BSc Zoology Series

Volume II

Cytology, Genetics and Molecular Genetics

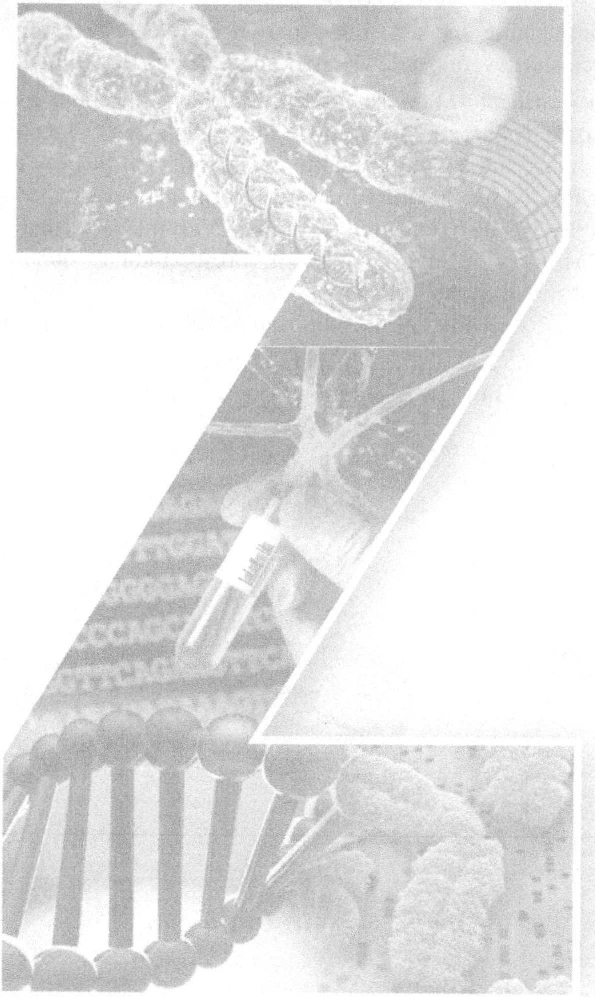

ABOUT THE AUTHOR

B N Pandey is Professor and Head at the Postgraduate Department of Zoology, Purnia College (affiliated to Bhupendra Narayan Mandal University, Madhepura), Purnia, Bihar. He obtained his BSc (Hons.) and MSc degrees in Zoology from Bhagalpur University, Bhagalpur and PhD from L N Mithila University, Darbhanga, Bihar. While working for his PhD, Dr Pandey collaborated with Prof. L I Korochkin of Institute of Cytology and Genetics, Novisibrisk, Moscow (USSR). He has also worked in collaboration with Prof. Susan Dawson of Utah State University, USA, on Population Health.

Dr Pandey has extensive teaching and research experience. Twelve research scholars have successfully completed their PhD degrees and several students are doing research under his learned supervision.

He has been awarded a Certificate of Merit for Excellence in Teaching by the Vice Chancellor, B N Mandal University, Madhepura, Bihar; another certification as Teacher of Outstanding Merit by the Ministry of Youth Affairs, Govt. of India; and a Gold Medal by the Zoological Society of India, for his contribution to research in this field. Dr Pandey has organised national seminars and symposia as well as the reputed All India Zoological Congress. He has also delivered lectures in Vidyasagar virtual classrooms. Major research projects sanctioned by the Ministry of Environment and Forests, Indian Council of Medical Research, etc., have benefited from his contributions. He is also a fellow of the Zoological Society of India and the Society of Environmental Sciences, and a member of the editorial boards of many journals.

Dr Pandey has published about 100 research papers in national and international journals. His areas of research interest include: Aquatic Biology, Population Genetics, and Population Health.

BSc Zoology Series
Volume II
Cytology, Genetics and Molecular Genetics

B N Pandey
Professor and Head
Postgraduate Department of Zoology, Purnia College
(Affiliated to Bhupendra Narayan Mandal University, Madhepura)
Purnia, Bihar

Tata McGraw Hill Education Private Limited
NEW DELHI

McGraw-Hill Offices
New Delhi New York St Louis San Francisco Auckland Bogotá Caracas
Kuala Lumpur Lisbon London Madrid Mexico City Milan Montreal
San Juan Santiago Singapore Sydney Tokyo Toronto

Tata McGraw-Hill

Published by the Tata McGraw Hill Education Private Limited,
7 West Patel Nagar, New Delhi 110 008

BSc Zoology Series: (Volume II)—Cytology, Genetics and Molecular Genetics

ISBN (13) : 978-1-25-902877-9
ISBN (10) : 1-25-902877-1

Vice President and Managing Director: *Ajay Shukla*

Head—Higher Education Publishing and Marketing: *Vibha Mahajan*
Publishing Manager—SEM & Tech Ed.: *Shalini Jha*
Editorial Executive—Sponsoring: *Smruti Snigdha*
Development Editor: *Renu Upadhyay*
Sr Copy Editor: *Nimisha Kapoor*
Sr Production Manager: *Satinder S Baveja*
Proof Reader: *Yukti Sharma*

Marketing Manager—Higher Education: *Vijay Sarathi*
Product Specialist: *Sachin Tripathi*
Graphic Designer— Cover: *Meenu Raghav*

General Manager—Production: *Rajender P Ghansela*
Production Manager: *Reji Kumar*

Typeset at Print-O-World, 2579, Mandir Lane, Shadipur, New Delhi 110 008, and printed at

Cover Printer :

The McGraw Hill Companies

CONTENTS

PREFACE

This *BSc Zoology Series* of five volumes will be useful for all undergraduate students of life sciences. The series has been developed to follow a unique test-friendly approach to especially assist undergraduate-level students in exam preparation. Besides, the applicants of CSIR-NET, GATE, Civil Services and other competitive examinations will also find this series very helpful.

About The Series

The following five volumes collectively structure this series:

Volume 1: Animal Diversity

Volume 2: Cytology, Genetics and Molecular Genetics

Volume 3: Biochemistry, Physiology and Endocrinology

Volume 4: Ecology and Animal Behaviour

Volume 5: Evolution, Comparative Anatomy, Biometry, Economic Zoology and Animal Development

These volumes cover the latest syllabi, as per the UGC curricula, of BSc courses taught across different Indian universities. Each part of a volume in the series contains a synopsis which briefly introduces the theme and then details important features topic-wise. This is followed by a comprehensive section on objective-type questions which includes short-answer questions, long-answer questions, multiple-choice questions, fill in the blanks, true or false questions, and questions based on reasoning and diagrams.

This arrangement has been ideated to first get the students acquainted with a chapter by going through the synopsis and then attempt to answer different sets of questions based on that chapter. Such a flow seeks to encourage self-study and aids quick revision of the topics in a lesson. While the synopsis provides a clear framework and considerable depth to topic-wise study of the syllabi, the stupendous variety in exercises covers a broad spectrum of learning tools.

What Makes This Series Unique?

The changing pattern of syllabus of academic life-science courses has induced a change in the type of questions appearing in undergraduate-level examinations of major universities and noted competitive tests. A distinct alteration in the nature of objective questioning has been identified. Objective questions, now part of compulsory questions, include the variations mentioned above. It then becomes imperative that the students be made fully conversant with this new pattern.

However very few books, adequately containing the required pedagogical features, are available to facilitate such a pattern of study. Recognising the growing interest of students and a need for a comprehensive yet basic-level text, I have authored this *BSc Zoology Series* to aid test-ready academic study.

Besides students, this series will amply assist various faculty members in the design and preparation of periodical tests for internal evaluation, question papers for undergraduate-level university examinations as well as CSIR-NET, GATE and Civil Services examinations, etc.

Salient Features Of The Series

- Apposite theory to aid quick revision for examinations
- Wide range of chapter-end exercises designed as per undergraduate examinations
- Surplus artwork to help develop a holistic understanding of concepts

Volume II : Cytology, Genetics and Molecular Genetics

Introduction

Cytology is the study of all aspects of cells, *Genetics* is the science of genes and heredity and *Molecular Genetics* is the study of genes at a molecular level. Cell biology, genetics and molecular genetics are interrelated. This volume covers the basic concepts, explains the interrelation and describes recent developments in the field of the cell biology, genetics and molecular genetics. A comprehensive comparison of historical findings and recent developments has been included to provide up-to-date knowledge and promote interest levels of students in these fields.

Highlight

This volume elucidates all the important topics such as Cell Organelles, Genetics of Cell Division, Mendelism, Chromosomal Aberrations, Mutation, Modern Concept of Gene, Human Genetics, Protein Synthesis, Gene Regulation, Human Genome Project, Genetics of Cancer, Immunogentics, and Genetic Engineering.

Organisation of Volume II

This volume has been classified into two parts, viz., *Cytology and Genetics and Molecular Genetics*. The first part, *Cytology* covers topics such as **Cell**—discusses characteristics, shape, size and types; **Cell Organelles**—details structure and various functions; **Cell Cycle**—describes the various phases and emphasises on the checkpoints such as DNA replication, DNA repair, spindle assembly, separation of chromatids, etc.; and **Cell Division**—elucidates the two types of divisions and their significance.

The second part, *Genetics and Molecular Genetics* covers topics such as **Mendelism**—deals with the principle of inheritance, transfer of characters from parent to the offspring and suppression and expression of these in the offspring; **Multiple Alleles**—discussed as an exception to Mendel's law; **Sex Determination**—describes the mechanism by which the sex of an offspring is determined; **Sex-linked Inheritance**—explains the mechanism by which characters are passed to the offspring; **Linkage**—discussed as an exception to Mendel's law of inheritance; **Crossing Over**—deals with the exchange of non-sister chromatid parts between homologous chromosomes and the various results; **Chromosomal Aberrations**—describes nature, occurrence and result; **Mutation**—explains the evolution of new characters; **Modern Concept of Gene**—traces origins of the concept and compares work done by Mendel with recent research on genes at a molecular level; **Human Genetics**—describes the branch of genetics that deals with the inheritance of human traits and also discusses how mutation causes diseases or genetic defects in human beings and their transmission to the offspring; **Cytoplasmic Inheritance**—explained as the mode of inheritance under control of the plasma genes and a comprehensive differentiation from Mendelian inheritance has also been presented; **Nucleic Acid**—describes nature, types and functions in living organisms; and **Protein Synthesis**—explained as the expression of characters in all living organisms under the precise control of the genetic material.

Further, **Genetic Code**—explained through codons and coding processes; **Gene Regulation**—discusses the activation and suppression of genes at different times; **Human Genome Project (HGP)**—describes its scale, beginnings and outlook; **Immunogenetics**—discussed as a branch of molecular genetics concerned with the relation between immunity and genetic factors in diseases along with its numerous applications; **Genetics of Cancer**—discusses symptoms, causes and treatment of cancer; **Prions**—describes structure and functions and their ability to cause fatal diseases; **Transposons**—discusses structure, use, role in gene regulation and related genetic diseases; **Apoptosis**—explained as programmed cell death, and discusses occurrence and result, especially in patients suffering from AIDS; **Genetics of Ageing**—described as a universal phenomenon, and covers its occurrence, progression, causes of ageing (including genetic causes) and related diseases; **Genetic Engineering**—elucidated as a highly specialised technique, discussing the process and the resultant 'recombinant DNA' in detail and its use in isolation and cloning; and **Genetics of Bacteriophages**—discusses nature and use, especially for engineering genes through gene cloning.

Online Learning Centre

For further interesting resources and supplements, please visit *http://mhhe.com/pandey/cgmg/vol2*

Acknowledgements

Writing this series has been a tremendous yet fulfilling endeavour. All the volumes have taken a final shape after endless inputs of time and effort. Though many teachers and students assisted me in compiling this book, I must especially mention the effort made by my colleague, O P Ambasta who extended immense support in myriad ways for bringing out the series in its present form. I am also indebted to A K Jha for his many valuable contributions.

I am grateful to the following reviewers for their helpful suggestions for improving the contents of this series.

A K Ojha – *Rajendra College, Balangir, Odisha*

P N Pandey – *SSPG College, Ayodhya, Uttar Pradesh*

S P Sinha – *Bhagalpur University, Bhagalpur, Bihar*

Budhadeb Manna – *University of Calcutta, Kolkata, West Bengal*

I am thankful to the team at Tata McGraw Hill Education, most notably Smruti Snigdha for giving me the opportunity to author this series and Renu Upadhyay for helpful suggestions to improve the quality of the content and regular reminders for timely completion of the project. It has been a pleasure to work with Nimisha Kapoor and Yukti Sharma, who took great care during the copy-editing and production processes of all the volumes.

I welcome all feedback, criticisms and suggestions for improvements in all the volumes from teachers, students and all other readers of this series. You can write to me at *b.n.pandey@hotmail.com*.

B N Pandey

Publisher's Note

Do you have a feature request? A suggestion? We are always open to new ideas (the best ideas come from you!). You may send your comments to *tmh.sciencemathsfeedback@gmail.com* (Don't forget to mention the title and author's name in the subject line).

CYTOLOGY

Introduction

Cell Biology deals with the study of cells. It is also known as Cellular Biology or Cytology.

Important Events/Discoveries in Cell Biology

1595	Jansen discovered the first compound microscope.
1665	Robert Hooke described cells in a cork.
1674	Leeuwenhoek discovered Protozoa and saw bacteria some nine years later.
1772	Alfsno Corti observed living matter in cells.
1781	F Fontana discovered nucleolus in the skin cell of an eel.
1833	Brown described cell nucleus in the cells of an orchid.
1835	Dujardin named living matter in cells as sarcode.
1838	Schleiden and Schwann proposed the cell theory.
1839	Purkinje named jelly-like substance in cells as protoplasm.
1840	Albrecht von Roelliker pointed out that sperm cells and egg cells are also cells.
1841	Robert Remark described amitotic cell division in the RBC of a chick's embryo.
1856	N Pingshem observed how a sperm cell penetrates an egg cell.
1857	Kolliker described mitochondria.
1858	Rudlof Virchow gave his conclusion *Omnis cellula e cellula*, that is, cells develop only from pre-existing cells.
1866	Haeckel established that the nucleus is responsible for storing and transmitting hereditary characters.
1861	Schultzee proposed the protoplasm theory.
1873	Anton Schneider described chromosomes.
1875	Van Beneden observed centrioles.
1879	Flemming described the behaviour of chromosomes during mitosis.
1881	Balbiani discovered giant salivary gland chromosomes.
1882	Knock identified bacteria.
1882	Flemming described cell division (mitosis).
1882	W Pfitzner discovered chromomeres.
1883	Schimper named chloroplasts.
1884	Mobius first discovered structures that were later named cell organelles.
1885	Hertwig and Strasburger described the role of nucleus in heredity.
1886	C A Mac Munn discovered cytochrome.

1888 Benden discovered centrosome.
1898 Golgi discovered Golgi apparatus.
1898 Waldeyer described chromosomes.
1902 Mc Clung discovered sex chromosomes.
1905 Farmer coined the term meiosis.
1906 M Tswett discovered chromatography.
1912 A Carrel discovered the technique of tissue culture.
1924 A Feulgen developed the test for identifying DNA in the cell.
1931 W H Lewis discovered pinocytosis.
1931 Ruska build the first Transmission Electron Microscope.
1938 T Svedberg developed the technique for ultracentrifugation.
1938 Behrens used differential centrifugation to separate nuclei from cytoplasm.
1939 First commercial transmission microscope was produced by Siemens.
1945 K R Porter discovered endoplasmic reticulum.
1952 De Duve identified lysosomes.
1952 Grey and co-workers established a continuous cell line.
1953 Watson and Crick proposed the structure of DNA.
1955 Nutritional need of animal cell was defined by Eagle.
1957 Meselson, Stahl and Vinogard developed density gradient centrifugation in cesium chloride solution for separation of nucleic acids.
1963 Chance and Parsons, Smith and H Fernandez–Moran discovered elementary particles in the mitochondrion.
1965 Cambridge instruments produced the first commercial scanning electron microscope.
1972 Singer and Nicolson proposed the fluid mosaic model.
1974 Claude and Palade described the ultra structure of a cell.
1976 Keith et al. described the microtrabecular system in the cytoplasm.
1976 Sato and coworkers described that different cell lines have different requirements of a mixture of hormones and growth factors in serum-free media.
1978 Mitchell discovered chemiosmotic mechanism of ATP synthesis.
1981 Transgenic mice and fruitflies were produced and mouse embryonic stem cell line was established.
1997 First sheep cloned.
1998 Mice cloned from stem cells.
1999 Hamilton and Baclcombe discovered SiRNA as post-transcriptional gene silencing in plants.
2001 Hunt and Nurse discovered cell cycle regulation by cyclin and cyclin-dependent kinase.
2007 Craig venter made synthetic chromosome.

Some Facts about Cells

- Cell is the structural and functional unit of life.
- The term cell was coined by Robert Hooke (1665).

- Robert Hooke was a mathematician and physicist.
- Robert Hooke actually observed empty cell walls of dead plant tissue under a microscope.
- He published his work in the form of a book named 'Micrographia' in London.
- A V Leeuwenhoek (1683) was the first to observe free cells (Protozoa, bacteria, sperms and erythrocyte).
- Alfosno Corti (1772) observed living matter in cells.
- F Fontana (1781) discovered nucleolus in the skin cells of an eel.
- Robert Brown (1831) discovered nucleus in the cells of an orchid root.
- F Dujardin (1835) named living matter in cells as sarcode.
- The jelly-like substance of cells was named as protoplasm by J E Purkinje (1839).

Characteristics of Cell

- Cell is the smallest unit of life.
- All cells have common ancestry.
- All cells have similar physical and chemical composition.
- Each cell may exist independently or as an integral part of an individual.
- Cells are capable of self-division and self-regulation.
- Each cell has its own metabolism.
- Cells are units of physiological activities.
- Cells contain genetic information and are the means to use it.
- Each cell has a definite lifespan.
- Life is not possible without cells.
- Cells are responsible for transferring life from one generation to the other.
- An organism lacks an activity which is not expressed by its cells.
- Cells have the ability to respond to stimuli.
- Each cell is totipotent, i.e., capable of forming an entire organism.
- Cells taking part in protection are without protoplast (e.g., skin surface cells and cork cells).
- In the human body, there are 260 types of cells.
- Humans have 100 trillion cells.
- Damaged or infected cells are destructed by apoptosis.

Size

- The size of cells is variable.
- Smallest cell is that of PPLO (Pleuro Pneumonia-Like Organism) while the egg of an ostrich is the largest cell.
- The size of a typical cell is 10 µm and its mass is 1 ng.
- The size of a cell is controlled by the:
  (a) Nucleo-cytoplasmic ratio
  (b) Ratio of cell surface to the cell volume
  (c) Rate of the metabolism

Shape

The shape of the cell may be:

- Variable (constantly changing) (e.g., *Amoeba* and *leucocytes)*
- Elongated (e.g., nerve cells)
- Spherical (e.g., eggs of many animals)
- Cuboidal (e.g., thyroid gland)
- Discoidal (e.g., RBCs)
- Flattened (e.g., squamous epithelium)
- Branched (e.g., pigment cells of the skin)

Cell Theory

- Cell theory was given by Schleiden and Schwann (1838).
- According to cell theory:
  (a) Cell is the structural unit of life
  (b) Body of living organisms is made up of cells
- Rudlof Virchow (1855) pointed out that new cells arise from pre-existing cells.
- Louis Pasture gave experimental prove that life originates from existing cells.
- Viruses are an exception to cell theory.

Types of Cells

Basically, there are two types of cells – prokaryotic cells and eukaryotic cells. The basic differences between prokaryotic and eukaryotic cells are given below.

Table 1

S. No.	*Feature*	*Prokaryotic cell*	*Eukaryotic cell*
1.	Size	Generally 1–10 μm	Generally 10–100 μm
2.	Organisation	Generally unicellular	Generally multicellular
3.	Cell wall	Present in most cells (Noncellulosic)	Present in plants (cellulosic)
4.	Nuclear membrane	Absent	Present
5.	Nucleolus	Absent	Present
6.	DNA	Naked	Combined with histone
7.	RNA polymerase	Sensitive to rifamycin	Nonsensitive to rifamycin
8.	Mitotic apparatus	Absent	Present
9.	Cellular organelles	Lack membrane	Bound with membrane
10.	Mitochondria	Absent	Present
11.	Endoplasmic reticulum	Absent	Present
12.	Chloroplast	Absent	Present in plant cells
13.	Ribosomes	70S	80S
14.	Lysosomes	Absent	Present
15.	Centrioles	Absent	Present

Contd.

Table 1 *Contd.*

16.	Microtubules	Absent	Present
17.	Vacuoles	Absent	Present
18.	Cardiolipin	Present	Absent
19.	Cyclosis	Absent	Cyclosis is common
20.	Flagella	Single stranded	Flagella are 11(9+2) Stranded
21.	Division	Amitosis	Mitosis or meiosis

Table 2 *Differences between Animal and Plant Cells*

S. No.	*Feature*	*Plant cell*	*Animal cell*
1.	Shape	Rigid	Irregular
2.	Cell wall	Present	Absent
3.	Nucleus	Peripheral	Central
4.	Plastid	Present	Absent, except in few Protozoans
5.	Vacuoles	Large, central	Small
6.	Centrosome	Absent	Present
7.	Glyoxysomes	Absent	Present

Cell Organelles

Protoplasm

- Protoplasm is the living substance of which a cell is made.
- It includes all parts of a cell.
- It is a transparent and jelly-like material.
- Huxley has defined protoplasm as the physical basis of life.
- O Hertwig (1892) proposed protoplasmic theory which states that living matter of plants and animals is the protoplasm.
- Protoplasm is divided into cytoplasm and nucleoplasm.
- Carbon, hydrogen, nitrogen and oxygen are the main constituents of protoplasm.
- Protoplasm contains both inorganic and organic substances.
- Inorganic substances include water that forms 90 per cent of the protoplasm and mineral salts like sodium chloride and gases like oxygen and carbon dioxide.
- Organic substances include proteins, carbohydrates, lipids, nucleic acid and enzymes.
- It has ability of nutrition, respiration, excretion, reproduction, metabolism and growth.
- It shows irritability and conductivity.
- It is heavier than water.
- Its viscosity is more than water.
- It contains positively (+) and negatively (–) charged particles.

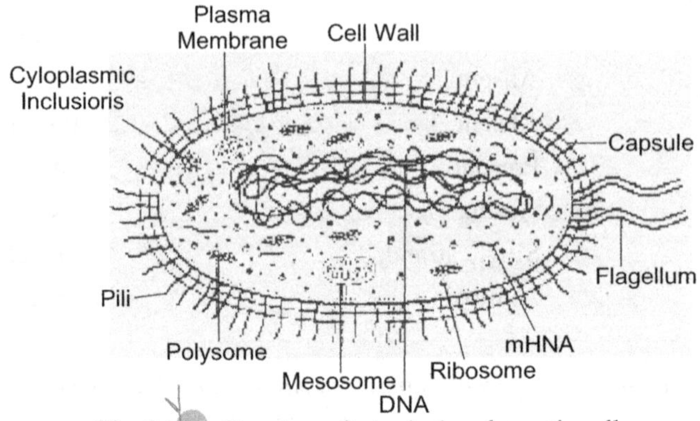

Fig. 1 Structure of a typical prokaryotic cell

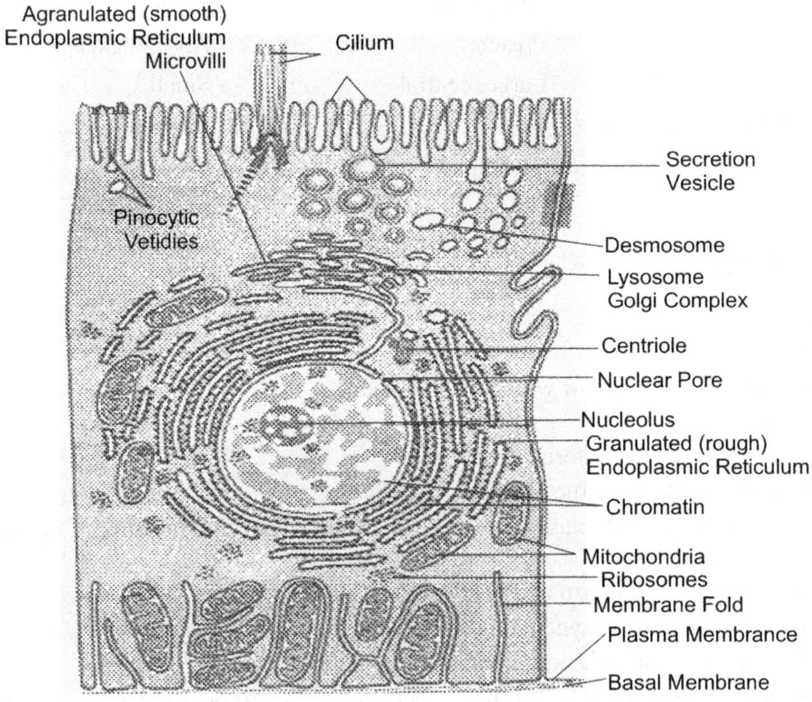

Fig. 2 Structure of an eukaryotic (animal) cell

- Protoplasm is colloidal in nature.
- The colloidal particles of protoplasm may exist both as sol and gel states.
- The movement of molecules of protoplasm is known as the Brownian movement.
- The substances formed by protoplasm are known as 'Deutoplasm'.

Cell Wall

- Cell wall is a rigid or semirigid envelope lying outside the cell membrane of plants, fungal and most of the prokaryotic cell.
- It is the cell wall that provides the most remarkable difference between plant cells and other eukaryotic cells.
- The cell wall is outside the protoplast and is thus part of the apoplast.
- It provides structural support, protection and also acts as a filtering device.
- Most of the carbon in the terrestrial ecosystem is located in the cell walls of plants.
- In plants, it is made up of cellulose and often lignin; in fungi, it is composed mainly of polysaccharides; in bacteria, it is composed of peptidoglycan. Generally, diatoms have a cell wall composed of silicic acid.
- The cell wall consists of middle lamella, primary wall and secondary wall. Often a tertiary cell wall is also present.

  (a) Middle Lamella
  - It is an extra or the outermost component that is formed between adjacent cells during cell division.
  - It is made up of calcium and magnesium pectate compounds.
  - It is viscous and acts as a cementing material between adjacent cells.

  (b) Primary Cell Wall
  - A primary cell wall is generally thin, flexible and extensible layer, formed while a cell is growing.
  - The cellulose in the primary wall is stabilised by hydrogen bonds.
  - The cellulose molecules form long chains that are cross networked with polysaccharides units called pectin.

  (c) Secondary Wall
  - As the cell matures, it lays down an inner secondary wall layer.
  - The secondary wall is just inside the primary wall.
  - It is approximately 5–10 μm thick.
  - It has an additional compound called lignin which makes cell walls waterproof.
  - It is more rigid than the primary cell wall.
  - It gives additional support to plants.
  - It is the main defence against harmful microbes like fungi or bacteria.
  - Secondary wall is the main constituent of the wood.
  - In some cases, a tertiary wall is present beneath the secondary wall as in the gymnosperm tracheids. It is very thin and is composed mainly of hemicellulose xylan. It lacks cellulose microfibrils.

Plasmodesmata

- Primary wall contains small pores through which cytoplasm of adjacent cells communicate by means of small cytoplasmic channels. These small cytoplasmic channels are called plasmodesmata.
- Majority of plasmodesmata contain a narrow tube-like structure called 'desmotubule' which is formed from endoplasmic reticulum of the connected cells.
- Plasmodesmata are typically formed during cell division.
- Plasmodesmata are approximately 50–60 nm in diameter.
- A typical plant cell may have 103–105 plasmodesmata.

- Plasmodesmata allow free movement of small metabolites as well as growth hormones between cells.
- Due to presence of plasmodesmata and intercellular spaces, the plant body is divided into two parts:
  - **(a) Symplast:** It is the living part of a plant body. It comprises protoplast bounded by a single plasma membrane. It stores energy reserves.
  - **(b) Apoplast:** Apoplast is the nonliving part of a cell. It is external to the plasma membrane. It comprises cell wall, spaces between cells and dead lumens. It stores bound water. Aqueous solution can move in the apoplast and symplast.

Protoplast The mass of protoplasm present in a cell is called protoplast. It consists of plasmalemma and everything inside it. Individual protoplasts are connected by plasmodesmata and comprise the symplast.

Cell Membrane

- Every cell is surrounded by a thin and elastic membrane which isolates cytoplasm from the external environment. This membrane is known as cell membrane.
- This membrane is selectively permeable and is also known as plasma membrane or plasmalemma.
- This membrane is not visible under light microscope.
- Cell membrane is responsible for controlled entry and exit of ions like sodium, potassium and calcium.
- This membrane is found both in prokaryotes and eukaryotes.
- In bacteria, yeast and plants, this membrane is bounded by the cell wall.
- Generally, thickness of this membrane is approximately 70–75Å.
- Chemically, this membrane is made up of lipid, protein and a little amount of carbohydrate.
- The organisation of phospholipid, proteins and carbohydrate in the plasma membrane is described with the fluid mosaic model.

Fluid Mosaic Model

- According to this model, the cell membrane is a fluid mosaic of lipids, proteins and carbohydrates.
- This model was proposed by Singer and Nicolson (1972).
- In the cell membrane, there is a continuous layer of phospholipid in which proteins are embedded.
- The phospholipids are arranged in a bilayer.
- The polar hydrophilic phosphate heads face outwards and their nonpolar hydrophobic fatty acid tails face each other in the middle of the bilayer.
- The proteins are inserted in many ways forming a mosaic of proteins.
- Proteins are of two types, viz., extrinsic (peripheral) and intrinsic (central).
- Extrinsic proteins are easily separable, whereas intrinsic proteins are not easily separable.

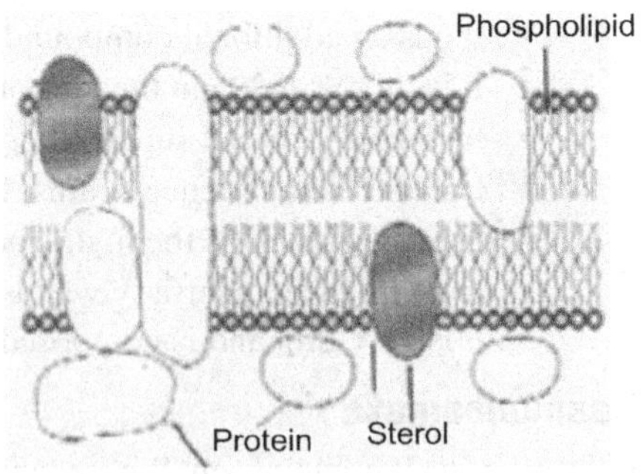

Fig. 3 Fluid mosaic model

- As there is rapid movement of protein and lipid molecules, so this membrane is considered to be highly fluid.
- The extracellular surface of a plasma membrane is covered with a carbohydrate group (oligosaccharides) that forms a loose coat called glycocalyx.

Role of Lipids

- Lipid molecules play a role in maintaining fluid property of the membrane.
- Due to the absence of covalent bonds between the lipids in the bilayer, the cell membrane has fluidity.
- Generally, the flipflop movement of lipid molecules from one monolayer to the other is very rare. But the exchange of places with their neighbours in monolayer takes place very rapidly resulting in lateral diffusion.
- The double bonds in unsaturated hydrocarbons tend to increase fluidity of phospholipid.
- Inositol plays a role in cell signalling.
- Glycolipid assists in cell recognition.
- Sterol inhibits phase transition.

Role of Proteins

- Protein forms the main bulk of a cell membrane.
- Transport proteins help in the transport of specific substances across the membrane.
- Channel proteins form open pores through the membrane, allowing the passage of the molecules.
- Carrier proteins selectively bind and transport small molecules.
- Many proteins function as enzymes.
- Some peripheral proteins serve as anchor points for the cytoskeleton or extracellular fibres.

Role of Carbohydrates

- Oligosaccharides give a cell identity (i.e., distinguish self from nonself) and are distinguishing factor in human blood types and transplant rejection.
- They help in maintaining asymmetry of a membrane.
- It has been suggested that oligosaccharides are negatively charged, so the positively charged protein may remain bounded to a cell membrane due to electrostatic attraction.

Specialised Structure

- Plasma membrane is modified to form some specialised structures to perform various functions such as absorption, fluid transport, electrical coupling, cell adherence, etc.
- These structures are microvilli (absorption), tight junctions, desmosomes (intercellular attachments), interdigitations (increase adherence and surface area for exchange of materials) and gap junctions (intercellular communications).
- Tight junctions serve to seal intercellular spaces.

Function

- Plasma membrane provides shape to the cell as well as protection to cytoplasmic organelles.
- It permits movements of certain substances in and out of the cell.
- The transportation of molecules across the membrane takes place by active or passive transport.
- The plasma membrane contains specific sites which help in the recognition of specific hormones.
- It regulates fusion of the membrane with other membranes in the cell via specialised junctions.
- It provides special site for binding and catalysis of enzymes.
- It helps in the release of secretary products of the cell.

- The in-folding of the plasma membrane in bacteria is called 'mesosome' which contains a chain of respiratory enzymes where the process of oxidative phosphorylation takes place.
- Plasma membrane provides a site for attachment of cytoskeleton filaments or components of extracellular matrix.

Cytoplasm

- It is the part of the cell between the cell membrane and the nuclear membrane.
- It consists of cytosol (matrix) and organelles.
- The cytosol is a transparent semifluid substance that is not held within organelles.
- Organelles are membrane-bound structures in a cell.
- Organelles are found embedded in the cytoplasm.
- Cell organelles have characteristic shape, specific chemical composition and functions.
- Cell organelles may be extra cytoplasmic (nucleus) or cytoplasmic (mitochondria, plastids, Golgi complex, lysosomes, ribosomes, endoplasmic reticulum, centrosomes, cytoskeleton, cilia, flagella, etc.) structures.
- Cytoplasm is the medium for chemical reactions.

Cytoskeleton

- The internal framework of a cell that supports its organelles cell contents and assists their movement is known as cytoskeleton.
- The cytoskeleton is composed of three types of fibrous components which are microtubules, microfilaments and intermediate filaments.
- The cytoskeleton also contains fibrous components called microtrabeculae.
- The three-dimensional structure of a cytoskeleton can be studied with the help of fluorescent labels or by electron microscopy in fixed preparations.

(a) Microtubules

- Microtubules are filamentous intracellular structure, responsible for various kinds of movement in all eukaryotic cells.
- They are one of the components of the cytoskeleton.
- Microtubules were discovered by de Robertis and Franchi (1953) in the axoplasm of nerve fibres.
- Slautterback (1963) coined the term microtubules.
- Microtubules may be labile structure (pseudopodia in *Amoeba*) or stable structure (those of cilia and flagella).
- Microtubules are long, hollow tubular cylinders having an outer diameter of 250 Å and a wall thickness of approximately 50 Å.
- The transverse section of cytoplasmic microtubules shows 13 sub-units (profilaments) which lie parallel to the long axis of microtubules.
- The profilaments are made up of alpha and beta tubulins which are arranged alternately.

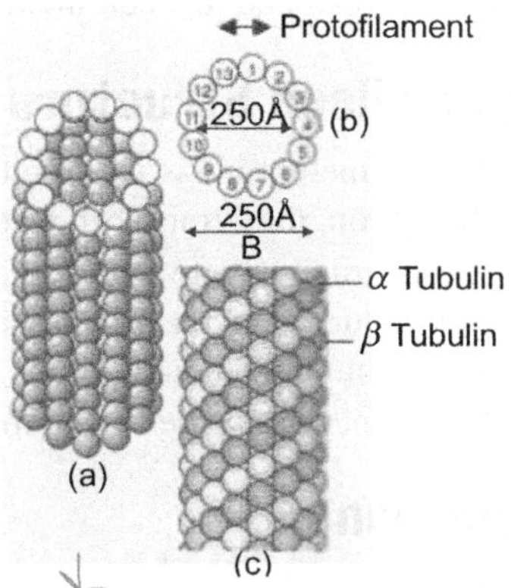

Fig. 4 (a) Three-dimensional model of microtubule structure. (b) Cross section of microtubule consisting of 13 sub-units (c) Helical structure of microtubule consisting of α and β tubulins.

- The basic arrangement of tubulin appears to be helical with 13 tubulin molecules per turn of the helix.
- Microtubules exhibit rapid assembly and disassembly at their ends.
- Ca^{++}, Mg^{++} and calmodulin-bound microtubular-associated protein are required for assembly.
- Colchicine prevents assembly of microfilaments.
- Microtubules provide shape to the cell.
- Microtubules function as an 'intracellular engine'.

(b) Microfilaments

- Microfilaments are fine unbranched proteinaceous structures which are generally arranged in bundles.
- Microfilaments are found in all eukaryotic cells.
- The thickness of microfilaments is approximately 6–10 nm and their length may be indefinite.
- A microfilament has a beaded appearance.
- Actin and myosin are the main constituents of microfilaments.
- Growth of microfilaments occurs through aggregation of actin molecules by nucleation.
- Microfilaments assist in cytokinesis in animal cells, help in the movement of microvilli and cause formation and retraction of pseudopodia.

(c) Intermediate Filaments

- Intermediate filaments are unbranched and noncontractile structures having a diameter of 100Å.
- They are intermediate between microtubules and microfilaments.
- They disorganise during cell division and organise again after cell division.
- They are found in several types of cells such as nerve cells, neuroglial cells, fibroblasts and epidermal cells.
- They maintain the shape of the cell.
- Depending on their biochemical composition and distribution, they show considerable tissue specificity. Their five major classes have been recognised:

1. Keratin-containing Intermediate Filaments

- They are generally known as cytokeratins.
- They are synthesised in living layers of epidermis.
- They connect cell membrane with desmosomes.

2. Glial Filaments

- They are present throughout the cytoplasm and are formed of acid protein.

3. Neurofilaments

- They are formed of three polypeptides.
- They form loose bundles of 100 Å which run parallel to the nerve axon.

4. Desmin-containing Filaments

- Desmin filaments are characteristics of all types of muscle cells (cardiac, smooth and skeletal).
- They are composed of desmin protein.
- They are concentrated in Z lines.

5. Vimetin-containing Filaments

- They are present in most cells of the body.
- They are wavy and made up of single polypeptide.
- They provide support to the nucleus and keep it in position.

Microtrabecular System

- Keith et al. (1976) described the microtrabecular system in cytoplasm.
- It is also termed as microtrabecular lattice.
- The microtrabecular filaments are 2–3 nm in diameter and 300 nm long and are made up of actin.
- Microtrabecular filaments form a link with all cell organelles, microtubules and microfilaments.
- They establish a true link between cytoskeleton components.

Plastids

- Plastids are large cell organelles found only in autotrophic cells.
- The term 'plastid' was coined by Schimper (1885).
- They are generally spherical or ovoid in shape.
- All plastids are derived from proplastids.
- They can be easily seen through a light microscope.
- Once formed, one type of plastids can be converted into other types.
- There are three types of plastids:

1. **Leucoplasts**
    - Leucoplasts are colourless plastids.
    - They are abundantly found in the cells of fruits, seeds, tubers and rhizomes.
    - They store nutrients.
        (a) Amyloplasts: Store starch
        (b) Elaioplasts: Store fat
        (c) Aleuroplasts: Store protein
2. **Chromoplasts**
    - Chromoplasts are red, yellow or orange in colour.

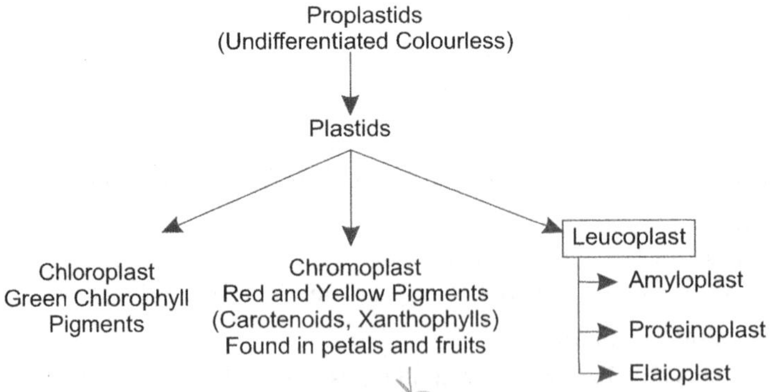

Fig. 5

    - Their colour is due to the presence of two pigments, viz., carotene and xanthophylls.
    - They are found in petals of flowers and fruits.
3. **Chloroplasts**
    - Chloroplasts are the second largest and semi-autonomous cell organelle found in plant cells and other eukaryotic organisms that conduct photosynthesis.

- Chloroplast converts light energy (from the sun) to chemical energy via the process of photosynthesis.
- The shape, size and volume of chloroplasts are altered by sunlight.
- The main pigment located in chloroplasts is chlorophyll.
- In green plants, there are two types of chlorophyll, viz., chlorophyll 'a' and chlorophyll 'b'.
- In chlorophyll 'b', one of the methyl groups of chlorophyll 'a' is replaced by a formyl group.
- Chlorophyll contains Mg^{++} and has four pyrrole rings.
- Chloroplasts are surrounded by an outer membrane and an inner membrane, separated by an inter-membrane space.
- The outer membrane of chloroplasts is freely permeable to small molecules.
- The inner membrane of chloroplasts is impermeable to small molecules, even to H^+.
- The inner membrane contains many translocators that regulate the passage of substances in and out of the chloroplast.

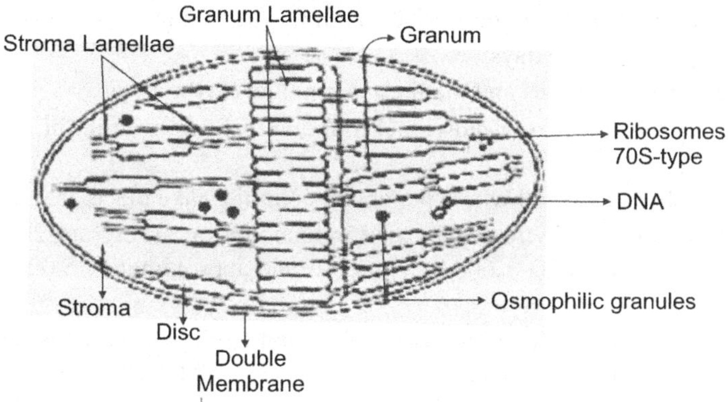

Fig. 6 Sectional view of plant chloroplast

- The fluid inside the inner membrane of the chloroplast is called stroma.
- Stroma contains ribosomes, circular DNA, RNA and soluble enzymes of Calvin cycle. It also contains *ribulose biphosphate carboxylase* (RuBPCarboxylase).
- Within the fluid there is an interconnected system of stacks of discs and each sac is called thylakoid.
- Thylakoids are the structural and functional components of the chloroplast.
- Thylakoids are the sites of photosynthesis. Their membranes contain all the enzymes required for photosynthesis.
- Interaction between chlorophyll, electron carriers, coupling factors and other components takes place within the thylakoid membrane.
- A stack of thylakoids is called granum.
- The inner surface of the thylakoid membrane bears small spherical structures called quantasomes.
- Quantasomes are photosynthetic units.
- Quantasomes consist of two photosystems called PS-I and PS-II, containing approximately 300 chlorophyll molecules. Cytochromes *b* and *f* as well as ATP synthetase are present.
- It has been suggested that chloroplasts may have originated from endosymbiotic cyanobacteria.

Mitochondria

- Mitochondria are the most important cell organelle present in all eukaryotic cells.
- They are descendents of a free-living bacterium and occupy 20 per cent volume of the cytoplasm.
- Mitochondria are the chief site of cellular respiration and oxidative phosphorylation.
- Mitochondria were first observed by Kolliker (1850).
- Flemming (1882) named them as fila, whereas Altmann (1892) called them bioblasts.
- Benda (1897) gave the term mitochondria (Mito-thread; chondrion-granule).
- Michaelis (1905) stained them with Janus Green B.
- Kingsburry (1912) pointed out that mitochondria are the site of cellular respiration.
- Palade and Sjastrand (1940–1950) studied the fine structure of mitochondria.
- In animal cells, mitochondria are the second largest organelle, whereas in plant cells it is the third largest organelle.
- Mitochondria are semi-autonomous cell organelle.
- Mitochondria are the cell organelle containing electron transport chain.
- The lifespan of mitochondria is 5–10 days.
- Mitochondria are absent in prokaryotes and mature mammalian erythrocytes.
- The number of mitochondria varies from individual to individual and even in different cells of the same individual.
- There are more mitochondria in the cell where metabolic activities take place.
- In *Micromonas* and *Trypanosoma*, there is only one mitochondrion. There are 20–24 mitochondria; in sperm; 300–400 in kidney cells; 1,000–1,600 in liver cells and approximately 5,00,000 in the flight muscles of insects.
- The number of mitochondria in a cell can increase by fission (following mitosis) and decrease by their fusing together. Defect in either process can produce serious, even fatal illnesses.
- The shape of mitochondria is variable. Generally, the shape of mitochondria is sausage-like.
- The size is also variable. Generally, the length of mitochondria is approximately 1.5–10 µm, whereas diameter is approximately 0.25–1 µm.
- Mitochondria are covered by a double membrane called outer membrane and inner membrane.
- Outer membrane is permeable to many metabolites.
- Outer membrane contains a protein called porin and has low content of cardiolipin as well as carrier proteins.
- The inner membrane contains five complexes of integral membrane proteins:
  (a) NADH dehydrogenase (Complex I)
  (b) Succinate dehydrogenase (Complex II)
  (c) Cytochrome *c* reductase (Complex III)
  (d) Cytochrome *c* oxidase (Complex IV)
  (e) ATP synthase (Complex V)
- Space between the outer and inner membranes is called the outer chamber.

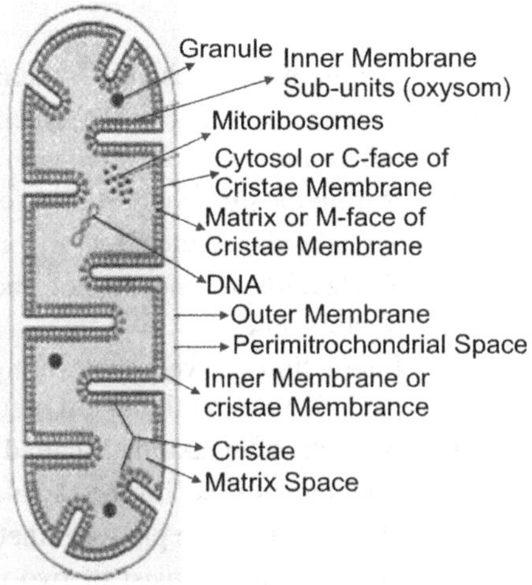

Fig. 7 Sectional view of mitochondrion

- Space enclosed by the inner membrane is called the inner chamber.
- The inner chamber contains a jelly-like substance called matrix. In matrix, 70S ribosome and circular DNA are present. Enzymes of Krebs cycle are also present in the matrix (except succinate dehydrogenase).
- The inner membrane bears many finger-like projections called cristae.
- Each crista has two faces called M (matrix) face and C (cytoplasmic) face.
- The shape of cristae is variable and the number of cristae is directly associated with the efficiency of mitochondria.
- The inner membrane contains many stalked particles called elementary particles or sub-units of Fernande–Moran or F_0–F_1 complex.
- Each elementary particle consists of a head, a stalk and a base.
- The head is approximately 75–100 Å in diameter, the length of stalk is approximately 50 Å and the size of base is 40 × 110 × 115 Å.
- The space between two elementary particles is approximately 100 Å and the number is approximately 104–105 per mitochondrion.
- Chemically, the head is made up of *ATPase*, stalk of oligomycin sensitivity protein (OSCP) and base of proton channel.
- Chemically, mitochondrion is made up of protein and lipid. Besides, it also contains enzymes of respiratory chain complex.

Mitochondrial DNA (mtDNA)

- The first mitochondrial DNA was isolated from chicken in 1966.
- Mitochondria contain 5–100 copies of DNA molecules.
- Mitochondrial DNA is circular or linear.
- Mitochondrial DNA is more stable and of higher density than nuclear DNA.
- Mitochondrial DNA is rich in guanosine–cytosine base.
- There are only 16,596 bases in human mtDNA.
- Mitochondrial DNA does not follow the usual rule of genetic inheritance.
- Mitochondrial DNA is inherited only from the mother.
- Mutations in mt genes may lead to diseases.
- Some of the diseases known to be associated with mtDNA are Parkinson's disease, Huntigton's disease, Leber's optic neuropathy and Kearns–Sayre syndrome.
- mtDNA plays a role in ageing.

Origin There are three views regarding the origin of mitochondria:

(a) Origin from various cell membranes
(b) By division of fully formed mitochondria
(c) *de novo* origin

Endoplasmic Reticulum

- Branching off and continuous with the outer membrane of the nucleus, there is a double-walled space which is zigzag throughout in the cytoplasm. This is called endoplasmic reticulum.
- The term 'endoplasmic reticulum' was coined by Porter (1953).
- Endoplasmic reticulum is found in all eukaryotic cells except mature mammalian erythrocytes.

- It forms 30–60 per cent of the total endomembrane system.
- In muscle cells, the endoplasmic reticulum is known as the sarcoplasmic reticulum, which is mainly concerned with storage and release of Ca^{++}.
- Endoplasmic reticulum of retinal cells is called myeloid bodies.
- Endoplasmic reticulum is both mobile and elastic.
- Structurally endoplasmic reticulum consists of cisternae, vesicles and tubules.
- Cisternae are flattened, tubular structures arranged parallel to each other. They are connected with each other and are more abundant in rough endoplasmic reticulum.
- Vesicles are oval to rounded structures and abundant in secretary cells.
- Tubules are unbranched or branched structures and abundant in cells synthesising lipids.
- Endoplasmic reticulum is of two types, viz., rough endoplasmic reticulum and smooth endoplasmic reticulum.

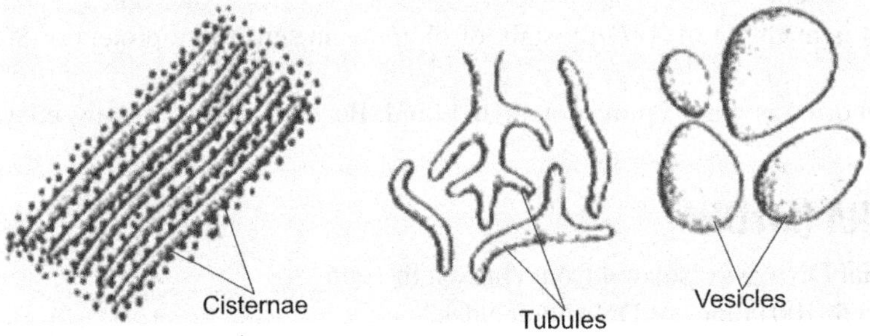

Fig. 8 Components of endoplasmic reticulum

Table 3

S. No.	*Rough Endoplasmic Reticulum (RER)*	*Smooth Endoplasmic Reticulum (SER)*
1.	It is also known as granular endoplasmic reticulum.	It is also known as agranular endoplasmic reticulum.
2.	It bears ribosomes.	It lacks ribosomes.
3.	It contains ribophorins.	It lacks ribophorins.
4.	Cisternae are abundant.	Tubules and vesicles are abundant.
5.	It is involved in the synthesis of protein.	It is involved in the synthesis of lipids.

- Endoplasmic reticulum provides mechanical support to the cells as well as forms a nuclear envelope.
- SER contains cytochrome P-450 which detoxifies pollutants, carcinogens, drugs, etc.

Ribosomes

- Ribosomes are ribo-nucleo-protein particles and one of the smallest nonmembranous cell organelle.
- They are so named due to presence of high amount of RNA.
- They are popularly known as the 'protein factory of the cell'.
- Ribosomes were first seen by Claude (1941) who called them microsomes.

- Palade (1958) coined the term 'ribosomes'.
- Ribosomes are found both in prokaryotes and eukaryotes.
- Ribosomes are also known as organelle within organelle as they are also found in mitochondria and chloroplasts.
- They may be found either in a free state or attached to cytoplasmic membranes like endoplasmic reticulum.
- The average size of ribosomes is approximately 150–200 Å.
- The number of ribosomes is high in cells involved in protein synthesis (pancreatic cells, liver cells, etc.).
- Each ribosome has two functional sites, viz., 'A'(amino acyl) site and 'P' (peptidyl) site. The 'A' site receives tRNA while 'P' site binds with the growing polypeptidal tRNA.
- Basically there are two types of ribosomes, viz., 70S and 80S.
- The main differences between 70S and 80S ribosomes are as follows:

Table 4

S. No.		*70S*	*80S*
1.	Distribution	Prokaryotes, mitochondria, chloroplasts	Eukaryotes
2.	Size	Smaller	Larger
3.	Molecular weight	3 million	4–5 million
4.	Sub-units	Smaller sub-unit of 30S and larger sub-unit of 50S	Smaller sub-unit of 40S and larger sub-unit of 60S
5.	RNA	3 molecules 30S sub-unit– 16S 50S sub-unit – 23S and 5S	4 molecules 40S sub-unit – 18S 60S sub-unit – 28S, 5.8S and 5S
6.	Nuclear proteins	30S – 21 50S – 31	40S – 33 60S – 49
7.	Total proteins	50–60	70–80

- The association and dissociation of two sub-units of ribosomes depend on Mg^{++} ion concentration.
- In prokaryotes, biogenesis of ribosomes occurs in the cytoplasm, whereas in eukaryotes it occurs in the nucleolus.
- Ribosomes are the site of protein synthesis.

Lysosomes

- Lysosomes are tiny cell organelles of eukaryotes which are popularly known as the 'suicidal bags' of the cell due to their autolysis activity.
- Lysosomes have been described as the 'recycling centre' because they free the metabolites from the worn out cell organelles by digesting them.
- In many organisms, lysosomes are involved in programmed cell deaths.
- The term 'lysosome' was coined by de Duve (1955).
- Lysosomes are rich in hydrolytic enzymes and are bounded by a single membrane.
- These enzymes are capable of digesting nucleic acids, polysaccharides, fats and proteins.
- Most of the enzymes of lysosome work in acidic medium.
- Lysosomes are stable in living cells.

- The lysosomal membrane is resistant to enzymes enclosed by it.
- Cholesterol, cortisone, cortisol and heparin act as membrane stabilisers for lysosomes, whereas steroids, sex hormones and lipid-soluble vitamins (A, D, E and K) act as membrane labilisers.
- Lysosomes show polymorphism. The four different types of lysosomes are:
  (a) **Primary Lysosomes:** They are also known as storage granules and contain digestive enzymes.

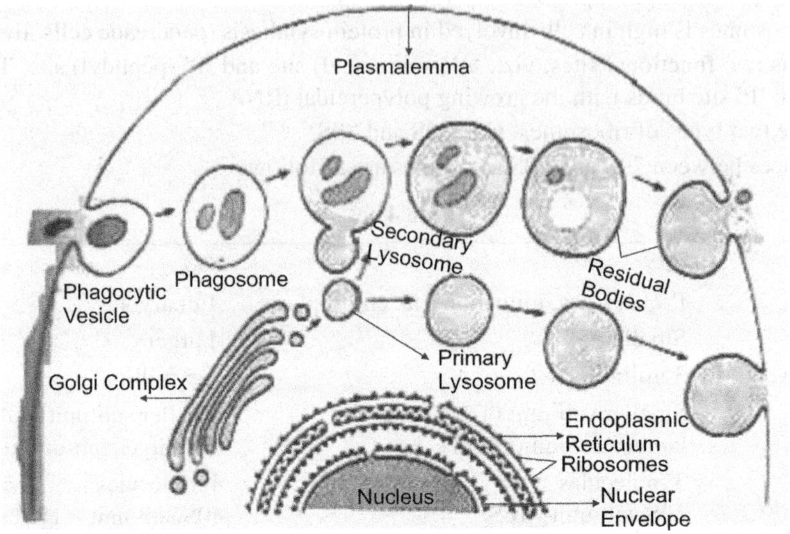

Fig. 9 Polymorphism in lysosome

  (b) **Secondary Lysosomes:** They are also known as heterophagosomes and are formed by the fusion of primary lysosome and engulfed materials.
  (c) **Autophagic Vacuoles:** They feed on their own cell organelle.
  (d) **Residual Bodies:** They contain undigested materials.
- The various forms of lysosomes are due to different stages of digestion of a particle.
- Malfunctioning of lysosomes causes a large number of diseases in humans such as Tay–Sachs disease, Nieman–Pick disease, Farber's disease, Hunter's syndrome, Hurler syndrome, Scheic syndrome, Sanfilippo syndrome, etc.
- Lysosomes are involved in extracellular as well as intracellular digestion.

Origin Primary lysosome originates from the Golgi complex.

Microbodies

- Microbodies are diverse group of organelles found in cytoplasm of almost all cells.
- Microbodies are roughly spherical structures bound by a single membrane.
- They have the ability to absorb oxygen to carry out direct oxidation.
- Rohdin (1954) coined the term 'microbodies'.
- Microbodies are mainly of two types, viz., peroxisomes and glyoxysomes.

Peroxisomes

- Peroxisomes are oval-to round-shaped microbodies and found in all eukaryotic cells.
- The diameter of peroxisomes is approximately 0.5–1 μm.
- They are self-replicating bodies and bud off from the endoplasmic reticulum.
- Like mitochondria, peroxisomes are the major site of oxygen utilisation.
- They contain oxidative enzymes such as catalase and urate oxidase.
- Peroxisomes use molecular oxygen and hydrogen peroxide to perform oxidative reactions.
- In photosynthetic cells, they are involved in photorespiration.
- In animal cells, they perform lipid metabolism.
- Peroxisomes protect cells from the toxic effect of hydrogen peroxide.
- They are numerous in liver and kidney cells.
- Absence or reduced number of peroxisomes result in a serious disease called Zellweger syndrome.

Glyoxysomes

- Glyoxysomes are specialised form of peroxisomes found in some plant cells, especially in the cells of germinating seeds and filamentous fungi.
- They are oval, rounded or polygonal microbodies having a diameter of approximately 0.5–1 μm.
- They were first discovered and named by Briedenbatch (1967).
- Glyoxysomes contain enzymes for beta oxidation of fatty acids and glyoxylate cycle.
- They are involved in gluconeogenesis.

Spherosomes

- Spherosomes are the smallest cell organelle.
- They are spherical structures found in some plant cells.
- They have a diameter of approximately 0.5–1 μm with a single membrane boundary.
- They have fairly granular matrix containing triglycerides.
- Spherosomes are abundant in cells in which lipids are stored.
- They contain hydrolytic enzymes.
- They originate from the endoplasmic reticulum.

Golgi Complex

- Golgi complex is the cell organelle which exhibits structural variation and is primarily related with the cell secretion.
- It was first reported by the Italian neurologist Camillo Golgi (1898) in the nerve cell of barn owl.
- Golgi forms a part of the endo-membrane system and constitutes 2 per cent of the total cytoplasmic volume.
- Golgi complex is found in all the eukaryotic cells. They are lacking in prokaryotic cells, mature mamalian RBCs and sperms.
- The shape and size of Golgi complex are variable and depend on the physiological condition of the cells.
- Golgi complex is well developed in secretary cells.

- The specific density of Golgi complex is less than that of mitochondria and endoplasmic reticulum.
- The ultra structure of Golgi complex has been studied by Dalton and Felix (1954) in the epididymis of rat.
- Golgi complex is made up of three components:

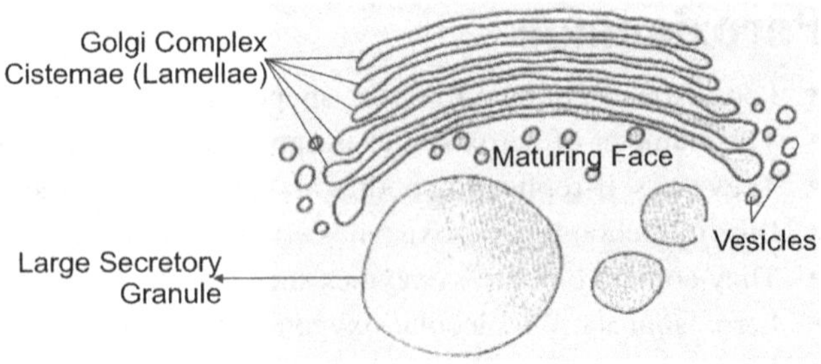

Fig. 10 Ultrastructure of Golgi complex

(a) Cisternae

- These are flattened tubular sac-like structures which occur in stacks.
- These are the most constant element of the Golgi complex.
- Each cisternae has two faces, viz., forming face (F-face) and maturing face (M-face).
- These two faces differ in the staining property.
- Cisternae lack ribosomes.

(b) Vacuoles

- Vacuoles are large rounded clear structures which are formed by the cisternae.

(c) Vesicles

- These are small droplet-like structures having a diameter of 4,000–8,000 Å and are formed by the cisternae.
- Golgi complex is surrounded by a clear zone of exclusion in which mitochondria, ribosomes, plastids and granules are lacking.
- Chemically, Golgi complex is made up of proteins and lipids. Besides, it also contains many enzymes.
- Golgi complex is formed from:
  (a) Plasma membrane (b) Endoplasmic reticulum
  (c) Nuclear membrane (d) Annulate lamellae
- Golgi complex is involved in the synthesis, packing and secretion of many substances.
- Golgi complex helps in cytokinesis.

Centrosome

- It is a thin, clear and homogenous structure located outside the nucleus.
- Centrosome is also called the 'microtubule-organising centre'.
- Centrosomes are the self-regulating bodies and form astral ray and spindle during cell division.
- Centrosome was discovered by Edouard Van Benden (1888) and the term 'centrosome' was coined by T Boveri (1888).
- Centrosome is lacking in prokaryotes, *Amoeba* and seed plants.
- Each centrosome consists of two centrioles called diplosomes.
- The clear cytoplasmic area around centrioles is called the centrosphere.
- Centrioles are cylindrical bodies lying perpendicular to each other.
- The wall of the centriole consists of parallel tubular structure.
- The cylinders are formed by nine groups of microtubules and each group contains three microtubules.

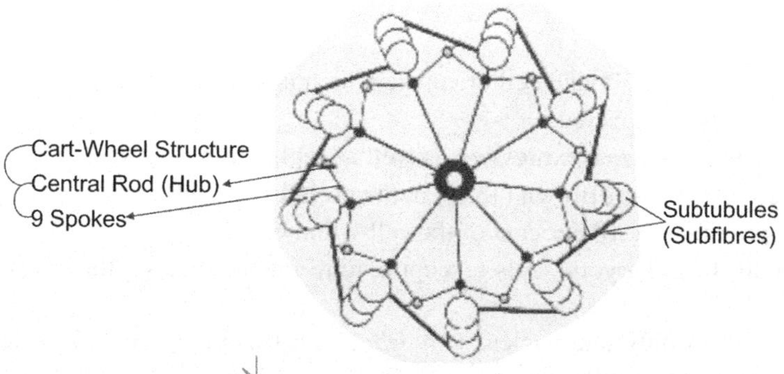

Fig. 11 Fine structure of centriole in TS

- The microtubules are made up of tubulin protein.
- The centrioles are morphologically identical to the basal bodies.
- Replication of centrosome takes place in the late S phase or G_2 phase of the cell cycle.

Cilia and Flagella

- These are hair-like projections protruding from the outer surface of the cell bounded by the cell membrane.
- Cilium is the projection shorter than the cell, whereas flagellum is the extension larger than the cell.
- Cilia are shorter but more in number, whereas flagella are longer but fewer in number in comparison to flagella.
- Cilia beat in a coordinated manner, whereas flagella beat independently.
- They are found both in animal and plant groups.
- A cilium or flagellum consists of three parts, viz., basal body, basal plate and shaft.
- Basal body is embedded in the outer part of the cytoplasm below the cell membrane.
- Basal body is a microcylinder structure consisting of microtubules having 9+2 arrangements.
- 9+2 arrangement is characteristic of eukaryotic cilia and flagella.
- Cilia and flagella in prokaryotes (bacteria and cyanobacteria) lack 9+2 arrangement.
- They contain tubulin, dynenin and nexin proteins. Dynenin has *ATPase* activity.
- They help in locomotion, nutrition, respiration, circulation, excretion, etc. They also serve as sensory structures.

Vacuoles

- Vacuoles are fluid-filled small cavities bounded by a thin membrane called tonoplast.
- They are a characteristic of plant cells.
- They are special storage organs and maintain turgidity of the cell.
- The vacuole sap contains gases, acids, sugars, salts, etc.
- In lower organisms, contractile vacuole is present which assists in osmoregulation and excretion.
- Anthocyanin (water-soluble pigment) is present in the cell sap of flower petals.
- In a mature plant cell, all the small vacuoles unite to form a centrally located single large vacuole called the central vacuole.
- In mature animal cells, vacuoles are small and many in number.

Nucleus

- Nucleus is the largest membrane-bound cell organelle which is known as the 'master cell organelle'.
- It is the first organelle to be discovered.
- It is mainly concerned with the gene expression as well as replication of DNA.
- Nucleus was discovered by Robert Brown (1831) in the orchid root cells.
- Nucleus occupies approximately 10 per cent of the cell volume.
- Nucleus is found in all the eukaryotic cells (except mature mammalian erythrocytes and sieve tubes of seed plants).
- Generally in a cell, there is only one nucleus (*Amoeba*). But there may be two nuclei (*Paramecium*) or there may be many nuclei (*Opalina*).
- A typical nucleus consists of following components:

1. **Nuclear Envelope**
    - Nucleus is surrounded by an envelope called nuclear envelope which isolates it from the cytoplasm.
    - It is a double-membrane structure called outer and inner nuclear membranes.
    - These membranes are lipoprotein membranes.
    - The outer membrane may contain ribosomes and is continuous with the endoplasmic reticulum.
    - The outer membrane bears small pores called nuclear pores.
    - Each nuclear pore is surrounded by a ring-like structure called annulus.
    - The pores and annuli together constitute the pore complex.
    - Through nuclear pores, exchange of material occurs between the cytoplasm and nucleus.
    - In between two membranes, there is a perinuclear space of approximately 10–70 nm.
    - The inner surface of the inner membrane bears meshwork of fibrous protein called nuclear lamina.
    - Nuclear lamina connects inner nuclear membrane with chromatin.
    - Nuclear laminas regulate assembly and disassembly of nuclear membrane during cell division.
    - Nuclear laminas are made up of three principal proteins, viz., laminas A, B and C.
    - Nuclear membranes are impermeable to most molecules.
    - Nuclear membranes disappear during late prophase and reappear during telophase.
2. **Nucleoplasm**
    - Nucleus contains a jelly-like fluid called nucleoplasm.

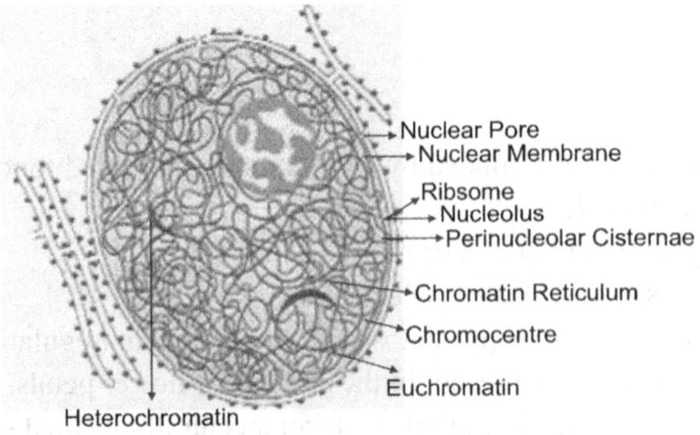

Fig. 12 Ultrastructure of nucleus

- Nucleoplasm contains various enzymes involved in metabolic pathways as well as replication of DNA and transcription of RNA.
- Some proteins present in the nucleoplasm are involved in the regulation of chromatin structure and function.

3. **Nucleolus**
   - It is a large spherical structure present in the nucleoplasm.
   - It was discovered by Fontana (1728) and the term 'nucleolus' was coined by Bowman (1840).
   - Nucleolus contains 80 per cent proteins and 20 per cent mixture of DNA and RNA.
   - Nucleolus is formed by a special region of the chromosome called 'Nucleolar Organiser Region' (NOR).
   - In eukaryotes, nucleolus is the site of synthesis of ribosomes and its assembly.
   - A cell may contain up to four nucleoli, but within each species the number of nucleolus is fixed.
   - Nucleolus is renewed at each cycle.
4. **Chromatin Fibres**
   - The thread-like filamentous structure present in the nucleoplasm is termed as chromatin fibres.
   - Chromatin fibres are complex of DNA and proteins.
   - Chromatin fibres are observed at interphase stage.
   - During cell division, these chromatin fibres become short and thick thread-like structures called chromosomes.
   - Chromatin materials are of two types, viz., euchromatin and heterochromatin.
   - The basic differences between euchromatin and heterochromatin are as follows:

Table 5

S. No.	*Euchromatin*	*Heterochromatin*
1.	It stains lightly.	It stains deeply.
2.	It is granular.	It is fibrous.
3.	It occurs in the diffused region.	It occurs in the condensed region.
4.	It is genetically active.	It is relatively inert genetically.
5.	It shows normal crossing over.	The frequency of crossing over is less.
6.	It replicates during early S-phase.	It replicates during late S-phase.
7.	It is found in the acetylated form.	It is found in the non-acetylated form.
8.	It does not exhibit heteropycnosis.	It exhibits heteropycnosis.
9.	It is less affected by temperature, sex, age, etc.	It is more affected by temperature, sex, age, etc.

Heterochromatin is of two types:

1. **Constitutive Heterochromatin**
   - It is found in all cells and all stages of life cycle.
   - DNA of this type of chromatin is permanently inactive.
   - Constitutive DNA is highly repetitive.
   - It remains in condensed state throughout the cell cycle.

- It is never transcribed.
- Most of the chromatin occurring around centromere, in the telomeres, in C bands of chromosomes is constitutive heterochromatin.

2. **Facultative Heterochromatin**
   - It develops during the development of the organisms.
   - It has no permanent condensation.
   - It follows periodic dispersal and in dispersal state it is actively transcribed.
   - It may have extrachromosomal inheritance.
   - It results from the inactivation of one of the two X chromosomes in females.

Chromosome

- Chromosomes are nuclear components having the power of self-duplication.
- They carry all information for a cell to grow, thrive and reproduce.
- They are popularly known as 'hereditary vehicles' as they carry genetic information from generation to generation.
- A chromosome is not visible in the nucleus when a cell is not dividing.
- The term 'chromosome' was coined by Waldeyer (1898).
- The shape and size of chromosomes are variable. However, the size and number of chromosomes remain constant for a particular species.
- A typical eukaryotic chromosome consists of:
- **Chromonemata:** These are filamentous thread-like structures. They represent chromatids in the early stages of condensation.
- **Chromomeres:** These are bead-like structures which are linearly arranged on the chromosomes.
- These are tightly folded regions of the DNA and more clearly visible in the polytene chromosome.
- They are not visible during metaphase as the chromosome is tightly coiled.

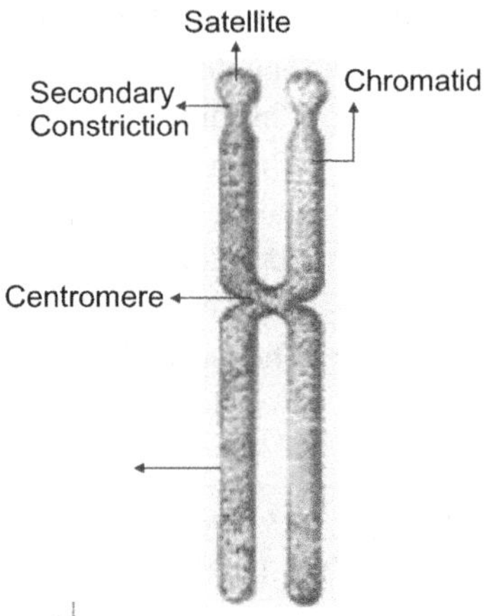

Fig. 13 Generalised structure of a eukaryotic chromosome

Chromatids

- A chromosome has two symmetrical structures at metaphase called chromatids which are held together by the centromere.
- Chromatids from one mother chromosome are called sister chromatids and chromatids from two different chromosomes are called nonsister chromatids.

Primary Constriction

- It is a narrow and constricted area of the chromosome which contains the centromere.
- Centromere is concerned with the movement of chromosomes during cell division.
- Fibrils of microtubules attach to it during cell division.
- Centromere contains a cup-like structure called kinetochore (0.20–0.25 μm).
- Kinetochore is the implantation site to which spindle microtubules are attached.
- The position of centromere on the chromosome is fixed and determines the shape of the chromosomes.

Depending on the position of centromere, a chromosome may be:

(a) **Telocentric:** Centromere is terminal in position.
(b) **Acrocentric:** Centromere is subterminal in position.
(c) **Metacentric:** Centromere is located in the middle of the chromosome.
(d) **Submetacentric:** Centromere is located slightly away from the middle point.

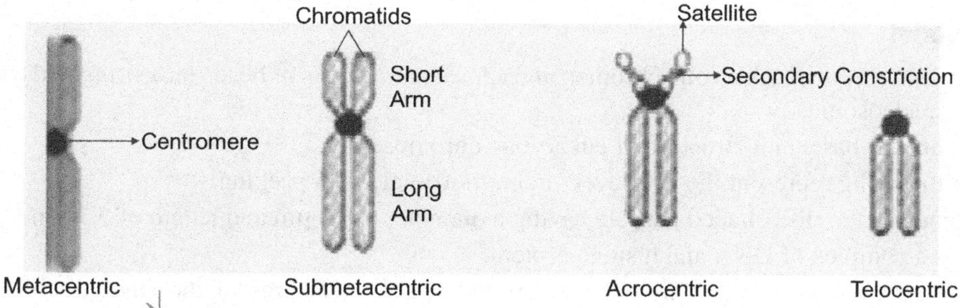

Fig. 14 Different types of chromosomes according to position of centromere

- Depending on the number of centromere, a chromosome may be:
  (a) **Acentric:** Without centromere
  (b) **Monocentric:** With one centromere
  (c) **Dicentric:** With two centromeres
  (d) **Polycentric:** With more than two centromeres
  (e) **Diffused:** Centromeres are indistinct

Secondary Constriction

- In addition to primary constriction, there is a secondary constriction.
- The difference between primary and secondary chromosomes can be observed during anaphase as a chromosome can bend only at the site of primary constriction.
- Secondary constrictions are constant in position and hence are useful in identifying a particular chromosome in a set.
- Secondary constriction may arise because the rRNA genes are transcribed very actively and thus interfere with chromosomal condensation.
- It contains the genes coding for 5.8S, 18S and 28S rRNA which induce formation of nucleolus. Hence, is named as nucleolar organiser region.
- In human beings, chromosomes 13, 14, 15, 21 and 22 are nucleolar chromosomes.

Satellite

- Satellites are round or elongated or knob-like appendages of chromosomes.
- Satellite is produced if secondary constriction is present in the distal region of the chromosome.
- The shape and size of satellites remain constant.
- It is attached with rest of the chromosome by a thin chromatin filament.

Telomeres

- Telomeres are the specialised ends of a chromosome which exhibit structural and physiological polarity.
- They are nonsticky ends of chromosomes.

- They confer stability to chromosomes.
- They are synthesised by the enzyme telomerase.
- Blackburn and Gall (1978) sequenced first telomere from *Tetrahymena thermophilia.*
- Each species has characteristic telomeric repeat sequences. However, widely divergent species may have the same telomeric unit.
- The telomeric DNA in eukaryotic cells is gradually lost in successive generations.

Nucleosome Model

- Under electron microscope, eukaryotic chromosome appears as a series of beads on a string and each bead is known as a nucleosome.
- Nucleosomes are the basic unit structure of eukaryotic chromosomes.
- The beads on the string represent the first level of chromosomal DNA packing.
- The nucleosome is a flat disc-shaped particle having a diameter of 10 μm and length of 5.7 μm.
- Nucleosome is a complex of DNA and histone protein.
- Each nucleosome consists of a histone octamer around which 1.75 turns of the DNA double helix is wrapped, which are 146 nucleotide pairs long.
- Each histone octamer consists of two flat tetramers of H-2A, H-2B, H-3 and H-4.
- Besides H-2A, H-2B, H-3 and H-4, one more histone is found called histone, H-1, which holds the two ends of the DNA double helix around the histone octamer.
- Histone H-1 is not conserved and is tissue-specific.

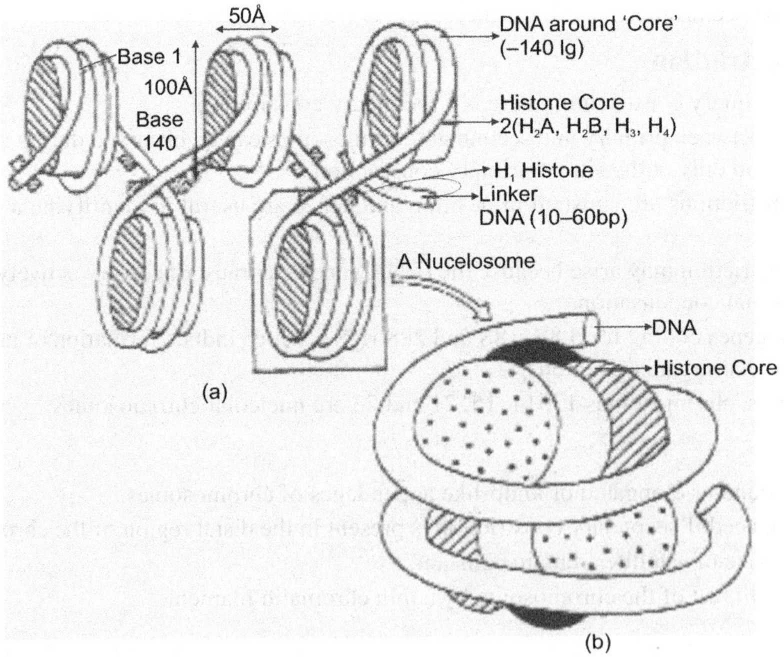

Fig. 15 Nucleosome model: (a) Repeating unit of nucleosome (b) Single nucleosome

- H-1 is chemically active and may react with H-1 of adjacent nucleosome to assist in coiling.
- Each nucleosome core particle is separated from the next by a region of linker DNA that varies in length from a few nucleotide pairs to 80 nucleotide pairs in different species.
- In each nucleosome, 142 hydrogen bonds are formed between the DNA and the histone core.
- A diploid human cell contains approximately 30 million nucleosomes.
- It has been suggested that the formation of nucleosome converts a DNA molecule into a chromatin thread about one-third of its initial length and this gives a first level of DNA packing.
- Several models have been proposed to explain how nucleosomes are packed in 30 μm chromatin fibre. One of these and the most consistent with the available data is a series of structural variations known as the zigzag model. This model suggests the following:
  (a) The 30 μm structure found in chromosome is probably a fluid mosaic of different zigzag variations.
  (b) There is variation in the length of linker DNA, and these differences in linker length may introduce further local perturbations into the zigzag structure.
  (c) Lastly, the presence of other DNA-binding proteins and DNA sequence that are difficult to fold into nucleosome punctuate the 30 μm fibre with irregular features.
- **Chemical composition:** Chemically, a chromosome is made up of DNA, RNA, histone and nonhistone proteins.
- **Functions**
  1. Chromosomes control all cellular activities.
  2. They carry transmission of characters from one generation to another.
  3. They control development and differentiation of characters.
  4. They help in the formation of nucleolus.
  5. Any change in the structure or number of chromosome leads to the formation of new characters which acts as raw material for evolution.

Polytene Chromosome

- Polytene chromosomes are the giant chromosomes which are easily visible under light microscope.
- They are found in the tissues of dipteran larvae.
- They were first observed by Balbiani (1882) in the salivary gland of the *Chironomous*.
- They are formed by endomitosis, i.e., duplication of DNA without nuclear division.
- They are visible during interphase.
- The cells of polytene chromosome do not undergo mitosis.
- Polytene chromosomes show somatic pairing, i.e., paternal and maternal chromosomes lie side by side which permits identification of deletions, duplications and inversions.
- All the polytene chromosomes may remain attached to a common chromocentre.
- Polytene chromosomes contain darkly stained dark bands and clear zones of interbands.
- Bands are Feulgen positive and are regions of high DNA concentration.
- Interbands are Feulgen negative and are regions of low DNA concentration.
- There are approximately 5,000 bands and 5,000 interbands in the genome of *Drosophila*.
- The banding patterns of polytene chromosome of *D. melanogaster* were studied by C B Bridges in 1935.
- At certain times, bands become enlarged to form swellings called puffs (Balbiani rings).
- Puffing is due to uncoiling of individual chromosomes in a band.
- Puffs represent active sites of RNA synthesis.
- Polytene chromosomes are very suitable for *in situ* hybridisation.

Lampbrush Chromosome

- Lampbrush chromosomes are found in the diplotene stage of meiosis of all animals and are the largest known chromosome.
- They are so named because of their brush-like appearance.
- They were first observed by Flemming (1882) in the oocytes of salamander and the name Lampbrush was given by Ruckert (1992).
- Lampbrush chromosomes are extensible and elastic. They can be stretched to 2×2 of their original length.
- They are best seen in the oocytes of salamander due to their high DNA content.
- They are present in the form of bivalents in which maternal and paternal chromosomes are held together by chiasmata.
- Each bivalent contains a centromere.
- Each bivalent contains four chromatids which are represented by axial filaments.
- Axial filaments consist of DNA.
- The axial filaments become tightly coiled to form chromomeres.
- Lateral loops come out from chromomeres.
- There are approximately 10,000 loops per chromosome set.
- Loops are symmetrical and each loop appears at constant position.
- Each loop has an axis made up of DNA.
- Loop DNA appears to be thick as it is covered with nonhistone proteins as well as nascent RNA molecules.
- Telomeres are lacking in lateral loops.
- Lampbrush chromosomes are an excellent material for *in situ* hybridisation of cloned DNA to RNA.
- Loops are the sites of RNA synthesis.

Cell Cycle

- A series of events that occur in a cell and results in its division and duplication is known as cell cycle.
- Each cell has a cell cycle clock which determines that a cell should or should not divide.
- A cell cycle takes 12–24 hours for most mammalian cells and approximately 20–30 minutes in *E. coli* cells.
- Nondividing cells are not considered to be in the cell cycle.
- The time required to complete one cell cycle (from the starting one cell division to the beginning of the next) is known as the generation time.
- The cell cycle is divided into two parts, viz., interphase and dividing phase.

Interphase

- It is the longest phase of the cell cycle and is the period between two consecutive cell divisions.
- Mammalian nerve cells have the longest interphase as they do not divide after birth.
- During interphase, the cell becomes enlarged.
- This phase is characterised by high rate of metabolism.
- Interphase is divided into three phases:

1. G_1 Phase

- It is the first growth phase which is also known as the first gap phase.

- It is a crucial decision point.
- The timing of G_1 phase is most variable even in different cells of the same species.
- During this phase, synthesis of RNA and proteins takes place.

2. **S Phase**
   - It is the synthetic phase during which replication of DNA occurs.
   - Histone proteins are synthesised during this phase.
   - Duplication of centrioles takes place.
   - Each chromosome is made of two chromatids.
   - It takes approximately 10 hours to replicate the 3 billion bits of information contained in the nucleus of a single human cell.

3. **G_2 Phase**
   - It is the second growth phase which is also known as the second gap phase or premitotic phase.
   - Synthesis of RNA and proteins occurs during this phase.
   - Replication of cell organelles and condensation of chromosomes takes place.
   - The cell continues to increase in size.
   - During G_2 phase, a cell contains two times (4C) the amount of DNA present in the original diploid cell (2C).
   - The lengths of S phase and G_2 phase are almost equal.

G_0 State

- A cell after cell division may withdraw from the cell cycle and enter into the resting phase called G_0 state or it may enter into the G_1 phase of the cell cycle.
- Cells in the G_0 state are viable and metabolically active.
- G_0 cells not only simply represent the absence of signals for mitosis but also an active repression of genes needed for mitosis.
- Most of the lymphocytes in human blood are in the G_0 state.
- Cells in culture can also be in the G_0 state.
- Cancer cells cannot enter the G_0 state.
- On stimulation, G_0 cells enter the G_1 phase.

Cell Cycle Checkpoints

- In a cell cycle, the following three checkpoints have been identified:

1. **G_1 Checkpoint**
   - It is the most important checkpoint in the cell cycle which is also known as restriction point.
   - It detects damaged DNA and prevents entry into the S phase.
   - G_1 checkpoint blocks entry into the S phase by inhibiting S–Cdk complex.
   - p53 levels are increased in damaged cells and block entry into the S phase.
   - A p53 mutation is the most frequent mutation that leads to cancer.
   - A p27 is a protein that binds to cyclin and CDK blocking entry to the S phase.

2. **G_2 Checkpoint**
   - It prevents the cell from entering mitosis (M phase).
   - It is triggered by Maturation-Promoting Factor (MPF) which is a cyclin CDK complex.

- The cdc 2 cyclin B kinase is a key molecule in regulating this transition.
- Damage to DNA after the S phase inhibits the action of CDK 1, thus preventing the cell from proceeding from G_2 to mitosis.

3. **Spindle Checkpoint (M-phase Checkpoint)**
   - It detects any failure of spindle fibres to attach to kinetochores and arrest the cell in metaphase.
   - It detects incorrect alignment of the spindle itself and blocks cytokinesis.
   - It triggers apoptosis, if the damage is irrepairable.
   - In mammals and in yeast, the spindle checkpoint is inactivated by mutants in MAD and BUB genes.
   - Colchicine, which inhibits spindle assembly, manifests the presence of the spindle check point.
   - A cell cycle is mainly controlled by two classes of regulatory proteins, viz., cyclins and cyclin-dependent kinases. Leland, Hartwell Hunt and Nurse were awarded the Nobel Prize in Physiology/Medicine in 1902 for their discovery of these key molecules.

Cell Division

- The process by which a cell divides to form new cells is known as cell division.
- Cell division is the basic feature of life.
- Mature new cells arise from the pre-existing cells, so a cell divides to form new cells.
- Cell divisions take place for growth, reproduction and repair work of the body.
- The main aim of cell division is to maintain the original genome.
- A human body may undergo approximately 10,000 trillion divisions in an entire lifetime.
- Cell division is mainly of two types, viz., mitosis and meiosis.

Mitosis

- Mitosis is a type of cell division in which two daughter cells are formed from a single cell, having the same number of the chromosomes as found in the original mother cell.
- Mitosis takes place both in somatic as well as reproductive cells.
- The term 'mitosis' was given by Flemming (1882).
- A cell undergoing mitosis is called mitocyte.
- Mitosis may be acentric (without centromere) as in plants or centric (with centromere) as in animals.
- Generally, the process of mitosis is completed within 1–2 hours.
- The steps of mitosis are controlled by different genes and if mitosis is not regulated properly, it may result in health problems such as cancer.
- The process of mitosis involves two steps:

1. **Karyokinesis**
2. **Cytokinesis**

   1. **Karyokinesis:** Karyokinesis is the division of the nucleus. It involves the following stages:

      (a) Prophase
      - Prophase is the first and longest stage of mitosis.
      - The cell becomes spheroid and there is an increase in viscosity and refractivity.
      - Each chromosome consists of two chromatids jointed together by a centromere.
      - The two centrioles start to move in opposite poles of the cell.
      - Around each centriole, radiating fibres appear called asters.

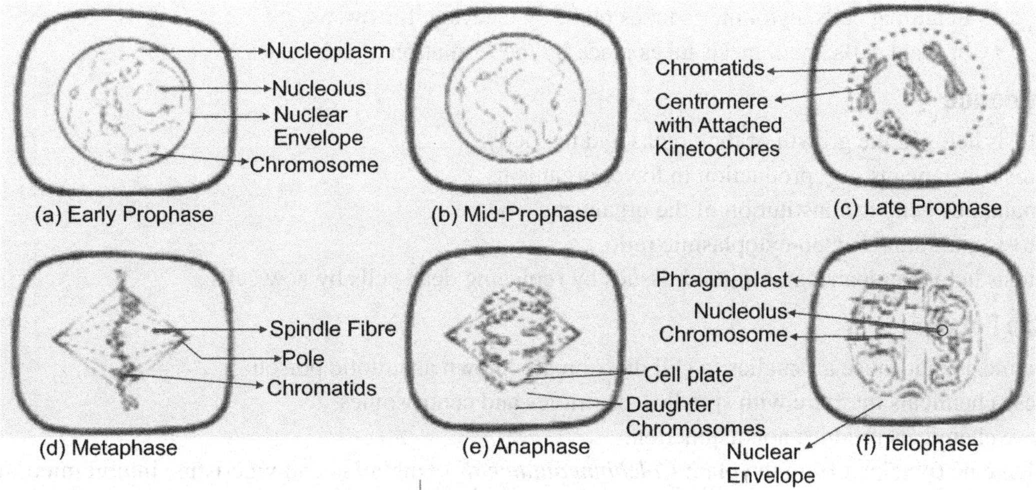

Fig. 16 Different stages of mitosis

- Asters are cytoplasmic in origin.
- The two asters are connected by spindle fibres.
- Spindles are formed from microtubules and are made up of tubulin.
- The aster, centrioles and spindle together form the mitotic apparatus.

(b) Metaphase

- Nuclear membrane and nucleolus disappear.
- The chromosome becomes arranged on equatorial plate.
- The chromosomes attain their full contraction.
- Metaphase is the most suitable stage for studying the morphological characterisation, banding and counting of the chromosomes.

(c) Anaphase

- Anaphase is the most dynamic and shortest stage of mitosis.
- The centromere divides. The sister chromatids separate and move towards opposites poles.
- Each chromatid has its own centromere.
- At the end of anaphase, there is a group of chromosomes at each pole.
- Mitotic anaphase has the same number of chromosomes as metaphase but half number of chromatids.
- The chromosomes may assume shape of 'J', 'V' or rod-like.

(d) Telophase

- It is the last stage of mitosis.
- Chromosomes at each pole begin to uncoil.
- Nuclear membrane and nucleolus appeared around each group of chromosomes, resulting in the formation of two daughter nuclei.

2. Cytokinesis

- Cytokinesis is the division of the cytoplasm.

- In animal cells, cytokinesis takes place by cleavage furrow.
- In plant cells, cytokinesis takes place by the formation of cell plates.

Significance

- Mitosis helps in the growth of the organs and the body.
- Mitosis is a means of reproduction in lower organisms.
- It maintains genetic constitution of the organisms.
- Mitosis maintains nucleo-cytoplasmic ratio.
- Mitosis helps in repairing of injured tissues by replacing dead cells by new cells.

Mitotic Poison

- Chemicals that cause anomalies in cell division are known as mitotic poison.
- These chemicals interfere with spindles, centrioles and centrosomes.
- These chemicals produce nondisjunction.
- Colchicine (obtained from the plant *Colchium autumnale*) vinblatine and vincristine inhibit microtubules assembly, whereas diezepam prevents separation of centrioles.
- Microorganisms remain unaffected by mitotic poison as they lack spindle apparatus.

Meiosis

- Meiosis is a special type of cell division as a result of which four daughter cells are formed, in which the number of chromosomes is reduced to half of that in the original cell.
- Meiosis occurs in reproductive cells.
- A cell undergoing meiosis is called meiocytes.
- Meiosis is also known as reductional division as it reduces the chromosome number to half.
- Meiosis involves two nuclear divisions with only one replication of DNA.
- Meiosis is essential for sexual reproduction.
- Meiosis was first discovered and described in the eggs of sea urchin by Oscar Hertwig in 1876.
- The term 'meiosis' was coined by Farmer and Moore (1905).
- Bdelloid rotifers have lost the ability to perform meiosis division, while meiosis does not occur in archaea or bacteria.
- Meiosis may be:
- **Gametic:** It is also known as terminal meiosis. It occurs at the time of gamete formation and is found in animals.
- **Zygotic:** It is also known as initial meiosis and occurs immediately after zygote formation. It is found in lower plants.
- **Sporic:** It is also known as intermediate meiosis. Meiosis division occurs in between the formation of zygote and gamete, resulting in the formation of haploid megaspores and microspores. Such type of meiosis is a characteristic of higher plants and some thallophytes.

Meiosis involves two divisions:

1. **Heterotypic division:** It is the first division as a result of which the chromosome number is reduced to half. It is also known as reductional division.
2. **Homotypic division:** It is the second division which is totally mitotic in nature.

1. **Heterotypic division:** It involves the following stages:

(i) Interphase: Before meiosis begins, the genetic material is duplicated.

(ii) Prophase I: It is the longest and most important phase of the meiosis. It is divided into six substages:

(a) Leptotene

- Chromosomes become more distinct.
- Chromosomes bear bead-like structures called chromomeres.
- The number, size and position of chromomeres are characteristic on a chromosome.
- Chromosome may develop a basket-like arrangement called bouquet and this stage is called bouquet stage.
- Some plant cells form a tangle of threads called synizetic knot (e.g., *Trillium*).

(b) Zygotene

- Chromosomes become shorter and thicker.
- Pairing of homologus chromosomes takes place.
- Paired chromosomes are called bivalents and the process of pairing is called synpasis.
- The number of bivalents is half of the number of diploid chromosomes.
- The pairing of homologous chromosomes may begin from the centromere (procentric), or from either ends (proterminal), or from anywhere (at random).
- Of the two homologous chromosomes, one is derived from the male parent and other from the female parent.
- As a result of pairing, a tripartite structure is formed called syneptonemal complex.
- Syneptonemal complex first appeared during zygotene.
- Syneptonemal complex is lacking in the meiosis of male diptera.

(c) Pachytene

- Each bivalent splits longitudinally, so in each bivalent the number of chromatids is four (tetravalent). This stage is called tetrad stage.
- Exchange of chromatids takes place between nonsister chromatids. This process of exchange of chromatids is known as crossing over.
- As a result of crossing over, a cross-shaped (X) structure is formed, called chiasmata.
- Pachytene is the longest stage in mammalian spermatogenesis.
- Pachytene may last for days, weeks or even years.

(d) Diplotene

- Separation of homologous chromosomes begins. This process is known as terminalisation.
- Due to contracting tendency of chromosomes, the chiasmata are pulled towards the ends of paired homologous chromosomes.
- Syneptonemal complex begins to break down.
- Diplotene may last for months or years.

(e) Diakinesis

- The process of terminalisation is completed.
- Chromosomes condense and thicken.
- Nuclear membrane and nucleolus begin to disintegrate.

(iii) Metaphase I

- Formation of spindle begins.

- Bivalents become arranged on equatorial plate.
- Bivalents are arranged in such a way that their arms lie over the equator while the centromeres are directed towards the poles.
- There is a 50–50 chance for the daughter cells to get the mother's or father's homologue as orientation is random, with either homologue on a side.

(iv) Anaphase I

- The homologous chromosomes separate from each other and move to opposite poles.
- Each pole has haploid number of chromosomes. In this stage, reduction in the number of chromosomes takes place.
- There is no division of centromeres.

(v) Telophase I

- Telophase I is similar to telophase of mitosis except that there is only one set of (replicated) chromosome in each cell.
- Nuclear membrane and nucleolus reappear around each group of the haploid set of chromosomes at each pole.
- Thus, two daughter haploid nuclei are formed; each with two chromatids.
- Sometimes telophase is absent.

Cytokinesis

- Cytokinesis results in the formation of two daughter cells having haploid number of chromosomes.

Interkinesis

- The period between telophase I and prophase II is called interkinesis.
- It is quite a short period or completely absent.
- In *Trillium* and odonata, there is no telophase I and interphase.

Homotypic Division

It is similar to mitosis but lacks the S phase. It involves the following stages:

1. **Prophase II**
   - The chromosomes condense and each chromosome consists of two chromatids, viz., one parental and the other recombinant.
   - Nuclear membrane and nucleolus disappear.
   - DNA does not replicate.
2. **Metaphase II**
   - Spindle fibres are formed.
   - Chromosomes arrange on the equatorial plate.
3. **Anaphase II**
   - The two chromatids of chromosome separate and move towards the opposite poles.
   - Centromere divides, so each chromatid has its own centromere.
4. **Telophase II**
   - Nuclear membrane and nucleolus reappear around each group of chromosomes.
   - Thus, four haploid daughter nuclei are formed.
   - Lastly, cytokinesis takes place as a result of which four haploid daughter cells are formed.

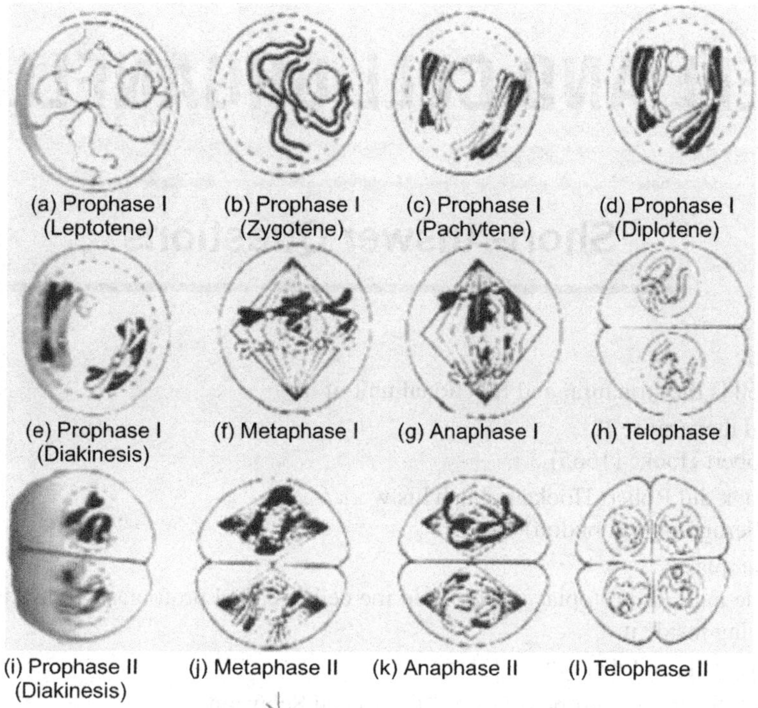

Fig. 17 Different stages of meiosis

Significance

- Meiosis is a means of gametes formation.
- Meiosis maintains constancy in chromosome number from generation to generation.
- It produces genetic variation (due to crossing over) in organisms which acts as raw material for evolution.

CELL AND CELL ORGANELLES

Short-Answer Questions

1. What is cell?
   Answer: Cell is the structural and functional unit of life.
2. Who coined the term cell?
   Answer: Robert Hooke (1665)
3. In which book did Robert Hooke publish his work?
   Answer: Micrographia (London)
4. What is protoplast?
   Answer: The mass of protoplasm present in the cell is called protoplast. It consists of plasmalemma and everything inside it.
5. Who proposed the cell theory?
   Answer: Cell theory was proposed by Schleiden and Schwann.
6. What are organelles?
   Answer: Organelles are membrane-bound structures present in a cell. They remain embedded in the cytoplasm.
7. What are microfilaments?
   Answer: Microfilaments are fine unbranched proteinaceous structures which are generally arranged in bundles. They are present in all eukaryotic cells.
8. Name three types of plastids.
   Answer: (a) Leucoplast
   (b) Chromoplast
   (c) Chloroplast
9. Which ion is present in chloroplast?
   Answer: Mg^{++}
10. What are thylakoids?
    Answer: Thylakoids are the structural and functional units of the chloroplast. They are the site of photosynthesis.
11. What are quantasomes?
    Answer: Quantasomes are the photosynthetic units.
12. Where mitochondrial proteins are synthesised?
    Answer: Mitochondrial proteins are synthesised by cytosolic as well as matrix ribosomes.
13. Name the specialised protein present in the outer membrane of the mitochondria.
    Answer: Porin

14. Write characteristics of the human mitochondrial genomes.
Answer: (a) Size – 16.6 kbp
(b) Nature of DNA molecule – ds circular DNA molecule
(c) Number of DNA molecules – More than one
(d) Repetitive DNA – Very little
(e) Introns – Absent
(f) Inheritance – Maternal

15. Name some diseases known to be associated with mitochondrial DNA (mtDNA).
Answer: Some of the diseases known to be associated with mtDNA are Parkinson's disease, Huntigton's disease, Leber's optic neuropathy and Kearns–Sayre syndrome.

16. Which enzyme of Krebs cycle is not present in the mitochondrial matrix?
Answer: Succinate dehydrogenase

17. Name two biochemical events that do not occur in the cytoplasm.
Answer: (a) Krebs cycle
(b) Oxidative phosphorylation

18. Name the largest electron transport complex.
Answer: Complex I NADH dehydrogenase, containing FMN and Fe-S.

19. Define plasmodesmata.
Answer: Primary cell wall contains small pores through which cytoplasms of adjacent a cells communicate by means of small cytoplasmic channels. These small cytoplasmic channels are called plasmodesmata.

20. What is glycocalyx?
Answer: Glycocalyx (cell coat) is the carbohydrate coating of the cell surface which protects a cell from mechanical and chemical damage and also mediate cell–cell adhesion.

21. What is transverse diffusion?
Answer: The movement of molecules from one membrane surface to another is called transverse diffusion (flip flop).

22. Distinguish between 70S and 80S ribosomes.
Answer: The main differences between 70S and 80S ribosomes are as follows:

Table 6

	70S	*80S*
Distribution	Prokaryotes, mitochondria and chloroplasts	Eukaryotes
Size	Smaller	Larger
Molecular weight	3 million	4–5 million
Sub-units	Smaller sub-unit of 30S and larger sub-unit of 50S	Smaller sub-unit of 40S and larger subunit of 60S
RNA	3 molecules 30S sub-unit of 16S 50S sub-unit of 23S and 5S	4 molecules 40S sub-unit of 18S 60S sub-unit of 28S, 5.8S and 5S
Nuclear proteins	30S – 21 50S – 31	40S – 33 60S – 49
Total proteins	50–60	70–80

23. Lysosomes have been regarded as a recycling centre. Why?

Answer: Lysosomes have been described as a recycling centre because they free metabolites from the worn out cell organelles by digesting them.

24. Name some diseases caused by malfunctioning of lysosomes.
Answer: Malfunctioning of lysosomes causes a large number of diseases in humans such as Tay–Sachs disease, Nieman–Pick disease, Farber's disease, Hunter's syndrome, Hurler syndrome, Scheic syndrome, Sanfilippo syndrome, etc.

25. What are peroxisomes?
Answer: Peroxisomes are oval to round-shaped microbodies and are found in all eukaryotic cells. Peroxisomes protect cells from the toxic effect of hydrogen peroxide.

26. Name the largest cell organelle.
Answer: Nucleus

27. Name the second largest cell organelle.
Answer: Mitochondria

28. Name three nonmembranous cell organelles.
Answer: (a) Ribosomes
(b) Cilia
(c) Flagella

29. How acidic pH is maintained in lysosome?
Answer: An H^+ pump in the lysosomal membrane uses the energy of ATP hydrolysis to pump H^+ into lysosome and thus maintains the acidic pH of the lumen.

30. Name the cell organelle that recycles decomposed products of proteins, lipids, carbohydrates and forms urea.
Answer: Mitochondria

31. Name the cells that contain myofilaments.
Answer: Muscle cells

32. Name the principal microtubule-organising centre of the cell.
Answer: Centrosome

33. Which organelle is known as the protein-packing plant?
Answer: Golgi complex

34. Name the proteins that form gap junctions.
Answer: Transmembrane proteins called connexins form gap junctions.

35. Name the first organelle to be discovered.
Answer: Nucleus

36. Name the largest chromosome known.
Answer: Lampbrush chromosome

37. In which stage of meiosis is the lampbrush chromosome found?
Answer: Dipoltene

38. Who discovered polytene chromosome?
Answer: Balbiani (1882) in the salivary gland of *Chironomous*.

39. Which enzyme helps in the formation of telomere?
Answer: Telomerase

40. Name three organisms in which the chromosome number is 46.
Answer: (a) Sable antelope (*Hippotragus niger*)

(b) Reeves's Muntjac (*Muntiacus reevesi*)
(c) *Homo sapiens*

41. What are glyoxysomes?
Answer: Glyoxysomes are specialised peroxisomes found in some plant cells. They contain enzymes of glyoxylate cycle.

42. Who proposed the Fluid mosaic model?
Answer: Singer and Nicolson (1972)

43. What is the function of tight junctions?
Answer: Tight junctions bind cells together to form a leak-proof sheet.

44. Name the animal having only one chromosome.
Answer: (a) *Parascaris univalens* (nematode)
(b) *Myrmecia pilosula* (Ant)

45. What is nucleolus?
Answer: Nucleolus is a darkly stained body found in the nucleus. It is involved in the formation of ribosomes.

46. How is the size of a cell controlled?
Answer: The size of a cell is controlled by the:
(a) Nucleo-cytoplasmic ratio
(b) Ratio of cell surface to the cell volume
(c) Rate of the metabolism

47. When is a centriole termed as a basal body?
Answer: When a centriole forms a cilium or flagellum, it is termed as basal body.

48. What is GERL?
Answer: A complex of Golgi complex, endoplasmic reticulum ribosome and lysosome.

49. Name the cell organelle which forms acrosome.
Answer: Golgi complex

50. Write two similarities between cilia and flagella.
Answer: (a) Both develop from the basal bodies
(b) Both have similar chemical composition

51. Name the eukaryotes that lack microtubules.
Answer: *Amoeba* and slime molds

52. How are ribosomes classified?
Answer: Ribosomes are classified on the basis of sedimentation rate.

53. What is tonoplast?
Answer: The membrane surrounding a sap vacuole is called tonoplast.

Long-Answer Questions

1. Draw a labelled diagram of an animal cell (no description). How does it differ from a plant cell?
2. Describe Fluid mosaic model. Mention important functions of cell membrane.

3. Describe the structure of chromosomes in the light of recent researches.
4. Justify the statement: Mitochondria are the powerhouse of the cell.
5. Describe ultrastructure of ribosomes and mention their role in protein synthesis.
6. Give an account of structure and function of the nucleus. Why is nucleus regarded as the controlling centre of a cell?
7. Write short notes on:
   (a) Role of lysosomes in cell pathology
   (b) Mitochondrial DNA
   (c) Rough endoplasmic reticulum
   (d) Nucleosome
   (e) Polytene chromosome
   (f) Centrioles
8. Describe the structure of nucleolus and discuss its role in the formation of ribosomes.
9. Distinguish between:
   (a) Heterochromatin and euchromatin
   (b) Smooth and rough endoplasmic reticulum
   (c) Prokaryotic cell and eukaryotic cell
   (d) Animal cell and plant cell
   (e) Ribosome and lysosome
   (f) Primary cell and secondary cell wall
   (g) Cilia and flagella
10. Describe the ultrastructure and function of the Golgi complex.
11. Describe the mechanism of electron transport chain.

CELL CYCLE

Short-Answer Questions

1. What is cell cycle?
   Answer: A series of events that occurs in a cell and results in its division and duplication is known as cell cycle.
2. Name the two main phases of cell cycle.
   Answer: (a) Interphase
   (b) M Phase
3. Give three processes that take place in interphase.
   Answer: (a) Synthesis of energy-rich compounds
   (b) Synthesis of DNA, histone and other proteins
   (c) Duplication of centrioles
4. What is generation time?
   Answer: The time required to complete one cell cycle (from the starting one cell division to the beginning of the next) is known as generation time.
5. Which cells have the longest interphase?
   Answer: Mammalian nerve cells have the longest interphase as they do not divide after birth.
6. During which phase of the cell cycle are histone proteins synthesised?
   Answer: S (synthetic) phase

7. What is G_0 state?
Answer: A cell after cell division may withdraw from the cell cycle and enter into resting phase, called G_0 state or it may enter the G_1 phase of the cell cycle.
8. Name the cells that never enter the G_0 state.
Answer: Cancer cells
9. What is the average cell cycle span for mammalian cells?
Answer: 24 h
10. How many checkpoints have been identified in cell cycle?
Answer: Three, which are:
(a) G_1 checkpoint (b) G_2 checkpoint
(c) M (metaphase) phase checkpoint
11. How is a cell cycle controlled?
Answer: A cell cycle is mainly controlled by two classes of proteins called cyclins and cyclin-dependent protein kinases.
12. When cells are stimulated to divide which cyclin appears first?
Answer: G1cyclin–CDK (cyclin-dependent protein kinase) complexes

Long-Answer Questions

1. What is cell cycle? Describe the process of cell cycle.
2. What are cell checkpoints? Describe the regulation of activity of cyclin-CDK (cyclin-dependent protein kinases) complexes.
3. Write short notes on:
(a) Cell cycle checkpoints (b) Role of Rb protein in cell cycle regulation
(c) Role of p^{53} in cell cycle regulation

CELL DIVISION

Short-Answer Questions

1. What is mitosis?
Answer: Mitosis is a type of cell division in which two daughter cells are formed from a single cell having the same number of the chromosomes as found in the original mother cell.
2. Mitosis is divided into how many phases?
Answer: Mitosis is divided into four phases:

(a) Prophase (b) Metaphase
(c) Anaphase (d) Telophase

3. During which phase of cell division, spindle is formed and nucleolus disappears?
Answer: Formation of spindle begins in prophase and is completed in metaphase. Nucleolus disappears in telophase.

4. Which stage of mitosis is most suitable for studying the morphological characterisation, banding and counting of the chromosomes?
Answer: Metaphase

5. Name the stage of cell cycle during which:
(a) Chromosomes moved to spindle equator
(b) Centromere split and chromatids separate
Answer: (a) Metaphase
(b) Anaphase

6. What is endomitosis?
Answer: Endomitosis is the process of chromosome duplication without cell division.

7. Which is the most dynamic and shortest stage of the mitosis?
Answer: Anaphase

8. What is called uncontrolled mitotic cell division?
Answer: Neoplasia that occurs due to mutation in the genetic material.

9. Name the adult tissues where mitosis is almost absent.
Answer: Mitosis is almost absent in the adult nervous tissues and striated muscle (skeletal and cardiac).

10. Give examples where mitosis is more frequent and less frequent.
Answer: In vertebrates, mitosis is more frequent in tissues that require intense renewal like epithelial cells and bone marrow, whereas it is slow in those tissues that require less renewal like adult bone marrow and connective tissue.

11. Why is mitosis also called equational division?
Answer: Because the number of chromosomes is the same in the parent and daughter cells.

12. What is the significance of mitosis?
Answer: Mitosis is essential for:
(a) Growth (b) Asexual reproduction of eukaryotes
(c) Embryonic development (d) Renewal of tissues

13. What is mitotic poison?
Answer: (a) Chemicals that cause anomalies in cell division are known as mitotic poison.
(b) These chemicals interfere with spindles, centrioles and centrosomes.
(c) These chemicals produce nondisjunction.

14. Why do microorganisms remain unaffected by mitotic poison?
Answer: Because they lack mitotic poison.

15. What happens if errors occur in mitotic division?
Answer: Errors in mitotic divisions can either kill a cell through apoptosis or cause mutations that lead to cancer.

16. Write three main differences between mitosis and meiosis?
Answer:

	Mitosis	Meiosis
(a)	Mitosis takes place in somatic cell.	Meiosis takes place in reproductive (gametic) cells.
(b)	The number of chromosomes in daughter cells remain the same.	The number of chromosomes is reduced to half in daughter cells.
(c)	As a result of mitosis, two daughter cells are produced.	As a result of meiosis, four daughter cells are produced.

17. What type of cell division produces haploid cells?
Answer: Meiosis

18. Name the stages during which the following events occur:
(a) Pairing of homologous chromosome
(b) Crossing over
(c) Reduction in chromosome number
Answer: (a) Zygotene
(b) Pachytene
(c) Anaphase I

19. What is synapsis?
Answer: During zygotene stage of prophase I, homologous chromosomes are paired which are termed as bivalent and the process of pairing is known as synapsis.

20. Define chiasmata.
Answer: During pachytene stage, as a result of exchange of chromosomal segments between homologous chromosomes (crossing over), a cross-shaped structure is formed which is known as chiasmata.

21. During which stage of meiosis did synaptonemal complex first appear?
Answer: Zygotene

22. Which is the longest stage of meiosis?
Answer: Prophase I

23. During which stage of meiosis does separation of homologous chromosomes takes place?
Answer: Anaphase I

24. Define meiosis.
Answer: Meiosis a special type of cell division as a result of which four daughter cells are produced, in which the number of chromosomes is reduced to half of original cell.

25. Who discovered and described meiosis for the first time?
Answer: Oscar Hertwig (1876) discovered and described meiosis for the first time in the eggs of sea urchin.

26. Name the three different ways through which synapsis takes place during zygotene?
Answer: (a) Procentric – Pairing begins from centromeres
(b) Proterminal – Pairing begins from either end of the chromosome
(c) At random – Pairing takes place from anywhere

27. In which stage of meiosis does division of the centromere take place?
Answer: Anaphase II

28. Distinguish between karyokinesis and cytokinesis.
Answer: Karyokinesis is the division of the nucleus, whereas cytokinesis is the division of the cytoplasm.

29. Name the stage of meiosis where disjunction takes place.
Answer: Anaphase I

30. How does anaphase II differs from anaphase I?
Answer: Anaphase II differs from anaphase I in having single-stranded chromosomes and separated chromatids.

31. Bouquet stage is the characteristic of which stage of cell cycle in animal cells?
Answer: Leptotene stage (Prophase I)

32. Name the stage of meiosis in which foundation of variation is laid down.
Answer: Pachytene

33. Name the stages of meiosis during which:
(a) Terminalisation of chiasmata begins –
(b) Terminalisation is completed –
Answer: (a) Diplotene
(b) Diakinesis

Long-Answer Questions

1. Draw a well-labelled diagram of mitosis (no description). What is its significance?
2. Give an account of meiosis in animal cells.
3. Give a comparative account of mitosis and meiosis.
4. Write short notes on:
(a) Pachytene (b) Diplotene
(c) Anaphase II (d) Synaptonemal complex
(e) Mitotic apparatus
5. Draw diagrams of different phases of meiosis I. Give the significance of meiosis.

CELL AND CELL ORGANELLES

Multiple-Choice Questions

1. Which one of the following statements is incorrect?
   (a) The DNA of mitochondria is rich in G:C ratio.
   (b) Ribosomes are produced in nucleolus.
   (c) RBC and mature sperm contain 80S ribosomes.
   (d) Size of nucleus depends on the number of chromosomes.
2. Which one of the following is an incorrect match.
   (a) Tonoplasts – Vacuoles (b) Histones – Prokaryotes
   (c) Mitochondria – Beta-oxidation of fatty acids (d) Golgi complex – Traffic police of the cell
3. Cell organelle with single membrane is:
   (a) Endoplasmic reticulum (b) Glycosome
   (c) Lysosome (d) All of the above
4. Endosymbiotic theory is related with the origin of:
   (a) Nucleus (b) Mitochondria (c) Golgi complex (d) Centrosome
5. Cell coat is also known as:
   (a) Glycocalyx (b) Microsome (c) Glycosome (d) Pectofibrin
6. The side of Golgi complex closest to the nuclear envelope and endoplasmic reticulum is known as:
   (a) *Cis* or forming face (b) *Trans* or maturing face
   (c) Preparing face (d) Terminal face
7. What is correct about cilium?
   (a) Short microtubular organelle (b) Used for locomotion
   (c) Helps in creating currents in the surrounding fluid (d) All of the above
8. Lysosomes of leucocytes and monocytes are important in defence against:
   (a) Bacteria (b) Viruses (c) Both (a) and (b) (d) None of the above
9. Compartment where uncoupling of receptor and ligand occurs is:
   (a) Endosome (b) Oxysome (c) Basal body (d) Crista
10. Which one of the following drugs disrupts actin microfilaments without affecting the microtubules?
    (a) Colchicine (b) Chloroquinine (c) Cytochalasin B (d) Streptomycin
11. Electrical coupling between cells depends on:
    (a) Tight junctions (b) Gap junctions (c) Desmosomes (d) None of these
12. Mitoplast is related with:
    (a) Mitochondria (b) Chromosome (c) Golgi complex (d) Plastid

13 Permeability is regulated by:
(a) K^+ (b) OH^- (c) Na^+ (d) Ca^{++}

14. Nuclear region within a prokaryotic cell is known as:
(a) Nucleosome (b) Nucleoid (c) Viroid (d) Chromocentre

15. Largest cell organelle is the:
(a) Nucleus (b) Mitochondrion (c) Golgi complex (d) Centrosome

16. An exception of cell theory is:
(a) Virus (b) Viroids (c) Prions (d) All of the above

17. Ribosomes are absent in:
(a) RBC (b) Sperms
(c) Both RBC and sperms (d) Chloroplasts and mitochondria

18. Which one of the following enzymes destroys H_2O_2?
(a) Catalase (b) Urate oxidase
(c) D-Amino oxidase (d) α-Hydroxylic acid oxidase

19. Most of the cytosolic H_2O_2 is produced by:
(a) Peroxisomes (b) Golgi complex
(c) Mitochondria (d) Mitochondria and membranes of ER

20. Match column I with column II and select the correct answer using answer codes.

Column I		Column II
(A) Lysosome	1.	Photorespiration
(B) Golgi complex	2.	Protein synthesis
(C) Peroxisome	3.	Formation of acrosome
(D) Rough endoplasmic reticulum	4.	Converts cellular polymers to monomeres

Answer codes:

	A	B	C	D
(a)	4	1	3	2
(b)	3	4	2	1
(c)	4	3	1	2
(d)	1	3	4	2

21. Nucleosome lacks:
(a) H_1 (b) H_2A and H_2B (c) H_3 (d) H_4

22. Hyperactivity of lysosome results in:
(a) Gout (b) Lung fibrosis (c) Arthritis (d) All of the above

23. Which one of the following cell organelles is related with glycosylation?
(a) Mitochondria (b) Lysosome
(c) Golgi complex (d) Endoplasmic reticulum

24. Specific density of Golgi complex is:
(a) More than the mitochondria (b) More than the endoplasmic reticulum
(c) Less than the mitochondria and endoplasmic reticulum (d) Equal to the mitochondria and endoplasmic reticulum

25. Consider the following statements:
(A) Duplication of mitochondrial DNA occurs during the S phase
(B) Human erythrocytes have well-developed mitochondria
(C) Mitochondria are stained with Sudan Black B
(D) Mitochondria are autonomous cell organelle

The correct statements are:
(a) A and B (b) A and D (c) B and C (d) None of the above

26 What is common between mitochondria and chloroplasts?
(a) Naked DNA (b) Semi-autonomous nature
(c) Production of ATP (d) All of the above

27. Mitochondrial swelling can be induced by:
(a) Ca^{++} (b) Thyroxin (c) Phosphate (d) All of the above

28. What is correct about microtubules?
(a) Microtubules help in maintaining the shape of a cell.
(b) They help in intracellular transport.
(c) Along with microfilaments, they assist in cell movement.
(d) All of the above

29. Nucleoli are larger and more in number in cells that are actively involved in:
(a) Lipid synthesis (b) Protein synthesis
(c) Carbohydrate synthesis (d) Beta-oxidation of fatty acids

30. Lomasomes are the:
(a) Projections from adjacent cells
(b) Foldings of plasmalemma found in fungal cells
(c) Thickened area of plasma membrane having hair-like structures
(d) Infoldings of plasma membrane having respiratory enzymes

31. Which one of the following is reduced during active stage of mitochondria?
(a) Krebs cycles (b) Oxidative phosphorylation
(c) Electron transport chain (d) None of these

32. Which form of DNA lacks guanine base?
(a) B and D (b) Z (c) A and Z (d) D and E

33. Peroxisomes are concentrated in:
(a) Brown fat (b) Liver (c) Kidney (d) Adipose cells

34. Which one of the following organelles is responsible for the disappearance of tail in tadpole larva during metamorphosis?
(a) Lysosome (b) Endoplasmic reticulum (c) Golgi complex (d) Mitochondria

35 Match column I with column II and select the correct answer using answer codes.

Column I	Column II
(A) Osteogenesis	1. Peroxisome
(B) Secretion	2. Chloroplast
(C) Breakdown of H_2O_2	3. Lysosome
(D) Synthesis of food in the presence of sunlight	4. Golgi complex

Answer codes:

	A	B	C	D
(a)	3	4	1	2
(b)	3	1	4	2
(c)	4	3	2	1
(d)	1	3	4	2

36. Deficiency of the lysosomal enzyme, alpha *iduronidase* causes:
(a) Hurler syndrome (b) Hunter syndrome

(c) Lesch–Nyhan syndrome (d) Beckwith–Wiedmann syndrome

37. Which one of the following is a lysosomal storage disorder?
    (a) Hunter syndrome (b) Tay–Sachs disease
    (c) Noorie's disease (d) Osteogenesis imperfecta
38. Kearns–Sayre syndrome is caused due to delection of:
    (a) A part of mitochondrial DNA (b) Chromosome 5
    (c) Chromosome 9 (d) Chromosome 15
39. The Leber's blindness is caused by:
    (a) Hypoactivity of lysosome (b) Hyperactivity of lysosome
    (c) A point mutation in mitochondrial DNA (d) A point mutation in nuclear DNA
40. Photophosphorylation coupling factor is found in:
    (a) Mitochondria (b) Endoplasmic reticulum
    (c) Chloroplasts (d) Glyoxysomes
41. Which one of the following histones is not conserved between species?
    (a) H_1 (b) H_2A and H_2B (c) H_3 (d) H_4
42. Humans have approximately ______________ pg of DNA per haploid genome.
    (a) 2 (b) 3 (c) 30 (d) 84
43. Which one of the following diseases is associated with mitochondrial DNA?
    (a) Chronic opthalmoplegia (b) Parkinson's disease
    (c) Kearns–Sayre syndrome (d) All of the above
44. Corneal clouding develops in:
    (a) Hurler syndrome (b) Hunter syndrome
    (c) Parkinson's disease (d) Sickle cell anaemia
45. Which one of the following is a lysosomal storage disease?
    (a) Scneic syndrome (b) Sanfilppo syndrome (c) I-cell disease (d) All of the above
46. Which one of the following statements is incorrect?
    (a) Opthalmoplegia is caused by the spontaneous deletion of chunk of mitochondrial DNA.
    (b) Opthalmoplegia involves maternal inheritance.
    (c) Hunter syndrome is due to deficiency of lysosomal sulphoiduronate sulphate.
    (d) Mitochondrial DNA plays a role in the ageing process.
47. Secondary lysosomes contain:
    (a) Hydrolytic enzymes (b) Undigested food
    (c) Primary lysosomes and food particles to be digested (d) Enzymes which digest own cell organelles
48. Mitochondria are absent in human:
    (a) Liver cells (b) Brain cells (c) Erythrocytes (d) Osteoblasts
49. Pg value is the:
    (a) Amount of DNA present in the genome (b) Amount of DNA present in the gamete
    (c) Total number of chromosomes in a cell (d) Length of DNA in a cell
50. Which one of the following proteins is responsible for loosening of cell wall and cell expansion?
    (a) Expansion (b) Chaperons (c) Both a and b (d) Spectrin
51. Which one of the following acts as a stabiliser for lysosomes?
    (a) Cortisone (b) Cholesterols (c) Antihistamine (d) All of the above

52. In mitochondrion, lowest pH occurs in:
(a) Matrix (b) Intermembrane space (c) Cristae (d) Ribosomes
53. Shape of the cell is controlled by:
(a) Viscosity of cytoplasm (b) Skeleton and function
(c) Cell wall, position and age (d) All of the above
54. The cell organelle related with the movement of materials to the exterior of the cell is:
(a) Chromosome (b) Golgi complex
(c) Endoplasmic reticulum (d) Mitochondria
55. The part of sperm formed by Golgi complex is the:
(a) Head (b) Middle piece (c) Axial filament (d) Acrosome
56. The ability of cell membrane is determined by:
(a) Proteins (b) Carbohydrates (c) Phospholipids (d) None of these
57. Match column I with column II and select the correct answer using answer codes.

Column I	Column II
(A) Smooth endoplasmic reticulum	1. Protein units of microtubules
(B) Nucleosomes	2. Junction of cell adherence
(C) Desmosomes	3. Synthesis of lipids
(D) Tubulin	4. Basic structure of chromatin fibres

Answer codes:

	A	B	C	D
(a)	3	4	2	1
(b)	3	1	2	4
(c)	2	1	3	4
(d)	4	2	1	3

58. Centrioles are not present in:
(a) *Amoeba* (b) Some algae and fungi
(c) Higher gymnosperms (d) All of the above
59. Maintenance of the Golgi complex in healthy condition depends on the presence of ______________ in the cell.
(a) Mitochondrion (b) Centrosome (c) Nucleus (d) Ribosome
60. Which one of the following histones is rich in lysine?
(a) H_1 (b) H_2A (c) H_3 (d) H_4
61. Arginine-rich histones are:
(a) H_2A and H_2B (b) H_1 and H_3 (c) H_3 and H_4 (d) H_2B and H_3
62. An organelle without cell membrane:
(a) Centrosome (b) Ribosome (c) Nucleolus (d) All of the above
63. Centriole is rich in:
(a) Tubulin (b) *ATPase*
(c) Both tubulin and *ATPase* (d) *ATPase* and *catalases*
64. Which one of the following statements is incorrect?
(a) C-value is characteristic of species.
(b) C-value is expressed in picograms or base pairs.
(c) Mitochondria are yellowish due to riboflavin.
(d) Lifespan of mitochondria is about 50–100 days.

65. Which one of the following is not present in a complete cell?
(a) Ribosome (b) Lysosome (c) Chromosome (d) Microsome

66. Which one of the following ion is related with the formation of cross bridges?
(a) K^+ (b) Na^+ (c) Mg^{++} (d) Ca^{++}

67. In the fluid mosaic model, the phospholipids bilayer:
(a) Has protein embedded in it (b) Is covered by outer and inner layers of protein
(c) Is covered by a layer of protein (d) Has carbohydrate embedded in it

68. The enzyme glucose-6-phosphatase is found in the endoplasmic reticulum of:
(a) Brain (b) Kidney (c) Liver (d) Heart

69. The number and rate of function of the nucleolar genes are controlled by:
(a) Slave genes (b) Master genes (c) C G genes (d) Jumping genes

70. A sudden increase in which one of the following ions in the surrounding environment acts as a trigger for causing contraction of the microvilli?
(a) K^+ (b) Na^+ (c) Mg^{++} (d) Ca^{++}

71. Presence of which one of the following in the cell membrane is essential for cell recognition?
(a) Proteins (b) Carbohydrates
(c) Steroids (d) Phospholipids and steroids

72. Which one of the following statements is incorrect?
(a) Degradation of bone is activated by vitamin A and parathyroid hormone.
(b) Lysosomes help in germination of seeds.
(c) Lysosomes of leucocytes and monocytes are essential in defence against bacteria and viruses.
(d) In rheumatoid arthritis, lysosomal enzymes inhibit development of cartilage.

73. Clathrin pits and vesicles help in the:
(a) Transfer of immunoglobulins (b) Recycling of membranes
(c) Uptake of yolk (d) All of these

74. Classification of ribosomes is based on:
(a) Shape (b) Size (c) Weight (d) Sedimentation rate

75. Consider the following statements:
(A) Mitochondria and chloroplasts are energy-transducing and semi-autonomous cell organelles
(B) Mycoplasma is the smallest cell with cell wall
(C) Virus lacks cell wall
(D) Nucleolus has porous membrane

The correct statements are:
(a) A and D (b) A, C and D (c) A, B and D (d) A and C

76. Which one of the following is an incorrect match?
(a) Oxysome – Site of phosphorylation (b) Nucleolus – Synthesis of chromatin
(c) Fo-sub-units – Spherical head of the oxysome (d) Spherosome – Fat digestion

77. Which one of the following functions of mitochondria occurs only in animals?
(a) Beta oxidation (b) TCA cycle
(c) ATP synthesis (d) Electron transport chain

78. Duplication of mitochondrial DNA takes place during:
(a) G_1 phase (b) S phase (c) G_2 phase (d) G_0 state

79. Knowledge of which one of the following is required to study permeability?

(a) Chemical organisation of the cell membrane (b) Chemical organisation of the cell
(c) Both (a) and (b) (d) Nature of the cell coat

80 Which one of the following cells do not divide after birth in humans?
(a) Neuron cells (b) Liver cell (c) Kidney cell (d) Muscle cell

81. In humans, the nuleolar organisers are located in the secondary constrictions of chromosomes:
(a) 13 and 14 (b) 15 and 21 (c) 22 (d) All of these

82. The cell organelle related with the formation of nematocysts and trichocysts is the:
(a) Endoplasmic reticulum (b) Lysosome
(c) Golgi complex (d) Centrosome

83. Consider the following statements:
(A) Endoplasmic reticulum is highly developed in green cells
(B) *Apparato reticulo interno* was discovered by C Golgi in 1898
(C) Chloroplasts are rich in manganese
(D) Quantasomes are found in chloroplasts

The correct statements are:
(a) A, B and D (b) B and D (c) C and D (d) A, C and D

84. Pyrenoids accumulate:
(a) Respiratory enzymes (b) Glycogen
(c) Starch (d) ATP

85. The protein present in the arms of microtubules is:
(a) Nexin (b) Dynein (c) Desmin (d) Vimentin

86. Desmin filaments are found in:
(a) Astrocytes (b) Axons (c) Cardiac muscles (d) None of these

87. DNA duplication is not shown by differentiated:
(a) Liver cells (b) Muscle cells (c) Erythrocytes (d) Nerve cells

88. Match column I with column II and select the correct answer using answer codes:

Column I		Column II
(A) Cancer cells	1.	Foxglove
(B) Digitation	2.	Microtubulins
(C) Desmosomes	3.	Glycosamino glycans
(D) Tubulin	4.	Cell adhering junctions

Answer codes:

	A	B	C	D
(a)	4	1	2	3
(b)	3	2	4	1
(c)	3	1	2	4
(d)	3	1	4	2

89. What is common between microbodies and mitochondria?
(a) Respiratory enzymes (b) Production of ATP
(c) Use of oxygen (d) Release of oxygen

90. In which one of the following do anaemia ferritin molecules accumulate in mitochondria?
(a) Sickle cell anaemia (b) Cooley's cell anaemia
(c) Pernicious anaemia (d) None of these

91. The flow cytometry technique allows detection of difference of ______mega base pairs.
(a) 1.2–2 (b) 1.5–4 (c) 2.5–5.5 (d) 4.5–10.5

92 In which one of the following syndromes do the liver and kidney lack peroxisomes?
(a) Beckwith–Wiedmann syndrome (b) Cerebro–hepato– renal syndrome of Zellweger
(c) Prader–Willi syndrome (d) Lesh–Nyhan syndrome

93. Smaller cells are more active because they have:
(a) Smaller surface area
(b) Lower nucleo-cytoplasmic ratio and larger surface area
(c) Larger surface area
(d) Larger surface area and higher nucleo-cytoplasmic ratio

94. Which one of the following is mismatched?
(a) Hurler's disease – Absence of lysosomes
(b) Periplasm – Space between the outer and inner nuclear membrane of nucleus
(c) Uncontrolled autophagy – Causes cancer
(d) Chromosomal puff – Transcriptional activity

95. Consider the following statements:
(A) Anoxygenic bacteria are photosynthetic
(B) They lack chlorophyll a
(C) They have PS II
(D) The anoxygenic bacteria do not evolve O_2 during photosynthesis
The incorrect statements are
(a) All (b) B and C (c) B and D (d) C

96. Both transcriptional and translational processes in mitochondria and chloroplasts are inhibited by
(a) Chloramphenicol (b) Rifampicin
(c) Both chloramphenicol and rifampicin (d) α-Amanitin and cyclo-hexamide

97. Mitochondrial respiratory enzyme, *cytochrome 'c' oxidase*, is made up of seven sub-units in which:
(a) Two are coded by nuclear genes and five by mitochondrial genes
(b) Four are coded by nuclear genes and three by mitochondrial genes
(c) Six are coded by nuclear genes and one by mitochondrial genes
(d) All sub-units are coded by mitochondrial genes

98. Mitochondrial and chloroplast DNA lack:
(a) Histones (b) Thymine (c) Adenine (d) Guanine

99. P-450 and P-448 present in smooth endoplasmic reticulum help in the detoxification of:
(a) Pollutants (b) Carcinogen (c) Drugs (d) All of the above

100. Chlorophylls trap light energy for the production of:
(a) ATP (b) NADPH
(c) CO_2 (d) Both ATP and NADPH

101. Freeze fracture technique is used to study:
(a) Membrane-bound organelles (b) Nuclear membrane
(c) Plasma membrane (d) Both plasma membrane and nuclear membrane

102. Which one of the following statements is correct?
(a) The positive charges on the histones enable them to bind to DNA.
(b) The five histones are present in the molar ratios of approximately 2 : H_1 : $2H_2A$: H_2B : $2H_3$: $2H_4$.

(c) The inner membrane of mitochondria is highly permeable to H^+ ions.
(d) Y- chromosome bears satellite.

103. Karyotype is used to determine the:
(a) Cause of multiple miscarriages (b) Cause of abnormal newborns
(c) Cause of individual malignancy (d) All of the above

104. The higher levels of chromatin are maintained by histone:
(a) H_1 (b) H_2A and H_2B (c) H_3 (d) H_4

105. _______________is the best material for the study of cell membrane.
(a) Liver (b) Kidney (c) Erythrocyte (d) All of these

106. Cells response to change in:
(a) Temperature (b) pH (c) Nutrient levels (d) All of the above

107. Endosymbiotic theory was first postulated by:
(a) Lynn Margulis (b) Y Daskal (c) L C Moore (d) Callan and Randall

108. Cytorrhysis is the complete collapse of a/an:
(a) Cell membrane (b) Animal cell
(c) Plant cell's cell wall due to loss of water (d) Mitochondria due to loss of water

109. Cytorrhysis causes much greater loss of _________cells.
(a) Number (b) Shape and structure
(c) Structure (d) Size, structure and number

110. A_{549} cells are:
(a) Highly dividing cells (b) Degenerated cells
(c) Human basal epithelial cells (d) Cells unable to divide

111. Which one of the following statments is incorrect?
(a) Cell contents are contained within a cell surface that contains proteins and a lipid bilayer.
(b) The functioning of a cell depends on its ability to extract and use chemical energy stored in organic molecules.
(c) The two membranes of mitochondria have same properties.
(d) Human mitochondrial DNA consists of 5–10 rings of DNA and appears to carry 16,568 base pairs.

112. Nuclear DNA changes by _______________ with each generation:
(a) 20 per cent (b) 45 per cent (c) 50 per cent (d) 75 per cent

113. Humans have an estimated _______________ cells.
(a) 10^4 (b) 10^8 (c) 10^{15} (d) 10^{20}

114. Humans body contains at least _______________different types of cells:
(a) 100 (b) 150 (c) 200 (d) 400

115. The human mitochondrial genome encodes only:
(a) 13 proteins (b) 22 tRNAs (c) 2 rRNAs (d) All of the above

116. What is incorrect about mitochondrial DNA?
(a) The DNA is not packed with proteins. (b) All genes are carried in a single circular DNA.
(c) DNA is not bounded by a nuclear envelope. (d) There is 50 per cent change in mitochondrial DNA from parent to offsprings in each generation.

117. Mitochondrial numbers are controlled by:
(a) Apoptosis (b) Apolysis (c) Autophagy (d) Cell function

118. In mammals, __________ mitochondrial DNA is inherited from the mother:
(a) 25 per cent (b) 50 per cent (c) 75 per cent (d) 99.99 per cent

119. Consider the following statements:
(A) Histones tend to suppress genetic activity
(B) Nonhistone proteins stimulate genetic activity
(C) Nonhistone proteins are synthesised throughout the cell cycle
(D) Nonhistone proteins show variation in structure in different species and even tissues

The correct statements are:
(a) A (b) A, B and C (c) A, C and D (d) All of these

120. In which one of the following is the diploid number of chromosomes 46?
(a) Humans (b) Evening bats (c) Red squirrels (d) All of the above

121 Which one of the following inhibits synthesis of DNA but does not interfere with other cell functions?
(a) Colchicine (b) Amethopterin
(c) Ammonical silver nitrate (d) Phytohaemagglutinin

122. Which one of the following enzymes is synthesised by the cooperation of both nuclear as well as chloroplast genes?
(a) *Ribose biphosphate carboxylase* (b) *ATPase*
(c) *Pyruvic kinase* (d) *Arginase*

123. Match column I with column II and select the correct answer using answer codes:

Column I	Column II
(A) Nucleosome	1. Photorespiration
(B) Lysosome	2. Microtubules
(c) Tubulin	3. Chromatin
(d) Peroxisome	4. Autolysis

Answer codes:

	A	B	C	D
(a)	3	4	1	2
(b)	2	4	3	1
(c)	3	1	2	4
(d)	3	4	2	1

124. The only region of the chromosome where DNA is single stranded is:
(a) Telomere (b) Centromere (c) Secondary constriction (d) Satellite

125 Telomere is rich in:
(a) Guanine (b) Guanine and cytosine
(c) Adenine and guanine (d) Cytosine and thymine

126 Which one of the following statements is correct?
(a) Replication of DNA at the telomere is regulated by the enzyme *telomerase*.
(b) In cancer cells, shortening of chromosome is prevented by *telomerase*.
(c) Constitutive heterochromatin and satellite bodies have repetitive sequences of DNA.
(d) All of these

127. Which one of the following statements is not true?
(a) New organisms generally get DNA from their mother.
(b) Mitochondrial DNA is used in the forensic science for identifying corpses or body parts.
(c) Mitochondrial DNA changes at a faster rate.

(d) Peroxisomes utilise oxygen for carrying out their activities.

128. Peptidoglycan is responsible for:
(a) Rigidity of the bacterial cell wall
(b) Determination of shape of the cell
(c) Both (a) and (b)
(d) Attachment of two cells

129. In Pomper's disease:
(a) Number of peroxisomes decreases
(b) Mitochondria become overactive
(c) Liver cells become engorged with glycogen
(d) Glycogenolysis increases

130. Zellweger syndrome is related with:
(a) Lysosome (b) Ribosome (c) Spherosome (d) Peroxisome

131. Humans contain approximately ___________picogram of DNA per haploid genome
(a) 1 (b) 2 (c) 3 (d) 5

132. Endoplasmic reticulum participates in the formation of:
(a) Vacuoles (b) Glyoxysomes (c) Protein bodies (d) All of the above

133. Consider the following statements:
(A) Intercellular coupling is universally shown by all cancer cells
(B) Junctional interconnections are less abundant during embryonic development
(C) The coupling between cells is electrical as well as metabolic
(D) Viroids contain single RNA covered by a protein capsid
The correct statement is:
(a) A (b) B (c) C (d) D

134. Consider the following statements:
(A) The cell coat is negatively charged
(B) The cell coat may bind with Na^+ and Ca^{++}
(C) The strength of the cell coat varies from cell to cell
(D) Usually malignant cells have higher glycolytic activity
The incorrect statements are:
(a) A and D (b) C and D (c) B and C (d) None of these

135. Peptidoglycan is not present in:
(a) Prokaryotes
(b) Eukaryotes
(c) Both prokaryotes and eukaryotes
(d) None of these

136. Fanconi anaemia is due to:
(a) Long telomere (b) Lack of centromere (c) Short telomere (d) Short satellite

137. Which cell organelle is present in all the five kingdoms?
(a) Nucleus
(b) Mitochondria
(c) Endoplasmic reticulum
(d) Ribosome

138. Cells that commit suicide:
(a) Shrink
(b) Develop blebs on the surface
(c) Release cytochrome C due to mitochondrial break
(d) All of the above

139. Familial myalgia syndrome is a recessive mitochondrial syndrome associated with complex:
(a) I (b) II (c) III (d) IV

140. What is true about replication of mitochondrial DNA?
(a) Rapid rate (b) Lack of proof reading (c) Asynchronic (d) All of the above

141. Match column I with column II and select the correct answer using answer codes.

Column I	Column II
(A) Mitochondria	1. *Monoamine oxidase*
(B) Chloroplast	2. High concentration of cardiolipin
(C) Outer membrane of the mitochondrion	3. Tend to pump out H^+ and retain OH^-
(D) Inner membrane of the mitochondrion	4. Tend to pump out OH^- and retain H^+

Answer codes:

	A	B	C	D
(a)	3	4	1	2
(b)	4	1	2	3
(c)	2	4	1	3
(d)	3	2	4	1

142. Multivesicular bodies are applicable to:
(a) Primary lysosomes (b) Secondary lysosomes
(c) Residual bodies (d) Autophagic vacuoles

143. The first disease caused by mutation in mitochondrial DNA was reported in:
(a) 1970 (b) 1978 (c) 1988 (d) 1992

144. Mitochondrial diseases can originate from:
(a) DNA damage (b) Oxidant damage to protein and lipid in the mitochondria itself
(c) Both (a) and (b) (d) None of these

145. Damage of mitochondria can be controlled by:
(a) Antioxidants (b) *Superoxide dismutase enzyme*
(c) Uncoupling proteins (d) All of the above

146. Cytology is related with:
(a) Cell structure (b) Cell composition
(c) Interaction of cell with other cells (d) All of the above

147. Which one of the following syndrome is due to deletion in the mitochondrial DNA and can be maternally inherited?
(a) Zellweger syndrome (b) Aicardi–Sayre syndrome
(c) Oscar syndrome (d) All of these

148. Eukaryotic cells arose during:
(a) Coenozoic era (b) Mesozoic era (c) Palaeozoic era (d) Proterozoic era

149. The chromosome number is 48 in:
(a) Rhesus monkey (b) Chimpanzee
(c) Gorilla and deer mouse (d) All of the above

150. Marker enzyme for mitochondria is:
(a) *Pyruvate dehydrogenase* (b) *ATPase*
(c) *Succinate dehydrogenase* (d) *NADH dehydrogenase*

151. Which one of the following statements is correct?
(a) Mitochondrial inheritance is non-Mendelian.
(b) Mitochondria are transmitted through nuclear DNA.
(c) All mitochondria are inherited from the mother's ovum.
(d) All of the above

152. Absence of dyenin in cilia and flagella causes:

(a) Immobile cilia and flagella
(b) Infertility in males
(c) Chronic respiratory infection
(d) All of the above

153. The largest human chromosome is approximately ____________ million base pairs in length.
(a) 150 (b) 175 (c) 247 (d) 380

154. Consider the following statements:
(A) FISH (Fluoresence in situ hybridisation) has become a commonly used staining technique
(B) It is now used for microdetection of syndrome
(C) FISH uses specific DNA probes, repetitive DNA
(D) Through this staining technique, unknown sequences of DNA are fluorescently labelled and hybridised to chromosomes

The incorrect statement is
(a) A (b) B (c) C (d) D

155. 9+2 cilia/flagella arrangement is not found in:
(a) Monera
(b) Monera and protista
(c) Protista
(d) Monera, protista and fungi

156. Mitochondria tend to:
(a) Pump out H^+ and OH^-
(b) Retain H^+ and OH^-
(c) Pump out H^+ and retain OH^-
(d) Pump out OH^- and retain H^+

157. What is incorrect about cell ghosts?
(a) They are animal cells with their internal contents.
(b) They are animal cells without their internal contents.
(c) It is difficult to observe them under a microscope.
(d) Cell ghosts of erythrocytes have been favourable material for the study of plasma membrane.

158. Which one of the following statements is incorrect about gas vacuoles?
(a) Gas vacuoles are found in prokaryotes.
(b) Gas vacuoles store metabolic gases.
(c) They regulate buoyancy and dilute the intensity of harmful radiations.
(d) They do not provide mechanical strength.

159. Which one of the following techniques is used to measure cell activity like a change in membrane potential?
(a) Patch clamp (b) Phagocytosis (c) Apoptosis (d) Ion channel

160. Which one of the following is responsible for coiling of nucleosome strand?
(a) H_1 (b) H_2A and H_2B (c) H_3 and H_4 (d) All of these

161. Cart wheel constitution is related with:
(a) Lysosome (b) Centrosome (c) Ribosome (d) Nucleolus

162. The most abundant lipid in a cell membrane is:
(a) Glycolipid (b) Phospholipid (c) Sulpholipid (d) Steroid

163. Which one of the following cell organelles participates in cyclophasmic inheritance?
(a) Nucleus (b) Ribosome (c) Mitochondrion (d) Centrosome

164. Intermediate fibres which form a part of cytoskelton are composed of:
(a) Keratin (b) Vimentin (c) Dermin (d) All of these

165. The major mass of cytoskeleton is represented by:
(a) Spectrin (b) Ankryn (c) Synemin (d) Glycophorin

166. Mitochondria are capable of ______________ in the cell.
(a) Movement (b) Changing position (c) Changing set (d) All of these

167. In prokaryotes, plasma membrane is the site of:
(a) Respiration (b) Photosynthesis (c) Synthesis of lipid (d) All of the above

168. Satellite region of chromosome contains:
(a) DNA and RNA (b) Centromere (c) No DNA (d) No RNA

169. Pyronine stain is used for:
(a) Nucleus (b) Nucleolus (c) Chromosome (d) Mitochondria

170. Which one of the following statements is incorrect?
(a) In rapidly dividing cells, the endoplasmic reticulum is poorly developed.
(b) The size of endoplasmic reticulum is relatively constant in different types of cells.
(c) The reticulocytes, which produce haemoglobin, have poorly developed endoplasmic reticulum.
(d) A rapidly dividing cell may be basophilic.

171. Which one of the following is applicable to lysosome?
(a) Autolysis (b) Autophagy (c) Apoptosis (d) All of the above

172. Which one of the following statements is correct about centromere?
(a) Centromere is a constricted region, containing no genes.
(b) Centromere appears only during cell division.
(c) Centromere joins two sister chromatids of the chromosome.
(d) All of the above

173. The shape of cell depends on its:
(a) Location (b) Cell type (c) Structure (d) Function

174. Gap junctions play a key role in:
(a) Electrical coupling
(b) Metabolic cooperations between adjacent cells
(c) Both (a) and (b)
(d) Electrical coupling, oxidative phosphorylation and metabolic cooperation

175. The first organelle to be discovered was:
(a) Nucleus (b) Mitochondria (c) Ribosome (d) Chloroplast

176. Which one of the following is not found in eukaryotes?
(a) Lysosome (b) Ribosome (c) Mesosome (d) Peroxisome

177. Location of plasmodesmata is determined by the:
(a) Spindle
(b) Cell plate
(c) Endoplasmic reticulum
(d) Activity of the cell

178. Intermediate filaments are sensitive to:
(a) Proteolysis (b) Colchicine (c) Both (a) and (b) (d) None of the above

179. *Glycosyl transnsferases* are concentrated in:
(a) Microbodies
(b) Golgi complex
(c) Mitochondria
(d) Endoplasmic reticulum

180. The histones present in chromosomes help in:
(a) Condensation process
(b) Regulation process
(c) Both (a) and (b)
(d) None of the above

181. Consider the following statements:
(A) Histones H3 and H4 are rich in arginine
(B) Histones are positively charged

(C) Histones are negatively charged

(D) Histones bind tightly to the negatively charged phosphate in DNA

The correct statements are:

(a) A, B and D (b) A and B (c) B and D (d) C and D

182. Which one of the following is the most variable histone?

(a) H_4 (b) H_3 (c) H_2A (d) H_1

Answers to Multiple-Choice Questions

1. (c)	2. (b)	3. (d)	4. (b)	5. (a)	6. (a)	7. (d)	8. (c)
9. (a)	10. (c)	11. (b)	12. (a)	13. (d)	14. (b)	15. (a)	16. (d)
17. (c)	18. (a)	19. (a)	20. (c)	21. (a)	22. (d)	23. (c)	24. (c)
25. (d)	26. (d)	27. (d)	28. (d)	29. (b)	30. (b)	31. (a)	32. (d)
33. (a)	34. (a)	35. (a)	36. (a)	37. (b)	38. (a)	39. (c)	40. (c)
41. (a)	42. (b)	43. (d)	44. (a)	45. (d)	46. (b)	47. (c)	48. (c)
49. (a)	50. (a)	51. (d)	52. (b)	53. (d)	54. (c)	55. (d)	56. (b)
57. (a)	58. (d)	59. (c)	60. (a)	61. (c)	62. (d)	63. (c)	64. (d)
65. (d)	66. (d)	67. (a)	68. (c)	69. (b)	70. (d)	71. (b)	72. (d)
73. (d)	74. (d)	75. (d)	76. (b)	77. (a)	78. (c)	79. (c)	80. (a)
81. (d)	82. (c)	83. (b)	84. (c)	85. (b)	86. (c)	87. (c)	88. (d)
89. (c)	90. (b)	91. (b)	92. (b)	93. (d)	94. (b)	95. (d)	96. (c)
97. (b)	98. (a)	99. (d)	100. (d)	101. (d)	102. (a)	103. (d)	104. (a)
105. (c)	106. (d)	107. (a)	108. (c)	109. (b)	110. (c)	111. (c)	112. (c)
113. (a)	114. (c)	115. (d)	116. (d)	117. (c)	118. (d)	119. (d)	120. (d)
121. (b)	122. (a)	123. (d)	124. (a)	125. (b)	126. (d)	127. (c)	128. (c)
129. (c)	130. (d)	131. (c)	132. (d)	133. (c)	134. (d)	135. (b)	136. (c)
137. (d)	138. (d)	139. (b)	140. (d)	141. (b)	142. (b)	143. (c)	144. (c)
145. (d)	146. (d)	147. (c)	148. (d)	149. (d)	150. (c)	151. (d)	152. (d)
153. (c)	154. (d)	155. (a)	156. (c)	157. (a)	158. (d)	159. (a)	160. (a)
161. (b)	162. (b)	163. (c)	164. (d)	165. (a)	166. (d)	167. (d)	168. (c)
169. (b)	170. (b)	171. (d)	172. (d)	173. (d)	174. (c)	175. (a)	176. (c)
177. (c)	178. (a)	179. (b)	180. (c)	181. (a)	182. (d)		

Fill in the Blanks

1. Action potential on the outer surface of plasma membrane is ______________.
2. The association and dissociation of ribosomal sub-units depends on the concentration of ______________.
3. The Fo-Fi combination functions as ______________.
4. In adipose cells, endoplasmic reticulum is represented by a few ______________.
5. Phagosomes and pinosomes are collectively known as ______________.

6. The organelle related with the production and decomposition of H_2O_2 and beta-oxidation of fatty acids is ______________________________.
7. Replicas of chromosomes are visible as ____________________.
8. The animal cells are interconnected by ______________________.
9. Centromeric index is the ratio of _____ of the two arms of a chromosome.
10. Lomasomes are more common in ________________________.
11. Each nucleosome is made up of __________________________ and _________.
12. From evolutionary point of view, prokaryotes are considered to be the ancestors of ____________.
13. Histone ____________________________has tissue-specific forms.
14. The total amount of DNA per genome of an organism (expressed as pico gram or base pairs) is known as ________________________.
15. The smallest cell without cell wall is ____________________.
16. Extrachromosomal DNA is found in __________ and ____________________.
17. One DNA molecule is equal to______________________chromatid.
18. Cell polarity is determined by ______________________.
19. In humans, centromere contains__________________________ million base pairs.
20. Most human cells contain 23 pairs of chromosomes except______________.
21. ________________________ is the largest cell.
22. The term 'totipotency' was given by__________________________________.
23. Mesosomes are the extensions of ____________________________________.
24. Mesosomes contain ____________________enzymes.
25. The membrane separating vacuoles from cytoplasm is known as _________.
26. Sandwich model of plasma membrane was given by ________________.
27. The larger sub-unit of 70S ribosome is ______ and smaller sub-unit is________.
28. Chemically, ribosomes are made up of _______ and __________________.
29. The arrangement of fibril in centrioles is _____________, whereas cilia and flagella have ___________ arrangement.
30. In animal cells, the multinucleate condition without cell membrane between adjacent cells is known as ____________________.
31. ______________of prokaryotes are analogous to mitochondria of eukaryotes.
32. Folds of inner wall of mitochondria are called __________________________.
33. Golgi complex is mainly related with cell ______________________________.
34. Cilia and flagella are similar in structure but differ in ______________________, ______________ and ______________.
35. Two nucleosomes are joined together by ___________________________.
36. Spherosomes are related with the synthesis and storage of ____________________.
37. Glyoxysomes generally found in ____________________rich plant cells.
38. The formation of primary lysosomes can be blocked by __________________.
39. The glyoxylate cycle allows Protozoa, fungi, and plants to convert fats into________________.
40. Glyoxysomes take part in _________________________cycle.

41. Out of two centrioles present in a spermatozoon, the distal centriole gives rise to ______________.
42. The elastic nature of cell membrane in animal cells is due to the presence of __________________.
43. The pH of cytoplasm is __________________________________.
44. In humans, the nucleolar organisers are located in the secondary constrictions of _____________, _______________, _____________, _______________and_______________ chromosomes.
45. Telomere prevents attachment of two_______________________.
46. Kinetochores can be separated by _________________________ technique.
47. Nuclear DNA differs from mitochondrial DNA in having ____________________.
48. There are ___________________ nucleotide pairs in a diploid cell of humans.
49. ___________ is an undifferentiated cell which is capable of giving rise to specialised daughter cells.
50. The array of plaque-like connections between plasma membrane of adjacent cells is known as ________________.
51. Intracellular attachments include________, _____________ and___________.
52. Peroxisomes are rich in ______________________.
53. There is a __________________________in the chromosome of yeast.
54. Prokaryotes may carry extrachromosomal _______________ called plasmid.
55. Mitochondria of humans contains _______________distinct types of protein.
56. Mitochondrial DNA shows substantial similarity with ______________ genomes.
57. The simplest chromosomes are found in _____________________________.
58. Heterochromatin consists mainly of ______________________________________.
59. The number of chromosomes is 46 in ______________and _________________.
60. Ca^{++} ATPase of plasma membrane is a ___________ protein and was first discovered in the membranes of ________________.
61. In eukaryotes, DNA is always present in combination with _________________protein.
62. _____________ is a type of lipid that stiffens the membranes of animal cells.
63. _________________________________ acts as garbage disposal of the cell.
64. The acidity of the lysosome is maintained with the help of ________________.
65. The origin of mitochondria is explained by the _______________________theory.
66. Cytoplasmic bridges present between adjacent cells are known as ______________.
67. Telomerase synthesises only the _____________________-rich strand of telomere.
68. Microbodies are of two types, viz., _______________ and ________________________.
69. A centriole contains _________________________ microtubules.

Answers to Fill in the Blanks

1. Negative
2. Mg^{++}
3. *ATP synthetase*
4. Tubules
5. Endosomes
6. Peroxisome
7. Chromatides
8. Desmosomes
9. Lengths
10. Fungi
11. DNA, histones—H_2A, H_2B, H_3 and H_4
12. Eukaryotes
13. H_1
14. C-value
15. Mycoplasma or PPLO
16. Mitochondria, chloroplasts
17. One

18. Microtubules
19. One
20. Red blood cells
21. Egg of ostriches
22. Morgan
23. Plasma membrane
24. Respiratory
25. Tonoplast
26. Danielli and Davson
27. 50S, 30S
28. Protein, rRNA
29. 9+0, 9+2
30. Syncitium
31. Mesosomes
32. Cristae
33. Secretion
34. Number, length, motion
35. Linker DNA
36. Fat
37. Fat
38. Promycin
39. Carbohydrate
40. Glyoxalate
41. Axial filament
42. Lipids
43. 6.8
44. 13, 14, 15, 21 and 22
45. Chromosomes
46. CREST
47. Intrones
48. $2 \times 4 \times 10^9$
49. Stem cells
50. Gap junction
51. Tight juctions, belt desmosomes, spot desmosomes
52. *Peroxidase*
53. Single microtubule
54. DNA
55. 615
56. Bacterial
57. Viruses
58. Inactive DNA
59. Hare and humans
60. Transport, red blood cells
61. Histone
62. Cholesterol
63. Lysosome
64. Hydrogen ion pumps
65. Endo symbiotic
66. Plasmodesmata
67. G
68. Peroxisomes, Glyoxysomes
69. 27

True or False

1. Prokaryotic cells multiply more rapidly in comparison to eukaryotic cells.
2. A few prokaryotic cells are much larger than eukaryotic cells.
3. In the absence of a nucleus, Golgi complex decreases in size and disappears.
4. The mitochondrial DNA is not capable of mutation.
5. RNA is Feulgen positive.
6. Flagella fuse to form cirri.
7. No cell organelle can survive outside the cell.
8. Cilia and flagella arise from basal bodies.
9. Oxysomes are found in peroxisomes.
10. Water and dissolved materials are unable to pass through a lignified wall.
11. The G:C ratio of mitochondrial DNA is poor.
12. Mature sperms and RBCs lack ribosomes.
13. Cell wall prevents bursting of cells in a hypotonic solution.
14. Cancer cells get differentiated in culture.
15. Size of a nucleus depends on the number of chromosomes present in it.
16. The F_o-F_1 combination functions as *ATP synthetase*.
17. The buoyant density of mitochondrial DNA is low.
18. Multiplication of mitochondria occurs by binary fission.
19. Mitochondria lack an adequate DNA repair mechanism.

20. Mesosomes are quite prominent in Gram-negative bacteria.
21. Phagosomes and pinosomes are collectively known as endosomes.
22. When ATP concentration is high, the respiratory chain is inhibited.
23. Each chromatid consists of single chromatin fibre and each chromatin fibre contains a single DNA double helix.
24. C and R bands have been used in plant karyotyping.
25. In an adult human, there may be 10^7 mitochondrial DNA molecules.
26. Protoplasm is the part of cell that contains different cell organelles excluding nucleus.
27. Nerve cell is the longest animal cell.
28. Dyenin protein is found in centrioles.
29. Microfilaments are hollow structures.
30. Both microfilaments and microtubules are contractile structures.
31. Distribution of microtubules determines cell polarity.
32. Crossing over is common in heterochromatin.
33. Histone proteins have not undergone any change during evolution and are identical in all plant and animal cells.
34. The mitochondrial wall is not permeable to CoA.
35. Oxysome is the centre of phosphorylation.
36. Cells contain 75–85 per cent water.
37. Water exists in cells in two forms, viz., free and bound.
38. Calmodulin is found in the cytoplasm of tissue cells.
39. Mitochondria are one of the most sensitive indicators of injury to a cell.
40. Presence of a specific carbohydrate at the cell membrane is essential for the cell recognition phenomenon.
41. The cell coat is positively charged.
42. The melting points of mitochondrial DNA and nuclear DNA are the same.
43. In the spermatozoa of some fish, the negative charge of DNA is neutralised by protamines.
44. Nonhistone chromosomal proteins (NHC) stimulate gene activity.
45. Supernumerary chromosomes are present in all the individuals of a species.
46. Deletions in mitochondrial DNA may occur all the time.
47. Microtubules function as an intracellular engine.
48. Cancer cells are mortal.
49. Eukaryotic cells are ten times the size of prokaryotic cells.
50. In prokaryotes, cell size is limited by efficient metabolism.
51. Mitochondria evolved more than 10^9 years ago.
52. Some mitochondrial genes lack codon.
53. Apoptosis causes swelling of mitochondria.
54. Mitochondrial DNA denaturates at a higher temperature than nuclear DNA.
55. The length of each nucleosome is constant in different tissues.
56. Mitochondrial DNA follows the Mendalian rules of genetic inheritance.

57. Mitochondria mainly synthesise hydrophilic proteins.
58. In an adult human, 50–70 billion cells die each day.
59. Gap junctions are called nexus.
60. Microtubules take part in endocytosis.
61. Lowering of temperature increases active transport.
62. Protein constitutes approximately 35 per cent of cytosol.
63. In humans, chromososome 2 is telocentric.
64. In 80S ribosome, RNA/protein ratio is 1:1.
65. Fluidity of membrane decreases at 0°C.

Answers to True or False

1. True	2. True	3. True	4. False	5. False	6. False	7. True	8. True
9. False	10. False	11. False	12. True	13. True	14. False	15. True	16. True
17. False	18. True	19. True	20. False	21. True	22. False	23. True	24. False
25. True	26. False	27. True	28. False	29. False	30. False	31. True	32. False
33. True	34. True	35. True	36. True	37. True	38. True	39. True	40. True
41. False	42. False	43. True	44. True	45. False	46. True	47. True	48. False
49. True	50. True	51. True	52. True	53. False	54. True	55. False	56. False
57. False	58. True	59. True	60. False	61. False	62. False	63. False	64. True
65. True							

Give Reasons

1. A typical cell is lacking in eukaryotes.
   — Because in higher organisms, specialisation of labour exists and accordingly different groups of cells are modified in different ways.
2. Mitochondria are semi-autonomous cell organelle.
   — Because biogenesis of mitochondria is due to the involvement of two genetic systems, i.e., mitochondrial and nuclear genes.
3. Mitochondrial genome is genetically not self-sufficient.
   — Because most proteins required for the structure and function of this organelle are controlled by the nuclear DNA.
4. DNA molecules having high guanine and cytosine content are more resistant to temperature in comparison to adenine- and thymine-rich molecules.
   — Because guanine and cytosine are held together by three hydrogen bonds, whereas adenine and thymine are held together by two hydrogen bonds.
5. Mitochondrial DNA plays a role in the ageing process.
   — Because a specific decrease in the amount of *cytochrome c-oxidase* (an enzyme involved in respiration) occurs, which is encoded by mitochondrial DNA.

6. Histones are basic proteins.
   — Because they have high content of arginine and lysine which are basic amino acids.
7. The proof that nuclei contain constant amount of DNA was a landmark achievement in cell biology.
   — Because it shows that DNA is the genetic material and genetic information is not destroyed during the differentiation of the various somatic tissues.
8. In some regions, endoplasmic reticulum shows a three-layered unit membrane structure.
   — Because such a structure allows drastic changes in the surface of the membrane.
9. Secondary constrictions can be used as a marker.
   — Because the position of secondary constriction is always constant.
10. If a chromosome breaks, the broken ends may fuse.
    — Because broken ends of a chromosome lack telomeres.
11. The outer surface of plasma membrane is negatively charged.
    — Due to the presence of sialic acid residues, carboxyl and phosphate groups.
12. Svedberg units are not added.
    — Because loss of area occurs when two sub-units are bound. Besides, the shape of a fully assembled ribosome has different aerodynamic properties in comparison with two unbound sub-units.
13. Cancer cells are immortal.
    — Because cancerous cells contain *telomerase* enzyme in large quantities which reconstruct the telomeres and permit cell division to continue.
14. Animal cells generally have irregular shape.
    — Because they lack a cell wall.
15. A close relationship has been developed between mitochondria and nucleus.
    — Because majority of enzymatic proteins of mitochondria are synthesised by the nucleus.
16. Mitochondrial DNA has a higher buoyant density and denaturates at higher temperature.
    — Because mitochondrial DNA has higher content of guanine and cytosine.
17. Nucleolar chromatin segments are condensed and replicate late.
    — Because of the presence of inactive rDNA cistron.
18. Animals ribosomes have lower sedimentation coefficient, still they are slightly larger than 70S ribosomes of bacteria.
    — Because animal ribosomes contain more protein.
19. Plant cells have few mitochondria in comparison to animal cells.
    — Because in plant cells, chloroplasts also synthesise ATP.
20. Lysosome provides defence against certain bacteria.
    — Because it hydrolyses the peptidoglycan of the cell wall.
21. Cells are small in volume.
    — Because they have to maintain a large surface area to volume ratio in a balanced state.
22. Mitochondrial ribosomes are rarely affected by antibiotics.
    — Because mitochondria are covered by a double membrane due to which any antibiotic is not able to enter easily into the organelle.
23. Mitochondrial diseases are transmitted only by females to their offspring.
    — Because in sexually reproducing organisms, mitochondria are maternally inherited.
24. Cells stop dividing.

— Because telomeres become shorter with each division, and are unable to protect the chromosome.

25. Study of chromosome behaviour is very important in biology.
    — Because its study helps us to follow the behaviour of DNA molecules and genes in a visual function.
26. Mucolipidoses are classified as lysosomal storage disease.
    — Because they involve increased storage of substances in the lysosome.
27. Peroxisomes protect the cell from the hydrogen peroxide produced by them.
    — Because the oxidative enzymes in peroxisomes breakdown the hydrogen peroxide into water and oxygen.
28. White blood cells have more lysosomes.
    — Because they have to digest more materials than other type of cells in their quest with bacteria, viruses and other foreign intruders.
29. The *lysosomal hydrolases* are active at acidic pH which is beneficial.
    — Because if hydrolases leak lysosomes, they are not likely to do damage, unless the cell becomes acidic.
30. Mitochondrion is well suited for cellular respiration.
    — Because:
    (a) It is double-membraned and smooth but permeable to CO_2 and pyruvic acid
    (b) Inner membrane bears cristae which provide a large surface area for reaction to take place and cristae contain co-enzymes of respiration
    (c) Stalked particles present on the inner surface of the membrane seem involved in the conversion of ADP to ATP
    (d) Matrix contains enzymes for Krebs cycle and oxidative phosphorylation
31. Mitochondria have their own DNA.
    — Because they descended from a proto-bacterium that merged with eukaryotic cells over 2 billion years ago.

Questions based on Diagrams

1. Select the answer which gives the correct combination of letters pointing to the organelles of the cell shown in the diagram below with the numerical order of their functions.
   1. Protein synthesis
   2. Seed germination
   3. Cell division
   4. Packing
   5. Synthesis of ATP

	1	2	3	4	5
(a)	D	E	C	A	B
(b)	D	B	C	A	E
(c)	E	A	D	C	B
(d)	C	E	B	A	D

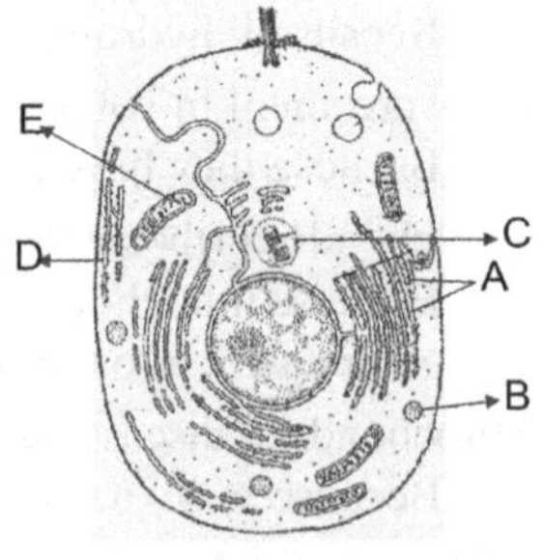

2. In the diagram of specialised structure of plasma membrane below, which part is involved in cell adherences?
   (a) I
   (b) II
   (c) III
   (d) IV

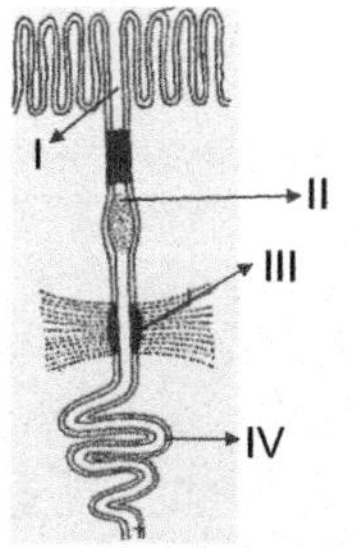

3. In the diagram of mitochodrion below, identify the part in which respiratory chain complexes are located.
   (a) I (b) II (c) III (d) IV

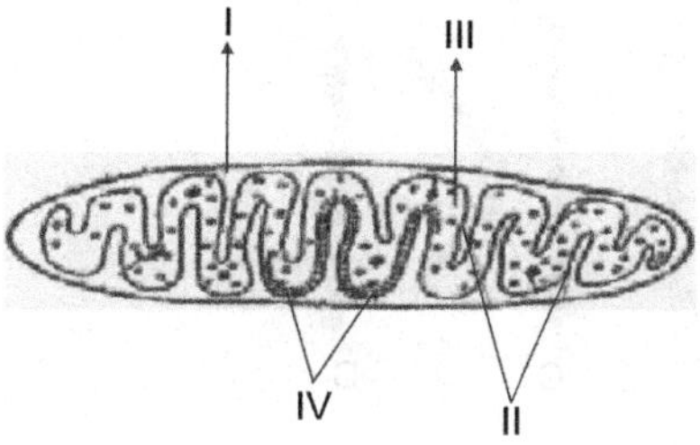

4. In the diagram above, which part is involved in oxidative phosphorelation?
   (a) IV (b) III (c) II (d) I
5. Identify the labelled part which is formed by Golgi complex in the diagram below.
   (a) I (b) II (c) III (d) IV

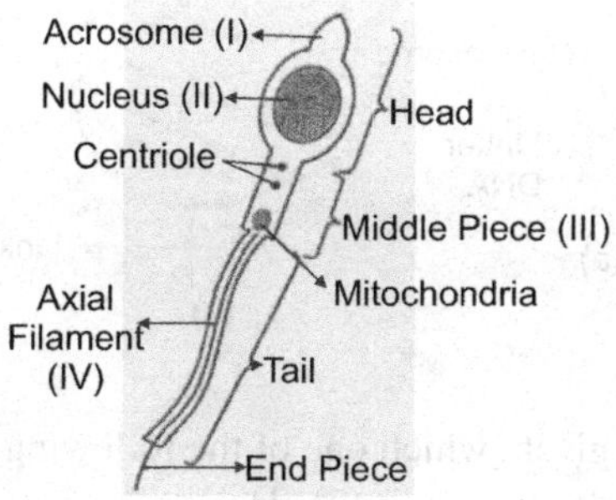

6. In the diagram above, which labelled part is known as 'Engine room' of the sperm?
   (a) IV (b) III (c) II (d) I
7. Identify the labelled part in the given diagram of chromosome that synthesises enzyme *Telomerase*.
   (a) I (b) II (c) III (d) IV

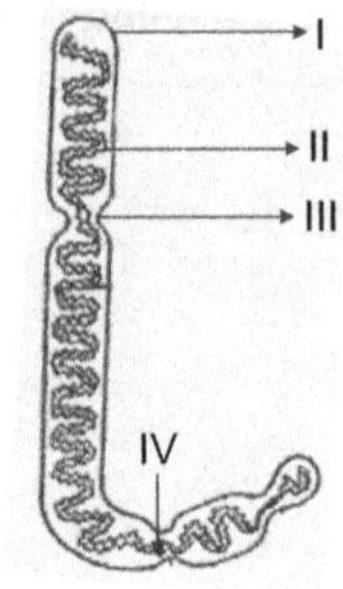

8. In the diagram above, which region is characterised by large amounts of tendemly repeated satellite DNA and specific proteins in mammals?
(a) IV (b) III (c) II (d) I

9. In the diagram below, which one is a telocentric chromosome?
(a) A
(b) B
(c) C
(d) D

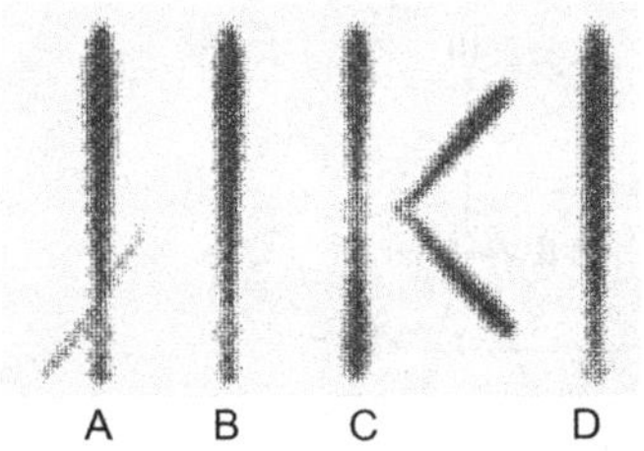

10. The diagram below is associated with:
(a) Ribosome
(b) Lysosome
(c) Chromosome
(d) Centrosome

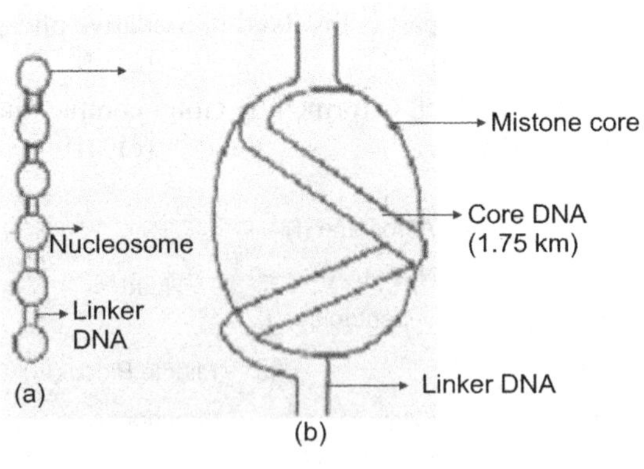

11. In the diagram of nucleosome above given, which one of the following is absent?
(a) H_1 (b) H_2A (c) H_3 (d) H_4

Answers to Questions based on Diagrams

1. (b) 2. (c) 3. (b) 4. (a) 5. (a) 6. (b) 7. (a) 8. (b)
9. (d) 10. (c) 11. (a)

GENETICS OF CELL DIVISION

Multiple-Choice Questions

1. Consider the following statements about a phase of the cell cycle:
   (A) Takes approximately 30–50 per cent part of entire cell division
   (B) Transcription of rRNA, tRNA, mRNA and protein synthesis occurs
   (C) Chromatin is fully extended
   (D) Sometimes, fast dividing cells lack this phase (striated muscle cells and neurons)
   This is:
   (a) G_0 phase (b) S phase (c) G_1 phase (d) M phase
2. The rates of RNA transcription and protein synthesis are very low during:
   (a) G_1 phase (b) S phase (c) G_2 phase (d) G_0 phase
3. Which one of the following functions to block the cell cycle, if the DNA is damaged?
   (a) CDK (b) p^{27} (c) p^{53} (d) Cyclin
4. Experimental evidence of cell cycle checkpoint was given first by:
   (a) Rao and Johanson (1970) (b) Karlin and Nevo (1986)
   (c) Weiner and Lee Hartwell (1988) (d) Von Holde (1988)
5. Most of the histone production occurs during:
   (a) G_1 phase (b) S phase (c) G_2 phase (d) G_1 and G_2 phases
6. Which one of the following is a protein that binds to cyclin and CDK, blocking entry into the S phase?
   (a) p^{27} (b) p^{53} (c) p^{15} (d) p^{21}
7. Which one of the following is regulated by checkpoints?
   (a) Initiation of DNA synthesis (b) Assembly and integrity of the mitotic spindle
   (c) Attachment of chromosome to mitotic spindle (d) All of these
8. Most cells in the human body are in:
   (a) G_0 phase (b) G_1 phase (c) S phase (d) G_2 phase
9. Cyclins accumulate during _____ phase of the cell cycle.
   (a) G_1 (b) S (c) G_2 (d) All of these
10. Tissues that require fast cellular renovation (mucosa and endomaterial epithelia) have:
    (a) No G_1 phase (b) Shorter G_1 phase (c) Longer G_1 phase (d) All of the above
11. Pascal triangle is a model of:
    (a) Prokaryotic cell (b) Eukaryotic cell (c) Cell division (d) DNA distribution
12. The correct sequence of phases of cell cycle is:
    (a) $G_1 \rightarrow G_0 \rightarrow S \rightarrow M$ (b) $G_1 \rightarrow S \rightarrow G_2 \rightarrow M$ (c) $G_0 \rightarrow S \rightarrow G_1 \rightarrow G_2 \rightarrow M$ (d) $G_1 \rightarrow S \rightarrow G_0 \rightarrow G_2 \rightarrow M$
13. Consider the following statements:
    (A) Cdc 25 is a protein phosphatase responsible for dephosphorylatoin and activation of Cdc2

(B) Cdc 25 is phosphorylated throughout interphase but not in mitosis
(C) p^{53} is a transcription factor whose activity is regulated by phosphorylation
(D) The ataxia telangiectasia mutated gene (ATM) encodes a protein kinase which acts as a tumour suppressor

The correct statements are:
(a) All of these (b) A, B and C (c) A and D (d) B and C

14. Cyclin was first discovered in:
(a) *Escherichia coli* (b) *Drosophila melanogaster*
(c) Sea urchin (d) *Neurospora*

15. The cells with high levels of Maturation Promoting Factor (MPF) are present in:
(a) G_0 phase (b) M phase (c) G_1 phase (d) S phase

16. Which one of the following is applicable to Epidermal Growth Factor (EGF)?
(a) Stimulate cell proliferation (b) Small protein
(c) Single peptide chain (d) All of these

17. Replication of centriole takes place during:
(a) G_0 phase (b) G_1 phase (c) S phase (d) G_2 phase

18. Which one of the following statements is incorrect about Cdc2 (cell division control) protein?
(a) First identified in yeast (*S. pombe*) (b) Equivalent to Cdc28 in budding yeast
(c) Equivalent to Cdk1 in mammals (d) None of these

19. Consider the following statements:
(A) In a normal cell, the p^{53} protein level is low
(B) p^{53} is a tumour-suppressor protein and is also known as the 'Guardian of the Genome'
(C) The major regulator of p^{53} is Mdm2
(D) p^{53} is an anti-transcriptional activator

The correct statement is:
(a) All of these (b) A, B and C (c) B and C (d) C and D

20. Of the 22,000 genes in each human cell, nearly _________ play a part in mitosis.
(a) All (b) 1,000 (c) 600 (d) 4

21. Which one of the following statements is incorrect?
(a) Synthesis of DNA occurs during S phase. (b) M phase is the shortest stage of cell cycle.
(c) Synthesis of histone occurs during S phase. (d) G_0 stage is characterised by significant metabolic activities.

22. In yeast, mitotic entry of G2 cells is delayed by the:
(a) *BUB* gene (b) *rad* 9 gene (c) p^{53} gene (d) p^{21} gene

23. Cyclin degradation and MPF inactivation result in:
(a) Nuclear reformation and cytokinesis (b) Chromosome decondensation
(c) Chromosome segregation (d) All of these

24. Decision for cell division is made during:
(a) G_0 phase (b) G_1 phase (c) S phase (d) M phase

25. Match column I with column II and select the correct answer using answer codes

Column I (Gene)	Column II (Function)
(A) *wee* 1	1. Dephosphorylation of Cdk1
(B) *suc* 1	2. Phosphorylation of Cdk

(C) *nim* 1 — 3. Negative control on wee 1

(D) *cdc* 25 — 4. Converts M form of Cdk1 to S form

Answer codes:

	A	B	C	D
(a)	3	4	2	1
(b)	4	1	2	3
(c)	2	4	3	1
(d)	2	1	4	3

26. Active MPF induces:
(a) Breakdown of nuclear membrane
(b) Condensation of chromosome
(c) Assembly of spindle
(d) All of these

27. In mammals, which one of the following genes causes delay in entry of cells with damaged DNA into the S phase:
(a) p^{21}
(b) p^{53}
(c) *Hct*[1]
(d) *nim*[1]

28. Inactive M-CDK is present in large amount:
(a) In the beginning of G_1 phase
(b) At the end of G_2 phase
(c) In the S phase
(d) In the beginning of M phase

29. The cell size becomes double during:
(a) G_1 phase
(b) S phase
(c) G_2 phase
(d) M phase

30. The condition during cell cycle when cell has undergone differentiation is known as:
(a) S phase
(b) M phase
(c) G_2 phase
(d) G_0 phase

31. In a diploid cell, the amount of DNA content in S phase is:
(a) $4n$
(b) $6n$
(c) $8n$
(d) $2n$

32. In cell cycle, DNA polymerase is active in:
(a) G_1 phase
(b) S phase
(c) G_2 phase
(d) G_0 phase

33. G_0 state of cell refers to:
(a) Pre-mitotic gap phase
(b) Abnormal synthesis of DNA
(c) Entry of cell in cell cycle
(d) Exit of cell from cell cycle

34. Which one of the following is known as invisible stage of M phase?
(a) G_0 phase
(b) G_1 phase
(c) S phase
(d) G_2 phase

35. The period between the completion of DNA synthesis and the start of mitosis is known as:
(a) G_1 phase
(b) S phase
(c) G_2 phase
(d) G_0 phase

36. The two main phases of cell cycle are:
(a) Interphase and mitosis
(b) Prophase and anaphase
(c) Metaphase and anaphase
(d) Prophase and telophase

37. Tumour growth factor beta inhibits:
(a) Abnormal cell proliferation
(b) Proteins p^{16} and p^{53}
(c) p^{Rb}
(d) All of these

38. The period after DNA synthesis has occurred but before the start of prophase is called:
(a) G_1 phase
(b) S phase
(c) G_2 phase
(d) G_0 phase

39. Which one of the following statements is incorrect about metaphase?
(a) It is the most dynamic stage.
(b) Chromosomes have maximum contraction.

(c) Each chromosome consists of two sister chromatids.
(d) It is suitable for counting of chromosome number and banding.

40. Division of centromere occurs during.
(a) Prophase (b) Metaphase (c) Anaphase (d) Telophase

41. Biosynthesis of nucleolus occurs during:
(a) Telophase (b) Anaphase (c) Metaphase (d) Prophase

42. Which one of the following does not occur during interphase?
(a) Duplication of centriole (b) Division of centromere
(c) DNA duplication (d) Synthesis of RNA and proteins

43. The anaphasic chromosomes appear:
(a) J shaped (b) V shaped (c) L shaped (d) All of the above

44. Centromeres of all the chromosomes lie over the equator during:
(a) Prophase (b) Metaphase (c) Anaphase (d) Telophase

45. Which one of the following is of the shortest duration?
(a) Anaphase (b) Metaphase (c) Prophase (d) Telophase

46. Which one of the following promotes and stabilises the formation of microtubules?
(a) Taxol (b) Podophyllotoxin (c) Vinblastine (d) All of these

47. Consider the following statements:
(A) Kinesin and dynein proteins are associated with centrosome
(B) The forces generated by kinesin and dynein proteins are opposite in direction
(C) All microtubules located between the poles and kinetochores have the same polarity
(D) Condensin in is a multi-sub-unit protein

The correct statements are:
(a) All (b) A, B and C (c) B, C and D (d) B and D

48. In meiosis, division of centromere occurs in:
(a) Metaphase I (b) Anaphase I (c) Metaphase II (d) Anaphase II

49. Crossing over occurs in:
(a) Zygotene (b) Pachytene (c) Diplotene (d) Diakinesis

50. Which one of the following is an incorrect match?
(a) Bivalent – Zygotene
(b) Chiasmata formation – Pachytene
(c) Chiasmata are almost fully terminalised – Metaphase I
(d) Kinetochores – Form the mitotic spindle

51. Free nuclear divisions occur in:
(a) *Opalina* (b) Dinoflagellates (c) Diptera (d) All of these

52. Separation of chromosomes into chromatids occurs in:
(a) Metaphase I (b) Anaphase I (c) Anaphase II (d) Telophase II

53. Chromosome condensation and microtubules formation begin in the:
(a) G_1 phase (b) S phase (c) G_2 phase (d) Prophase

54. Which one of the following occurs in zygotene?
(a) Terminalisation (b) Synapsis (c) Crossing over (d) Chiasmata

55. Consider the following statements about syneptonemal complex:
(A) Bipartite structure (B) First discovered by Moses (1956)

(C) Its lateral elements lack DNA and RNA
(D) King (1970) proposed hypothesis for the formation of syneptonemal complex

The correct statements are:
(a) All of the above (b) A, B and C (c) B and D (d) C and D

56. When cell size increases, surface area of cell ____________ in the same proportion as the cell volume.
(a) Decreases (b) Increases
(c) Does not increase (d) Depends on physiological condition of the cell

57. Bivalents appear as darkly stained round bodies during:
(a) Diakinesis (b) Diplotene (c) Pachytene (d) Zygotene

58. Which one of the following is related with alternation of generation?
(a) Amitosis (b) Mitosis (c) Meiosis (d) None of these

59. During meiosis, separation of homologous chromosomes begins in:
(a) Pachytene (b) Diplotene (c) Diakinesis (d) Anaphase I

60. The fact that each association of chromosomes in pair at zygotene involves one paternal and one maternal chromosome was first pointed out by:
(a) Von Winiwarter (1900) (b) Montgomery (1901)
(c) Sutton (1902) (d) Elinors E. Carothers (1913)

61. Disjunction takes place during:
(a) Pachytene (b) Metaphase I (c) Anaphase I (d) Telophase I

62. If in a cell, the number of bivalents is 20, what would the number of chromosomes observed during anaphase II?
(a) 5 (b) 10 (c) 20 (d) 40

63. Match column I with column II and select the correct answer using answer codes:

Column I	Column II
(A) Bouquet stage	1. Prophase
(B) Pairing of homologous chromosome	2. Leptotene
(C) High refractivity and viscosity cytoplasm	3. Zygotene
(D) Lampbrush chromosome	4. Diplotene

Answer codes:

	A	B	C	D
(a)	2	3	1	4
(b)	4	3	1	2
(c)	3	4	2	1
(d)	2	4	3	3

64. Which one of the following is applicable to mitosis?
(a) Syneptonemal complex (b) Crossing over
(c) Tetravalent (d) None of these

65. Division of mitochondria and plastid occurs during:
(a) G_1 phase (b) S phase (c) G_2 phase (d) M phase

66. Telophase and interphase are omitted in the meiosis of:
(a) Diptera (b) Odonata (c) Hemiptera (d) *C.elegans*

67. Syneptonemal complexes are lacking in the meiosis of:
(a) Male diptera (b) Female diptera (c) Male *C.elegans* (d) Male odonata

68. Mitotic poison interferes with:
(a) Centrioles (b) Centromeres (c) Spindle (d) All of these

69. Valium:
(a) Prevents the separation of centrioles (b) Prevents formation of spindle microtubules
(c) Induces formation of multipolar spindle (d) Increases formation of spindle microtubules

70. How many meiotic divisions are required to produce 100 seeds?
(a) 25 (b) 50 (c) 125 (d) 150

71. In which one of the following has cell cycle regulation been studied in detail?
(a) Yeast (b) *E. coli* (c) *Drosophila* (d) *Neurospora*

72. According to dynamic equilibrium hypothesis, ________ of microtubules is directly responsible for the movements of chromosomes.
(a) Polymerisation (b) Phosphorylation
(c) Polymerisation and depolymerisation (d) Phosphorylation and dephosphorylation

73. In eukaryotes, mitosis is mainly concerned with:
(a) Growth (b) Repair of the tissue (c) Asexual reproduction (d) All of these

74. In primary spermatocyte, there is a ________ amount of DNA.
(a) Tetraploid ($4n$) (b) Diploid ($2n$) (c) Haploid (n) (d) Octaploid ($8n$)

75. Chromomeres become larger and fewer in numbers in:
(a) Zygotene and pachytene (b) Pachytene
(c) Diplotene (d) Diplotene and diakinesis

76. The two lateral arms of syneptonemal complex are separated by an axial space of lower density in most plants and animals, except:
(a) Nematodes (b) Insects
(c) *C. elegans* (d) *Ascaris megalocephala*

77. DNA content is highest in:
(a) Primary spermatocyte (b) Secondary spermatocyte
(c) Spermatid (d) Spermatozoon

78. Which one of the following stages may last for days, week or even years?
(a) Leptotene (b) Zygotene (c) Pachytene (d) None of these

79. Segregation of homlogous chromosome takes place during:
(a) Diakinesis (b) Metaphase I (c) Anaphase I (d) Anaphase II

80. In which one of the following does syneptonemal complex start at the telomeres?
(a) *Ascaris megalocephala* (b) *Bombyx mori*
(c) *Drosophila melanogaster* (d) *Chironomous*

81. Which one of the following is the main component of syneptonemal complex?
(a) Protein (b) Lipid (c) DNA (d) Lipid and RNA

82. Centriolar cycle does not involve:
(a) Disorientation (b) Depolymerisation (c) Nucleation (d) Elongation

83. Chromosomes are shortest and thickest in:
(a) Prophase (b) Metaphase (c) Anaphase (d) Telophase

84. Which one of the following statements is incorrect about zygonemal DNA?
(a) Has a lesser buoyant density than other DNA (b) Rich in guanine–cytosine
(c) Has highly repeated sequences of nucleotides (d) None of these

85. Which one of the following is difficult to identify during mitosis?
    (a) Mitochondria (b) Golgi complex (c) Chloroplast (d) Centrioles
86. The best stage for cytogenetic study is the:
    (a) Prophase (b) Metaphase (c) Anaphase (d) Telophase
87. There is no crossing over in males of:
    (a) *Phryne fenestralis* (b) *Tipula caesia*
    (c) *Drosophila melanogaster* (d) All of these
88. The number of meiotic divisions required to produce 40 speramatids is:
    (a) 8 (b) 10 (c) 20 (d) 40
89. A cell having 40 chromosomes is undergoing mitotic division. What will be the number of chromosomes during anaphase?
    (a) 10 (b) 20 (c) 40 (d) 80
90. Chiasmata formation is the result of:
    (a) Terminalisation (b) Synapsis (c) Crossing over (d) Nondisjunction
91. Which one of the following inhibits microtubule assembly?
    (a) Colchicine (b) Vinblastine (c) Podophyllotoxin (d) All of these
92. Match column I with column II and select the correct answer using answer codes:

Column I	Column II
(A) Schneider (1887)	1. Lajtha (1963)
(B) Whiteman (1887)	2. Mitosis
(C) Flemming (1872)	3. Karyokinesis
(D) G_0 phase	4. Cytokinesis

Answer codes:

	A	B	C	D
(a)	4	3	2	1
(b)	2	4	1	2
(c)	2	1	2	4
(d)	3	4	2	1

93. Which one of the following statements is incorrect about telophase?
    (a) There is increase in the viscosity and refractivity of cytoplasm.
    (b) It is almost reverse of prophase.
    (c) The two centrioles organise themselves into centrosome.
    (d) Chromosomes undergo decondensation.
94. Mid body occurs in _________ of animal cell.
    (a) Cytokinesis (b) Telophase (c) Anaphase (d) Metaphase
95. Amitosis was first described by Robert Remark (1855) in:
    (a) Liver cells of chick embryo (b) Red blood corpuscles of chick embryo
    (c) Red blood corpuscles of frog (d) Kidney cells of chick embryo
96. Cell lineage theory was proposed by:
    (a) Remark (1841) (b) Rudolf Virchow (1855)
    (c) Van Benden (1887) (d) Winiwater (1900)
97. In mitosis, there is no pairing of homologous chromosomes except in:
    (a) Amphibian oocytes (b) Somatic cells of salivary glands of *Drosophila*
    (c) Kidney cells of chick embryo (d) Somatic cells of salivary glands of *Ascaris*

98. Which one of the following is applicable to mitosis?
(a) Reduction in the number of chromosomes (b) Synapsis
(c) Terminalisation (d) None of these
99. Reduction in the number of chromosomes takes place during:
(a) Diakinesis (b) Metaphase I (c) Anaphase I (d) Anaphase II
100. Tetrad formation occurs during:
(a) Zygotene (b) Pachytene (c) Diplotene (d) Metaphase I
101. The progress of cell cycle is regulated by:
(a) Actins (b) Myosins (c) Cyclins (d) All of these
102. At the end of cytokinesis, the daughter cells enter ____________ phase.
(a) G_0 (b) G_1 (c) S (d) None of these
103. Cells divide for:
(a) Growth (b) Repair (c) Reproduction (d) All of these
104. Chromosomes become visible during:
(a) Prophase (b) Metaphase (c) Anaphase (d) Telophase
105. A small amount of crossing over has been reported in the males of:
(a) *Drosophila melanogaster* (b) *D. ananassae*
(c) *D. malerkotliana* (d) *D. bipectinata*
106. Which one of the following is a checkpoint in the cell cycle?
(a) DNA damage (b) DNA replication (c) Spindle (d) All of these
107. Consider the following statements:
(A) Mutants of *MAD* and *BUB* genes in mammals and yeast activate the spindle checkpoint
(B) G_2 checkpoint monitors unreplicated and damaged DNA
(C) G_1 checkpoint stimulates progression into S phase by activating S-CDK complex
(D) In yeast, gene *Cdc*13 activates *Cdk*1 for S phase

The incorrect statements are:
(a) A, B and C (b) A and C (c) B and D (d) None of these
108. Synzetic knot is applicable to:
(a) *Drosophila* (b) *Chironomous* (c) *Lilium* (d) All of these
109. Spindle is without:
(a) Asters (b) Centrioles (c) Blepharoplast (d) All of these
110. Which one of the following appears to have a role in the assembly and disassembly of spindle microtubules?
(a) Calcium ion (b) Calcium-binding protein
(c) Calmodulin (d) All of these
111. Which one of the following statements is incorrect about spindle?
(a) It is essential for mitosis. (b) It is mainly protein gel.
(c) It undergoes cycle of dissolution and reformation. (d) None of these
112. Syneptonemal complex first appear during:
(a) Leptotene (b) Zygotene (c) Pachytene (d) Diplotene

Answers to Multiple-Choice Questions

1. (c)	2. (b)	3. (c)	4. (c)	5. (b)	6. (a)	7. (d)	8. (a)
9. (d)	10. (b)	11. (c)	12. (b)	13. (a)	14. (c)	15. (b)	16. (d)
17. (d)	18. (d)	19. (b)	20. (c)	21. (d)	22. (b)	23. (d)	24. (b)
25. (c)	26. (d)	27. (b)	28. (b)	29. (a)	30. (d)	31. (a)	32. (b)
33. (d)	34. (c)	35. (c)	36. (a)	37. (d)	38. (c)	39. (a)	40. (c)
41. (a)	42. (b)	43. (d)	44. (b)	45. (a)	46. (a)	47. (b)	48. (d)
49. (b)	50. (d)	51. (a)	52. (c)	53. (c)	54. (b)	55. (c)	56. (c)
57. (a)	58. (c)	59. (b)	60. (b)	61. (c)	62. (c)	63. (a)	64. (d)
65. (c)	66. (b)	67. (a)	68. (d)	69. (a)	70. (c)	71. (a)	72. (c)
73. (d)	74. (a)	75. (a)	76. (b)	77. (a)	78. (c)	79. (c)	80. (b)
81. (a)	82. (b)	83. (b)	84. (a)	85. (d)	86. (b)	87. (d)	88. (b)
89. (d)	90. (c)	91. (d)	92. (d)	93. (a)	94. (a)	95. (b)	96. (b)
97. (b)	98. (d)	99. (c)	100. (b)	101. (c)	102. (b)	103. (d)	104. (a)
105. (b)	106. (d)	107. (c)	108. (c)	109. (d)	110. (d)	111. (d)	112. (b)

Fill in the Blanks

1. ________ phase, _______ phase, and ________ phase are collectively called interphase.
2. The period between M phase and S phase is called _______ phase.
3. Each Cyclin-Dependent Kinase (CDK) consists of two parts, viz., an enzyme called _______ and a modifying protein called ________.
4. Mitotic cyclins accumulate gradually during ________ phase.
5. G_1 cyclins binds to CDK proteins during _______ phase.
6. The force acting on anaphase chromosome is about ______dynes.
7. In __________, there is duplication of chromosomes but the nucleus does not divide.
8. Terminal meiosis is also known as ________ meiosis.
9. The process of bringing of chromosome on equator is known as _________.
10. Chiasmata were first observed by _________.
11. Cell division was first studied by ___________.
12. Pairing of homologus chromosomes is known as __________.
13. Chiasmata first appeared in the region of ________.
14. The position of chiasmata may be _______, ________ or _________.
15. A diploid cell undergoing meiosis is called __________.
16. A cell undergoing mitosis is called _________.
17. The synapsis of homologous chromosomes results in the formation of _________.
18. The interzonal fibres are observed during ________ and _______ between the daughter chromosomes.

19. The site of implantation of microtubules in the chromosome is the __________.
20. Kinetochores can be isolated by the ___________ technique.
21. Separation of homologous chromosomes is called _________.
22. Congression occurs during _________.
23. The shape of a chromosome can be best studied in the ______ stage.
24. Oocytes contain ___________ chromosome.
25. Syneptonemal complex is formed during _________ of prophase I.
26. Four chromatids are found during _________ of meiosis.
27. The cell cycle phase between two mitosis is called __________.
28. Homologous pairs of chromosomes are identical in shape and appearance, except ______ chromosomes.
29. A tetrad is composed of two chromosomes, viz., one ___________ and one _________.
30. A tetrad is composed of two ___________.
31. The spindle is formed around a cytosolic structure called _________.
32. A _________ is the attachment point for the spindle fibres.
33. _______ phase is the time when a cell leaves the cell cycle and quits dividing.
34. Cohesion between sister chromatids is established during the ________ phase.
35. _________ is a type of cell division that forms egg and sperm cells.
36. __________ is the longest stage of cell division.
37. Successive S phases without entering into divisional phase result in endoploidy through _______.
38. In budding yeast, the cell cycle consists of three cycles, viz., ___________ cycle, _______ cycle and _________ cycle.
39. Duplication of chromosomes without cell division is called ___________.
40. During S phase, cyclins ___________ and _______ on making complex with Cdc28 lead to DNA replication.
41. In humans, on average after 52 divisions, the cells stop dividing. This is known as ________.
42. Mitosis and cytokinesis together constitute the ___________ phase of the cell cycle.
43. The cell cycle starts in _________, with active synthesis of proteins and RNA.
44. Before entering into the M phase, cells are in _________.
45. In meiosis, the first division is ___________, whereas the second division is _________.
46. Meiosis was first discovered in the eggs of _________.
47. __________ are eukaryote organisms, which do not have meiosis.
48. __________ meiosis occurs in lower plants.
49. The two copies of each chromosome formed during the S phase are called _________________.
50. A protein complex present inside each centromere is called _________.
51. The pedigree of cells related through divisions which enable one to trace a cell's predecessors and progeny is called__________.

Answers to Fill in the Blanks

1. G_1, S, G_2
2. G_1
3. Kinase, cyclin

4. G_2
5. G_1
6. 7×10^{-15}
7. Endomitosis
8. Gametic
9. Congression
10. Johanssen (1909)
11. Prevost and Dumas (1824)
12. Synapsis
13. Crossing over
14. Terminal, subterminal, interstitial
15. Meiocyte
16. Mitocyte
17. Syneptonemal complex
18. Anaphase, telophase
19. Kinetochore
20. CREST
21. Disjunction
22. Metaphase
23. Metaphase
24. Lampbrush
25. Zygotene
26. Pachytene
27. Interphase
28. Sex
29. Maternal, paternal
30. Dyads
31. Centrosome
32. Kinetochore
33. G_0
34. S
35. Meiosis
36. Interphase
37. Endomitosis
38. Chromosome, centrosome, cytoplasmic
39. Endomitosis
40. c1b5, c1b6
41. Hayflick limit
42. Mitotic (M)
43. G_1
44. Interphase
45. Reductional, equational
46. Sea urchin
47. Excavata
48. Zygotic
49. Sister chromatids
50. Kinetochore
51. Cell lineage

True or False

1. During interphase, the cell is metabolically inactive.
2. The steps of mitosis are controlled by a number of genes.
3. In meiosis, DNA is replicated only once but the nucleus divides twice.
4. Meiosis ensures that the haploids are a complete set of chromosomes.
5. Any circulating RBC that has a chromosome is very likely a sign of an RBC cancer.
6. A cell spends 90 per cent of its time in interphase.
7. Duration of G_1 is constant for all species.
8. Epithelial cells frequently enter the G_0 phase.
9. The cells that enter the G_0 phase are destined to die.
10. The length of S phase and G_2 phase, respectively, are almost equal in all cell types.
11. Cancer cells cannot enter the G_0 phase.
12. Errors in mitosis can kill a cell through apoptosis.
13. The G_1 cyclins are found in all eukaryotes.
14. The proteins encoded by the retinoblastoma susceptibility gene (*pRB*) and p^{53} proteins are tumour suppressors.
15. Cells in the G_0 phase may re-enter the G_1 phase.
16. Asymmetric cell division is also known as determinative cell division.
17. Muscles and connective tissues have shorter G_1 phase in comparison to mucosa and endometrial epithelia.
18. Activation of maturation-promoting factor causes stable interphase state.

19. Cyclin is a regulator of maturation-promoting factor.
20. Cells in culture cannot be in the G_0 state.
21. Cells in the G_0 state have duplicated DNA content.
22. Cyclin is destroyed during each mitosis.
23. Fall in activity of maturation-promoting factor initiates anaphase.
24. Cdc25 is inactivated by polykinase.
25. Generally, the number of chiasmata per bivalent depends on the length of chromosomes.
26. Amitosis was discovered by Carpenter.
27. Karyokinesis is a discontinuous process.
28. Mitosis is useful in regeneration.
29. In plant cells, cytokinesis is centrifugal.
30. In the meiotic division of *Trillium*, telophase I is absent.
31. Nerve cells of mammals have the longest interphase.
32. In mitosis, every chromosome behaves independently.
33. In meiosis, two replication cycles of DNA are followed by two divisions.
34. Two homologous chromosomes fuse during pairing.
35. During diakinesis, the number of chiasmata diminishes.
36. Diplotene may last for months or years.
37. The number of DNA strands during the G_2 phase is four.
38. Kinetochores lack tubulin.
39. Cohesin is a dimeric protein.
40. Prophase involves condensation of chromosomes and their splitting into chromatids.
41. Generally, chiasmata disappear during diplotene.
42. Syneptonemal complex plays a role in chromosome pairing.
43. The spindle structures of mitosis and meiosis differ in their structure.
44. Birefringence increases when the chromosomes are separating.
45. Root tip is the best material to study mitosis.
46. In anaphase I, chromosomes are double stranded.
47. Meiosis I separates sister chromatids.
48. Amitosis causes structural and functional abnormalities in cells.
49. In animal cells, microfilaments are involved in cytokinesis.
50. Menstruated oocytes are arrested in meiosis II.
51. Before a cell divides, it must make a copy of its DNA.

Answers to True or False

1. False	2. True	3. True	4. True	5. True	6. True	7. False	8. False
9. False	10. True	11. True	12. True	13. False	14. True	15. True	16. True
17. False	18. False	19. True	20. False	21. False	22. True	23. True	24. False

25. True	26. False	27. False	28. True	29. True	30. True	31. True	32. True
33. False	34. False	35. True	36. True	37. False	38. False	39. False	40. True
41. True	42. True	43. False	44. False	45. True	46. True	47. False	48. True
49. True	50. False	51. True					

Give Reasons

1. Eukaryotic cell division is more complicated than prokaryotic cell division.
   — Because eukaryotes possess:
   (a) A nucleus
   (b) More than one chromosome
   (c) Two types of nuclear division, viz., mitosis and meiosis
2. The two nuclei created by mitosis are genetically identical.
   — Because the sister chromatids at metaphase are identical.
3. Mitosis is often called copy division.
   — Because during this division, genetic material is copied.
4. Meiosis is different from mitosis.
   — Because meiosis involves two divisions as well as chromosomes line up differently at the two different metaphase plates (as tetrads in the first division and as dyads in the second division).
5. In second meiotic division, there is no duplication of genetic material.
   — Because second division follows without interphase or S phase.
6. Chromosomes are the most important part of mitosis.
   — Because during mitosis, the chromosomes are duplicated and then separated into two daughter cells.
7. Mutation in checkpoint genes may lead to cancer.
   — Because it causes deregulation of cell division.
8. Cyclins are so named.
   — Because of their constant cycle of synthesis and degradation during cell division.
9. Prokaryotic and eukaryotic cells do not divide in the same way.
   — Because of their structural differences.
10. Epidermal Growth Factor (EGF) and Insulin-like Growth Factor (IGF) are also known as progression factors.
    — Because they maintain the process of cellular progression to ongoing mitosis.
11. G_0 cells decrease in size.
    — Because of the rapid degradation of their protein and RNA content as well the slow rate of their synthesis.
12. The 2:0 errors are also known as nondisjunction errors.
    — Because of the failure of separation of sister chromatids during mitosis.
13. In microorganisms, the mitotic poisons have negative effect.
    — Because they lack spindle apparatus.

14. The distribution of genetic material among daughter cells is important.
    — Because their future development and generation of cells from that will be solely dependent on the genetic material received from the parent cell.
15. Chromomeres may be used to identify a specific chromosome of an organism.
    — Because they are characteristic in size, number and position for a particular chromosome.
16. In humans and animals, the type of meiosis that occurs is called gametic meiosis.
    — Because it occurs just before the formation of gametes.

GENETICS AND MOLECULAR GENETICS

History of Genetics

1866 Gregor Johann Mendel formulated the Laws of Heredity.

1869 Friedrich Miescher discovered a weak acid in the nuclei of white blood cells that is today called DNA.

1900 Carl Correns, Hugo de Vries and Erich Von Tschermak rediscovered Mendel's principles, independently marking the beginning of Modern Genetics.

1903 Chromosomes were rediscovered to be hereditary units.

1905 N Stevens and E Wilson independently described the behaviour of sex chromosomes as XX determines female and XY determines male.

1906 William Bateson coined the term genetics.

1908 A Garrod proposed that some human diseases are due to inborn errors in metabolism.

1908 Hardy–Weinberg law was derived.

1910 T H Morgan pointed out that genes are located on the chromosomes and proposed the theory of sex-linked inheritance.

1913 First genetic map of a chromosome was prepared by A Sturtvant.

1927 H J Muller used X-rays to produce artificial gene mutation in *Drosophila*.

1928 Frederick Griffith discovered a hereditary molecule that is transmissible between bacteria.

1931 Harriet Creighton and Barbara McClintock demonstrated cytological proof of crossing over in maize.

1933 Jean Brachet demonstrated that DNA is found in the nucleus and RNA in the cytoplasm.

1937 A E Blakeslee induced polyploidy with colchicine.

1948–1950 Barabara McClintock developed the hypothesis of transposable elements to explain colour variation in maize.

1941 E L Tatum and G W Beadle propounded 'one-gene–one-enzyme' theory.

1944 O T Avery, C McLeod and M McCarty isolated DNA as genetic material.

1949 L Pauling demonstrated that protein structure is under genetic control.

1950 E Chargaff demonstrated that adenine–thymine groups are always equal to cytosine–guanine groups.

1952 A D Hershey and M Chase proved that DNA is genetic material.

1953 J D Watson and F H C Crick demonstrated the double helical model for DNA.

1955 A Kornberg and S Ochoa synthesised nucleic acid in vitro.

1956 A Gierer and G Schramm demonstrated that RNA is the genetic material of TMV (Tobacco Mosaic Virus).

1956 J H Tijo and A Levan established that the correct number of chromosomes in humans is 46.

1957 S Benzer proposed the concept of cistron, recon, and muton.

1958 M Meseleson and F W Stahl experimentally proved that the mode of replication in DNA is semiconservative.

1958 F H C Crick proposed the central dogma.

1961 F H C Crick proved the triplet nature of genetic code.

1961 M W Nirenberg and J H Mathaei deciphered the genetic code.

1961 F Jacob and J Monod proposed operon concept.

1965 F H C Crick proposed the Wobble hypothesis.

1970 H Khorana synthesised an artificial gene from DNA nucleotides.

1970 H Temin and D Baltimore discovered reverse transcription.

1972 H Smith and K Wilcox isolated the first restriction enzyme, Hind II, which cut DNA molecules within specific recognition sites.

1972 W Fiers and his team were able to determine the sequence of a gene (gene for bacteriophage MS2 coat protein) for the first time.

1972 Paul Berg and Herb Boyer produced the first recombinant DNA molecules.

1972 S H Kim suggested three-dimensional structure (L-shaped model) of tRNA.

1973 E M Southern developed Southern blotting technique for analysing the related genes in DNA restriction fragment.

1973 Annie Chang and Stanley Cohen showed that a recombinant DNA molecule can be maintained and replicated in *E. coli*.

1977 P A Sharp and R J Roberts discovered split genes of adenovirus.

1979 Alwine and his team discovered the Northern blotting technique.

1980 Zambryski, Van Montagu and Schell developed transgenic plant (Agrobacterium mediated).

1980 F Sanger and his team sequenced the entire genome of bacteriophage φX174.

1981 Harbes, Jahner and Jaenisch developed transgenic mice.

1981 Three independent research teams announced the discovery of human oncogenes (cancer genes).

1983 K B Mullis discovered the polymerase chain reaction.

1985 A Jeffreys discovered DNA fingerprinting.

1987 S Tonegawa was awarded the Nobel Prize for discovering the rearrangements of DNA sequences of mammalian immunoglobulin genes to produce a large variety of antibodies.

1989 F Collins and L C Tusi identified the gene coding for the cystic fibrosis trans-membrane conductance regulator protein (CFTR) on chromosome 7, which when mutates causes cystic fibrosis.

1990 First gene therapy was performed.

1990 Genome project began.

1991 Lalji Singh developed a new technique of DNA fingerprinting using BKM-DNA probe.

1994 Genetically modified tomatoes were marketed.

1995 The genome of *Haemophilus influenza* (the first genome of free-living organism) was sequenced.

1997 The first cloning of a mammal (Dolly, a sheep) was performed by Ian Wilmut and his colleagues.

1998 The first genome sequence of a multicellular eukaryote *(Caenorhabditis elegans*) was released.

2001 First draft sequences of human genome were simultaneously released by the Human Genome Project and Celera Genomics.

2003 Successful completion of the Human Genome Project with 99 per cent genome sequenced to a 99.99 per cent accuracy.

2006 M Pembrey and O Bygren reported sex-specific male-line transgenerational response in humans.

2007 Controversies continue over human and animal cloning, research on stem cells and genetic modification of crops.

MENDELISM

- Gregor Johann Mendel was an Austrian priest in a monastery at Brunn.
- He conducted experiments on garden pea (*Pisum sativum*) for inheritance of characters.
- The conclusions were published in the annual proceedings of the Natural History Society of Brunn in 1865.
- The results remained obscure till a rediscovery by three scientists, namely, Hugo de Vries (Holland), Karl Correns (Germany) and Kris Von Tschermark (Austria) in 1900.
- On the basis of the results of the monohybrid cross and the dihybrid cross, Correns formulated four laws of inheritance, namely,
    1. Law of Unit Character
    2. Law of Dominance
    3. Law of Segregation
    4. Law of Independent Assortment
- Due to many exceptions in Law of Unit Character, this law was immediately dropped.
- Exceptions have also been reported in Law of Dominance and Law of Independent Assortment. Therefore, these laws are not perfect or universal laws.
- Law of Segregation (also called 'Law of Purity of Gametes') has no exception till date. Therefore, this law is called perfect law or universal law.

Mendel's History

1. Birth – 22 July 1822
2. Became Priest – October 1843
3. Became Gregor –1849
4. Experimentation period – 1856 to 1864
5. Publication of result –1865
6. Death – 1884
7. Father of Genetics – 1909

Selection of Material and Traits

Mendel selected Garden Pea (*Pisum sativum*) for his experiments because of the following reasons:

(a) It is an annual plant
(b) The flowers are bisexual
(c) Self-pollination
(d) Easy emasculation
(e) Large number of seed formation
(f) Large number of contrasting characters

He selected seven pairs of contrasting characters of the plants as given below.

Table 1

S. No.	*Characters*	*Dominant*	*Recessive*
1.	Plant length	Tall (T)	Dwarf (t)
2.	Flower position	Axial (A)	Terminal (a)
3.	Pod shape	Inflated (I)	Constricted (i)
4.	Pod colour	Green (G)	Yellow (g)
5.	Seed shape	Round (R)	Wrinkle (r)
6.	Seed colour	Yellow (Y)	Green (y)
7.	Flower colour	Red (R)	White (r)

Cross

Mendel studied the inheritance of unit character at a time. Crosses were done between parents of pure breeding line having contrasting characters. This experiment was carried out on F_1 and F_2 generations. The crosses were of the following two types:

1. Monohybrid cross
2. Dihybrid cross

1. Monohybrid Cross

- The cross between pure breeding line parents for a single pair of contrasting characters is called monohybrid cross.
- When a pure tall (TT) pea plant was crossed with a pure dwarf (tt) pea plant, in F_1 generation, all pea plants were tall (Tt).
- The F_1 plants are called hybrid and they were self-pollinated.
- In the F_2 generation, both tall and dwarf plants were found.
- The cross is presented below.

Results of the F_2 Generation

1. Phenotypic ratio = 3 tall : 1 dwarf (Phenotype ratio = 3:1)
2. Genotypic ratio = 1 Homozygous tall : 2 Heterozygous tall : 1 Homozygous dwarf (Genotype ratio = 1:2:1)

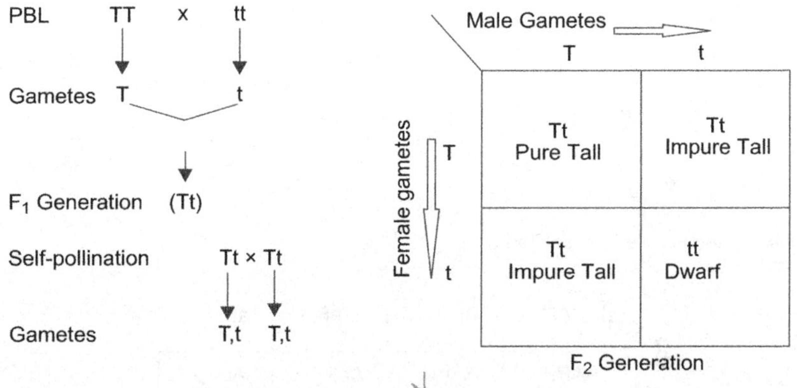

Fig. 1

Explanation of Monohybrid Cross

- Tallness and dwarfness are determined by a pair of contrasting factors or determiners. They are now called genes.
- These factors occur in pairs and are received from either parent.
- The factors are brought together by fertilisation and separate during gamete formation.
- These factors never contaminate each other.

2. Dihybrid Cross

- The cross between pure breeding line parents involving two pairs of contrasting characters is called dihybrid cross.
- When pure axial and purple-flowered pea plant was crossed with pure terminal and white-flowered pea plant in the F_1 generation, all pea plants were axial and purple flowered.
- In the F_2 generation, the self-cross of F_1 plants gave the following four types of combinations:
  - (a) Axial and purple flowered
  - (b) Axial and white flowered
  - (c) Terminal and purple flowered
  - (d) Terminal and white flowered
- The cross is represented below.

Results of the F_2 Generation

- Phenotypic ratio = 9 Axial and Purple : 3 Axial and White : 3 Terminal and Purple : 1 Terminal and White
- Genotypic ratio = 1:2:2:4:1:2:1:2:1

Explanation of Dihybrid Cross

- Axial, purple, terminal and white flower characters are determined by four different sets of factors.
- These factors occur in pairs and are received from either parent.
- These factors are brought together by fertilisation and separate during gametogenesis.
- These factors never contaminate.
- In the F_1 generation, cross-fertilisation leads to formation of hybrids, all having heterozygous axial and purple flowers.
- In the F_2 generation, the process of independent assortment takes place among factors, so new combinations

PBL, Axial and Purple Flower AAPP × aapp Terminal and White Flower

Gametes: AP, ap

Fertilisation → AaPp

All Axial and Purple Flower Plants

Self-pollination AaPp × AaPp

Gametes – 4 Types: AP, Ap, aP, ap × AP, Ap, aP, ap

♂ \ ♀	AP	Ap	aP	ap
AP	AAPP Axial & Purple	AAPp Axial & Purple	AaPP Axial & Purple	AaPp Axial & Purple
Ap	AAPp Axial & Purple	AApp Axial & White	AaPp Axial & Purple	Aapp Axial & White
aP	AaPP Axial & Purple	AaPp Axial & Purple	aaPP Ter. & Purple	aaPp Ter. & Purple
ap	AaPp Axial & Purple	Aapp Axial & White	aaPp Ter. & Purple	aapp Ter. & White

Fig. 2

like axial white-flowered and terminal purple-flowered pea plants arise.

- This cross suggests that factors express independently and they have the liberty to form new gene combinations in equal percentage.

Mendel's Law of Inheritance

On the basis of monohybrid and dihybrid cross, the following four laws were formulated:

1. **Law of Unit Character –** An individual's body contains many characters. Each of these characters behaves as a unit and its expression is controlled by a particular gene.
2. **Law of Dominance –** For one or more sets of contrasting characters, when two individual are crossed, the characters which appear in F_1 heterozygous are called dominant characters while those that do not appear are called recessive characters.
3. **Law of Segregation –** It is also called law of purity of gametes. 'Factors (genes) segregate at the time of gamete formation'.
4. **Law of Independent Assortment** – For more than one set of contrasting characters, the alleles segregate and assort independently.

Significance of Mendel's Laws

The significance of Mendel's laws is stated below.

1. The outcome of Mendel's laws is useful in Eugenics, which deals with the betterment of the human race.

2. The laws are useful in improvement of crops, cattle, fowl, etc.
3. Mendel's laws help in pedigree analysis for many diseases.

Backcross

The cross between the F_1 hybrid and any of the parent (back generation) is called backcross.

Test Cross

The cross between the F_1 hybrid and the recessive individual (parent) is called test cross. It is utilised to find out the heterozygosity in F_1 individuals.

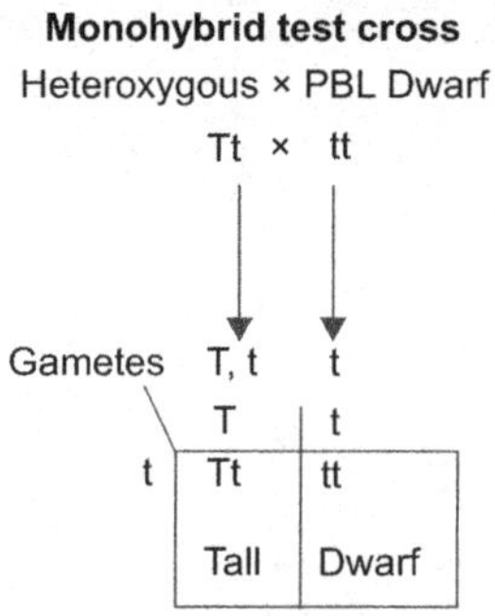

Result: Phenotypic and Genotypic Ratio = 1:1

Fig. 3

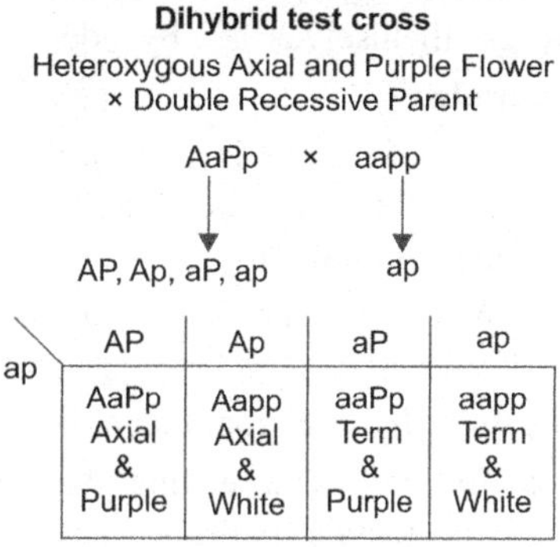

Fig. 4

Incomplete Dominance

It is a post-Mendelian cross. It is also known as blending inheritance. It is an exception to Mendel's findings.

Reference: When two contrasting alleles come together in the F_1 hybrid, the character expressed is neither of the parents' type, but a mixture of both.

Example

(i) Incomplete dominance is seen in the four o'clock plant (*Mirabilis jalapa*).

(ii) When a red-flowered (RR) plant is crossed with a white-flowered (rr) plant.

(iii) In the F_1 generation, all plants bear pink (Rr) flowers.

(iv) The self-cross in F_1 hybrids yields red-, pink- and white-flowered plants.

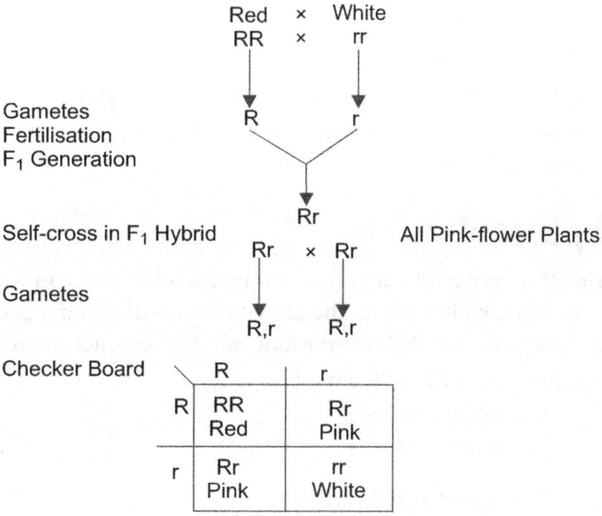

Fig. 5

Cross

Result

Phenotypic ratio = Genotypic ratio (1:2:1)

= 1 Red : 2 Pink : 1 White

Explanation – The allele R is incompletely dominant over allele 'r', so that all heterozygous (Rr) individuals produce pink flowers.

Co-dominance

It is a post-Mendelian cross and an exception to Mendel's findings. It is also known as Bateson factor hypothesis.

Definition – When two contrasting alleles come together in a heterozygous individual, both alleles express themselves side by side.

Example

(i) Co-dominance is seen in the inheritance of coat colour in cattle and blood group inheritance in humans.

(ii) When red-hair cattle is crossed with white-hair cattle.

(iii) In F_1 hybrids, the roan colour is expressed.

(iv) The cross (inbreeding) between such male and female hybrids give red-, roan- and white-hair-coloured cattle in the F_2 generation.

Result

Phenotypic ratio = Genotypic ratio

= 1 Red : 2 Roan : 1 White

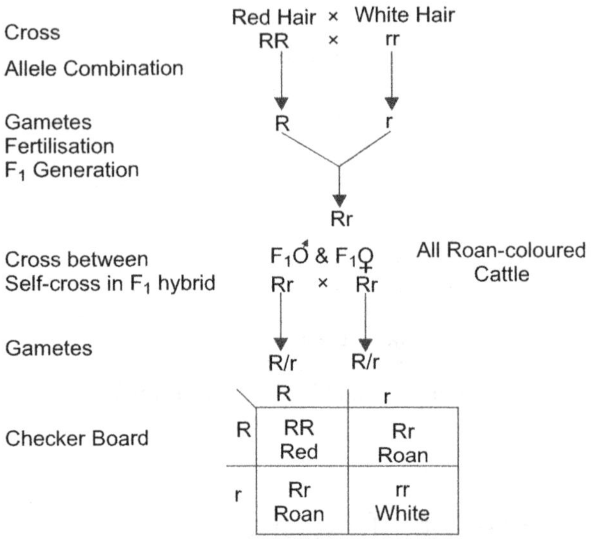

Fig. 6

Explanation – The allele for red hair colour (R) is equally expressed as the allele for white hair colour (r), so all heterozygous individuals have roan (Rr) colours.

Epistasis

Epistasis is the interaction of two non-allelic genes, in which, one gene masks the expression of another gene. The gene, which masks the activity of another gene, is called an epistatic gene. The gene, whose activity is suppressed, is called a hypostatic gene. The genes are located on different loci.

Epistasis is of the following two types:

1. Dominant epistasis
2. Recessive epistasis

1. Dominant Epistasis

In this case, a dominant gene suppresses the expression of another allele either dominant or recessive.

Example

(i) In poultry, there are two kinds of white breed, viz., the white leghorn and the white Plymouth rock.

(ii) When such homozygous poultry are left to interbreed, in the F_1 generation, all poultry become heterozygous white leghorn.

(iii) Interbreeding between the F_1 hybrid male and female produces 13 white and 3 coloured individuals.

(iv) The cross is given below.
(w = white; c = colour)_

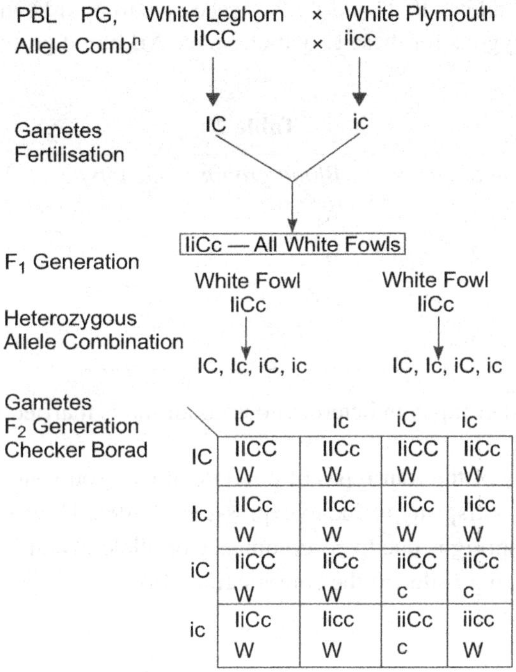

Fig. 7

Result

Phenotype ratio: 13:3

MULTIPLE ALLELES

- When a set of more than two alleles, occupying the same locus in the homologous chromosomes, affects the expression of phenotypic character, they are called multiple alleles and the phenomenon is called multiple allelism.
- The varied alleles in a set have arisen as a result of mutation in genes.

Characteristics of Multiple Alleles

(i) They are found on the same locus in the homologous chromosomes.

(ii) Only two genes of multiple alleles are found in an individual in diploid condition.
(iii) The gamete contains any one of the genes.
(iv) Crossing over is absent in multiple alleles.
(v) Multiple alleles influence only one character.
(vi) Their interaction shows dominant, recessive and co-dominance phenomena.

Example

The inheritance of ABO blood groups in humans is a very good example of multiple alleles. The four blood groups, viz., A, B, AB and O are due to the presence of three alleles occupying the same locus on the homologues chromosome. These alleles I^A, I^B and I^O, I^O is recessive to both I^A and I^B alleles. I^A and I^B show co-dominance. Individual heterozygous for these two alleles have AB blood group. The genotypes of four blood groups are as follows:

Table 2

S. No.	*Blood groups*	*Genotype*
1.	A	I^AI^A / I^AI^O
2.	B	I^BI^B / I^BI^O
3.	AB	I^AI^B
4.	O	I^OI^O

When an individual with blood group A in heterozygous condition is married to a woman with heterozygous B blood group:

- The offspring may have any of the four types of possible blood groups in ABO system.
- Blood group A and B in the offspring is due to expression of allele I^A over I^O and allele I^B over I^O.
- Blood group AB in the offspring is due to co-dominance of allele I^A and I^B.
- Blood group O in the offspring is due to the combination of two recessive alleles I° I°.

Cross

Result

The possibility of the blood group AB:A:B:O in the offsprings is 25 per cent each.

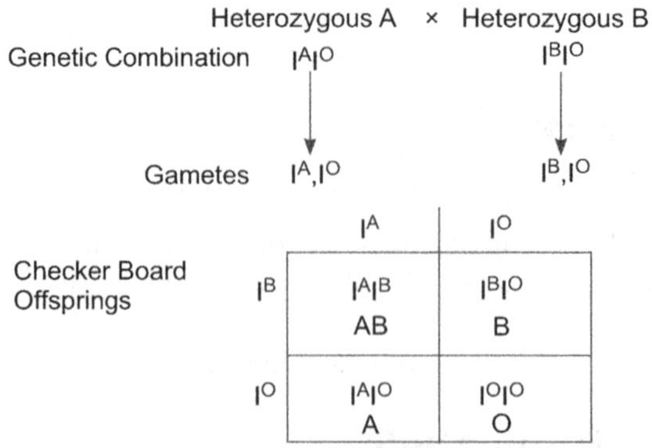

Fig. 8

POLYGENIC TRAITS AND QUANTITATIVE INHERITANCE

- There are certain characters in human population, which show phenotypic variations (e.g., height, weight, skin colour, etc.).
- These characters are expected due to cumulative effect of genes.
- These genes are called polygenes and their inheritance is called polygenic inheritance, while the characters are called polygenic traits.

Characters

(i) The effect of polygenes is cumulative.
(ii) All the alleles of polygenes have equal effect.
(iii) The inheritance is free from dominance, epistasis and linkage.
(iv) The environment has little role in the expression of characters.
(v) It was first found in humans by F C Galton in 1883.
(vi) It was first found in wheat by H Nilsson Enile in 1909.

Example

(i) When homozygous red wheat of pure breeding line (AABB) is crossed with homozygous white of pure breeding line (aabb).
(ii) In the F_1 generation, all plants produce wheat grains of intermediate red colour.
(iii) The self-cross among the F_1 hybrids produces five types of wheat grains.

Result

Dark red = $^1/_{16}$
Medium red = $^4/_{16}$
Intermediate red = $^6/_{16}$
Light red = $^4/_{16}$
White = $^1/_{16}$

LINKAGE

- The tendency of two or more genes located on the same chromosome to remain together, inherited from generation to generation, is known as linkage.
- Linkage is an exception to the principle of independent assortment.
- The concept of linkage was first introduced by T H Morgan (1910).
- The event was first observed by Bateson and Punnet in 1906 in the sweet pea plant (*Lathyrus odoratus*).

Characters

- Arrangement of all genes including linked genes is linear in chromosomes.

- Linked genes are found in the same chromosomes.
- Linked genes have the least chance of separation by crossing over called strong linkage.
- The inheritance of linked genes is of pure and combined form.

Example

- In the sweet pea plant (*Lathyrus odoratus),* the two flower characters are expressed by linked genes.
- The purple-coloured flower (PP) is dominant over the red-coloured flower and the long pollen (LL) is dominant over the round pollen grains.
- When pure purple-flowered pea plants having long pollen grains are crossed with red-flowered short pollen grains pea plant:
- In the F_1 generation, all sweet pea plants have purple flowers and long pollen grains.
- The test cross of such an F_1 hybrid with double recessive parent fails to produce the expected 1:1:1:1 ratio.
- The new ratio found due to linkage of two genes is 7:1:1:7.
- If the two alleles (PPLL) come from the same parents (PPLL X ppll), they tend to enter the same gamete. This is known as coupling.
- When the same allele (P and L) come from different parents (PPll X ppll), they tend to enter different gametes. This is known as repulsion.
- T H Morgan (1910) pointed out that coupling and repulsion are two aspects of the same phenomenon called linkage.

Types of Linkage

Linkage is of the following two types:

1. **Complete Linkage –** When genes are closely associated, they are always transmitted together. They do not undergo crossing over showing complete linkage. No new forms are formed.
   - It is rare and is found in male *Drosophila,* female silkworms, and a few others.
2. **Incomplete Linkage –** Incomplete linkage is the tendency of linked genes to separate and form recombinant types due to crossing over.
   - Incomplete linkage has been studied in *Drosophila*, other animals and maize.

Significance – Linkage maintains a specific trait from generation to generation.

Linkage Groups

- All the genes present on a chromosome constitute a linkage group.
- The number of linkage groups is equal to the haploid number of chromosomes.
- In *Drosophila*, the number of linkage groups is four; in maize, it is 10; and in humans, it is 23.

CROSSING OVER

- It is the process of exchange of segment between homologous pairs of chromosomes.
- It leads to the production of a new combination of genes. It is also called genetic recombination.
- It was first described by T H Morgan in 1911.

Time and Stage of Crossing Over

- Crossing over takes place during gametogenesis.
- Gametogenesis involves meiosis cell division.
- The actual time of crossing over is pachytene stage of meiosis I.
- Crossing over results in the formation of chiasma or chiasmata.

Significance of Crossing Over

- It confirms the linear arrangement of genes in chromosomes.
- It produces a new combination of genes.
- It causes appearance of new characters.
- It causes variation among offsprings of the same parents and among the same species.
- The process is applied for improvement of economically important varieties of plants and animals.

SEX DETERMINATION

- Sex is a unique feature of sexually reproducing organisms.
- Majority of the organisms show sexual dimorphism in which egg-producing organisms are known as female, while sperm-producing organisms are called males.
- Sex is inherited according to Mendelian laws.

Time of Sex Determination

Sex is determined at the following three times:

1. **Progamic** – Sex is determined before fertilisation.
2. **Syngamic** – Sex is determined at the time of fertilisation.
3. **Epigamic** – Sex is determined after fertilisation.

Types of Sex Determination

Sex is determined by genetically controlled mechanisms of the following types:

I. Sex Chromosome Mechanism

Dioecious organisms have two sets of chromosomes, viz., autosomes and allosomes or sex chromosomes.

- Allosomes are of two types, viz., X and Y.
- In dioecious and diploid organisms, the sex is determined by the following two members:
    1. Heterogametic male
    2. Heterogametic female

1. Heterogametic Male

- Females are homogametic (XX), while males have only one X chromosome and are heterogametic.

- Heterogametic males are of the following two types:
  - (i) XX–XO type
  - (ii) XX–XY type

(i) XX–XO Type

- In some insects (e.g., Grasshopper, Squash bug) males possess only one X chromosome.
- Presence of unpaired X chromosomes determines the male sex. For example, Squash bug

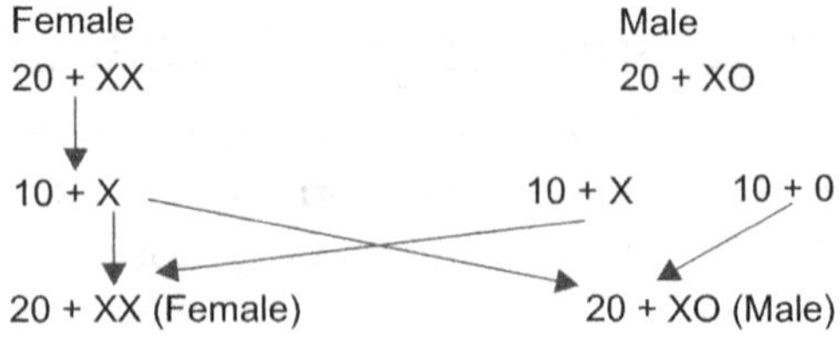

Fig. 9

(ii) XX–XY Type

- It was first studied by Wilson and Stevens (1902–1905) in milk weed bug (*Lygaeus turcicus*). Therefore, it is also known as *Lygaeus* type.
- It is found in many flies, beetles, plants and mammals including humans.
- Females are homozygous, i.e., XX, while males are heterozygous, i.e., XY.
- During fertilisation, females are produced when X chromosome containing sperm unites with the ovum.
- In this condition, the presence of Y chromosome determines the male sex offspring.
- In the plant *Cocclea indica* (family Cucurbitaceae) XX–XY mechanism was studied by Prof. R P Roy et al. of Patna University in 1974.

  For example, Human beings

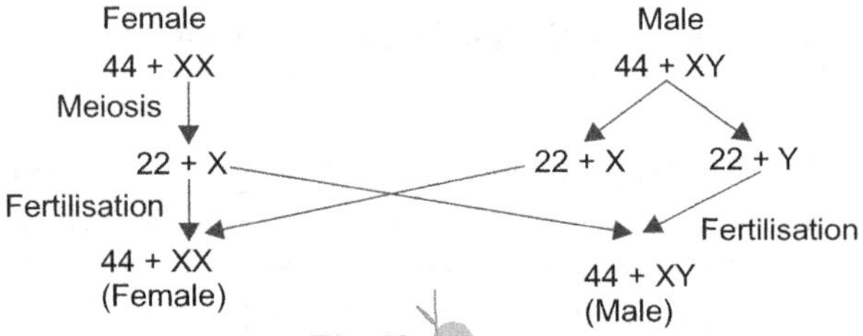

Fig. 10

2. **Heterogametic Female**
   - It is found in birds and some butterflies.
   - Females are heterogametic and produce two types of eggs, i.e., half containing X or Z chromosomes and half containing Y or W chromosomes.
   - Males are homogametic and all sperms contain X or Z chromosomes.
   - After fertilisation, two types of zygotes are formed, viz., XX or ZZ male and XY or ZW female.

Example: Silkworm

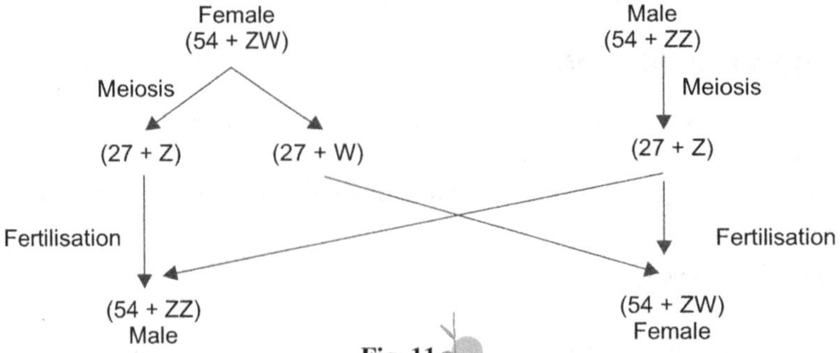

Fig. 11

II. Genic Balance Mechanism

1. This mechanism was put forward by C B Bridges in 1922.
2. According to Bridges, in *Drosophila*, sex is determined by the ratio of X chromosome and autosome, while Y chromosome remains insignificant.
3. Following table shows $^X/_A$ ratio responsible for determination of sex in *Drosophila*.

Table 3

S. No.	*X/A ratio*	*Sex*
1.	More than 1	Superfemale
2.	1	Female
3.	1 to 0.5	Intersex
4.	0.5	Male
5.	Less than 0.5	Supermale

III. Genotype Mechanism

- Sometimes a single gene present on autosome allosome affects the phenotypic expression of sex (e g., *Chalamydonomonas*, maize, *Drosophila* and several fishes).
- Sturtevant recognised a transformer gene (*tra*) in the autosome of *Drosophila.*
- Transfomer is a recessive gene which does not influence sex in heterozygous condition.
- In homozygous condition (*tra/tra*), it converts a female into a sterile male but has no influence on a normal male 2n *Drosophila.*

IV. Haploid-Diploid Mechanism of Sex Determination

This type of sex determining mechanism is found in bees, ants, wasps and some other hymenopterans.

In honeybees, the following three types of individuals are found:

1. Queens – Queens are diploid and fertile females and develop from fertilised eggs.
2. Workers – Workers are diploid, but sterile females and develop from fertilised eggs.
3. Drones – Drones are haploid fertile males. They develop from unfertilised eggs (parthenogenetically).

- In honeybees, the queen lays two types of eggs, viz., ferlilised and unfertilised.

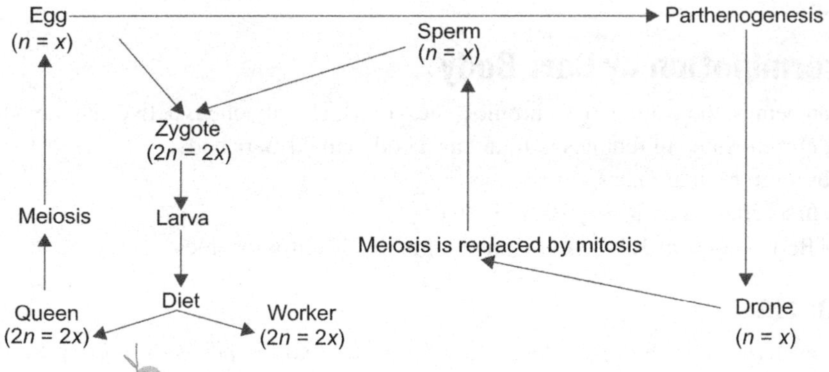

Fig. 12 Haploid-Diploid mechanism of sex determination in honeybees

- The larvae that hatch from unfertilised eggs develop parthenogenetically into drones.
- The development of diploid larvae into queen or worker depends on the quality of food.

The larva that get royal jelly as food throughout the larval period develop into queen, whereas the larva that does not get royal jelly as food throughout the larval period develop into a worker.

V. Sex Mosaic Mechanism

- In *Drosophila*, occasionally, some flies found are gynandromorphs, in which part of their body exhibits female characteristics and the other part exhibits male characteristics.
- Gynandromorphs develop due to the lack of X chromosomes in a particular cell.
- During the first zygotic division, two blastomeres with unequal number of X chromosomes are formed.
- The blastomeres with 2A+XX develop into the female half, while with 2A+X into the male half.

VI. Environmental Mechanism

- Environment plays an important role in the determination of sex in many plants and animals.
- When *Cannabis sativa* is grown in the summer season, the ratio of male and female flowers becomes 1:1, while in general, it gives male flowers only.
- When *Equisetum* plant is grown in the soil, it gives female flowers, while in bad soil, it gives only male flowers.
- In the animal kingdom, *Bonellia viridis* provides the best examples of environment-dependent sex determination.
- In this worm, all larvae are cytologically and genetically similar and hermaphrodite.
- If a single worm is reared from the eggs in isolation, it develops into an adult female.
- If newly hatched worms are reared on the proboscis of a female, they develop into adult males and migrate down to the nephridium of the female and stay as parasites.
- Young ones of a marine mollusk, *Crepidula,* if reared alone, always become females, but if reared in association with adults, they develop into males.
- In *Ophryotrocha,* it is the number of segments in the body which determines the sex. If the number of segments is up to 20, the individual is male and if the number of segments become more than 20, the individual is female. Quite interestingly, if due to some reason, the number of segments become less than 20, it again becomes transformed into a male.
- In American alligators, snapping turtles and some other reptiles, sex is determined by environmental temperature. In turtles, a temperature below 28°C leads to the production of more males and a temperature above 33°C produces more females. A temperature between 28°C and 33°C produces males and females in an equal ratio.

VII. Sex Determination by Barr Body

- In female human beings, there are two X chromosomes, in which only one is active and the other is inactive.
- The inactive X chromosome in females is found as a body called Barr body.
- Barr body is absent in normal males.
- Barr body was first reported by Murry Barr (1949).
- The number of Barr body is $n-1$, where n is the number of X chromosomes.

VIII. Freemartin

- In cattle, whenever twins of the opposite sex are born, the male is perfectly normal but the female is sterile. Such a sterile female is called a freemartin and the phenomenon is known as freemartinism.

- Freemartin is only produced when there is a vascular connection between two embryos.
- It is presumed that probably male hormones suppress the development of the ovary in the female.
- Sex determination is a complex phenomenon in which many mechanisms play an important role. The XX–XY mechanism is widely accepted.

SEX-LINKED INHERITANCE

In human beings, sex is determined by the XX–XY mechanism. These are called allosomes or sex chromosomes. Some of the characters inherited through the allosomes are called sex-linked characters.

The diseases whose genes are present on the sex chromosomes and inherited with sex chromosomes are called sex-linked diseases. The genes located on the X chromosomes are called sex X-linked genes and the genes present on the Y chromosomes are called holandric genes, because they are present in males only. Sex-linked inheritance was introduced by T H Morgan in 1910 in *Drosophila melanogaster*.

Sex-Linked Diseases in Humans

Genes for about 20 characters are found on X chromosomes and exhibit sex-linked inheritance.

Sex-Linked Diseases

In human beings, X and Y chromosomes are responsible for sex determination and are called sex chromosomes or allosomes. The characters whose genes are located on sex chromosomes are called sex-linked characters and their inheritance from one generation to another is called sex-linked inheritance. The concept of sex-linked inheritance was introduced by T H Morgan in 1910.

Inheritance through X Chromosomes

About 20 characters of human beings exhibit sex-linked inheritance and their genes are found located on X chromosomes. The most common examples of sex-linked diseases, inherited in the same pattern are given below.

(a) Colour-blindness
(b) Haemophilia
(c) White fore lock of hair
(d) Childhood muscular dystrophy
(e) Lesch–Nyhan Syndrome

(a) Colour-blindness

(i) It is a sex-linked recessive disease.
(ii) Colour-blind people are unable to distinguish between red and green colours.
(iii) It was first described by Horner in 1876.
(iv) Red colour blindness is called protanopia and green colour blindness is called deuteranopia.

- The inheritance and establishment of diseases in the progeny can be studies in the following three cases:

Case I – When a woman with normal vision is married to a colour-blind man, all her sons and daughters have normal vision.

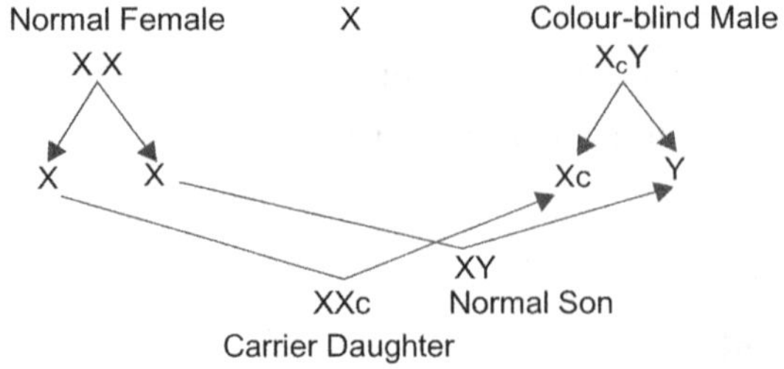

♀ \ ♂	Xc	Y
X	X_cX Carrier Daughter	XY Normal Son
X	X_cX Carrier Daughter	XY Normal Son

Fig. 13

Case II – When a woman, whose father was colour-blind, is married to a male with normal vision, half of her sons will have colour-blind vision.

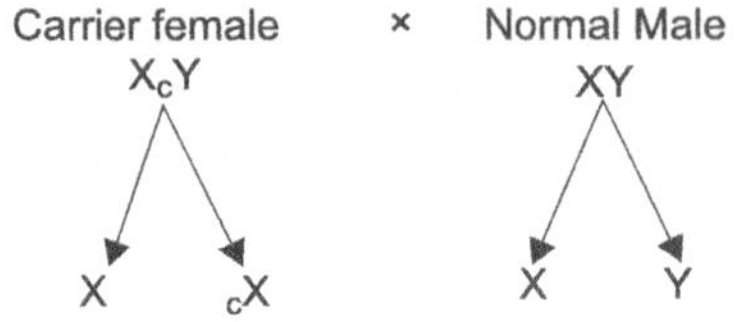

♀ \ ♂	X	Y
X	XX Carrier Daughter	XY Normal Son
Xc	X_cX Carrier Daughter	X_cY Colour-blind Son

Fig. 14

Case III – When a colour-blind woman is married to a man with normal vision, all her sons will have colour-blind vision, but all her daughters will have normal vision.

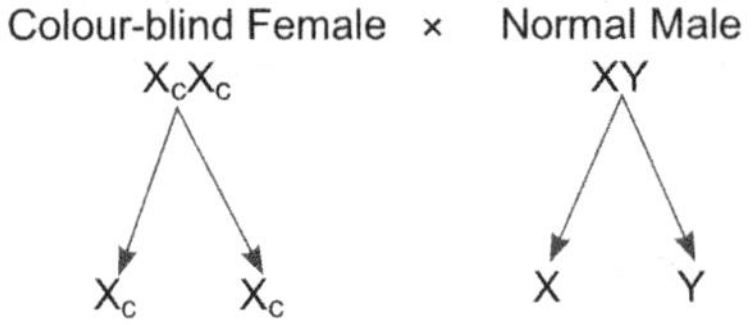

♀ \ ♂	X	Y
X_c	X_cX Carrier Female	X_cY Colour-blind Male
X_c	X_cX Carrier Female	X_cY Colour-blind Male

Fig. 15

Conclusion

The following conclusion may be drawn from the above three cases:

- Colour-blindness is more common in males because only X chromosomes carry the gene.
- In females, it is less frequent because two recessive genes are needed for expression.
- A female with only single colour-blind gene is called carrier, in which the disease does not appear.
- Females may be carriers but males are never so.
- A colour-blind father and a mother with normal vision have carrier daughters and sons with normal vision.
- A woman with normal colour vision, whose father was colour-blind, will give birth to normal and colour-blind sons in equal proportion.
- Colour-blind women always have colour-blind fathers and always give birth to colour-blind sons and carrier daughters.

- Colour-blind women give birth to colour-blind daughters only when their husband is also colour-blind.
- Women who have normal vision but can transmit colour-blindness are called carriers.

CHROMOSOMAL ABERRATIONS

- Chromosomal aberrations are abnormalities in the structure or number of chromosomes.
- They are often responsible for genetic disorders.
- They may occur spontaneously or can be induced by environmental agents such as chemicals, radiations and ultraviolet light.
- Generally, chromosomal abnormalities occur when there is error in cell division (mitosis or meiosis).
- Chromosomal aberration involves breaking of chromosome segments, their loss or union with the same (intrachromosomal aberration) or different chromosomes (interchromosomal aberration).
- Chromosomal aberrations can be recognised with the use of karyotype.
- Structural changes in chromosomes are of the following four types:
    1. Deficiency or deletion
    2. Duplication
    3. Inversion
    4. Translocation

1. Deficiency

- Deficiency or deletion is the loss of chromosomal segments.
- The lost segment may be terminal (terminal deletion) or intercalary (intercalary deletion).

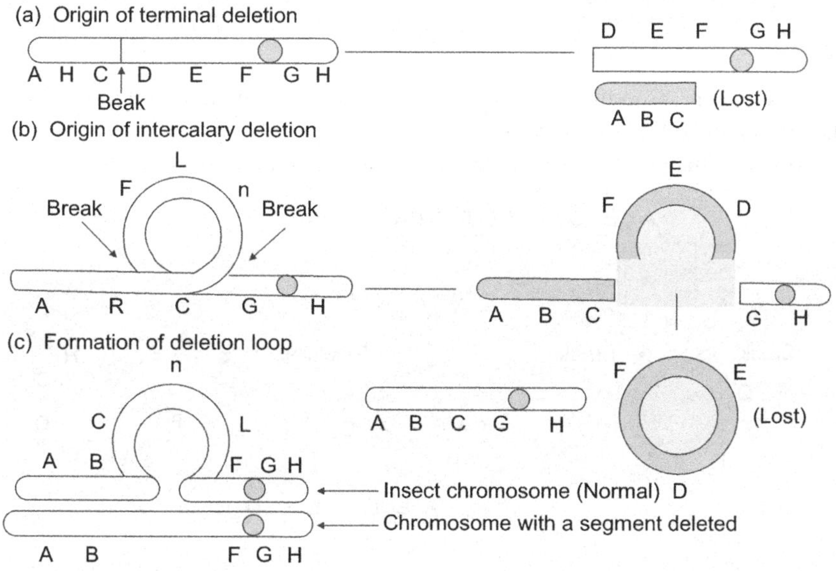

Fig. 16 Deletion: (a) Terminal deletion (b) Intercalary deletion (c) Formation of deletion loop

- Terminal deletion involves single break near the terminal end, while intercalary deletion involves two breaks; the separated segment is lost and there is reunion of the broken ends.
- In heterozygous deletion, the synapsed chromosome would show difference in length at one end (terminal deletion) or loop (intercalary deletion).
- Homozygous deletion may be lethal if some vital gene is involved.
- In the presence of deletion, a recessive allele behaves like a dominant allele. This phenomenon is known as pseudodominance.
- Heterozygous deficiencies lead to many genetical disorders such as retinoblastoma, myeloid leukaemia, cri-du-chat syndrome, etc.
- Deletion involves change in the amount of DNA.
- Genetically, deletion can be recognised by recessive lethality, pseudodominance and lack of revertability and cytologically by deletion loops.

2. Duplication

- When a set of genes are present twice in a chromosome, it is known as duplication.
- From evolutionary point of view, duplications are very important as they supply additional genetic material capable of evolving new functions.
- Duplication arises due to unequal crossing over.
- Adjacent duplicated segments may occur in tandem sequence with respect to each other (abcbcd) or in a reverse order (abccdb).
- Duplications are more frequent in nature and are less deleterious in comparison with deletions.
- Duplication involves change in the amount of genetic material.
- Repeated duplication of the same gene results in gene amplification as well as the formation of gene families.
- A small duplication is rarely lethal even when homozygous. However, sometimes it may be pathological.

3. Inversion

- Inversion involves two breaks in the chromosome followed by the reunion of the ends in a reverse manner, i.e., the broken segments rotate through 180°.
- Thus in inversion, only the arrangement of genes is changed and not the number.

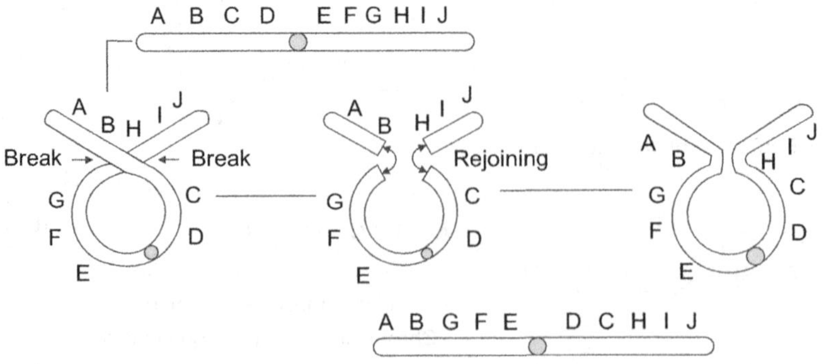

Fig. 17 Diagram showing the mechanism by which some of the inversion might occur

- Generally this change in the gene order does not produce clinical abnormality but it results in increased risk of generating abnormal gametes.

- Organisms may be either homozygous or heterozygous for an inversion.
- Inversions are of the following two types:
  - Paracentric inversion – The inverted segment does not include centromeres.
  - Pericentric inversion – The inverted segment includes centromeres.
- If crossing over takes place within the inverted segment of a paracentric inversion, then it results in the formation of acentic and dicentric chromosomes.
- If crossing over occurs within the loop of a pericentric inversion, then the resulting chromatids have a duplication and a deficiency.

4. Translocation

- Translocation is the exchange of chromosomal segments between the nonhomologous chromosomes.
- Translocation is of the following three types:
  - Simple Translocation – The broken segment gets attached to one end of the nonhomologous chromosome.
  - Reciprocal Translocation – It involves the exchange of chromosomal segments between members of two different pairs.
  - Shift Translocation – The broken segment gets inserted interstitially in a nonhomologous chromosome.

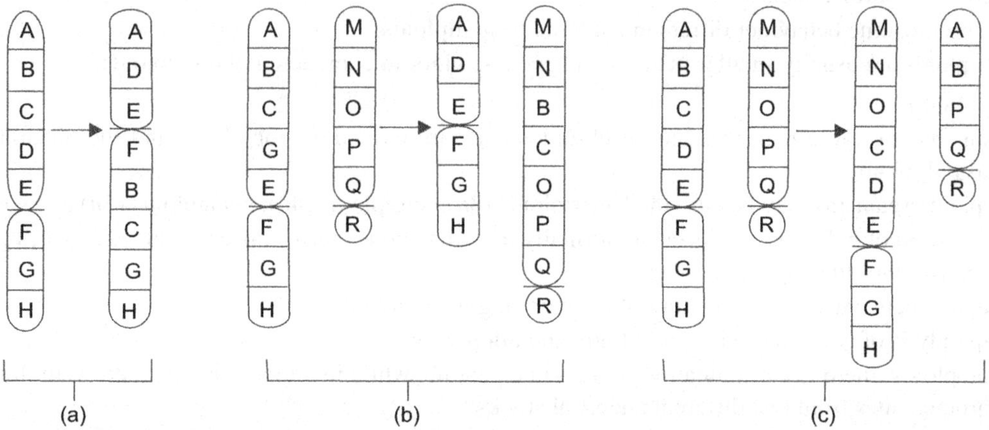

Fig. 18 Different types of translocation: (a) Simple translocation (b) Reciprocal translocation (c) Shift translocation

- Translocation occurs due to irregularities during crossing over.
- In translocation, there is no loss or addition of genetic material, only rearrangement of genes occurs. Thus, two new chromosomes are produced with an altered gene sequence.
- A balanced translocation carrier is at risk of producing chromosomally unbalanced offspring.
- Generally, translocation heterozygotes are less fertile in comparison with translocation homozygous. It is some of the gametes produced by them that give rise to unviable progeny due to deficiency or duplication of the chromosome segment.
- Translocation may alter the size of a chromosome as well as the position of the centromeres.

Changes in Chromosome Number

Variation in chromosome number is of the following two types:

1. Euploidy
2. Aneuploidy

1. Euploidy

- Euploidy involves variation of entire sets of chromosomes.
- It involves complete genomes.
- When one complete set of chromosome is present, it is called monoploidy (n). If two sets of chromosomes are present, it is called diploidy ($2n$) and if more than two sets of chromosomes are present, it is termed as polyploidy.

(a) Monoploidy

- Monoploidy is also termed as haploidy.
- Each chromosome is represented only once.
- Haploidy may arise as an abnormality in species which are diploid.
- It is common in plants but rare in animals.
- Among animals, haploidy occurs in male honeybees, ants and other insects which develop from unfertilised eggs (parthenogenesis).
- The chromosome behaviour differs in haploids from diploids.
- Monoploids are usually smaller in size and have less vigor in comparison with diploids.

(b) Polyploidy

Organisms having more than two sets of chromosomes are termed as polyploids and the phenomenon is known as polyploidy.

- Polyploid organisms may be classified as triploids ($3n$), tetraploids ($4n$), pentaploids ($5n$) and so on.
- A proportion of polyploidy cells occur normally in human bone marrow, because megakaryotes generally have 8–16 times the haploid number.
- Tetraploid cells are a normal feature of regenerating liver and many other tissues.
- Polyploidy is of two types, viz., autoploidy and alloploidy
- In autoploidy, there is multiplication of the same genome while in allopolyploidy, there is multiplication of chromosomes from two different ancestral stocks.
- Polyploidy arises due to error during meiosis or fertilisation. Sometimes abnormal mitosis also leads to polyploidy.
- Polyploidy is common in plants but rare in animals.
- Polyploidy is artificially induced with colchicine. Colchicine is an alkaloid derived from the autumn crocus (*Colchicum autumnale*) which disturbs spindle formation.

2. Aneuploidy

- Aneuploidy refers to variations in chromosome numbers involving individual chromosomes.
- Aneuploids have unbalanced chromosome sets, i.e., with extra or missing chromosomes.
- Aneuploids are generally produced by nondisjunction or some type of chromosome misdivision at either meiosis or mitosis.
- Aneuploids may be of the following types:

(a) **Monosomics** – Monosomics are individuals having one chromosome less than the diploid number ($2n$–1).
(b) **Double Monosomics** – There are two chromosomes less, that are different members of the chromosomal complement ($2n$–1–1).
(c) **Nullisomics** – They have two chromosomes less than the diploid complement ($2n$–2).
(d) **Trisomics** – Homolog of normal diploid complement is present three times ($2n$+1).
(e) **Double Trisomics** – There are two extra chromosomes, which are different members of the genome ($2n$+1+1).
(f) **Tetrasomics** – In a tetrasomy individual, one chromosome of the genome is present four times ($2n$+2).

GENE MUTATION

- Mutation is the sudden, discontinuous and inheritable change in genetic material.
- Mutation is the failure of the DNA repair mechanism.
- A mutation occurs when the DNA is damaged or changed in such a way as to alter the genetic message carried by that gene.
- Once the gene has been damaged or changed, the mRNA transcribed from that gene will carry an altered message.
- These changes could give rise to the following types of mutations:
  1. **Silent Mutation –** There is change in the nitrogenous base but there is no change in the amino acid.
  2. **Missense Mutation –** It involves changes in the nitrogenous base as well as in the amino acid sequence. Sickle cell anaemia is a good example of missense mutation, which is caused by the substitution of valine in place of glutamic acid in the haemoglobin chain.
  3. **Nonsense Mutation –** It changes the codon into terminating codon, resulting in the termination of the polypeptide chain. It results in the formation of short polypeptide chain which has little or no biological effect. Such mutations are also called terminator mutations.
- The term 'mutation' was coined by Hugo de Vries.
- Mutations may occur in somatic cells as well as germ cells.
- Mutations arising in somatic cells are not passed to the next generation, while those occurring in reproductive cells are passed onto the next generation.
- Mutations occurring in nature are called spontaneous mutations. They arise due to inherent errors in the DNA replication and transmission processes.
- Spontaneous mutation can occur at any point in the cell cycle.
- Mutation rate varies from 10^{-4} to 10^{-6} mutations per gene per generation. In humans, the mutation rate is 10^{-5} to 10^{-6} per gamete per generation.
- Mutations that occur due to or in response to some externally applied agents are called induced mutations.
- Depending upon the effect, mutation may be dominant or recessive.
- T H Morgan (1910) reported white eye mutation in *Drosophila melanogaster*.
- Mutations play a key role in speciation.
- A variety of agents cause mutations and they are called mutagens.
- Mutagens may be physical or chemical.

- Physical mutagens are X-rays, gamma rays, beta rays, ultraviolet rays, etc.
- There are varieties of chemical mutagens such as nitrogen mustard, sulfur mustard, dimethyl nitrosamine, ethylene oxide, di-ethyl sufonate, nitrous acid, hydroxyl amine, hydrazine, etc.

Molecular Basis of Mutations

There are two basic types of mutations.

1. **Substitution Mutations** – It involves replacement of one nitrogenous base of a triplet codon by another nitrogenous base. It may be of the following two types:
   - **Transition** – Transition is the replacement of a purine by another purine and replacement of one pyrimidine by another pyrimidine.
   - **Tranversion** – Transversion is the replacement of purine by pyrimidine and pyrimidine by purine.
2. **Frameshift Mutation** – Frameshift mutation is caused by the addition or deletion of a base pair in the gene, resulting in change in the reading frame of the DNA.
   - Addition or deletion of one or more nitrogenous bases results in a new codon sequence that code for quite different amino acids.
   - The change in amino acid sequence results in a change in the synthesised protein, which is generally nonfunctional.

Deletion

- Deletion is the removal of one or more nitrogenous bases from the DNA polynucleotide chain.
- Deletion results in the establishment of a new sequence which occurs by deletion of any number of bases, not divisible by three.
- Suppose the original reading frame is CAT GAT CAT GAT CAT GAT CAT, then deletion of C of the last codon will read as CAT GAT CAT GAT CAT GAT TA.

Insertion

- If one or more bases are added (provided it is not divisible by three), it will disturb the genetic message.
- +G CAT GAT GCA TGA TCA TGA TCA T
- If deletion and insertion take place simultaneously, then the message will be out of frame only in the triplet between the deletion and insertion.
- Deletion and insertion
- –C+C CAT GAT ATG ATC ATC GAT

Mechanism of Spontaneous Mutations

Spontaneous mutations arise by mutagens present in the environment, such as radiation, radioactive compounds, heat and naturally occurring base analogues like caffeine. Spontaneous mutations that arise by tautomerisation are described here.

Tautomerisation

- Isomerisation between tautomers is called tautomerisation.

- Tautomers are the two forms of the same compound.
- Normally, in a DNA molecule, adenine (purine) pairs with thymine (pyrimidine) while guanine (purine) pairs with cytosine (pyrimidine).
- All these four nitrogenous bases (adenine, thymine, cytosine and guanine) of DNA have rare tautomeric forms.
- These rare forms are called tautomers and are formed by the rearrangements of hydrogen atoms.
- The normal bases of DNA are generally present in the keto form and amino form.
- As a result of tautomeric arrangement, they can be transformed into rare enol form and imino form, in which distribution of electron is slightly different.
- In a tautomeric state, adenine pairs with cytosine and thymine pairs with guanine.
- The unusual pairing of adenine with cytosine results in G–C pairing in the next generation causing formation of mutant forms in some progeny.
- The rare base pairing result in misreplication of DNA leading to mutation.
- Ambiguity of base pairs during replication may also result in spontaneous mutations.

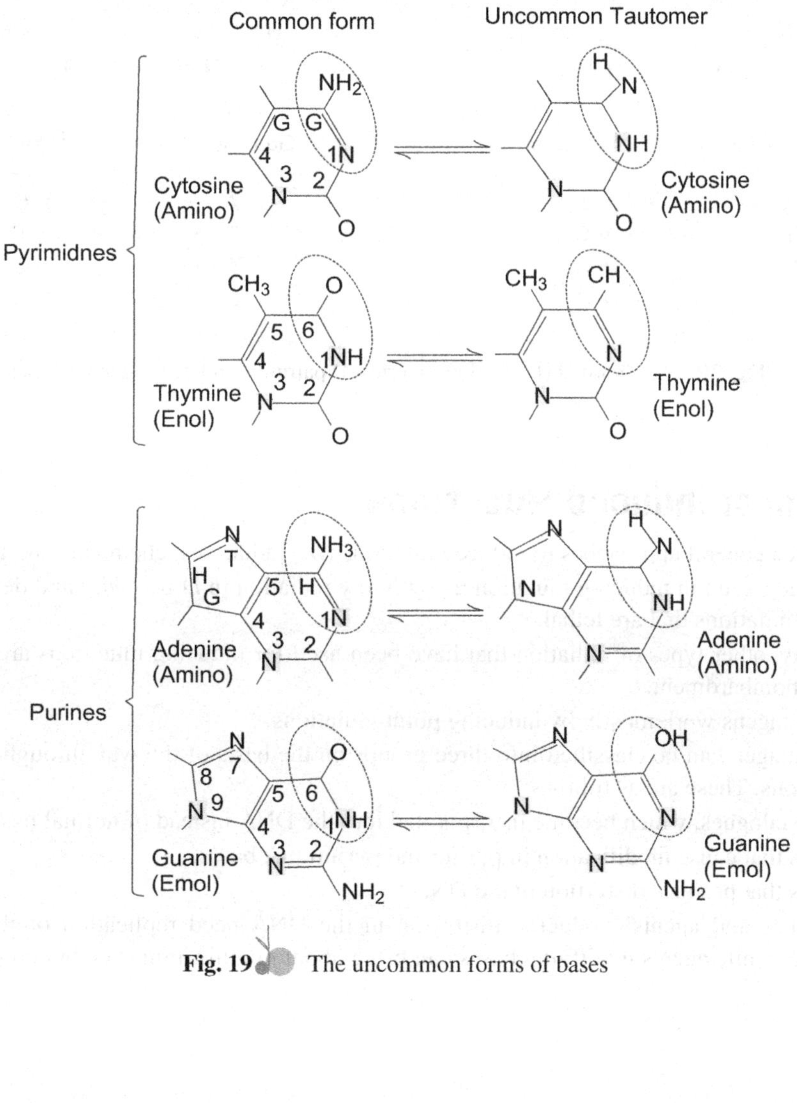

Fig. 19 The uncommon forms of bases

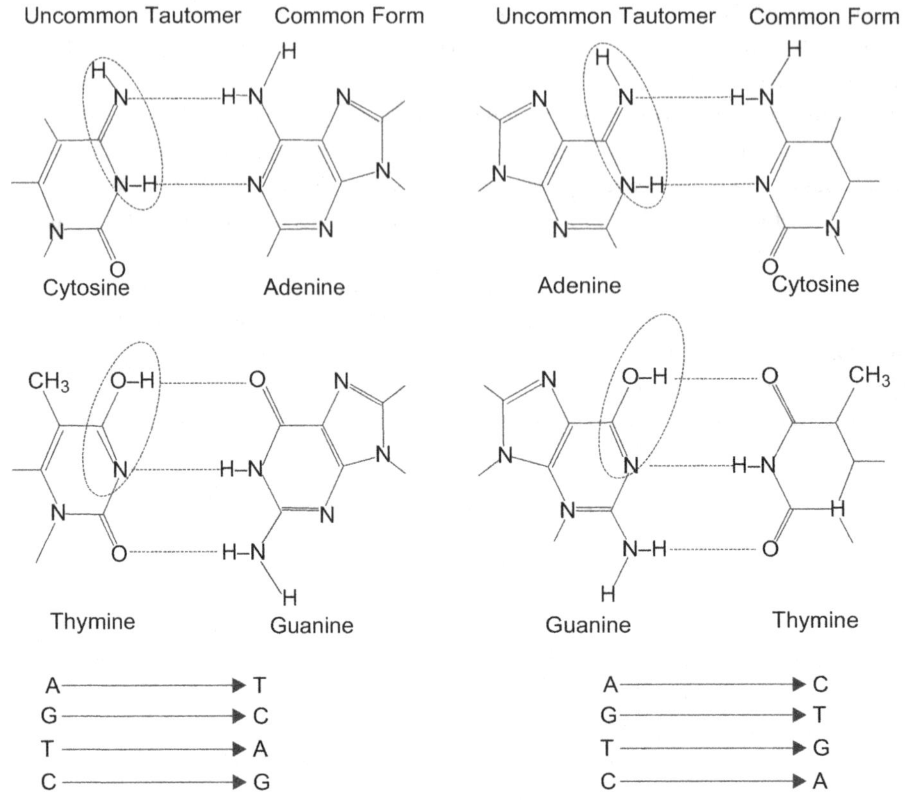

Fig. 20 Normal (I) and Tautomeric (II) pairing of four nitrogenous bases

Mechanism of Induced Mutations

- There are three general approaches to induce mutation, viz., radiations, chemicals and transposons.
- H J Muller pioneered in inducing mutation using X-ray radiation in *Drosophila* and developed a method of detecting mutations that are lethal.
- Besides X-ray, other types of radiation that have been used for inducing mutations are gamma rays and fast neutron bombardment.
- Chemical mutagens work mostly by inducing point mutations.
- Chemical mutagen can be classified into three groups on the basis of the way through which they bring about mutations. These are as follows:
  1. Base analogues, which become incorporated into the DNA instead of normal bases.
  2. Agents that cause modification in purine and pyrimidine bases.
  3. Agents that produce distortion in the DNA.
- Base analogues and agents producing distortion in the DNA need replication of the DNA for their incorporation, while agents modifying bases can bring about modification even in nonreplicating DNA.

1. Base Analogues

- Base analogues are chemicals having structures similar to nitrogenous bases.
- Base analogues sometimes become incorporated into the DNA in place of normal nitrogenous bases.
- 5-bromouracil (5-BU) is a pyrimidine analogue and is structurally very similar to thymine. It can pair with adenine or guanine.
- 2-aminopurine (2-AP) is a purine analogue that can pair with cytosine and thymine.
- Both 5-bromouracil and 2-aminopurine only mutate when they are incorporated in the replicating DNA.

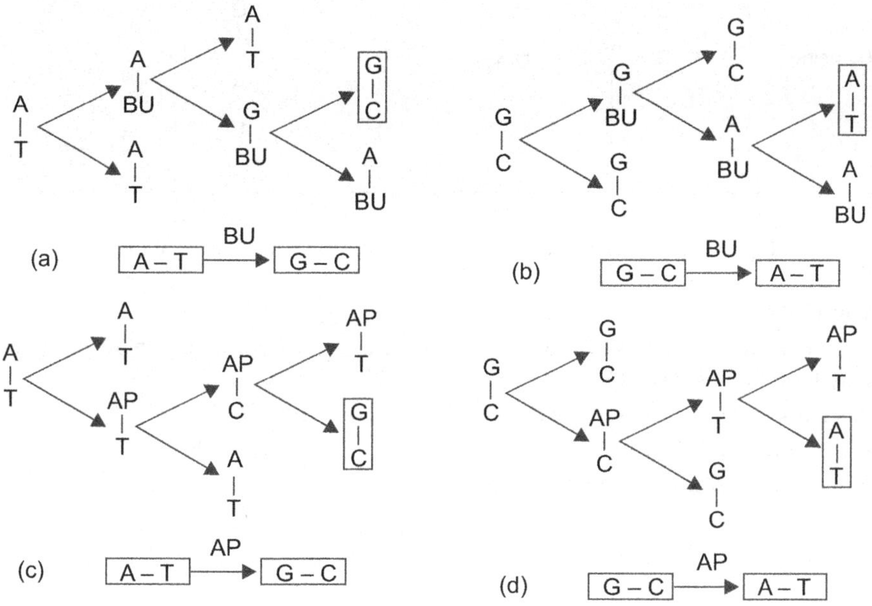

Fig. 21 Changes in base pairing due to the incorporation of base analogues

2. Base Modification

There are some mutagens which cause change in base pairing, resulting in incorrect pairing.

- Such modifications involve alkylation, hydroxylation, deamination, depurination, etc.

(a) Deamination

- Some of the chemicals (like nitrous acid and hydroxyl amine) cause deamination of nitrogenous bases. They replace amino group (–NH_2) group by hydroxyl group (–OH).
- Deamination of cytosine results in the formation of uracil; deamination of adenine leads to the formation of hypoxanthine (H); and that of guanine forms xanthine.
- Hypoxanthine shows similarity with guanine.
- During the course of DNA replication, uracil pairs with adenine while xanthine pairs with cytosine.
- Thus, it results in the substitution of A = T for G ≡ C and G ≡ C for A = T.

Normal base of DNA	Normal pairing	Altered base by deamination	Altered pairing
Adenine	A = T	Hypoxanthine (H)	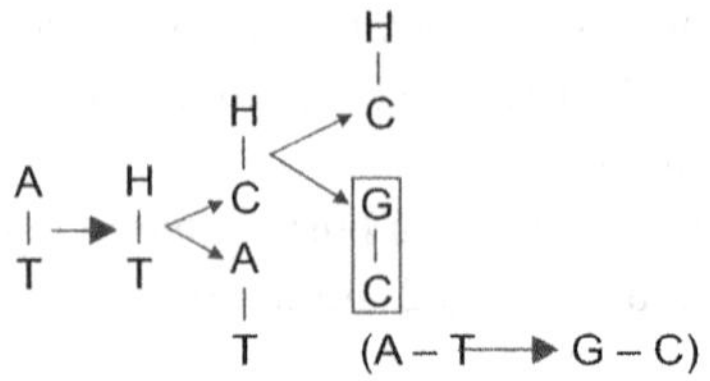
Cytosine	C≡G	Uracil	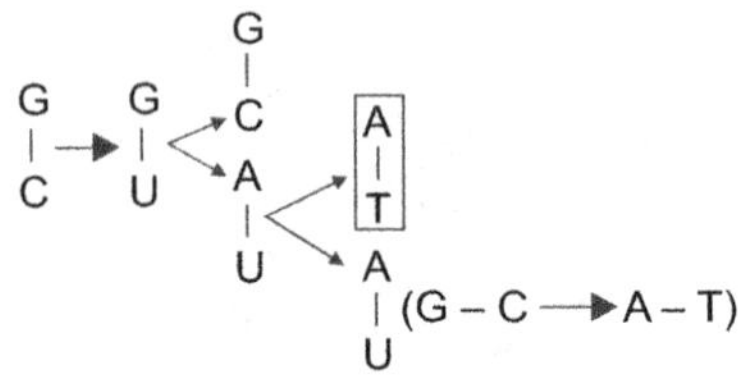

Fig. 22 Deamination caused by nitrous acid and abnormal pairing

(b) Hydroxylation

- Hydroxyl amine complexes with cytosine and causes its hydroxylation, resulting in the formation of hydroxyl cytosine (HC).
- This hydroxyl cytosine pairs with adenine instead of guanine.
- At the time of DNA replication, this introduces thymine at this level.
- Thus, G–C pairing changes to A–T pairing.

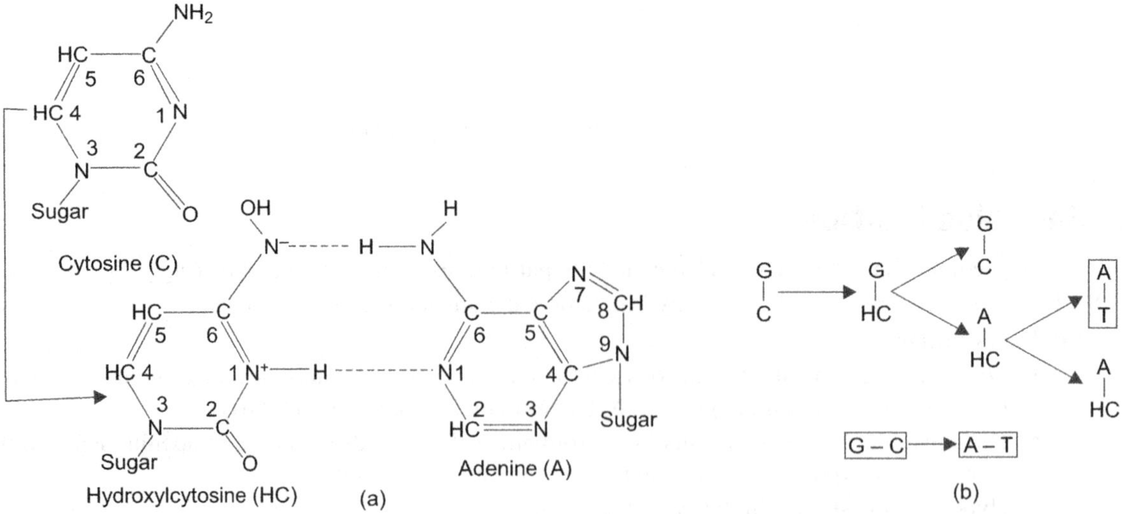

Fig. 23 Transition of C–G to A–T produced by hydroxyl amine due to conversion of cytosine to hydroxyl cytosine

(c) Alkylation

- Alkylation is caused by alkylating agents.
- Alkylating agents cause mutations by transitions, transversions, deletions and frameshifts.
- Alkylating agents (such as ethyl-methane sulfonate, ethyl-ethane sulfonate and mustard gas) can mutate both replicating and nonreplicating DNA.
- Alkylating agents bring about methylation and ethylation of nitrogenous bases.
- Alkylation of guanine results in mispairing with thymine and during the course of replication, it may result in $G \equiv C \longrightarrow A = T$ pairing.
- Ethyl methane sulfonate (EMS) removes guanine from the strand of the DNA and leaves a gap point.
- At the time of replication, any of the four nitrogenous bases become inserted in the gap.
- In the next replication, the gap is filled by a base, which is complementary to the inserted base.
- This may lead to transition or transversion.

3. Agents causing Distortion of DNA

There are certain acridine dyes (proflavin, acridine orange) that can be intercalated between bases of the DNA strand causing distortion in the DNA.

- This results in insertion or deletion of bases during replication, resulting in mutations.

Transposable Elements as Mutagens

- Scientists are now using transposable elements to create new mutations.
- Transposable elements are mobile pieces of DNA that can move from one location to another in a genome.
- Often when they move to a new location, the result is a new mutant.
- The mutant arises due to the presence of a new piece of DNA in a wild-type gene that disrupts the normal functioning of the gene.
- Thus, transposable elements are a powerful source of creating insertional mutations. This is known as insertional mutagenesis.
- Besides, prion replication has been shown to be subjected to mutation.

MODERN CONCEPT OF GENE

- The presence of genes was first proposed by Mendel in 1865. He called it factor. The term 'gene' was coined by Johannson in 1909.
- Gene is the basic fundamental unit of heredity and life.
- Genes control both structure and specific function of the cells and thus the entire organism.
- Genes are present in each and every cell of all organisms.
- Chemically, each gene consists of a specific sequence of DNA building blocks, called nucleotides. Each nucleotide is made up of pentose sugar, nitrogenous base and phosphoric acid.
- On an average, a gene consists of 1,500 nucleotide base pairs.
- The strands of DNA into which genes occur are organised into chromosomes.
- Each gene provides a blueprint for the synthesis of enzymes and other proteins.

- A gene expresses itself producing a specific protein through the process of transcription and translation.

$$\text{Gene} \xrightarrow{\text{Transcription}} \text{mRNA} \xrightarrow{\text{Translation}} \text{Protein}$$

- A single gene may occur in several forms called alleles. Generally a gene has two forms, viz., dominant and recessive. When a gene exits in more than two forms, it is known as multiple alleles.
- Sometimes two alleles of a character express equally. This is known as co-dominance.
- Genes can replicate and can produce its own copies.
- The expression of genes is influenced by environmental factors.
- The total set of genes of an organism is called a genome.
- Genes may vary in their make up from person to person.
- Each gene occupies a fixed position on the chromosome called locus.
- Genes are arranged in a single linear order on a chromosome.
- Alteration in the number and arrangement of genes may result in mutation.
- Changes that occur due to mutations in germ cells can be transmitted to the next generation.
- Mutations that affect somatic cells may result in various types of cancers.
- Many genes are present on each chromosome and they are inherited together. Such genes are called linked genes.
- Sometimes two or more genes interact to produce a particular trait. This is termed as 'interaction of genes'.
- Sometimes a single pair of genes produces two or more characters. This is known as pleiotropism.
- Transfer of alleles of genes occurs from one population to another, called gene flow, which leads to change in gene frequencies.
- The number of genes in a particular cell may be increased (by polyploidy or hyperploidy) or decreased (hypoploidy).
- Inbreeding leads to homozygosity while outbreeding results in heterozygosity within the gene pool.
- Fine structure of gene has revealed that a gene consists of cistron, recon and muton.
- Cistron is a portion of the DNA specifying a polypeptide chain.
- Recon is the smallest unit of DNA capable of recombination.
- Muton is the smallest unit of DNA capable of undergoing mutation.
- Some genes are naturally split. Such genes contain some sequences which do not code for amino acids (called introns) and some sequences that code for amino acid (called exons).
- Genes having only one intron are called monointron gene (RNA tyr gene) while genes having more than one introns are called multi-introns genes (rat muscle X-acting gene has six introns).
- There are certain genes, which code for more than one polypeptide. Such genes are called overlapping genes.
- Overlapping genes share some of the same sequences (In bacteriophage φX174, gene *E* lies entirely within gene *D*1, but they are translated in different reading frameworks).
- There are genes which move from one location to another on a chromosome.

Such genes are called jumping genes or transposons.

- A gene that produces two proteins simultaneously from a long transcript by changing the end point of protein synthesis is known as nested gene. The entire coding sequence of such a gene lies between the start codon and terminating codon of a larger external gene. Thus, the entire coding sequences are present in other genes.
- The genes that have lost their ability to perform a function due to mutations are called pseudogenes. They look like normal genes but do not express any RNA or protein. They are often termed as nonfunctional DNA and regarded as junk.

- Genes that lack well-defined exon or introns sequences, i.e., the exon in one instance might be introns in another case, are termed as complex genes. These genes manifest excessive rearrangement in sequence (mRNA), i.e., before post-transcriptional modifications.
- Genes that cause cancer are called oncogenes. Oncogenes are activated form of proto-oncogenes.
- The genes that code for proteins, which act as transcription factor, enhancing the rate at which certain DNA sequences are transcribed are termed as homeotic genes. Homeotic genes play a key role in the early development and differentiation of embryonic tissue in eukaryotic organisms. These genes were first identified in *Drosophila*.
- The group of genes showing similarity with each other is termed as gene family, arising by duplication. All genes in the family may occur on the same locus. For example, five members of growth hormone gene family are clustered on chromosome 17 in humans. Gene families provide information how new genes arise and diversify.

HUMAN GENETICS

- Human genetics is the branch of genetics that deals with the study of inheritance of human traits.
- Sir Archibald Garrod is generally described as the Father of Human Genetics.
- About 200 traits are known to be genetically transmitted in human beings. Some of these traits are transmitted through typical Mendelian pattern and some through non-Mendelian pattern.

Human Karyotype

- Karyotype refers to the complement of chromosomes, either at species level or at individual level.
- The correct number of human chromosomes (46) was given by Tijio and Levan (1956). Previously, the chromosome number of humans was considered to be 46.
- Normal human karyotype contains 22 pairs of autosomes and one pair of sex chromosomes.
- Karyotype of normal karyotype of men is 44+XY and that of women is 44+XX.
- The table given below shows the nomenclature of chromosome groups proposed by the *Denver Report and the London Report* (1963).

Table 4

S. No.	*Denver report*	*London report*	*Description*
1.	Group 1–3	Group 1–3 (A)	Large and metacentric chromosomes
2.	Group 4–5	Group 4–5 (B)	Large submetacentric chromosomes
3.	Group 6–12	Group X, 6–12 (C)	Medium-sized submetacentric chromosomes
4.	Group 13–15	Group 13–15 (D)	Large acrocentric chromosomes; all may have satellites
5.	Group 16–18	Group 16–18 (E)	Chromosome 16 is metacentric while 17 and 18 are submetacentric
6.	Group 19–20	Group 19–20 (F)	Small metacentric chromosomes
7.	Group 21–22	Group 21–22 +Y (G)	Short acrocentric chromosomes having satellites but Y chromosome lacks satellite

- Karyotypes are arranged with short arm of the chromosome on the top and long arm at the bottom.

Chromosome Banding

- A band is the part of a chromosome which is differentiated from its adjacent region by appearing darker or lighter due to staining property.
- The banded chromosomes appear as a continuous series of light and dark bands.
- There are no interbands.

G Banding

- G banding is one of the most popular banding methods used for karyotype analysis.
- G banding is obtained with Giesma stain following digestion of chromosome with trypsin (a proteolytic enzyme used for cell dispersal).
- G banding yields a series of darkly and lightly stained bands.
- The dark bands represent heterochromatic (rich in A–T base pairs), while the light bands represent euchromatic region (rich in C–G base pairs).
- Generally this method produces 300–400 bands in a normal human genome.

R Banding

- R banding pattern is the reverse of the G-banding pattern.
- It results in the staining of areas rich in G–C base pairs, which are typical for euchromatin, and light regions are heterochromatin (rich in T–A base pairs).

C Banding

- C banding is useful in staining the centromeres regions and the constitutive heterochromatic regions.
- This banding technique is very useful for the characterisation of plant chromosomes.

Q Banding

- Q bands are obtained by quinacrine staining and fluorescence.
- The distribution, number, size and intensity of each band is fixed for each chromosome pair.

T Banding

- T banding visualises telomere.

NOR Banding

- NOR banding is obtained by using ammonical silver reagent.
- The NOR bands differ from person to person and are helpful in studying chromosome polymorphism.

Human Chromosomal Abnormalities

1. **Wolf–Hirschhorn Syndrome**
    - This syndrome was independently reported by Wolf et al. (1965) and Hirschhorn et al. (1965).
    - This syndrome is due to partial deletion of the short arm of chromosome number 4.
    - It is characterised by severe psychomotor and retardation of growth.
    - The skull is microcephalic with cranial asymmetry.
    - Congenital heart malformation is common.

2. **Cri-du-chat Syndrome**
   - It was first described by Lejeune et al. (1963).
   - It is due to slight deletion of short arm of chromosome number 5.
   - The newborn cries like mewing of a cat, so it is also known as the cat-cry syndrome.
   - Such a syndrome is characterised by a small head, widely spaced eyes, moon-like face and receding chin. Dental malocclusion is common and congenital heart defects are common
   - Such syndromes are rare.
3. **Patau's Syndrome**
   - It was reported by Patau et al. (1960).
   - Patau's syndrome is due to trisomy of chromosome 13.
   - Such victims are characterised by a small head (microcephaly), widely spaced eyes, low set ears, clefts of the lips, palate and polydactyly.
   - Congenital heart disease, kidney disorders and malformed genitilia are common in such a syndrome.
   - One of the characteristics of such syndrome is the trigger thumb, i.e., the thumb and index finger overlap the third finger).
   - Life-threatening complications may develop during infancy or early childhood.
   - Meiotic nondisjunction is thought to be the cause of this chromosomal aberration.
   - The frequency of this syndrome is 1 in 10,000 live births.
4. **Edward Syndrome**
   - Edward syndrome is due to trisomy of chromosome 18.
   - Trisomy of chromosome 18 was independently described by Edwards et al. and Smith et al. (1960).
   - Among live-born children, trisomy of chromosome 18 is the most common autosomal trisomy after trisomy of chromosome 21.
   - Such a syndrome is characterised by severe psychomotor and growth retardation, microcephaly, microphthalmia, micrognathia, microstomia and heart defects.
   - Survival time is less than four months.
   - Edward syndrome is more frequent in females than in males.
   - Its incidence is 1 in 3,000 live births.
5. **Down Syndrome**
   - This is one of the most common and best-studied chromosomal anomalies in humans.
   - It was first described by Langdon Down (1866).
   - It is also known as mongolism due to facial appearance like mongoloids.
   - It is characterised by trisomy of 21 chromosome. There is a presence of an extra copy of genetic material on the 21 chromosome, either in whole (trisomy 21) or part (due to translocation).
   - Individuals with such syndrome are characterised by
     (a) Rounded face and broad forehead
     (b) Microgenia (abnormally small chin)
     (c) Hypotonia (poor muscle tone)
     (d) Protruding tongue
     (e) Height, weight and head circumference are smaller in children with such syndrome.
     (f) There is increased risk of developing epilepsy.
     (g) Mental retardation is common.
     (h) Usually both males and females are infertile.

(i) Prominent epicanthus fold in the eyelids is a common feature.

- Frequency of this syndrome is 1 in every 600–700 live births.
- More than 95 per cent cases of Down syndrome are caused by nondisjunction and remaining due to translocation.
- With an increase in the age of parents (35+ years mother and 42+ years father), there is increase in the frequency of Down syndrome.

Sex Chromosomal Abnormalities

1. **Klinefelter Syndrome**
    - It was described by Klinefelter et al. (1942).
    - Klinefelter syndrome is also known as XXY condition.
    - Chromosomal complement of such an individual is 44+XXY.
    - Such individuals are phenotypically male by virtue of having Y chromosome.
    - They have underdeveloped testes (hypogonadism) leading to sterility and ill-developed male secondary sexual characters.
    - They are taller individuals having less body muscles, less facial and body hair and hips wider than other boys, and have enlarged breasts (gynecomastia).
    - XXY constitution arises either by fertilisation of an XX egg by a Y sperm or an X egg by an exceptional XY sperm.
    - Frequency of such syndrome is 1 per 500 male live births.
    - The risk of Klinefelter syndrome increases with increasing maternal age.
2. **Turner Syndrome**
    - Turner syndrome is a condition associated with monosomy of an X chromosome.
    - It was first described by Turner (1938).
    - Their genomic complement is 44+XO, indicating that they have 45 chromosomes.
    - Such individuals are phenotypically female but are sterile due to the absence of ovarian function.
    - They are of a short stature having webbed neck, puffiness or swelling of both hands and feet, along with heart defects and kidney problems.
    - In Turner syndrome, there is deficient secretion of FSH and estrogen.
    - Epithelial cells lack Barr body.
    - Frequency of Turner syndrome is 1 in 2,500 live female births.
3. **XYY Syndrome**
    - It is a genetic condition in which males have an extra Y chromosome in each cell.
    - Affected males have 47 chromosomes (44+XYY).
    - This is not usually an inherited condition but a defect that occurs during cell division.
    - Such males tend to be tall having normal levels of testosterone.
    - Most of these individuals have normal sexual development and normal fertility.
    - The incidence of XYY syndrome is not affected by maternal or paternal age.
    - The frequency of XYY syndrome is 1 in 1,000 live births.
4. **Fragile X Syndrome**
    - Fragile X syndrome is the most common form of mental retardation.
    - Fragile X syndrome is due to a change in the gene that is inherited at the time of conception.

- This gene is termed as FMR1 gene, located on the X chromosome.
- FMR1 gene is responsible for making a protein responsible for proper development of the brain. So, when this gene is not functioning properly, functioning of the brain is affected.
- X chromosome is unusually fragile at one tip in the victim of such a syndrome.
- Generally most people have 29 repeats at the end of their X chromosome but those with fragile X have more than 700 repeats due to duplication.
- Its frequency is 1 in 1,500 births in males and 1 in 2,500 births in females.

Human Genetic Diseases

- Genetic or hereditary diseases are conditions caused by subtle damage to the DNA, which is passed from one generation to the next.
- The damaged gene is inherited according to rules of genetics.
- These diseases may be:
  1. Autosomal recessive
  2. Autosomal dominant
  3. Sex-linked recessive
  4. Sex-linked dominant

1. Autosomal Recessive Diseases

In autosomal recessive inheritance, the gene is located on one of the autosomes and both the sexes are equally affected. Recessive would mean that two copies of the gene are necessary to have the trait; one inherited from the mother and one from the father. A person with only one recessive gene is said to be a 'carrier' for the trait or disease, but they do not have any health problems from 'carrying' one copy of the gene. Most people do not know they carry this defective gene since it is recessive in nature until they have a child with the disease.

Some important autosomal recessive diseases are given below.

(a) Albinism

- Albinism is an autosomal recessive trait.
- An albino person lacks the dark pigment called melanin, which provides dark colour to skin, hair and iris.
- Albinism is due to the deficiency of the enzyme tyrosinase.
- The enzyme tyrosinase converts amino acid tyrosine to melanin through an intermediate product dihydro phenyl alanine (DOPA).
- The gene for albinism (a) does not produce tyrosinase. Its normal allele (A) produces tyrosinase. Thus, only a homozygous (aa) individual is affected.
- For heterozygous parents with normal pigmentation (Aa), two different types of gametes may be produced, viz., A or a. From such a cross, one-fourth of the children could be albino.

(b) Phenylketonuria (PKU)

- PKU is an inborn metabolic disease in which a homozygous recessive individual lacks the enzyme phenyl alanine hydroxylase.
- Enzyme phenyl alanine hydroxylase is required for the conversion of amino acid phenyl lalanine to tyrosine.
- Lack of this enzyme is due to an abnormal recessive gene with locus on chromosome number 12.

- The defective gene is due to substitution.
- There is a high level of phenyl alanine in the blood and tissue fluids.
- The high level of phenyl alanine in the blood damages the development of the brain. Muscles and cartilages of legs may be defective as a result of which the victim is unable to walk properly.
- The heterozygous individuals are normal but are carriers.
- Its frequency is 1 in 15,000 infants.

(c) Alkaptonuria (Onchronosis)

- Alkaptonuria was the first disease to be interpreted as a single gene trait.
- It was discovered by Archibald Garrod (1902).
- It is a rare condition in which patients have arthritis and excrete large amounts of homogentisic acid in their urine.
- The urine of such a person turns black on exposure to light.
- Alkaptonuria is due to the deficiency of an enzyme, oxidase, which is required for the breakdown of homogentisic acid.
- Lack of this enzyme is due to the absence of the normal form of a gene that controls synthesis of the enzyme.

(d) Tay–Sachs Disease

- It is an autosomal recessive disease resulting in degeneration of the nervous system.
- Children homozygous for this trait rarely survive past five years of age.
- Sufferers lack the ability to make enzyme N-acetyl-hexosaminidase, which breaks down the GM2 ganglioside lipid.
- The lipid accumulates in lysosomes in the brain cells, eventually killing the brain cells.
- Symptoms of the disease manifest after birth.
- It is rare in general population (i.e., 1 in 30,00,000).

(e) Galactosemia

- Galactosemia is an autosomal recessive disease in which the sufferer lacks the ability to convert galactose into glucose.
- It is due to the deficiency of the enzyme galactose phosphate uridyl transferase (GALT).
- GALT locus is on 9p and carrier detection is possible.
- Milk is toxic to galactosemic children and such children die around three years of age.
- Its incidence is 1 in 40,000.

(f) Cystic Fibrosis

- Cystic fibrosis is an autosomal recessive trait with locus at 7q21–31.
- Cystic fibrosis is caused by a mutation in the gene for the protein cystic fibrosis transmembrane conductance regulator (CFTR).
- There is excess production of glycoprotein which interferes with salt metabolism.
- Elevated levels of sodium and chloride ions in sweat.
- There is pancreatic insuffiency and chronic lung disease secondary to recurrent infection.
- The disease is common in Caucasian population.

(g) Wilson Disease

- It is an autosomal recessive trait with locus on 13q14.
- It is characterised by defective copper metabolism with reduced ceruloplasmin and increased hepatic copper.

- Its incidence is 1 in 2,00,000 in United Kingdom and 1 in 25,000 in Japan.

(h) Sickle Cell Anaemia

- Sickle cell anaemia is a recessive autosomal trait.
- Sickle cell anaemia is a molecular disease and is produced by the presence of an abnormal gene called haemoglobin S.
- Haemoglobin S is the result of an anomaly of the beta chain, wherein the sixth position glutamic acid is substituted by valine.
- The sickle cell trait was first discovered in an American of African ancestry by Herrick (1910).
- In sickle cell anaemia, the shape of RBCs becomes sickle-like, because of which the oxygen-carrying capacity is reduced.
- During reduced oxygen tension, red-cells sickle cell anaemia carriers may be sickle. Thus, this recessive trait may be co-dominant at the level of the gene product or dominant in altered environment.

2. Autosomal Dominant Trait

- Gene for such trait is located on one of the autosomes and both sexes are equally affected.
- If one parent has a dominant trait, there is a 50 per cent chance that any child will have this trait.
- Some important autosomal dominant diseases are as follows:

(a) Huntington's Disease

- It is an autosomal dominant disease resulting in progressive destruction of brain cells.
- Its gene is located on chromosome 4.
- If a parent has the disease, 50 per cent children will be victims of this disease.
- The disease does not manifest until after age of 30.

(b) Achondroplasia

- Achondroplasia is an autosomal dominant trait in which only a single copy of the abnormal gene is required to cause achondroplasia.
- The gene of achondroplasia fully penetrates, i.e., everyone who possesses it has achondroplasia.
- It results in an abnormally short stature and is the most common cause of short stature with disproportionately short limbs.
- It is one of the oldest known birth defects.
- An average worldwide occurrence is 1 in 25,000 births.

(c) Marfan Syndrome

- Marfan syndrome is an autosomal dominant trait and is a disorder of the connective tissue.
- The gene for Marfan syndrome, called fibrilin-1, is located on chromosome 15.
- It is characterised by arachnodactyly, i.e., long limbs with reduced upper or lower segment ratio. There is extreme looseness of joints.
- Lifespan of such an individual is 40–50 years.

(d) Congenital Spherocytosis

- Congenital spherocytosis is an autosomal dominant trait with locus in 8p.
- It is characterised by chronic hemolytic anaemia. There is reduced red blood survival and increased osmotic fragility.
- Its incidence is 1 in 4,500 births.
- Some other common examples are polydactyly, brachydactly, etc.

3. Sex-Linked Recessive Diseases

- The genes for sex-linked diseases are located on either X chromosome or Y chromosome.
- In X-linked inheritance, genes are located on the X chromosome.
- In females, X-linked recessive genes are expressed only during homozygous condition, so females suffer less in comparison with males.
- Important X-linked recessive diseases are colour-blindness, haemophilia, G-6PD deficiency, Lesch–Nyhan dystrophy, childhood muscular dystrophy, white forelock of hair, etc.

(a) Lesch–Nyhan Syndrome

- It is a rare inherited trait.
- Lesch–Nyhan syndrome is an inherited X-linked recessive trait.
- Mutations in HPRT1gene cause this syndrome, due to which there is severe deficiency of the enzyme hypoxanthine phosphoribosyltransferase1.
- Lesch–Nyhan syndrome is characterised by the overproduction and accumulation of uric acid.
- The overproduction of uric acid may cause gouty arthritis, kidney stones and bladder stones.
- Its frequency is 1 in 3,80,000 individuals.

(b) Glucose-6-Phosphate Dehydrogenase (G-6PD) Deficiency

- G-6-PD deficiency is the most common metabolic disorder.
- It is characterised by the deficiencies of the enzyme glucose-6-phosphate dehydrogenase, which catalyses oxidation of glucose-6-phosphate to 6-phosphogluconate with simultaneous reduction of NADP to NADPH.
- G-6-PD deficiency is an X-linked trait (chromosome location Xq28).
- G-6-PD deficiency was first identified in people of African origin in America by Beutler (1959).
- Beans and certain drugs induce hemolysis in G-6-PD deficient persons.
- G-6-PD deficiency is widespread in India.
- G-6-PD deficiency confers relative advantage against the human parasite *Plasmodium falciparum.*

X-Linked Dominant Inheritance

- In X-linked dominant inheritance, the gene is located on the X chromosome but acts in a dominant manner.
- Both males and females may display the trait by having only one copy of the gene.
- Examples of X-linked dominant disease are Rett syndrome and incontinentia pigment (IP).

Y-Linked Traits

- Y chromosome is small and carries few genes, so there are few Y-linked disorders.
- Hypertrichosis (tuft of hair on pinna) is a Y-linked trait.

Blood Groups

- Blood groups are complex chemical systems found on the surface of red blood cells.
- International Society of Blood Transfusion (ISBT) has recognised 29 blood groups.
- Across 29 blood groups, over 600 different blood groups' antigens have been found.
- Many of these blood groups are rare or found only in certain ethnic groups (e.g., Diego positive blood group is found only among East Asians and Native Americans).
- Of these blood groups, ABO and Rh blood groups are the most important for blood transfusion.

ABO Blood Groups

- Blood groups are useful genetic markers and excellent traits of population genetic studies.
- Blood grouping is a valuable trait in blood transfusion, forensic, medicine and paternity disputes.
- The ABO system is a polymorphic system of antigens carried on red cells and other tissues.
- Landsteiner (1900) recognised existence of the ABO blood system in humans comprising A, B and O groups.
- De Castello and Sturli (1902) discovered the fourth group AB of the system.
- The later studies subdivided the group into A_1, A_2 (Dungern and Hirszfeld, 1911) and A_3 (Friedenreich, 1936).
- The mode of inheritance of A, B and O was studied by Bernstein in 1924.
- Almost every individual has the same blood group for life, but very rarely, an individual's blood type changes through addition or suppression of an antigen, infection, malignancy or autoimmune disease.
- ABO system is the only blood group system in which antibodies are consistently, predictably and naturally present in the serum of the people who lack antigen.
- The ABO blood group expression is controlled by genes at the ABO locus located on chromosome 9.
- ABO blood types are also present in some animals like cows, sheep, apes, chimpanzees, banobos, gorillas, etc.
- The A and B allele differ from each other by seven nucleotide substitutions; four of which are translated into different amino acids in the gene product.
- On the basis of presence of antigens and antibodies, human blood groups have been grouped into four types:
  - Blood group A having antigen A and antibody β
  - Blood group B having antigen B and antibody α
  - Blood group AB having antigen AB and without antibody
  - Blood group O having no antigen but antibody α and β
- Blood group is inherited from both parents.
- The four blood groups—A, B, AB and O are controlled by three alleles—$I^A I^A$, $I^B I^B$ and $I^O I^O$.
- I^A and I^B are co-dominant while I^O is recessive to both. People homozygous for the recessive I^O allele have blood group O.
- People having blood group A have genotype I^AI^A/I^AI^O and likewise people having blood group B have genotype I^BI^B / I^BI^O.
- Heterozygous I^AI^B individuals have blood group AB.
- The parents of a person of blood group O must be I^OI^O or I^AI^O to produce one-fourth I^OI^O progeny or I^BI^O or one may be I^AI^O and other I^BI^O.
- It is not possible in blood grouping to distinguish I^AI^A from I^AI^O or I^BI^B from I^BI^O, but this may be possible through pedigree analysis.
- Thus for blood group O, the parents must be homozygous (I^OI^O) or heterozygous (I^AI^O/I^BI^O).

♀ \ ♂	I^A	I^B	I^O
I^A	I^AI^A (A)	I^AI^B (AB)	I^AI^O (A)
I^B	I^AI^B (AB)	I^BI^B (B)	I^BI^O (B)
I^O	I^AI^O (A)	I^BI^O (B)	I^OI^O (O)

Fig. 24

Blood Transfusion

- The first successful experiment in blood transfusion was conducted on dogs by an English anatomist, Richard Lowenin, in 1666. In 1667, French scientist, Jean Baptiste Denis performed the first transfusion of animal blood to a human (lamb blood).

- There should be correct knowledge of blood groups of donor and recipient at the time of blood transfusion because if the blood of a wrong group is transfused, it will lead to death of the recipient.
- I^A allele synthesise antigen A and in the blood group of person A, the serum contains naturally occurring antibodies against the antigen B.
- Antibodies against antigen B will agglutinate any RBCs which have the antigen B in them, which may lead to death.
- Persons having blood group AB have both antigens A and B on the RBCs. So neither anti α nor anti β occur in the serum of such persons. They can receive blood from any other group. Therefore, AB blood group is known as a universal recipient.

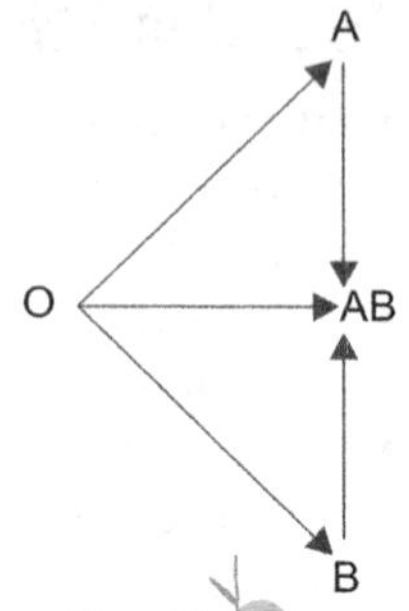

Fig. 25

- In blood group O persons, antigens A and B both are absent and the serum contains both anti α and anti β. Such a person can donate blood to any other group. Therefore, any person with blood group O is known as a universal donor.
- The diagram shown below describes the phenomenon of donors and recipients. The direction of arrow shows the recipients.
- Differences among ABO blood types occur because of the presence of different antigen and antibodies.
- Antigens of ABO blood systems are molecules composed of a protein–sugar combination that occur on the surface of red blood cells, while antibodies are substances that are produced by the immune system in response to external substances.
- ABO alleles determine the activity of specific sugar transfers.
- The A allele adds N-acetylgalactosamine to the precursor glycoprotein (H substance).
- The B allele adds D-galactose to the precursor glycoprotein (H substance).
- In the presence of O allele, the H substance is unchanged.

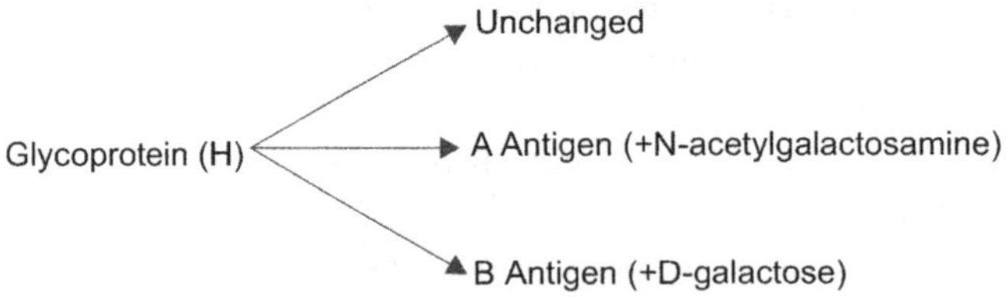

Fig. 26 Biosynthesis of ABO blood group substances

- The H substance is produced by the action of dominant allele at the locus, which is distinct from ABO locus.
- Individuals homozygous for recessive mutant allele H do not produce H antigen.
- Individuals with the rare 'Bombay phenotype' do not express substance H on the red blood cells and therefore, do not bind A or B antigens.
- A new technique has been developed by scientists for converting one blood group into another.
- The technique enables blood group A, B and AB to be converted into group O negative, which can be safely transfused to any person.
- Two bacteria, viz., *Elizabethkingia meningosepectium* and *Bacteroides fragilis* contain potentially useful enzymes that cut sugar molecules from the surface of red blood cells.
- The enzyme of *Bacteroides fragilis* can remove the antigen B and the enzyme of *Elizabethkingia meningosepectium* can destroy antigen A.

Rh Blood Group

- Rh blood factor was reported by Landsteiner and Weiner (1940) in the red blood corpuscles of rhesus monkey (*Rhesus macaca*).
- Rh antigen is a protein on the surface of red blood corpuscles in many persons.
- Those having this antigen are Rh^+ and those who lack this antigen are Rh^-.
- About 85 per cent of the population is Rh^+ and the remaining 15 per cent is Rh^-.
- Inheritance of Rh blood group is controlled by two alleles, viz., D and d.
- Individuals which are homozygous dominant (DD) or heterozygous (Dd) are Rh^+ and those who are homozygous recessive (dd) are Rh^-.
- Clinically, the Rh factor can lead to serious medical complications.
- Rh-type mother–foetus incompatibility takes place when an Rh^+ man fathers a Rh^- child with a Rh^- mother.
- Rh^+ father has genotype DD or Dd. There are two mating combinations possible:
- Only Rh^+ children are likely to have complications. When both the mother and foetus are Rh^-, the birth will be normal.
- In the first delivery usually there is no incompatibility. However, in the second, third and subsequent births, both are likely to have life-threatening problems.
- Normally, antiRh^+ bodies do not develop in the first pregnancy, unless the mother comes in contact with Rh^+ blood. So, her antibodies are not likely to agglutinate red blood cells of her Rh^+ foetus.

♀ \ ♂	D	D
d	Dd	Dd
d	Dd	Dd

♀ \ ♂	D	d
d	Dd	dd
d	Dd	dd

Fig. 27

- Placental rupture generally occurs at birth. Here some foetal blood gets mixed with the mother's blood, which stimulates the development of antibodies to Rh^+ blood antigens.
- Rh antibodies can cross the placenta.
- If a second pregnancy occurs, the antibodies present in the mother's blood start destroying the red blood cells of the foetus due to agglutination. Therefore, the baby is anaemic. The baby also suffers from jaundice. This is known as hemolytic disease of newborn (erythroblastosis fetalis).

CYTOPLASMIC INHERITANCE

- In cytoplasm, some self-reproducing particles are found having the ability of stability, mutability as well as influencing the transmission of characters. These particles are called plasmagenes. The mode of inheritance under the control of plasmagenes is termed as cytoplasmic inheritance.
- Cytoplasmic inheritance is also known as extranuclear, extrachromosomal, non-Mendelelian or uniparental inheritance.

- Cytoplasmic inheritance was first of all described by Correns (1908).

Characteristics of Cytoplasmic Inheritance

- Characters are controlled by plasmagenes and the sum total of plasmegnes is called plasmon.
- Plasmagenes are located in the cytoplasm, cytoplasmic organelles and on the cellular surfaces.
- Plasmagenes have the ability of self-duplication, stability as well as undergoing mutation.
- Plamagenes are inherited through cytoplasm.
- In cytoplasmic inheritance, the results of the reciprocal cross differ.
- The offsprings receive cytoplasm only from female gametes. As a result, in offsprings, plasmagenes of only the mother are manifested. Therefore, cytoplasmic inheritance is also known as maternal inheritance.

Examples of Cytoplasmic Inheritance

1. **Shell Coiling in Snails**
   - A good example of cytoplasmic inheritance was described by Boycott et al. (1930) in shell coiling of snails.
   - In snails, (*Limnaea peregra*) two types of shell coiling are found, viz., dextral and sinistral.

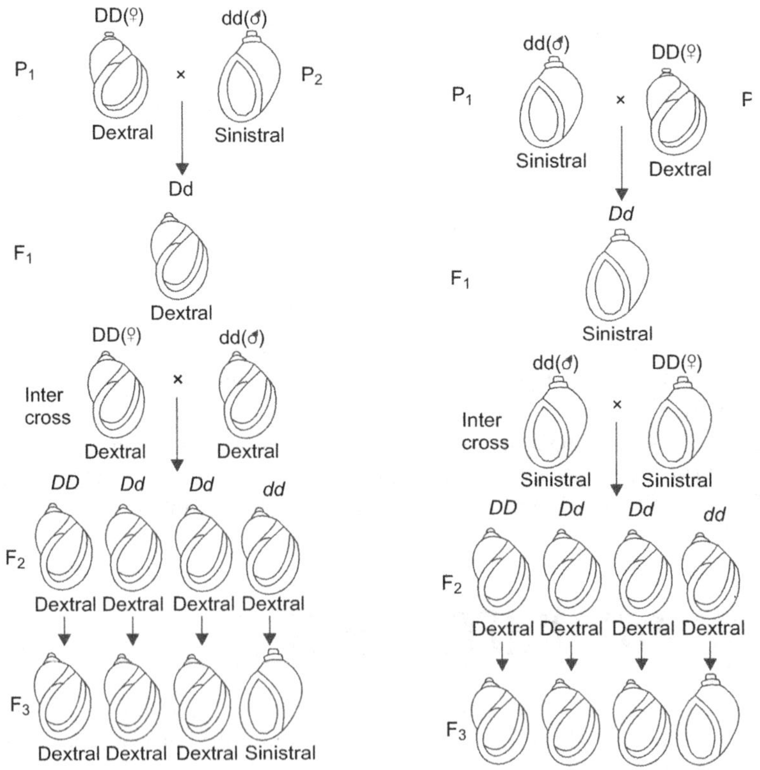

Fig. 28 Cytoplasmic inheritance in Limnaea peregra

- In dextral type, shell coiling is right handed (clockwise) and in sinistral type, it is left handed (anticlockwise).
- Inheritance of dextral coiling is controlled by dominant gene D and that of sinistral type by recessive gene d.
- Boycott et al. found that shell coiling is determined by the gene of the mother.
- When a dextral female is crossed with a sinistral male, all F_1 offsprings develop dextral coiling. Likewise, when a sinistral female is crossed with a dextral male, all F_1 offsprings develop sinistral coiling.
- Thus, it is clear that the genotype Dd may be dextral or sinistral depending on the genotype of the mother.
- Likewise, the genotype dd may be sinistral or dextral according to the genotype of the mother.
- Thus, it is clear that it is the genotype of mother, which determines the pattern of coiling in offsprings, irrespective of their own genotype.
- This is due to the presence of cytoplasmic particles (plasmagenes) in the cytoplasm.

2. Inheritance of Kappa Particles in *Paramecium*

- Inheritance of kappa particles in *Paramecium aurelia* is controlled by cytoplasmic inheritance (Sonneborn, 1938).

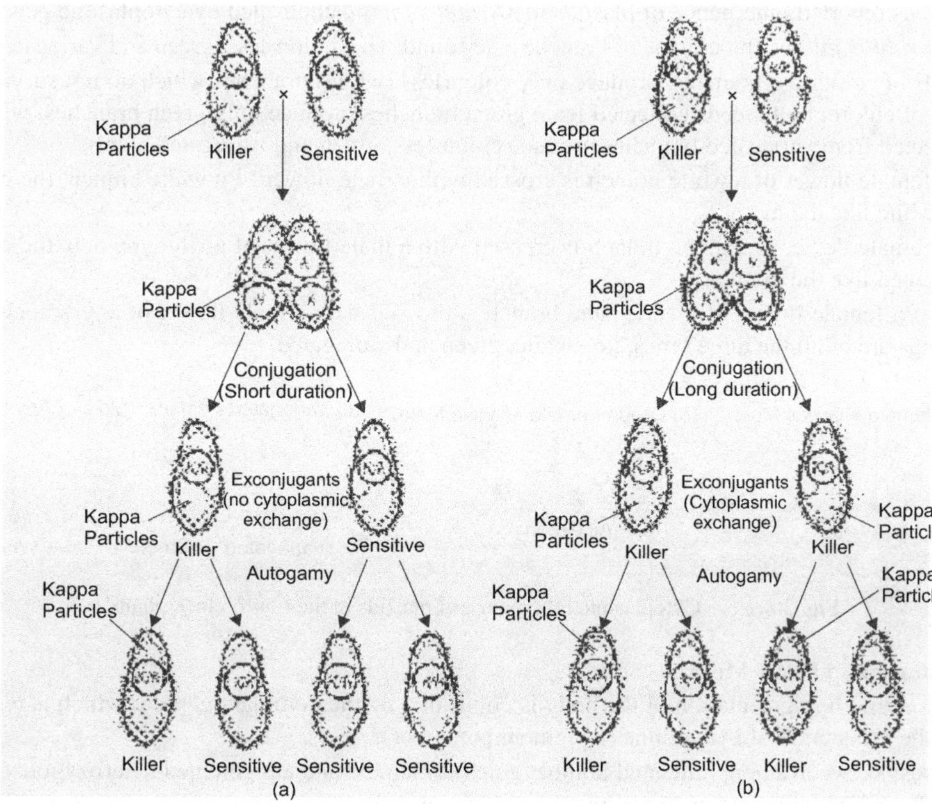

Fig. 29 Cross between killer and nonkiller strain of Paramecium: (a) Conjugation without cytoplasmic exchange (b) Conjugation with cytoplasmic exchange

- Kappa particles are found in a certain strain of Paramecium and are capable of producing a toxic substance called paramecin.
- Paramecin is lethal to strains that lack kappa particles.
- The production of kappa particles is dependent on a dominant allele K, so the killer strains are KK or Kk, while the nonkiller strains (that lack kappa particles) are kk.
- If the killer (KK) and nonkiller strains are allowed to conjugate and if the duration of conjugation is short, the ex-conjugate killer produces a killer strain and the ex-conjugate nonkiller produces a nonkiller strain.
- But if conjugation persists for a long duration, the ex-conjugate killer produces killer strains and quite surprisingly, the nonkiller also produces killer strains.
- It is so because if conjugation persists for a short duration, then only exchange of the nucleus occurs.
- But when conjugation persists for a long duration, it involves exchange of cytoplasm and the nonkiller strain receives plasma genes of kappa particles. Thus, the nonkiller strain, in spite of its genotype kk, starts producing kappa particles as it receives kappa particle K from the killer strain, due to the exchange of cytoplasm.

3. **Plastid Inheritance in *Mirabilis jalapa***
   - Correns reported inheritance of plastids in *Mirabilis jalapa* controlled by cytoplasmic genes.
   - In *Mirabilis jalapa*, three types of branches are found, viz., colourless, green and variegated leaves.
   - Seeds of colourless branches produce only colourless (white) branches which do not survive due to lack of chlorophyll. Seeds collected from green branches produce only green branches, while seeds collected from variegated branches produce colourless, green and variegated plants.
   - If a female flower of a white branch is crossed with a male flower of a white branch, the offsprings are white like the mother.
   - If a female flower of a green branch is crossed with a male flower of a white branch, the offsprings are green like the mother.
   - But if a female flower of a variegated branch is crossed with a male flower of any branch, the offsprings are of all the three types, i.e., white, green and variegated.

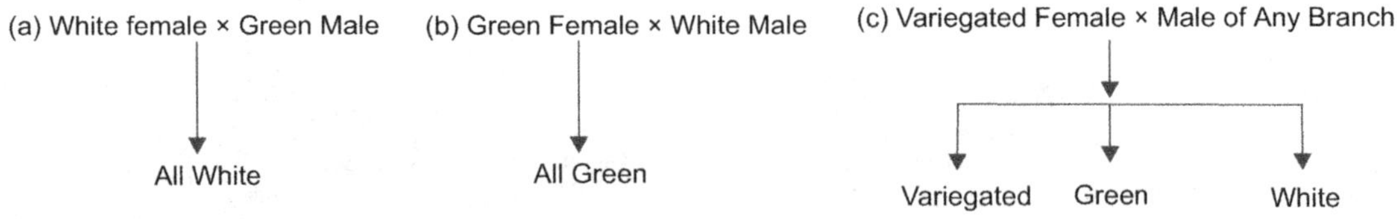

Fig. 30 Cytoplasmic inheritance of plastids in the four o'clock plant

4. **Pigmentation in Flour Moth**
   - In flour moth, pigmentation of the body is controlled by the dominant gene A, which is responsible for the production of kynurenine (a pigment precursor).
   - A cross between a nonpigmented homozygote (aa) female and a pigmented heterozygote (Aa) male results in both nonpigmented and pigmented larvae (1:1).
   - But a cross between heterozygote pigmented (Aa) female and homozygous (aa) nonpigmented male results in all pigmented larvae. However, when larvae mature, only half of them become pigmented (Aa) and the rest become nonpigmented (aa).

- It is so because nonpigmented (aa) larvae receive kynurenine pigment from the mother through the egg cytoplasm, during the early stages of development. However, they are unable to synthesise kynurenine pigment due to the absence of dominant allele A and this results in the subsequent loss of pigmentation, which occurs in half of the adult moths.

Nonpigmented (aa) Female × Pigmented (Aa) Male

↓

Nonpigmented (aa): Pigmented (Aa)

Pigmented (Aa) Female × Nonpigmented (aa) Male

↓

Pigmented (Aa): Pigmented (aa) (Larval Stage)

Fig. 31

5. Sigma Particles in *Drosophila*

- Heritier and Teisier reported a race of *Drosophila*, which was more sensitive to CO_2 than the others.
- This race is known as sensitive strain and it possesses a heat-liable substance called sigma factor.
- Sigma factor is present in the cytoplasm.
- Individuals of the sensitive race become anesthetised even at low concentrations of CO_2 and may be killed at higher concentrations of CO_2.
- If a CO_2 sensitive female is crossed with an insensitive male, all offsprings are CO_2 sensitive. However, if an insensitive female is mated with a sensitive male, either a few offsprings are sensitive or none of them is sensitive.
- It is due to the inheritance of sigma particles from the mother through the egg cytoplasm because of which insensitive individuals become sensitive. Thus, it indicates that the inheritance of sigma particles is under the control of cytoplasmic genes.

NUCLEIC ACID

Nucleic acids are of the following two types:

1. Deoxyribose Nucleic Acid (DNA)
2. Ribose Nucleic Acid (RNA)

Genetic Materials of Some Organisms

1. Double-stranded DNA – Protista, fungi, algae, plants, (dsDNA) higher animals, bacteria, small pox virus, animal virus, some bacteriophages T2, T4, T7, etc.
2. Single-stranded DNA – Bacteriophage φx174 (ssDNA)
3. Double-stranded RNA – HIV, reo viruses, (dsRNA)
4. Single-stranded RNA – TMV, influenza virus, (ssRNA), polio virus, bacteriophage F2, R17, etc.

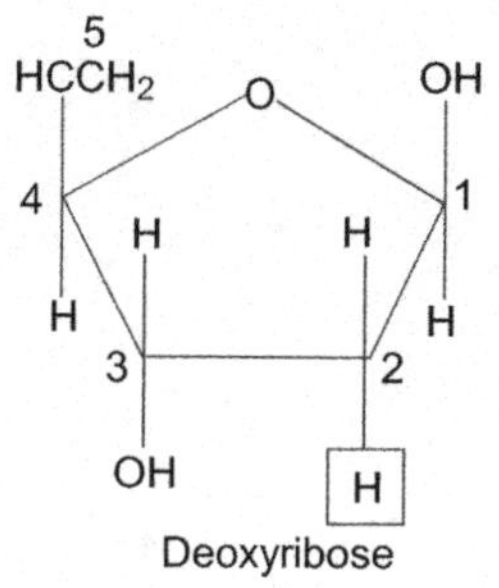

Fig. 32

Chargaff's Law

1. Base composition varies from species to species.
2. Different cells of the same species have the same base composition.
3. Base composition does not change with age, nutrition or change in environment.
4. For each species the number of A = T and the number of G = C.
5. A+T ratio is different in different species but similar in same species.

Chemical Composition

DNA contains the following three types of chemical compounds:

1. **Sugar**
   - Sugar found in DNA is of monosaccharide and of pentose type.
   - The pentose sugar has five-membered ring in which the second carbon is deoxidised.
2. **Phosphate**
   - In DNA, phosphate is found as phosphoric acid (H_3PO_4).

   Phosphoric acid contains three reactive hydroxyl (–OH) groups. Out of the three hydroxyl groups, two are involved in forming sugar phosphate back-bone of the DNA. The phosphate makes a nucleotide negatively charged.

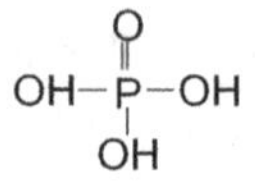

Fig. 33 *Phosphoric acid*

3. **Nitrogenous Bases**
   - Two types of nitrogenous bases are found in DNA, viz., purine and pyrimidine
   - Purine is a nine-membered, double-ring structure.

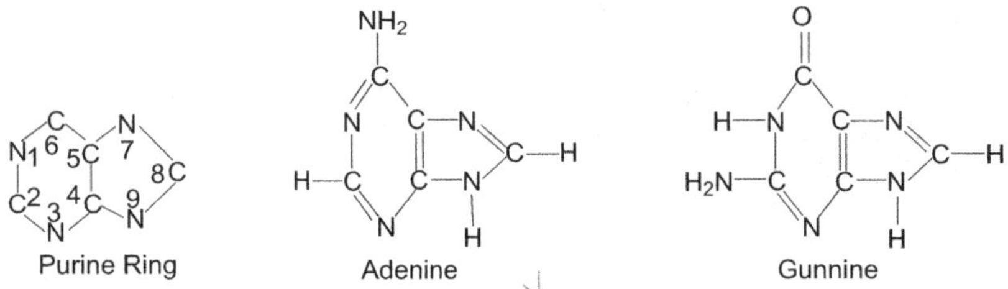

Fig. 34

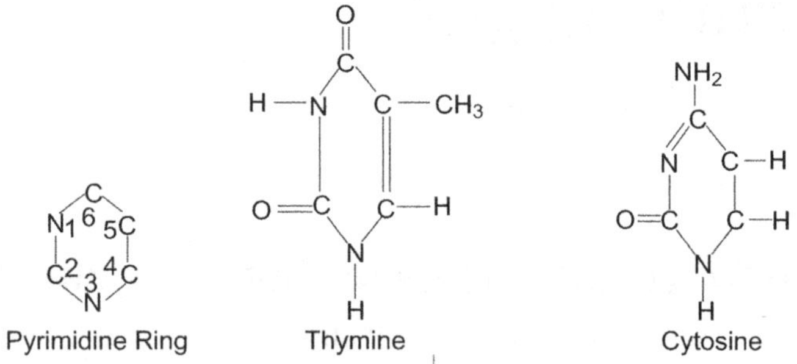

Fig. 35

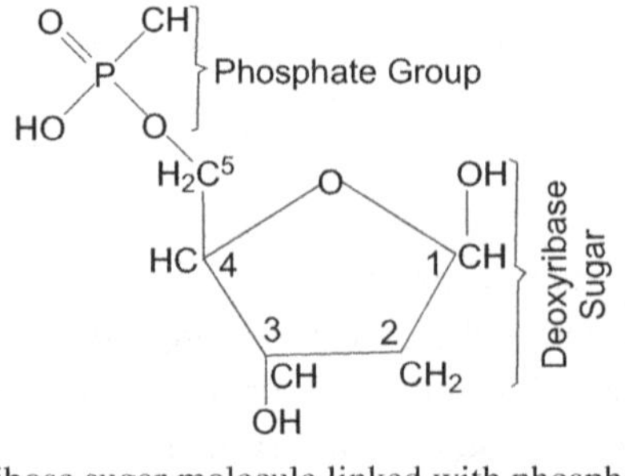

Fig. 36 A deoxyribose sugar molecule linked with phosphate group at 5' position

- In DNA, two types of purine are found, viz., adenine and guanine.
- Pyrimidine is a single-ring compound and in DNA.
- Two types of pyrimidine are found, viz., cytosine and thymine.

Nucleoside

- A combination of sugar and nitrogenous base is known as nucleoside.
- The nitrogenous bases unite with sugar through glycosidic bonds.
- In purine, the glycosidic bond forms between nitrogen no. 9 and carbon no. 1 of deoxyribose.
- In pyrimidine, the glycosidic bond is formed between nitrogen no. 3 and carbon no. 1 of deoxyribose.
- In DNA, the following four types of nucleosides are found:

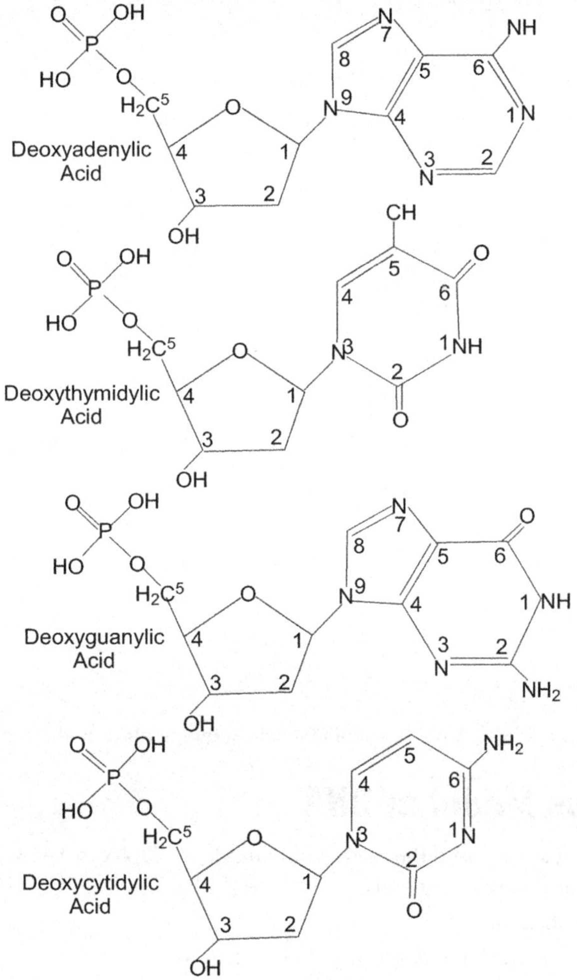

Fig. 37 Structure of four different 5'p 3' OH nuclectides found in DNA

Adenine + Deoxyribose = Deoxyadenosine
Thymine + Deoxyribose = Deoxythymidine
Cytrosine + Deoxyribose = Deoxycytidine
Guanine + Deoxyribose = Deoxyguanosine

Nucleotide

- A combination of sugar, nitrogenous base and phosphoric acid is known as nucleotide.
- In DNA, the following four types of nucleotides are found:
  Deoxyribose + Adenine + Phosphoric acid = Deoxyadenylic acid
  Deoxyribose + Guanine + Phosphoric acid = Deoxyguanylic acid
  Deoxyribose + Thymine + Phosphoric acid = Deoxythymidylic acid
  Deoxyribose + Cytosine + Phosphoric acid = Deoxycytidylic acid

Polynucleotide

- A combination of many nucleotides is known as polynucleotide
- In a polynucleotide chain, nucleotides are joined together by phosphodiester bonds (a diester bond involves two ester bonds).

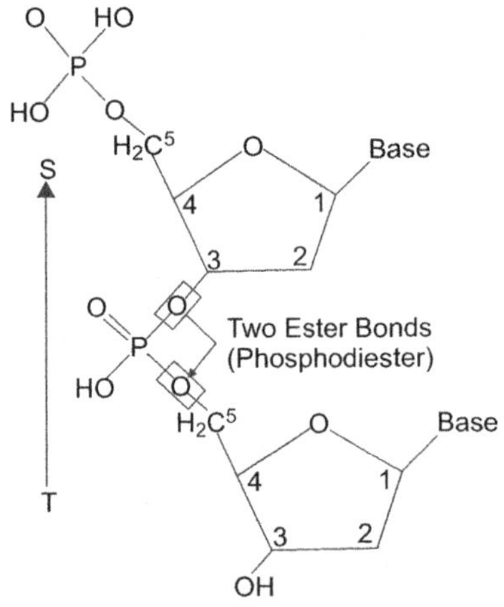

Fig. 38 A polynucleotide chain showing phosphodiester bond

Watson and Crick Model of DNA

- J D Watson and F S H Crick proposed the most stable model of DNA in 1953. According to the model, the structure of the DNA is as discussed ahead.
- DNA is a double helical structure.
- The helix contains two antiparallel and spirally coiled strands.
- The backbone of each strand is made up of deoxyribose sugar and phosphate joined by ester bonds.
- All the sugars in one strand are directed to one end, which is opposite to that of the sister strand.

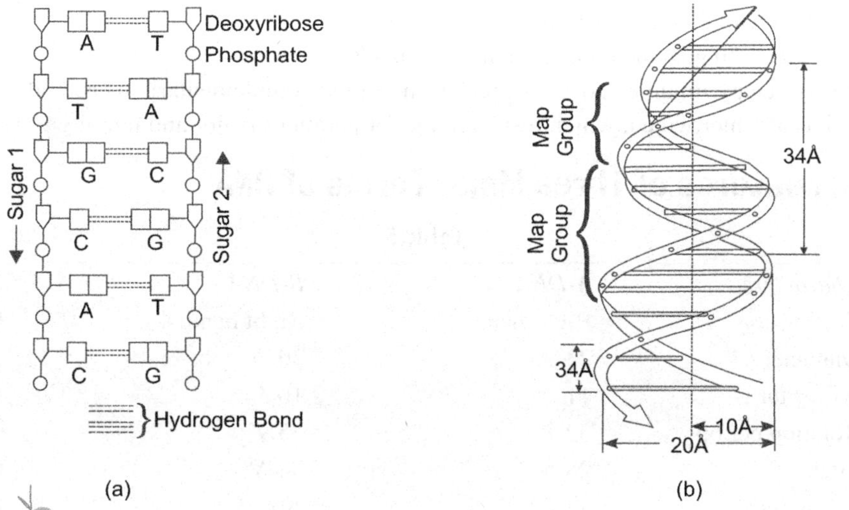

Fig. 39 (a) Watson and Crick model of DNA (b) Parts of two complementary chains of nucleotides of the DNA model

- One oxygen of phosphate group joins with carbon no. 3 of the deoxyribose.
- Another oxygen of the same phosphate joins with carbon no. 5 of the next sugar.
- Each strand completes a turn at 34 Å intervals called pitch.
- Each turn contains ten nucleotides; hence the distance between the two nucleotides is 3.4 Å.
- The width of the double helix is 20 Å.
- It contains two types of nitrogenous bases, viz., pyrimidine and purine.
- Purine (Adenine and Guanine) is a double-ring compound, while pyrimidine (Thymine and Cytosine) is a single-ring compound.
- Purine joins with deoxyribose sugar by glycoside bond that forms between C1 of deoxyribose and N9 of purine.
- Similarly, glycoside bond is formed between C_1 of deoxyribose and N_3 of pyrimidine.

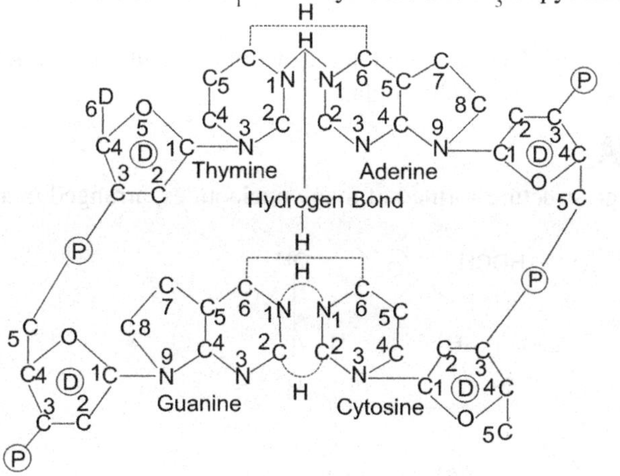

Fig. 40 DNA segment showing hydrogen bonds between two types of bases: Bonding between deoxyribose sugar and phosphate

- Adenine pairs join with thymine by two hydrogen bonds.
- Cytosine pairs join with guanine by three hydrogen bonds.
- The pairing between pyrimidine and purine makes the strand complementary to each other.
- Both the strands are intertwined, which results in the formation of major and minor grooves.

Structural Features of Three Major Forms of DNA

Table 5

S. No.	*Characters*	*A-DNA*	*B-DNA*	*Z-DNA*
1.	Coiling	Right handed	Right handed	Left handed
2.	Diameter	23°A	20°A	18°A
3.	bp per turn	11	10.5	12
4.	Rotation per bp	32.7°	35.9°	60°/2
5.	Pitch	28 A°	34 A°	69°A
6.	Mean twist	+18°A	+16°A	0°t
7.	Form of sugar			
8.	Phosphate bone	Regular	Regular	Zigzag

Functions of DNA

- DNA carries genetic characters from parents to offsprings.
- DNA controls all cellular activities.
- DNA brings about differentiation of cells during development.
- DNA contributes to evolution by undergoing gene mutations.
- It synthesises RNA.
- DNA controls post-natal development by its internal clock.

RNA

- RNA is a type of nucleic acid, chiefly fond in the cytoplasm. But it is also found in the nucleolus. Besides, it also occurs in mitochondria and chloroplast.

Structure of RNA

- RNA is a single-strand structure formed of many nucleotides, arranged in a linear fashion.

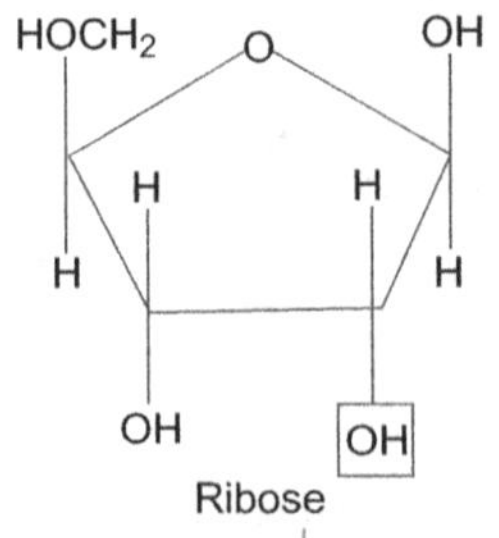

Fig. 41

- It contains ribose sugar, joined to phosphate with phosphodiester bond at carbon positions no. 3 and 5.
- The nitrogenous bases found in RNA are as follows:

  (a) **Purine** – Adenine and Guanine

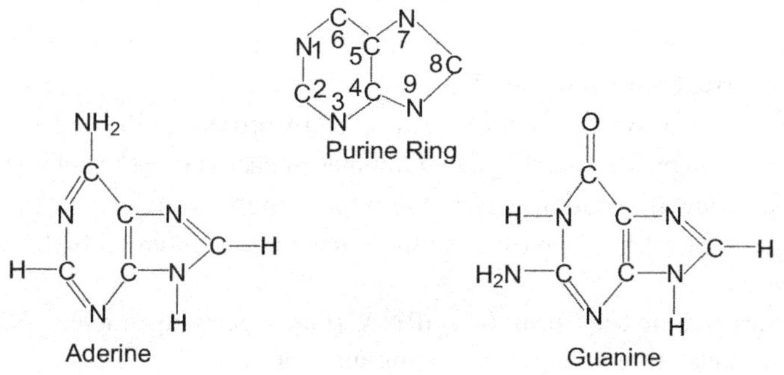

Fig. 42

  (b) **Pyrimidine** – Cytosine and Uracil

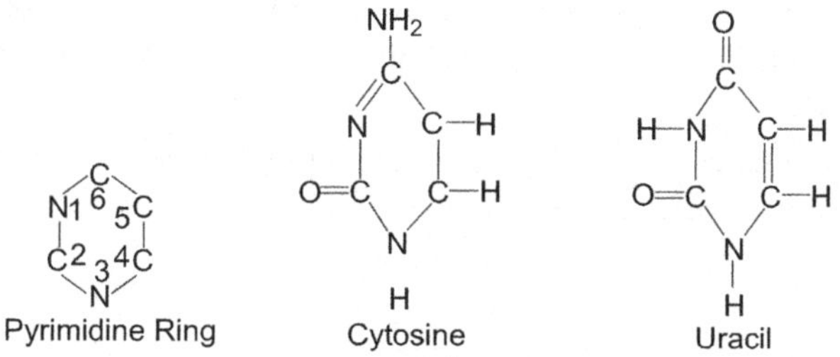

Fig. 43

- Nitrogenous bases unite with ribose by glycosidic bond.
- The glycosidic bond forms between C_1 of ribose and N_9 of purine.
- The glycosidic bond forms between C_1 of ribose and N_3 of pyrimidine.
- Purine and pyrimidine bases are not in equal number.
- RNA does not follow Chargaff's law.
- The nucleotides of a single-stranded RNA show intramolecular pairing, which provides stability to the RNA.

Types of RNA

In Eukaryotic cells, the RNA is not a genetic material and is synthesised from the DNA by a process called transcription. The following three types of RNA are found in the cytoplasm:

1. mRNA (Messenger RNA)
2. tRNA (Transfer RNA)
3. rRNA (Ribosomal RNA)

1. mRNA (Messenger RNA)

- It is synthesised inside the nucleus from one of the two strands of DNA and after synthesis it diffuses into the cytoplasm.

Fig. 44 *mRNA: A = Cap, B = Noncoding region, C = Initiation codon, D = Coding region, E = Termination codon, F = Noncoding region, G = Poly-A*

- It carries genetic information from chromosomal DNA to the cytoplasm for the synthesis of protein.
- The synthesis is carried out from 5' to 3' end.
- mRNA constitutes about five per cent of the total RNA present in the cell.
- It has a short lifespan in prokaryotes (e.g., two minutes in bacteria) but in eukaryotes, it survives for days.
- It is destroyed after few translations; therefore it has a high turnover.
- mRNA of all eukaryotes has a cap of 7-methyl Guanosine at 5' end which protects them against the action of nucleases.
- mRNA is synthesised in the form of hnRNA (heterogeneous nuclear RNA) which contains a sequence of both translating and nontranslating nucleotides.
- During maturation, the nontranslating sequence (introns) are removed by the process of RNA splicing.
- The sequence of nucleotides which code a polypeptide chain is called exon.
- Most of the mRNA contains a sequence of poly-A (polyAdynalic acid) tail attached to 3' end.
- The poly-A tail stabilises the mRNA and becomes shorter with the age of the mRNA.
- The coding region of mRNA starts with AUG and ends with UAA, UAG or UGA (termination codon).

2. **tRNA (Transfer RNA)**

- tRNA is a relatively smaller molecule containing 75 to 85 nucleotides.
- It is also called soluble or adaptor RNA.
- It constitutes 10 to 20 per cent of the total RNA of the cell.
- It is synthesised in the nucleus on DNA template (a part of DNA).
- It consists of a single-stranded polynucleotides chain, looped upon it to form a clover structure.
- In 1965, Holley et al. worked out the nucleotide sequence for yeast alanine tRNA and gave the clover leaf model.
- According to the clover leaf model, tRNA is folded to form five arms.
- The 3' end always terminates with C–C–A sequence, which is added post-transcriptionally.
- The activated amino acid joins at this C–C–A sequence.
- The 5' end terminates with G or C nucleotide.
- Loop IV interacts with the complementary region of the rRNA during protein synthesis.
- The variable loop differs greatly in length in different tRNA.
- Loop III contains anticodon side while loop I participates in amino acid activation.
- All tRNA show this characteristic of folding except mitochondrial tRNA.

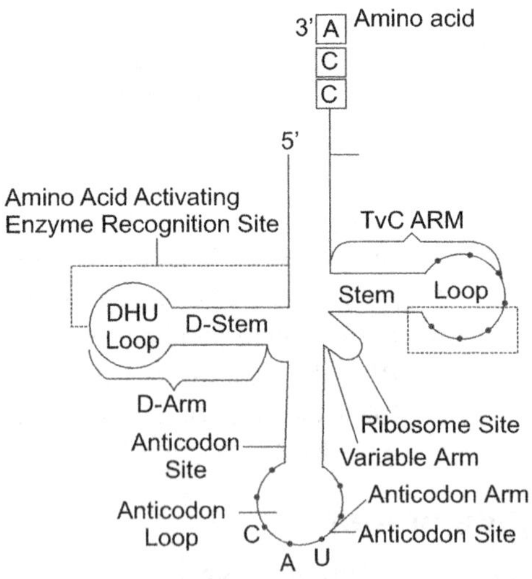

Fig. 45 tRNA

3. **rRNA (Ribosomal RNA)**

- Ribosomal RNA forms bulk (80 per cent) of the total cellular RNA.
- The RNA molecule is single polynucleotide helix, which is branched and flexible.
- In the helical region, base pairs are complementary and are joined by hydrogen bonds.
- Unfolded regions lack base complements.
- The rRNA helixes unfold on heating and refold upon cooling.
- The rRNA is stable for at least two generations.
- The synthesis of rRNA begins at gastrulation and increases with embryonic development.

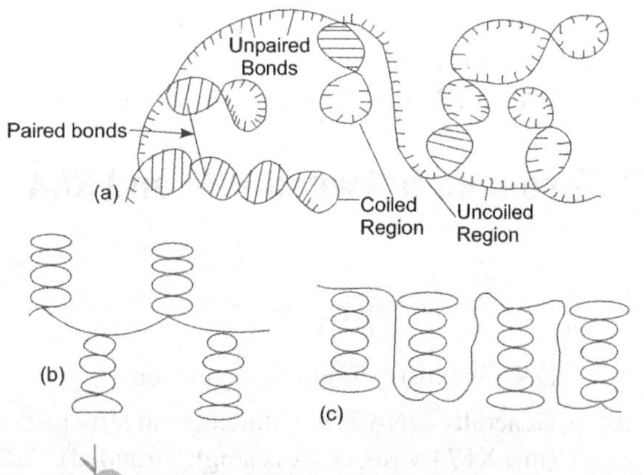

Fig. 46 Ribosomal RNA (rRNA): (a) As an extended strand (b) Without positive interaction (c) With positive interaction

- rRNA is the one of the only genes present in all cells.
- In prokaryotic cells, its number is about 15,000 per cell while in eukaryotic cells its number is numerous per cell.
- Ribosomes contain about 65 per cent rRNA and 35 per cent protein.
- Prokaryotic cells, mitochondria and chloroplast contain 70S ribosome which is made up of two sub-units; a larger sub-unit of 50S and a smaller sub-unit of 30S.
- The larger sub-unit (50S) consists of 23S and 5S rRNA and 34 different types of proteins, while the smaller sub-unit (30S) consists of 16S rRNA and about 21 different types of proteins.
- Eukaryotic cells contain 80S ribosome which consists of a larger sub-unit of 60S and smaller sub-unit of 40S. The larger sub-unit consists of 28S rRNA, 5S rRNA 5.8S rRNA and more than 45 types of different proteins, while the smaller sub-unit consists of 18S rRNA and 33 different types of proteins.
- The 3' end of 18S rRNA (16S rRNA in prokaryotes) contains a binding site for mRNA cap.
- The 5S rRNA contains binding site for tRNA.
- rRNA provides structural integrity to ribosome and also serves as the site for the attachment of mRNA.

Other Types of RNA

- **Complementary RNA (cRNA)** – cRNA is a viral RNA which is transcribed from negative sense RNA that serves as template for protein synthesis.
- **Small nuclear RNA (snRNA)** – snRNA is a class of eukaryotic small RNA molecules. It is usually found in nucleus as ribonucleoprotein and is apparently involved in processing of heterogenous mRNA.
- **Small nucleolar RNA (snoRNA)** – Small nucleolar RNAs are found in archaea and eukaryotes and are involved in nucleotide modifications of RNAs.
- **Genetic RNA (gRNA)** – In certain viruses, RNA acts as genetic material (e.g., TMV). Its molecule may be single stranded, double stranded and linear or circular.
- **Catalytic or enzymatic RNA** – These are the classes of RNA having the ability to participate in enzymatic reactions without the help of a protein. They are referred to as ribozymes and are found in many species. The first ribozymes were discovered the 1980s by Thomas R Cech and Sidney Altman.

- **Telomerase RNA** – Telomerase RNAs are found in majority of eukaryotes and are involved in the synthesis of telomere.
- **YRNA** – YRNA are found in animals and are involved in RNA processing and DNA replication.

Differences between DNA and RNA

Table 6

S. No.	*DNA*	*RNA*
1.	DNA is mostly found in the nucleus.	RNA is mostly found in the cytoplasm.
2.	Generally DNA is a double helical structure (in φX174 virus DNA is single stranded).	Generally RNA is a single helical structure (in some viruses like Reo virus, RNA is double helical).
3.	In DNA, sugar is pentose and deoxyribose type.	In RNA, sugar is pentose and ribose type.
4.	In DNA, nitrogenous bases are adenine, guanine, cytosine and thymine.	In RNA also four types of nitrogenous bases are found, but uracil is present instead of thymine.
5.	In DNA, base pairing is A=T/T=A: C=C/C=G.	In RNA, base pairing is A= U/U= A:C=G/G=C
6.	Base pairing occurs throughout the length.	Base pairing occurs only in the helical region.
7.	Generally DNA is genetic material.	RNA is nongenetic (except few viruses).
8.	DNA is of only one type.	RNA is of three types: mRNA, rRNA and tRNA.
9.	DNA is made up of a large number of nucleotides.	RNA is made up of a fewer nucleotides.
10. 11.	DNA can replicate to form its copy and forms RNA by transcription.	RNA lacks power of replication and transcription (in certain viruses it can synthesise RNA chains).
12.	DNA is feulgen positive.	RNA is feulgen negative.

REPLICATION OF DNA

- One of the most important properties of DNA is that it can synthesise an exact copy of itself. This property of DNA is known as replication.
- Replication of DNA takes place during interphase.
- Theoretcally, the following three types of replication are possible in DNA:

  1. Conservative Replication – According to the conservative mode of replication, out of the two DNA formed, one DNA has entirely new material and the other has entirely old material.

  2. Semiconservative Replication – According to this mode of replication, out of the two DNA stands-formed, in each DNA, there is one old and one new strand.

  3. Dispersive Replication – The dispersive mode of replication suggests that out of the two DNA formed, each strand of each DNA has patches of old and new strands.
- It was first proposed by Watson and Crick in 1953 that the replication is semiconservative, in which the new strand is synthesised on the parental strand.
- The semiconservative mode of replication suggests that half of the DNA molecules are conserved. The other half are synthesised as new strands.

- Meselson and Stahl (1958) gave strong experimental proof in favour of semiconservative mode of DNA replication.
- Meselson and Stahl grew *E. coli* in a medium containing heavy isotope ^{15}N for several generations, so that both strands of the DNA became fully labelled with ^{15}N.
- Thereafter, they transferred the bacteria in a medium containing ^{14}N.
- DNA was extracted from bacteria and tested for ^{15}N DNA in every succession generation through density gradient centrifugation (using cesium chloride).
- After the first generation, only hybrid molecules (^{14}N + ^{15}N) were present in the cells.
- After two generations, half (50 per cent) DNA molecules were light (^{14}N) and half (50 per cent) were hybrid (^{14}N + ^{15}N) and after the third generation, 75 per cent DNA molecules were light chain (^{14}N) and 25 per cent were hybrid (^{14}N + ^{15}N).
- After the first generation in ^{14}N medium, the bacteria settled at a level intermediate between light and heavy bands. This indicated that all DNA found after the first generation had intermediate density. This was only possible, if the mode of replication in DNA was semiconservative.

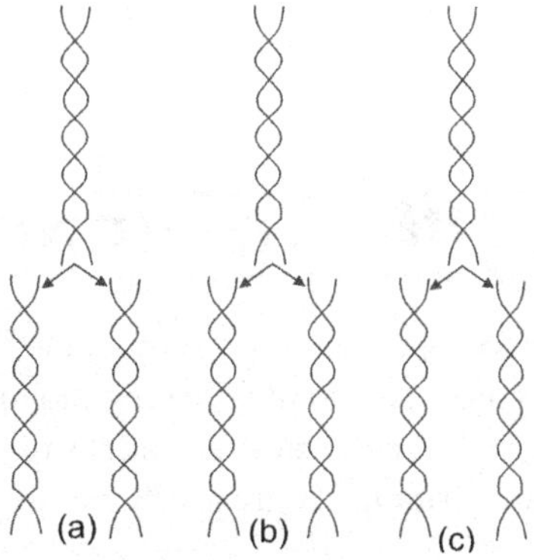

Fig. 47 Three possible modes of DNA replication: (a) Conservative (b) Semiconservative (c) Dispersive

Replication Process

The whole process of replication is completed in the following steps:

- The replication begins at a particular point called initiation point or origin.
- Origin is a specific point of 100–200 bp recognised by initiation protein.
- In prokaryotes, there is single origin while in eukaryotes multiple origins are found.
- The initiation leads to cut in the DNA strand called nick.
- Nick leads to unwinding of the double helix helped by unwinding proteins and superhelix relaxin proteins.
- Unwinding leads to separation of both strands; as a result replication fork is formed.
- The synthesis of new strand is called chain elongation which takes place in 5' ⟶ 3' direction.
- In one strand, it is continuous called leading strand, while in the other strand it is discontinuous called lagging strand.
- The lagging strand requires RNA primer for chain elongation.
- RNA primer is a short polyribonucleotide containing 9–50 N-bases.
- Discontinuous replication leads to synthesis of strand in fragments called Okazaki fragments.
- Okazaki fragments are short polydeoxyribonucleotides containing 200–2000 N-bases.

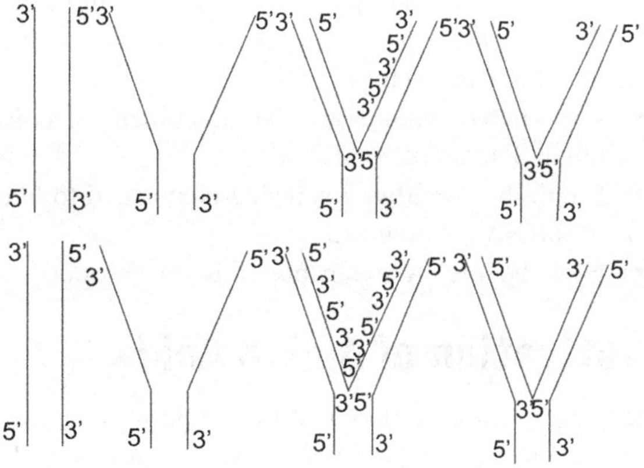

Fig. 48 Replication of DNA

- After the formation of Okazaki fragments, the RNA primer is excised by exonuclease.
- The Okazaki fragments join together by DNA ligase after the removal of RNA primer.

PROTEIN SYNTHESIS

- Proteins occupy central position in the architecture and functioning of living organisms.
- Proteins are made up of repeating monomeric units and each monomeric unit is known as amino acid.
- Amino acids are the building blocks of protein.
- There are 20 amino acids which are linked up by peptide bonds to form long chains called polypeptides.

Central Dogma

- Central dogma refers to the relationship between DNA, RNA and protein and it was originally proposed by Crick.
- Crick (1958) proposed one way flow of information according to which DNA transfers its information to RNA(transcription) and then RNA is translated to protein (translation).

$$DNA \xrightarrow{\text{Transcription}} RNA \xrightarrow{\text{Translation}} Protein$$

- Temin and Baltimore (1970) proposed reverse flow of information according to which DNA synthesises RNA and RNA can synthesise DNA in tumour viruses. This is also known as Teminism.

$$DNA \underset{\text{Reverse}}{\overset{\text{Transcription}}{\rightleftharpoons}} RNA \xrightarrow[\text{Translation}]{\text{Translation}} Protein$$

- Protein synthesis involves the following steps:

Transcription

- The direction of transcriptions is 5' ⟶ 3'
- Transcription takes place on the DNA template helped by different factors and RNA polymerase.
- Transcription results into the formation of hnRNA which splices to form mRNA.
- The mRNA matures by the attachment of poly-A tail at 3' end of the mRNA.
- Similarly, 7-methyl guanosine is attached to 5' end of mRNA forming cap.
- Now the mRNA is transported to the cytoplasm.

Activation of Amino Acids

Inactive form of amino acid in cytoplasm is activated by ATP and specific enzyme; as a result, aminoacyl adenylate enzyme complex is formed.

AA+ATP ⟶ AMP ~ AA + PP.

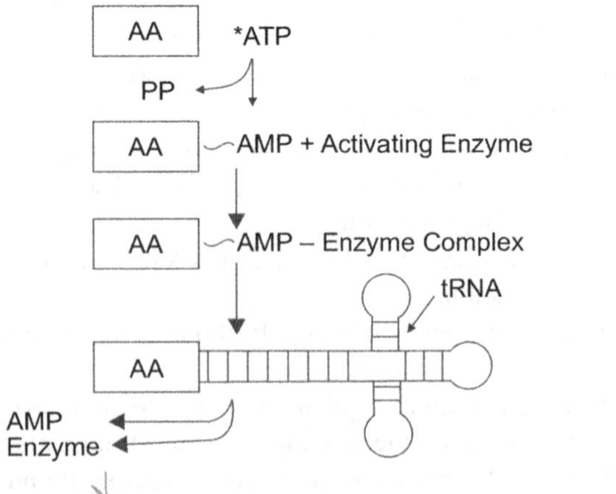

Fig. 49 Activation and attachment of amino acids

Transfer of activated amino acid to (tRNA)

Aminoacyl adenylate enzyme complex transfers the amino acid to tRNA in the presence of enzyme aminoacyl tRNA synthetase by which the amino acid gets attached to the tRNA.

AMP~AA Enzyme Complex + tRNA⟶AA-tRNA + Enzyme + AMP.

Initiation

In prokaryotes, the initiation of translation begins with tRNA f-met while in eukaryotes, it begins with tRNA met.

Attachment of tRNA with Ribosome

- **The larger sub-unit of ribosome has two sites called A site and P site.**
- **A site is also called acceptor site or aminoacyl site, which receives the activated aminoacyl complex.**
- **P site is also called donor site or peptidyl site, which helps in the formation of peptide bond and polypeptide chain elongation.**

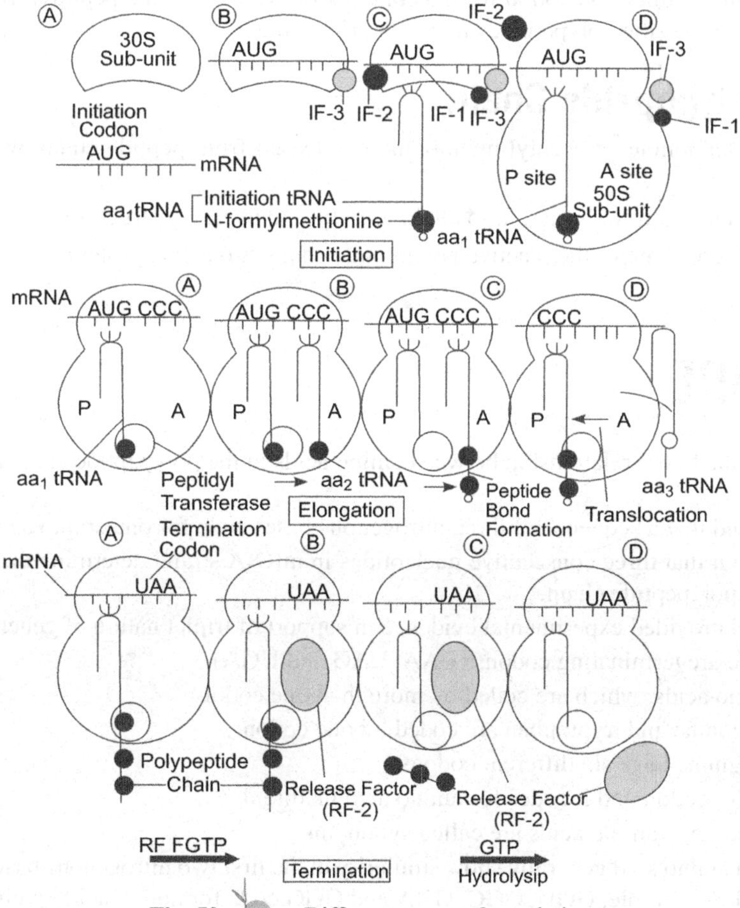

Fig. 50 **Different stages of translation**

Peptide Bond Formation

- The first peptide bond is formed between amino acid of the P site and amino acid of the A site.
- In this process, amino acid of the P site loses OH group and amino acid of the A site loses H group.
- After the formation of the peptide bond, tRNA of the P site is then released and the peptide bond moves to the A site.

Chain Elongation

The movement of tRNA from P site to A site is repeated. As a result, a long polypeptide chain is formed. During this process the ribosome shifts from 5' to 3' direction on mRNA

Chain Termination

- The termination of polypeptide chain is brought about by stop codons (UAA, UAG or UGA) of mRNA and releasing factors R1 and R2.
- The releasing factors and termination codon form a complex on which enzyme peptidyl transferase acts to terminate as well as release the polypeptide chain from ribosomes.

Activation of Polypeptide Chain

- The first amino acid Methionine or Formyl methionine is released from peptide chain by the action of enzyme deformylase.
- Now released polypeptide chain folds upon it to attain tertiary or quaternary structure.
- The attainment of structure changes the inactive polypeptide chain into active protein.

GENETIC CODE

- Genetic code refers to the linear relationship between amino acids in the polypeptide chain and triplets in DNA (or RNA).
- Gamow (1954) proposed that a sequence of three nitrogenous bases code for one amino acid.
- Crick (1961) pointed out that three consecutive nucleotides in mRNA strand determine the position of a single amino acid in a polypeptide chain.
- Nirenberg and Mathaei provided experimental evidence in support of triplet nature of genetic code.
- Out of 64 codons, three are terminating codons (UAA, UAG and UGA).
- Besides, there are amino acids, which are coded by more than one codon.
- Only amino acid methionine and tryptophan are coded by one codon.
- Leucine, serine and arginine have six different codons.
- Codon AUG acts as start codon and it codes for amino acid methionine.
- Codons that specify the same amino acids are called synonyms.
- Multiple codons for an amino acid generally show similarity in the first two nitrogenous bases and it is the third base that varies. For example, GGU, GGC, GGA and GGG code for amino acid glycine.

First Base	Second Base U	Second Base C	Second Base A	Second Base G	Third Base
U	UUU Phe	UCU	UAU Tyr	UGU Cys	U
	UUC	UCC Ser	UAC	UGC	C
	UUA Leu	UCA	UAA*	UGA*	A
	UUG	UCG	UAG*	UGG Try	G
C	CUU	CCU	CAU His	CGU	U
	CUC Leu	CCC Pro	CAC	CGC Arg	C
	CUA	CCA	CAA Gln	CGA	A
	CUG	CCG	CAG	CGG	G
A	AUU	ACU	AAU Asn	AGU Ser	U
	AUC Ile	ACC Thr	AAC	AGC	C
	AUA	ACA	AAA Lys	AGA Arg	A
	AUG Met	ACG	AAG	AGG	G
G	GUU	GCU	GAU Asp	GGU	U
	GUC Val	GCC Ala	GAC	GGC Gly	C
	GUA	GCA	GAA Glu	GGA	A
	GUG*	GCG	GAG	GGG	G

Fig. 51 Dictionary of genetic code
* GUG codes for methionine, When used as initiating codon
* UAA, UAG and UGA are termination codons

- The variation in the third nitrogenous base (known as wobble base) can be explained by the 'wobble hypothesis'.
- The wobble hypothesis was proposed by Crick (1965), which states that the first two nitrogenous bases of the triplet codon pair according to set rule, i.e., A with U, G with C but the third base has much more freedom of pairing. The wobble base permits more than one type of pairing.
- Crick (1965) found that if U is present at the first position of anticodon, it can pair with A or G at the third position of codon. Similarly, G present in anticodon can pair either with C or U.
- Thus, wobbling allows economy of tRNA molecules.

Table 7 Wobble pairing

S. No.	*Anticodon (First base)*	*Codon (Third base)*
1.	U	A, G
2.	C	G
3.	A	U
4.	G	U, C
5	I	U, C, A
6.	I	(Inosine resembles G)

Characteristics of Genetic Code

1. **Triplet** – Genetic code is triplet, i.e., a sequence of three nitrogenous bases in mRNA code for one amino acid.
2. **Non-overlapping** – Genetic code is non-overlapping, i.e., one base never participates in the formation of two adjacent codons.

3. **Commaless** – There is no punctuation mark between two adjacent codons, i.e., genetic code is commaless.
4. **Universal** – Genetic code is almost universal, i.e., one codon always codes for the same amino acid in all organisms.
5. **Degenerate** – There are amino acids, which are coded by more than one codon. Thus, genetic code is degenerate.
6. **Co-linear** – Sequence of nitrogenous bases in mRNA (DNA) corresponds to the sequence of amino acids in a polypeptide chain.
7. **Initiation or Start Codon** – AUG is the initiation codon, which codes for amino acid methionine. Rarely GUG acts as initiation codon.
8. **Terminating or Stop Codons** – Termination of polypeptide chain is brought about by specific codons called terminating codons. There are three terminating codons, viz., UAA (ochre), UAG (amber) and UGA (opal).
9. **Polarity** – Genetic code is always read in a fixed direction, i.e., in the $5' \longrightarrow 3'$ direction.

Deciphering Genetic Code

- Nirenberg and Mathaei (1961) synthesised RNA homopolymers comprising only one type, i.e., they produced mRNA either UUUUUU (poly-U), AAAAAA (poly-A), CCCCCC (poly-C) or GGGGGG (poly-G).
- Using in vitro protein synthesis, they found the following results:
    1. UUU codes for amino acid phenylalanine (UUU is the first code to be deciphered).
    2. AAA codes for amino acid lysine.
    3. CCC codes for proline.
    4. However, they could not find any result with poly-G as it did not serve as a template (it attains secondary structure and thus could not attach to ribosome). Thus, three of sixty four codons were easily accounted for.
- Further, work regarding deciphering genetic code took place by using synthetic mRNA containing two types of nitrogenous bases (heteropolymer) in the laboratories of Ochoa and Nirenberg. The synthetic mRNA contained two types of nitrogenous bases at random (random co-polymers). For example, in a random co-polymer using U and A nucleotides, eight triplets are possible—UUU, UUA, UAA, UAU, AAA, AAU, AUU and AUA. Theoretically, eight amino acids could be coded by these eight codons. However, the actual experiment yielded only six amino acids—phenylalanine, lysine, tyrosine, leucine, isoleucine and asparagine, by varying the relative composition, U and A, in the random co-polymer as well as determining the percentage of different amino acids in the protein obtained.
- Lederberg and Nirenberg (1964) used cellulose nitrate filter technique for determining codons of amino acids known as the filter binding technique.
- The ribosomes are left on the filter paper while tRNA washed through the filter in the absence of mRNA. But in the presence mRNA and ribosomes, the tRNAs stick to the filter.
- A mixture of synthetic poly-U mRNA and ribosome was placed on the filter and various tRNA carrying specific amino acids labelled by ^{14}C were separately passed through the filter.
- In this process, only amino acid phenylalanine was retained in the filter. Thus, it became clear that UUU is the code for phenylalanine.
- Lederberg and Nirenberg were able to determine a sequence of 54 codons out of 64 codons.

Exceptions to Genetic Code

Though genetic code is universal, there are a few exceptions:

1. In a few viruses, such as φX174 and SV-40, genes are known to overlap.
2. The codons AGG and AGA, code for arginine but in human mitochondria, both these act as terminating codons.
3. In *Paramecium* and some other ciliates, terminating codons UAA and UGA code for amino acid glutamine.
4. In *Euplotes octacarinatus* (ciliate), terminating codon UGA codes for cysteine. Only UAA is used as terminating codon while terminating UAG codon is absent.

GENE REGULATION

- The process by which cells translate their genetic information contained in their DNA into proteins is known as gene expression.
- These proteins have a variety of functions.
- But not all these proteins are needed at a time, i.e., different proteins are required at different times.
- Thus, all the functional genes do not simultaneously make protein.
- As the development proceeds, certain genes are activated while certain genes are suppressed.
- The activation and suppression of genes at different times is called gene regulation.
- Gene expressions are controlled at many levels to ensure the organism has appropriate response to its environment or internal changes.

Gene Regulation in Prokaryotes

- Gene regulation in prokaryotes is essential as they are single-cell organisms and they mainly depend on their environment for their activities.
- In bacterial cells, regulation of synthesis of enzymes occurs in a way that these enzymes are produced only when substrates, which are utilised by these enzymes, are present in the cell.
- Jacob and Monad (1961) proposed a model for gene regulation in prokaryotes, based on their study on the inducible operon system, for the synthesis of beta-galactosidase enzyme in *E. coli*, in order to explain the induction or repression of enzyme synthesis.
- This model is popularly known as the operon model.
- According to this model, control of gene expression in prokaryotes involves control of operon.
- An operon is a coordinate unit of gene expression.
- Operon consists of:
  1. **Promoter** – Promoter is the sequence of DNA where *RNA polymerase* enzyme binds.
  2. **Operator** – Operator is the sequence of DNA where active repressor binds. It controls the activity of a number of structural genes and itself is under the control of a repressor molecule synthesised by the regulator gene.
  3. **Structural Genes** – Structural genes are the DNA regions that code for protein. They synthesise mRNA under the control of an operator gene.

Regulatory Components

1. **Regulator Gene** – It is the sequence of DNA that codes for the production of a repressor protein (often located away from the operon it regulates).

2. **Repressor Protein** – In active form, it binds to the operator to prevent transcription of structural genes.

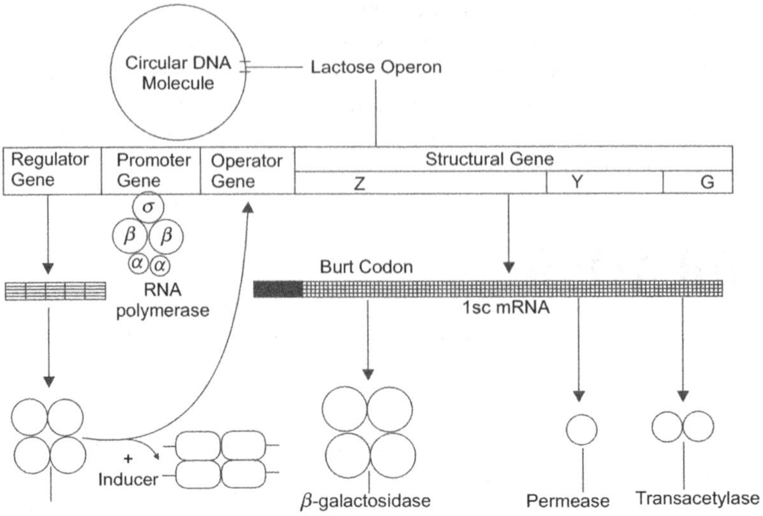

Fig. 52 The lac operon of the bacterium E. coli

Inducible Operon (Lac Operon)

- Inducible operons are generally in the off condition, i.e., their gene products are required occasionally or not at all by the cell.
- Repressor proteins are made in active form, and are capable of binding to the operator.
- Repressor proteins can be inactivated by binding to the inducer (in lac operon it is lactose) which is then activated.

Repressible Operon

- Repressible operons are generally in the on condition, i.e., the products of genes are generally required, most of the time, by cells for maintenance.
- Repressor proteins are made in inactive form, which bind to the operator.
- A repressor needs to bind to a co-repressor to make it an active repressor.
- A co-repressor is generally an end product of the metabolic pathway coded by the operon. In tryptophan operon, it is tryptophan.
- There must be high accumulation of co-repressor before it can bind to the repressor to make it active.

Negative versus Positive Controls

- Above examples simply show on and off at the operator representing a negative control.
- Some operons when turned on, do not transcribe at a sufficient rate, as *RNA polymerase* do not bind effectively to the promoter.
- These operons can be transcribed effectively by binding efficient binding of *RNA polymerase*. An activated helper is required for *RNA polymerase* binding.

Gene Regulation in Eukaryotes

- Gene regulation in eukaryotes is much more complex due to their multicellularity.
- In eukaryotes, regulation of gene is needed due to their cell specialisation.
- Each cell type differentiates by activating a different subset of genes.
- Gene regulation in higher eukaryotes may be either short term or long term.
- Gene regulation in eukaryotes occurs at the following levels:
  1. Transcription
  2. RNA processing
  3. mRNA lifetime (longevity)
  4. Translation
- One of the most popular models of gene regulation in eukaryotes is the Davidson–Britten's model known as 'Gene Battery Model'. According to this model, gene regulation occurs at the level of RNA processing.

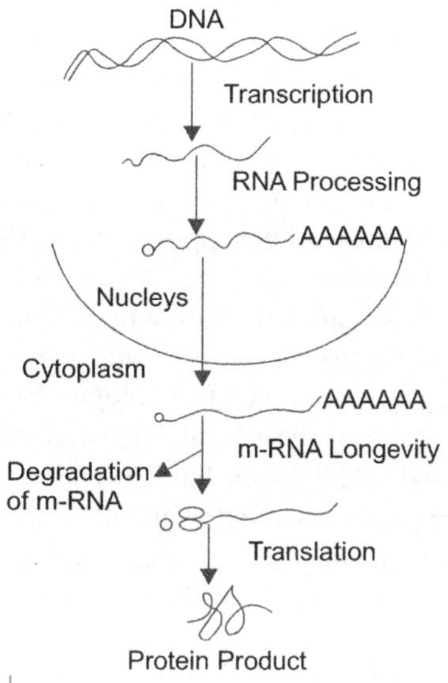

Fig. 53 Various levels of regulation in production of protein in eukaryotic cell

HUMAN GENOME PROJECT

- The Human Genome Project is a mega project that was started in October 1990 and was completed in April 2003. However, the announcement of sequencing of individual chromosomes was published in 2006 with the completion of assigning nucleotide sequences to chromosomes.
- The Human Genome Project made it possible, for the first time, to read nature's complete genetic blueprint for building a human being.
- The Human Genome Project was a 13-year-effort coordinated by the US Department of Energy and the National Institute of Health. Besides the United States, geneticists from United Kingdom (Welcome Group), Japan, Germany, France, China and India also joined the project.
- The total estimated cost of the Human Genome Project would be approximately nine billion US dollars.
- James Watson was the first director of the Human Genome Project. He was replaced by Francis Collins in 1993.
- Francis Collins (Director, Human Genome Project) and J Craig Venter (Founding President, Celera Genomics) are two important scientists associated with the Human Genome Project.
- The complete sequence of the first human chromosome (chromosome 22) was published in December 1999.
- The human genome reference sequences do not represent gene of a particular person's genome.
- The knowledge obtained from the sequences is applicable to every person as all humans share a basic set of genes and common genomic regions.
- Researchers collected blood (female) or sperm (male) samples from a large number of donors.

- It is easier to prepare DNA cleanly from sperms than other cell types as sperms have a higher ratio of DNA to protein. Although sperms contained all chromosomes necessary for study, scientists involved in the Human Genome Project also used white cells from the female donor's blood.
- Only a few samples were processed for DNA resources from a large number of donors. Thus, donor identities were protected and even scientists could not know whose DNA was sequenced.
- The Human Genome Project is an international scientific research project with a primary goal to
    1. Identify all the approximately 20,000–25,000 genes in human DNA.
    2. Determine sequences of the three billion base pairs that constitute human DNA.
    3. Store this information in databases.
    4. To improve tools for data analysis.
    5. Transfer related technologies to the private sector.
    6. Address the ethical, legal and social issues that may arise from the project.
- During decoding of the human genome, scientists also identified genes for cystic fibrosis, neurofibromatosis, Huntington's disease and an inheritable form of breast cancer.
- In addition, the project decoded genome of the bacterium *E. coli*, a fruit fly and a nematode worm, in order to study the genetic similarities among species. A mouse genome was also decoded.

Methodology of the Human Genome Project

- For sequencing, first complete DNA from a cell is isolated.
- The DNA is then broken into relatively small sizes and cloned into a suitable host using a specialised vector.
- The commonly used hosts were bacteria and yeast and the vectors were termed as BAC (Bacteria Artificial Chromosome) and YAC (Yeast Artificial Chromosome).
- The cloning resulted in amplification of each fragment of DNA.
- The fragments were sequenced using automated DNA sequences that worked on the principle developed by Frederick Sanger.
- The sequences were then arranged on the basis of some overlapping regions.
- It resulted in the generation of overlapping fragments for sequencing.
- The bases in the overlapping segments were identified and assembled in a linear order, using computer database.
- Throughout this process, the DNA fragments of a chromosome were sequenced to recreate its original nucleotide sequence.
- Such a study was conducted on all 23 chromosomes of the human genome to understand the exact genome structure of humans.

Salient Features of Human Genome

- There are 3.1647 billion base pairs in the human genome.
- The average gene size is 3,000 base pairs. Dystrophin is the largest human gene having 2.4 million base pairs.
- The human genome contains approximately 30,000 genes.
- Approximately 99.9 per cent base pairs are similar in all human beings.
- The function of over 50 per cent genes is unknown.
- Less than two per cent genes code for proteins.
- Chromosome 1 contains 2,968 genes (maximum) while Y chromosome contains 231 genes (minimum).

- Repeated sequences form the largest part of the human genome.
- A–T rich regions have poor genes while G–C rich regions have dense genes.
- Approximately 1.4 million locations have been identified by scientists in humans, where single base difference occurs (SNPs). They will help in localising the disease-associated sequences in the chromosomes.

Applications of the Genome Project

- The project provides database information of humans. Biotechnology based company may use data to manufacture protein which are used in the treatment of genetic diseases in humans.
- It helps in identifying genetic diseases.
- The project will help in knowing the real bases of human life.
- By matching genome of human and *Drosophila,* scientists have found that *Drosophila* contains remedial genes for 177 genetic diseases in humans. Thus, remedy is always around us.
- The action of harmful genes can be blocked by introducing antisense gene to stop genetic diseases.
- Efforts are in progress to determine genes that will cause reversion of cancerous cells to normal.

DNA FINGERPRINTING

- DNA fingerprinting is a technique to identify a person based on his/her DNA.
- DNA fingerprinting is specific to a person and it cannot be changed by any treatment.
- It is also known as genetic fingerprinting or DNA profiling.
- DNA fingerprinting is a very quick way to compare the DNA sequences of two living organisms.
- Sir Alec Jeffrey (1984) invented the technique of DNA fingerprinting.
- 99.9 per cent DNA is identical between individuals, but 0.1 per cent that differs can be used to distinguish one individual from another.
- DNA fingerprinting uses a specific type of DNA sequence, known as microsatellite.
- Microsatellites are short pieces of DNA, which repeat many times in a given person's DNA.
- In a given area, microsatellites tend to be highly variable that make them ideal for DNA fingerprinting.
- By comparing a number of microsatellites in a particular area, a person can be identified easily.

Technique of DNA Fingerprinting

- **Isolation of DNA** – DNA is obtained from tissues or cells of the body. Only a small amount of tissue like blood, semen, hair, or skin is needed.
- **Cutting, Sizing or Storing** – DNA is cut into small pieces at specific locations using restriction enzymes.
- DNA pieces are sorted according to size by a process called electrophoresis.
- In electrophoresis, particles are passed through seaweed agarose gel to determine particle sizes (DNA).
- **Transfer of DNA to Nylon Sheet** – The DNA is transferred to nylon sheet by placing the sheet on the gel and letting them soak.
- **Probing** – Radioactive or coloured probes are added to the nylon sheet. By doing so, a pattern is produced.
- **DNA Fingerprinting** – This pattern is called a DNA fingerprint.

Application of DNA Fingerprinting

- **Paternity** – The DNA pattern is inherited from both parents and therefore, it can be used for paternity and maternity. Thus, it determines the biological parenthood.
- **Criminal Investigation** – DNA fingerprinting can be used to determine whether a suspect was at the crime scene.
- **Diagnosis of Inherited Diseases** – DNA fingerprinting is an important advancement in diagnosis of inherited diseases, as early detection will allow the medical staff to provide medical treatment to the patient.
- **Identification of a Body** – This is useful if the body is badly decomposed or if only some body parts are available, e.g., in the aftermath of a natural disaster or war.
- **Breeding Programmes** – DNA fingerprinting is helpful in guiding breeding programmes for endangered animals.
- **Evolution** – DNA fingerprinting can solve problems of evolution.

MOLECULAR GENETICS OF CANCER

- Cancer may be defined as the uncontrolled growth of the cells (mitosis) that ultimately kills the organism.
- Cancer is a combination of many diseases with different types of causes, symptoms and treatments.
- Cancer cells are formed when the DNA is damaged and cannot be repaired.
- Cancer can be inherited through inheriting damaged DNA.
- Cancer cells are never in G_0 state.
- Hippocrates gave cancer its name.
- Hippocrates was the first to differentiate between benign and malignant tumours.
- Cancer is not a disease of the modern civilisation. The earliest recorded history describing cancer is between 3000 BC and 1500 BC.
- If there is abnormal growth and division of cells, without any controlling limit, it results in the formation of tumours in the body.
- Tumours are of the following two types:

1. **Benign Tumour**
    - Benign tumours are not cancerous.
    - They consist of well-differentiated cells.
    - Benign tumours can be treated easily
    - They do not spread to other parts of the body.
2. **Malignant Tumour**
    - Malignant tumours are cancerous.
    - They cause damage to adjacent organs by spreading cancer cells at a high rate.
    - The cells of this tumour invade the distant organs and form a new tumour, which has the same characteristics as the primary tumour.
    - They have infinite growth and are characterised by decrease in the cytoplasmic-nuclear volume ratio.

Classification

Cancer is grouped, based on its origin and histology, primarily in three major types.

1. **Carcinoma**
   - Carcinoma is the most prominent type of cancer. It accounts for 85 per cent of all types of cancers.
   - Different types of carcinoma develop from epithelial tissues and are ectodermal or endodermal in origin.
2. **Sarcoma**
   - Sarcoma is the rarest type of cancer. It is mesodermal in origin and mainly made up of connective tissue.
   - Different types of sarcomas are solid tumours growing from a muscle, bone or connective tissue.
   - All types of sarcoma account for 1 per cent of all cancers.
3. **Leukemia and Lymphoma**
   - Various types of these cancers comprise 7 per cent of all cancer cases.
   - They are grouped together because they develop in the cells of the immune system.
   - Types of leukemia develop in the cells of the myeloid system while types of lymphoma develop in the lymphatic system.
   - Hodgkin's disease is an example of lymphomas.
   - Besides the above mentioned, multiple myeloma (originates from plasma cells of the bone marrow) and mixed types of cancers are also found.

Characteristics of Cancer Cells

Tumour cells display characteristic set of features that distinguish them from normal cells. These traits allow the individual cell to form a tumour mass and eventually to metastasize to other parts of the body. Some important characteristics of cancer cells are as follows:

1. **Immortalisation** – Transformed cells are immortal and can grow at an uncontrollable rate, if nutrition is available.
2. **Angiogenesis** – Like any living cell, tumours need blood supply for food and oxygen. Cancer cells secrete growth factors that cause nearby blood vessels to produce branches that grow into cancerous tissues.
3. **Loss of Contact Inhibition** – Normal cells in culture stop growing when the plasma membranes come in contact with each other. This inhibition of growth after contact is called contact inhibition. However, transformed cells lack this property and they go on dividing continuously until they kill themselves.
4. **Lack of Differentiation** – Normally cells differentiate and become capable of specific functions (e.g., muscle cells are specialised for contraction but cancer cells are not specialised).
5. **Cytoplasmic Inclusions**
   - In transformed cells, microtubules and microfilaments are fewer in number and much thinner in comparison with normal cells.
   - The plasma membrane of cancerous cells contains predominantly GM_3 types of gangliosides but the plasma membrane of normal cells contains four types of gangliosides —GMIa, GM_1, GM_2 and GM_3.
   - There is slow disappearance of a major protein called LETS (Large External Transforming Sensitive Proteins).
   - Mitochondria become swollen and the number of cristae is reduced.

- The mitochondrial matrix becomes nonhomogenous.
- Ribosomes and endoplasmic reticulum are abundant.
- Generally, transformed cells have reduced Golgi complex.

6. **Genetic Changes**
   - The shape and organisation of the nuclei of the cancer cell may be markedly different from the nuclei of normal cells of the same origin. This change in appearance may be useful in the diagnosis and staging of tumours.
   - Some cancer cells have chromosomal mutations, either extra or missing chromosomes or part of chromosomes.
   - Cancer cells have extra copies of certain genes, a phenomenon known as gene amplification.
   - Cancer is also associated with gene mutation.
   - The DNA of cancer cells is different from normal cells, because they are damaged and cannot be repaired.
   - The nuclear membrane exhibits deep invaginations.
   - The nucleolus becomes irregular and enlarged in size due to polyploidy.

7. **Physiological Properties**
   - Cancer cells lack anchorage property.
   - Transformed cells require less amount of serum for their growth in comparison with normal cells.
   - In transformed cells, there is increased rate of sugar transport. This increases sugar intake by malignant cells.
   - In malignant cells, anodic movement is generally higher, indicating increase in negative surface charge.
   - Cancer cells are unable to oxidise NADH in the mitochondria, as they are deficient in glycerol phosphate dehydrogenase.
   - Cancer cells exhibit agglutination property with lectin.
   - Some cancer cells have defective electrical communication.
   - Tumour cells have high levels of protein survivin that inhibits apoptosis.
   - Cancer cells often produce enzymes that help them invade neighbouring tissues. These enzymes digest the barriers to migrate and spread the tumour cells.
   - The mobility of surface proteins increases in the transformed cells, thus permitting easier agglutination to tumour cells by lectins.
   - Plasma membranes of most transformed cells contain antigens which are absent in normal cells.

8. **Metastasis**
   - Metastasis is one of the important characteristics of cancer cells, i.e., they have the capacity to invade other organs. A cancerous tumour may shed cells into blood and lymphatic vessels and are dispersed to distant sites where they form secondary tumours. Due to this property, it is impossible to remove them surgically.

Genes and Cancer

- Cell division is a normal phenomenon of living organisms, which is dependent on the tightly controlled sequence of events.
- These events are based on the proper levels of transcription and translation of certain genes.
- Any shortage of disturbance in these processes results in unregulated growth that may lead to cancer.

- The genetic basis of cancer was first discovered through the work with retroviruses that were shown to cause cancer in monkeys, chicken, mice, etc.
- The viral genes identified, could be directly linked to the cause of cancer in experimental animals.
- The identified genes related to these processes are grouped into two categories:

Genes whose Protein Products Stimulate or Enhance the Division and Viability of Cells.

- It also includes genes that cause tumour growth by inhibiting cell death. The normal version of these genes is called proto-oncogenes and their mutated or otherwise damaged genes are called oncogenes. Oncogenes lead to cancer.
- Michaeal Bishop and Harold Varmus are pioneers of research on oncogenes and were awarded the Nobel Prize in 1989.
- More than 100 oncogenes have been identified and associated with some form of cancer. However, what activates these genes is not known. External factors, radiation, certain chemicals, etc., may cause activation of oncogenes and result in some types of cancer.
- These genes stimulate cell growth.
- Examples:
  1. MYC gene, which can result in lymphomas.
  2. RAF genes, which can result in stomach cancer.
  3. TRK genes, which can result in thyroid cancer.

Activation of Proto-Oncogenes

- The first oncogene was discovered in 1970 and was named Src. It was discovered in chicken retrovirus.
- Proto-oncogenes are activated in the following three ways:

(a) Gene Mutation

- Gene mutation can change proto-oncogenes to oncogenes.
- Such mutations can occur during normal cell division and can be caused by environmental factors such as chemicals, ultraviolet rays from the sun, X-rays and sometimes by DNA repair mechanism.
- For example, specific mutations in *ras* genes are frequently found in cells from varieties of tumours like colon, lung, breast and bladder cancers.
- Another example of proto-oncogenes by gene mutation is RET.

(b) Increase in Protein Concentration Leading to Activation of Proto-Oncogenes

- An increase in protein concentration may be caused by:
  1. Gene duplication
  2. An increased protein stability
  3. An increase in protein expression

(c) Translocation or Amplification

- Oncogenes can be activated by structural changes called translocation or amplification, which occurs in chromosomes.
- During chromosome translocation, a proto-oncogene on one chromosome may move to another chromosome, leading to structural alteration of genes.
- For example, translocation between chromosome 9 and 22, which is found in Chronic myeloid

lekumemia (CML), a proto-oncogene on chromosome 9 called c-Abl, is moved to chromosome 22, where it fuses with another gene called Bcr.

- Oncogenes can be activated by transposition to an active chromatin domain. In Burkitt's lymphoma (a childhood tumour, common in malarial regions of Central Africa and Pappua New Guinea), a characteristic translocation is seen between chromosomes 8 and 14. This translocation puts the MYC oncogene (located on chromosome 8) close to an immunoglobulin gene on chromosome 14.
- Proto-oncogenes are also activated by gene amplification. For example, breast cancer amplify ERBB2 and sometimes MYC.

Genes whose Protein Products Directly or Indirectly Prevent Cell Division

- They are called tumour-suppressor genes.
- Tumour-suppressor genes produce products that normally block the cell cycle, which would normally prevent the cancer.
- When this gene mutates, it causes a loss or reduction in its function, leading to cancer.
- When mutated, the mutant allele behaves as a recessive, i.e., as long as the cell contains one normal allele, tumour suppression continues.
- Because tumour-suppressor genes are recessive, cells that contain one normal and one mutated gene, i.e., heterozygous, still behaves normal (except X-linked tumour suppressor gene WTX).
- Tumour-suppressor genes are as follows:
  1. **RB Gene** – It is located on chromosome 13. If RB gene becomes bad, it may result in the development of retinoblastoma, bone, breast, lung, prostrate, bladder and other cancers.
  2. **p53 Gene** – p53 is a suppressor gene, located on chromosome 17.
- It can arrest replication of cells with damaged genes until the normal process has taken place.
- If cells with damaged genes grow and replicate, they may result in cancer.
- p53 gene suppresses growth of such cells.
- If this becomes bad, it may lead to the development of leukemia, breast, colon, soft tissues sarcoma and many other types of cancer.
  3. **BRCA1 Gene** – It is located on chromosome 17 and if it becomes bad, it may lead to the development of breast cancer.
  4. **BRCA2 Gene** – It is located on chromosome 13 and if it becomes bad, it may result in the development of breast cancer.

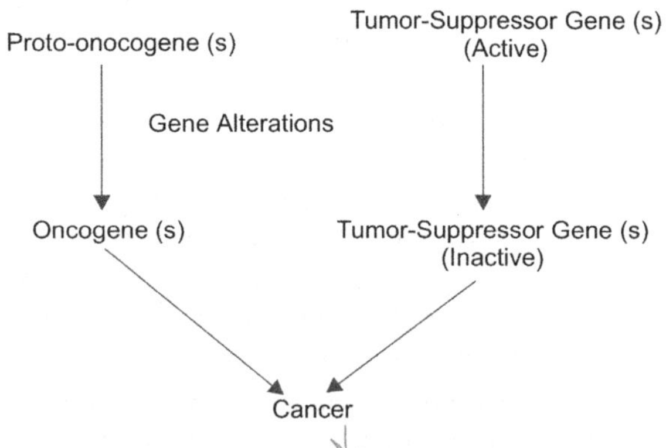

Fig. 54

  5. **APC Gene** – Its bad nature is associated with cancer of colon, pancreas and the stomach.
- The cancerous state results from alteration of several genes, both proto-oncogenes and tumour-suppressor genes.

Events that cause Cancer

- Carcinogens or other cancer-causing agents weaken the cell membrane and due to this weak cell membrane, a microbe is able to enter a normal cell.

- The microbe, after its entry, intercepts the glucose entering the cells.
- The microbe excretes myotoxins. As myotoxins are highly acidic, the inside of the cell becomes highly acidic, which is a characteristic of cancer cells.
- The cell's mitochondria that convert glucose into energy, receive very little glucose as the microbe intercepts most of the glucose.
- A lot of myotoxins and garbage, become available to the mitochondria that they cannot convert into energy.
- The mitochondria play a key role in providing energy to a cell (through the Krebs cycle and electron transport chain).
- To grab more glucose, signals are sent to the insulin receptors on the cell membrane.
- As a result, more glucose enters the cells. However, most of the glucose is utilised by the microbes and myotoxins increase in number.
- The cell is now cancerous as its energy levels drop and it is anaerobic.
- The microbes disturb the Krebs cycle and ETC as long as they are inside the cell.
- Each sick cancer cell contains healthy microbes, so it is not possible to kill microbes without killing the cells.

Carcinogens

- A carcinogen is a substance capable of causing cancer or aggravating cancer in humans and animals.
- Carcinogens include chemicals, radiations and some viruses.
- Sir Percival Potts (1775), a British Physician primarily suggested that the induction of cancer might be linked to agents in the environments. He suggested that exposure to soot caused high rates of cancer in chimney workers.
- Most carcinogens require promoters for the production of cancer.
- Benzopyrene applied on skin does not produce cancer. Croton oil when applied on skin also does not produce cancer. But when benzopyrene application is followed by croton oil, malignant growth occurs. Here croton oil acts as promoter.
- Many mutagens are carcinogens but some carcinogens are not mutagens.
- Alcohol and estrogen are examples of carcinogens that are not mutagens.
- The process leading to cancer by carcinogens is called carcinogenesis.
- The carcinogenic potency of chemical agents depends on their ability to bind to the DNA.
- It can be said that carcinogenesis is due to changes in DNA because carcinogens are mutagens.
- Carcinogens are generally electron deficient, so that they can bind certain regions of the DNA through covalent linkages and induce malignant growth.
- Carcinogens do not lead to cancer after every exposure.
- Some carcinogens result in cancerous changes following high-level prolonged exposure, while others may cause damage at lower levels and shorter exposure periods.
- However, the genetic make-up of an individual may influence the body's response to a carcinogen.

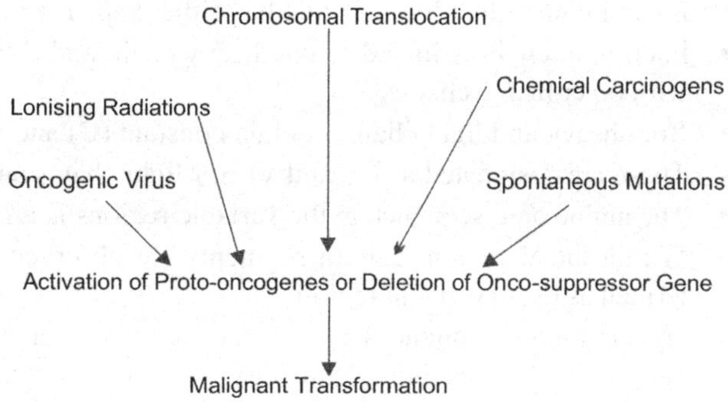

Fig. 55 Cancer causing sources/agents

- Some potent carcinogens are aflatoxin, vinyl chloride, alkylating agents like acridine dyes and polycyclic aromatic compounds such as benzopyrene, azodyes, etc.
- Besides ultraviolet radiations, sunlight is also capable of inducing malignant tumours.
- Viruses that are known to cause cancer such as HPV (cervical cancer), Hepatitis B (liver cancer) and EBV (a type of lymphoma) are all DNA viruses.
- Certain parasites may also induce cancerous growth, such as *Schistosoma haematobium* which induces cancer of the urinary bladder.

IMMUNOGENETICS

- Immunoglobulins are glycoproteins that function as antibodies.
- They are found attached with the cell membrane, secretions or circulating in the blood.
- They are produced in response to detection of foreign molecules in the body.
- These foreign bodies that trigger the formation of antibodies are called antigens.

Structure of Immunoglobulin

- In 1962, the structure of immunoglobulin molecule was independently proposed by Rodney and Gerald Edelman, for which they were awarded the Nobel Prize in 1972.
- IgG molecule is made up of two heavy chains and two light chains, joined together by disulphide bridges.
- In IgG, the heavy (H) chains are composed of 440 amino acids while light (L) chains are made up of 220 amino acids.
- Depending on heavy chain make-up, immunoglobulin are of the following five types:
    1. IgG having γ (gamma) heavy chain.
    2. IgM having μ (mu) heavy chain.
    3. IgA having α (alpha) heavy chain.
    4. IgD having δ (delta) heavy chain.
    5. IgE having ε (epsilon) heavy chain.
- Light chains are either kappa (k) or lambda (λ) in all classes.
- Each immunoglobulin molecule has either kappa (k) or lambda (λ) chain but not both.
- Each light chain is linked to one heavy chain and each heavy chain is associated with a light chain and with other heavy chains.
- Both heavy and light chains contain constant (C) and variable (V) regions.
- These are designated as CL and VL for light chains and VH and CH for heavy chains.
- The amino acid sequence in the variable regions is highly variable.
- Within the V region, certain segments are observed to be more variable than other segments and are termed as hypervariable regions.
- Hypervariable sequences are also called complementary determining regions (CDRs) as they form antigen-binding site complementary to the topology of the antigen structure.
- It is the CH region that determines the class of immunoglobulins.
- The V regions determine the antigen specificity of the antibody molecule.

- Papain cleaves the Ig molecule, so that two Fab (fragment antigen binding) fragments and one Fc (fragment crystallisable) fragment are produced.
- The antigen binding site is located in the Fab.
- The Fc portion is not able to bind to the antigen but it contains a complement binding site which is exposed when interaction between Fab fragment and the antigen occurs.
- This binding occurs through noncovalent interactions (Van der Waals force, hydrogen bonds, and hydrophobic interactions) and triggers conformational changes similar to that observed in enzyme-substrate interaction.

Immunoglobulin G (IgG)

- Immunoglobulin that contains ¥ (gamma) heavy chains are called IgG.
- IgG forms 70–80 per cent of the total immunoglobulins in circulation.
- IgG is also known as 7SIg due to its sedimentation coefficient.
- IgG promotes phagocytosis in plasma and activates the complement system.
- IgG are the only kind of antibodies that cross the placenta.

Immunoglobulin A (IgA)

- Immunoglobulins containing alpha (α) heavy chains are called IgA.
- IgA is basically made up of two light chains (k or λ) and two heavy chains.
- IgA is the second largest class of immunoglobulins, which accounts for about 10–15 per cent of the total immunoglobulins.
- They are found mainly in mucosal secretions, tears, colostrums and milk.
- They are the initial defence in mucosa against pathogens.

Immunoglobulin M (IgM)

- IgM contains mu (μ) heavy chains.
- IgM are also known as macroglobulins or 19S globulins.
- IgM is a polymer of five immunoglobulin molecules, each consisting of two light and two heavy chains.
- IgM antibodies are expanded in the surface of B cells and are formed primarily in the plasma.
- These are the first antibodies produced in significant quantities against an antigen.
- Natural antibodies are also IgM in nature.
- Due to multivalent sites, IgM is highly effective as an agglutinating antibody.
- IgM antibodies promote phagocytosis and activate complement system.

Immunoglobulin D (IgD)

- IgD contains delta (δ) heavy chains.
- IgD antibodies are found on the surface of B cells that have not been exposed to antigens.

Immunoglobulin E (IgE)

- IgE contains eplsion (ϵ) heavy chains.
- IgE is present in blood in very small amounts.
- Due to its cytophilic properties, it is mostly bound to cells and in circulation, it is present only in trace quantities.
- IgE is a monomer and plays an important role in allergic reactions.
- IgE levels increase in worm infections.

Biological Properties of Different Immunoglobulins

Table 8

S. No.		IgG	IgA	IgM	IgD	IgE
1.	Complement fixation	+	–	+	–	–
2.	Placental transfer	+	–	–	–	–
3.	Fixation to mast cells	–	–	–	–	+
4.	External secretions	–	–	+	–	–
5.	Receptor for binding to macrophages	+	–	–	–	–
6.	Natural antibodies	–	–	+	–	–
7.	Present in colostrums	+	–	–	–	–

Genetic Basis of Antibodies Diversity

- The Ig molecule is a tetramer having two types of polypeptide chains, in which, one is known as light (L) chain and the other is called heavy (H) chain.
- Each chain is coded by a single interrupted gene.
- Any type of light chain can associate with any type of heavy chain.
- Each protein chain is composed of two main regions, viz., the N-terminal variable or V region and the C-terminal constant or C region.
- There exists considerable variation in sequence from one protein to another in the variable region, while constant region exhibits considerable homology among the proteins.
- The V domain is generated by joining the variable regions of the light and heavy chains.
- There are many genes coding for variable regions while there are only few genes coding for constant regions.
- It is to be noted that neither V gene (coding for variable region) nor C gene (coding for constant region) is expressed independently.
- V gene must be joined physically with C gene to form a unit in the form of a light or heavy chain.
- Thus, two genes code for a single polypeptide chain.
- There are three immunoglobulin gene families:

Table 9

S. No.	Gene	Chromosome Number (Human)	Chromosome number (Mice)
1.	k-light chain	2	6
2.	L chain	22	16
3.	H chain	14	12

- The following two theories have been proposed to explain the ability of B cells to produce immunoglobulins:

1. **Somatic Variation Theory**
    - This theory states that germ line contains a limited number of variation genes, which are diversified by mutation or recombinational events during embryonic development of the immune system.

2. **Germ Line Theory**

- This theory states that all the genes are inherited and antibody diversity is a result of mutations and selection.
- However, neither of these theories gave a reasonable explanation of the structure of immunoglobulins.
- In 1976, Tonegawa and Hozumi reported the first evidence of V and C regions of immunoglobulins being encoded by different genes and the genes being rearranged in the course of B cell differentiation.
- There are about 200 different variable (V) region genes in germ line DNA.
- There are 12 diversity (D) region genes and four joining (J) region genes.
- Rearrangement of DNA takes place during differentiation of the cell into the B cell.
- This rearrangement leads to the formation of a functional VDJ recombinant gene.
- As any of the genes may recombine with any others, so this rearrangement has a capacity to produce $200 \times 12 \times 4 = 9{,}600$ possible combinations.
- Likewise, the 200 different V regions of light chains may recombine with five J regions, generating $200 \times 5 = 1{,}000$ possible light chains.
- Since the antibody combining site is formed from the variable region of one light chain and one heavy chain combination, the total possible immunoglobulin combinations is about $9{,}600 \times 100 = 10^7$.
- In addition, variations in the precise length of V-, D- and J-coding sequences and joints between them also lead to increase in diversity.

Allelic Exclusion

- Like other somatic cells, B cells are diploid and contain both maternal and paternal chromosomes.
- Though B cell is diploid, it expresses the rearranged heavy chain genes from only one chromosome and the rearranged light chain genes from only one chromosome.
- The process by which this is done is known as allelic exclusion.
- This ensures that functional B cells never contain more than one VH–DH–JH and one VL–JL unit.
- It is essential for antigenic specificity of the B cell because the expression of both alleles would make the B cell multispecific.

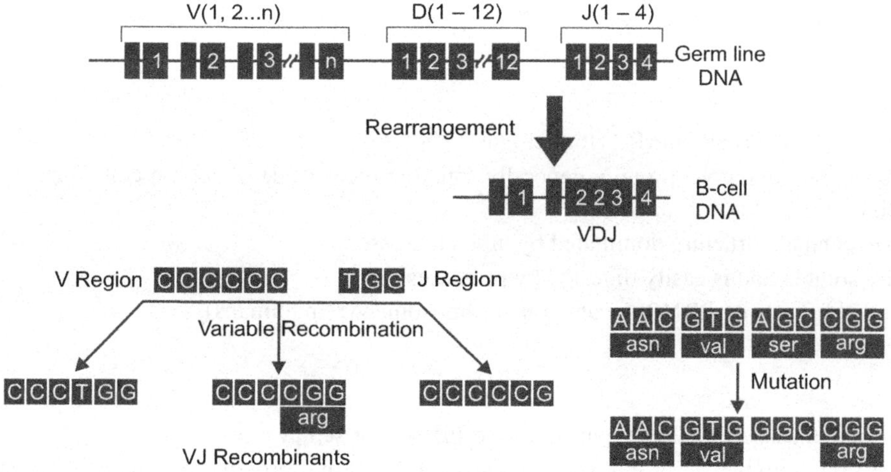

Fig. 56

- The phenomenon of allelic exclusion suggests that once a VH–DH–JH and VL–JL rearrangement has occurred, the recombination machinery is turned off leading to the suppression of expression of heavy and light chain genes on the homologous chromosomes.

PRIONS

- Prions are infectious agents, composed of proteins that cause fatal brain diseases.
- Prions lack nucleic acid (DNA or RNA) but are yet infective, i.e., they can multiply.
- A prion is a mutated protein rather than a mutated gene, which means there is a defect in the end product, and not a defect in the code.
- Prions are 100 to 1,000 times smaller than the smallest organism and as such would be the smallest living form on the earth.
- Prion itself is capable of modifying protein content of the entire organism.
- According to recent researches, the instructions of making prion proteins are on a gene that all mammals have and is mainly active in nerve cells.
- Prions are a product of normal human genes, termed as the PrP gene, found on chromosome 20. This gene contains two exons separated by single introns. The PrP protein is the precursor of the prion protein.
- Prions do not obey the central dogma principle.
- Two possible mechanisms have been suggested for their multiplication that violates the central dogma. The first possibility is of reverse translation that creates a nucleic acid sequence of amino acids in PrP and the second one is that PrP might be acting as template for its own synthesis.
- They are rod-shaped particles and may consist of one, two or three protein molecules (Pr P).
- The term 'prion' was coined by Stanley B Prusiner (1982). He received the Nobel Prize in Medicine/ Physiology, in 1997, for the discovery of prion.
- According to Prusiner, ceratin proteins come in two varieties, viz., cellular form designated as PrPC and infectious misfold form called PrPSc. The two forms are identical in their amino acid sequences and differ in their three-dimensional structure.

PrPC

1. The normal protein is called PrPC (for cellular).
2. It is a transmembrane glycoprotein, generally found at the surface of certain cells (e.g., hematopoietic stem cells).
3. It has its secondary structure dominated by alpha helices.
4. It is easily soluble and is easily digested by proteases.
5. It is encoded by a gene (PRNP) located on chromosome 20 (in humans).

PrPSc

1 The abnormal disease-producing gene is called PrPSc (for scrapie).

2. It has the same number of amino acid sequences as a normal protein, i.e., the primary structures are identical but the secondary structure is dominated by beta conformation.

3. It is insoluble in all the strongest solvents and is highly resistant to digestion by proteases.
4. When PrPSc comes in contact with PrPC, it converts PrPSc into more of itself.
5. These molecules bind each other forming aggregations.
   - According to the prion model, the disease progresses when a misfold PrPSc molecule comes in contact with cellular form PrPC . The appearance of misfold molecule (PrPSc) may be either spontaneous or caused by feeding an animal some food containing meat of diseased animals.
   - Prion diseases include scrapie in sheep, mad cow disease (bovine spongiform encephalopathy or BSE) in cattle, Kuru and Cruetzfeldt–Jakob disease in humans.
   - BSE was first recognised in 1986, whereas Kuru had already been identified in 1957. Scrapie is known for at least 200 years.
   - A prion disease, also called transmissible spongiform encephalopathies, is transmissible host to host in a single species and sometimes even from one species to another. In this disease, the brain tissue is destroyed, giving it a spongy appearance.
   - All prion diseases affect the structure of the brain or the neural tube and are universally fatal.
   - Prion diseases may be present as genetic, infectious or sporadic; all of which involve variation of prion protein.
   - More than 20 mutations of the *Prp* gene are known to cause inheritable human prion diseases.
   - Like other life forms, prions have the ability to evolve through mutations and natural selection.
   - Prions have been discovered in fungi. It is theorised that they may play a symbiotic role in the fungi immune system.
   - Prions can be detected with high sensitivity by Protein Misfolding Cyclic Amplification (PMCA) technology.
   - On the whole, the prions are against the basic tenet that all forms of life must contain nucleic acid.

TRANSPOSONS

- Transposons are the segments of DNA that move around different locations in the genome of a single cell.
- Transposons are also termed as transposable elements, mobile elements, translocatable elements, insertion elements and selfish genes or hopping genes.
- Transposons are also known as jumping genes as they are segments of DNA that are able to jump from location to location within a genome.
- Transposons have been called selfish DNA as their only function is (or appears so) to make more copies of themselves and junk DNA because there is no obvious benefit to their host.
- Transposons are long DNA sequences generally 10^2 to 10^4.
- Each transposon carries the gene that encodes enzyme transposae, which are required for its own transposition.
- Transposons form a large fraction of C value in eukaryotic cells.
- The first transposons were discovered in maize by McClintock in 1940 for which she was awarded the Nobel Prize in 1983. She noticed insertion, deletion and translocation by these transposons.
- Transposons produce gene products and are often involved in gene regulation.
- Larger transposons carry one or more genes in addition to those necessary for transposition.
- Transposons may grow to include more genes.

- In many cases transposons carry a sequence that acts as promoter, altering the regulation of genes at the site of insertion.
- The ability of transposons to increase genetic diversity along with the ability of genomes to inhibit most transposable elements activity results in a balance that makes transposons an important part of evolution as well as gene regulation.
- In the process of moving from location to location, transposons may cause mutations and increase or decrease the amount of DNA in the genome.
- It has been estimated that 80 per cent of spontaneous mutations are caused by transposons.
- Transposon elements are found both in prokaryotes and eukaryotes but are more abundant in eukaryotic genes
- There are three types of transposons:

1. **Class II Transposons**

   Consist only of DNA that moves directly from one location to another.

2. **Class III Transposons**
   - Also known as Miniature Inverted repeats Transposable Elements (MITEs).
   - The genomes of rice and *C. elegans* contain thousands of copies of recurring motif, consisting of almost identical sequence of about 400 base pairs flanked by characteristic inverted repeats of about 15 base pairs.
   - The genome of rice contains about 1,00,000 MITEs.
   - MIETEs have also been reported in mice, humans and *Xenopus*.

3. **Retrotransposons (Class I)**
   - Retrotransposons first transcribe DNA into RNA and then use reverse transcriptase to form a DNA copy of the RNA to insert it in a new location.
   - Many retrotransposons have long terminal repeats (LTRs) at their ends. There may be over more than 1,000 base pairs in each.
   - Retrotransposons also generate direct repeats at their new site of insertion.
   - HIV-1 and other human retroviruses (HIL-1, the human T-cell leukaemia virus) behave like retrotransposons.
   - The RNA genome of HIV contains one gene for transcriptase and one for integrase. The integrase serves the same function as the transposases of DNA transposons.
   - 42 per cent of the entire human genome consists of retrotransposons.
   - Transposable sequences identified both in prokaryotic and eukaryotic organisms have been grouped into two categories, viz., insertion sequences and longer transposons.
   - Insertion sequences (IS) are short and of about 1,000 bases while longer transposons may be longer than even thousand bases.

Insertion Sequences

- Insertion sequences are small compared with other transposable elements.
- They only code for proteins involved in transposition activity.
- They can insert at different regions of bacterial chromosomes and plasmid by the process of illegitimate recombination.
- The coding region of an insertion sequence is generally flanked by inverted repeats.

- Insertion sequences move by conservative transposition.
- The insertion sequences (IS elements) are categorised according to form IS_n, where *n* is the number (e.g., IS_1, IS_2, IS_3 IS_4, IS_{10}, etc).
- The mutagenic activity of IS elements is due to their insertion in the DNA and subsequent errors in the process of transcription.
- IS elements play a key role in the re-orientation and joining the genetic material.

Prokaryotic Transposons

- The prokaryotic transposons are composite and noncomposite.
- Composite transposons (such as Tn_{10}) consist of a central element, which contain several genes that provide resistance to antibiotics. It is mobilised through the activity of a transposase encoded within the insertion sequences flanking the element.
- Certain composite transposons are flanked by two copies IS either in direct or in inverted orientation.
- The frequency of transposition of composite transposons declines with increasing distance between IS sequences.
- Complex transposons may be replicated by either conservative or replicative transposition.
- Composite transposons can be easily recognised when they carry identifiable gene markers as those provide antibiotic resistance and produce toxin.
- Noncomposite transposons (such as Tn_3) lack insertion sequence at their extremities. They have inverted repeats necessary for transposition.

Eukaryotic Transposons

- Eukaryotic transposons were first discovered by Barbara Mc Clintock for maize in the 1940s.
- Most transposons found in eukaryotic cells transpose via RNA intermediates including those found in humans.
- Eukaryotic transposons resemble prokaryotic transposons in many features.
- Some important eukaryotic transposons are P elements in *Drosophila melanogaster*, TY elements in yeast, IAP and VL elements in rodents. The most common form of transposons in humans is the A/U sequence, which is about 300 bases long.

Significance

Transposons:

- Cause mutation and chromosomal break
- May be used as genetic markers
- Regulate gene expression
- Form a large part of constitutive heterochromatin region of the chromosomes that are permanently condensed and for most part generally inactive in every cell.
- Some have been found to cause genetic diseases such as haemophilia, severe combined immune deficiency, Duchenne muscular dystrophy, etc.
- May be used as transformation vectors.
- Used for tagging desirable genes.

APOPTOSIS

- Apoptosis is the programmed cell death, which occurs in the life cycle of all multicellular organisms.
- It is a genetically determined process.
- Apoptosis regulating genes have been found in every organism.
- It occurs during normal cell turnover, development and in the immune system.
- It is an active process that requires energy in the form of ATP.
- It may occur when a cell has become mutated and is on the verge of becoming cancer.

  Chief features of apoptosis are as follows:
  (a) Cell shrinkage
  (b) Blebbing
  (c) Chromatin condensation
  (d) Preservation of organelles and cell membrane
  (e) DNA fragmentation
  (f) Rapid engulfment by neighbouring cells prevent inflammation
- The term 'apoptosis' was used by Kerr et al. (1972).
- Apoptosis differs from necrosis in which cellular debris can damage the organism.
- Apoptosis involves only single scattered cells.
- In general, apoptosis is advantageous in the life cycle of an organism. For example, differentiation of fingers and toes in a developing embryo occurs because the fingers apoptose, resulting in separation of the digits.

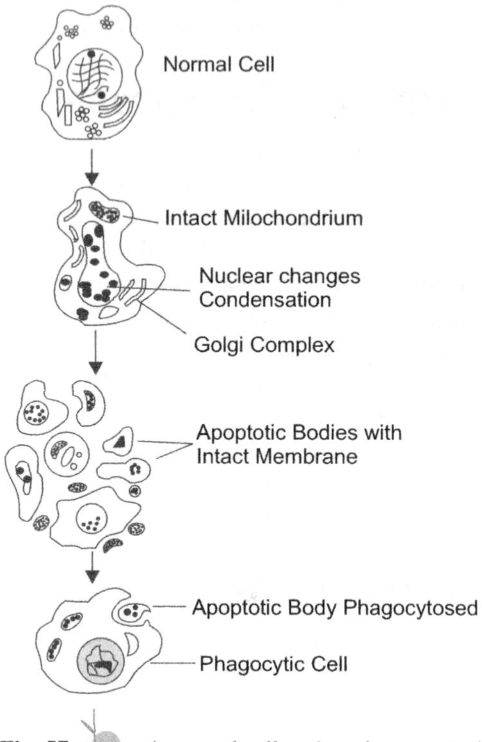

Fig. 57 A normal cell undergoing apoptosis

- Excessive apoptosis may cause atrophy, such as in neurodegenerative diseases, while too less apoptosis results in uncontrolled cell proliferation in cancer.
- Apoptosis is highly related with caspases, Bcl-2 proteins and granzyme B.

1. Apoptosis Triggered by Mitochondrial (Internal Signals) Pathway

- Bcl-2 protein is present on the outer surface of the mitochondria of a healthy cell that prevents apoptosis.
- Internal damage to the cell results in the migration of the Bax protein to the surface of the mitochondrion.
- Here it inhibits the protective effect of Bcl-2 and it also inserts itself in the outer membrane resulting in the formation of a hole causing release of cytochrome c.
- This cytochrome 'c' binds with the protein Apaf-1 (apoptotic protease activating factor-1) resulting in the formation of apoptosomes.
- This binding requires energy which is provided by ATP.
- The apoptosomes binds with caspase-9 and activates it.

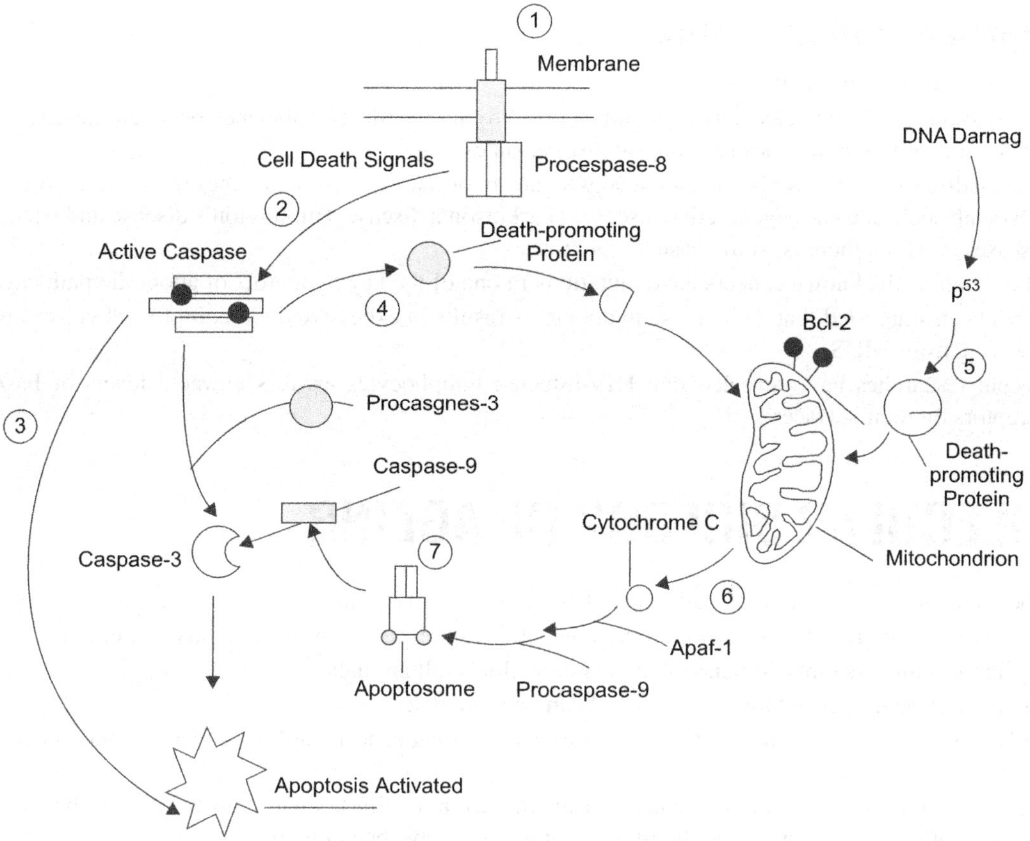

Fig. 58 Events showing apoptosis

- Caspase-9 is cleaved and doing so it activates caspases 3 and 7.
- The activation of these caspases results in the digestion of structural proteins in the cytoplasm as well as degradation of chromosomal DNA and lastly phagocytosis of the cell.

2. **Apoptosis Triggered by External Signals**
   - Fas and TNF receptors are integral membrane proteins.
   - Their receptor domains are exposed to the cell surface.
   - Binding of FasL and TNF results in the transmission of a signal to the cytoplasm that activates caspases.
   - Caspase-8 activates cascade of caspase activation resulting in phagocytosis of the cell.
3. **Apoptosis-Inducing Factor (ANF)**
   - Apoptosis-inducing factor is a protein which is generally found in the membrane space of the mitochondria.
   - When the cell receives a signal relating to death, the apoptosis-inducing factor is released from the mitochondria.
   - It then reaches to nucleus and binds with the DNA
   - Binding results in the fragmentation of the DNA and ultimately death of the cell.

Apoptosis and Diseases

Apoptosis is genetically regulated.

Their exists a balance between cell death and cell proliferation and any imbalance between these two results in abnormal expression of genes resulting in disease states.

- Several disease states have been associated with an abnormal expression of genes involved in programmed cell death such as neurodegenerative diseases (Parkinson's disease, Huntington's disease and Alzheimer's disease) in which there is an increase in apoptosis.
- About half of the known cancers have mutations in one of the key regulators of apoptotic pathway.
- It has been suggested that an increase in apoptosis results in excessive depletion of CD4 cells in patients suffering from AIDS.
- Recent researches have revealed that HIV-infected lymphocytes express elevated levels of Fas/Apo-1 receptors on their surfaces.

MOLECULAR BIOLOGY OF AGEING

- The collection of changes that lead human beings progressively more likely to die is termed as ageing.
- Ageing is a complex multifactorial process, which is influenced both by genes and the environment.
- Ageing is a universal phenomenon that occurs in almost all animals.
- Ageing can be influenced by our genes and can be regulated.
- Old age is a major factor in causing many diseases including cancer, cardiovascular and neurodegenerative diseases.
- The molecular and physiological causes of ageing are not fully known. However, a number of genetic factors that assist in determining the lifespan of animals have been identified.
- Genetics of ageing is the science of heredity for traits related to ageing such as lifespan, age at menopause, age at the onset of specific diseases later in life (e.g., Alzheimer's disease, prostrate cancer, etc.).
- The genetics of ageing is closely related to the biology of lifespan.
- A large number of structural and functional changes occur both at cellular and tissue levels with ageing of human beings and animals. Some of these important changes are as follows:
  (a) There is a progressive increase in oxidative stress.
  (b) Activation of inflammatory mediators.
  (c) Decrease in the number of post mitotic cells.
  (e) Decrease in the levels of hormones.
  (f) Decrease in the levels of certain enzymes and increase in the levels of few enzymes.
  (g) Decrease in the permeability of cell membranes.
  (h) Accumulation of age pigments.
  (i) Decrease in immunocompetence.
  (j) There is significant change in Apolipoprotein D gene expression (Apoplipoprotein D gene is involved in a mechanism that protects cells from ageing).
  (k) DNA methylation (leading to deactivation of gene) and histone acetylation (leading to activation of dormant gene).
  (l) Increase in chromosomal abnormalities and somatic mutations.

(m) Increase in cross linking and tensile strength of collagen.

(n) Increase in common disorders such as cardiovascular disease, cancer, dementia and diabetes.

- According to the genetic theory of ageing, lifespan is mainly determined by the genes we inherit.
- This theory suggests that our potential age is determined at the time of conception.
- The number of ageing genes (gerontogenes) remains to be established. However, there are no doubts about their existence. For example, in humans one of the forms of a gene coding, apolipoprotein E (ApoEe2), is associated with exceptional longevity and decreased susceptibility to Alzheimer's disease.
- Every human cell (except egg and sperm) contains two sets of 23 chromosomes (one set from the mother and the other set from the father). However, the relative proportion of aberrant cells with unusual number of chromosomes increases with age, and the process may cause cancer and other diseases.
- All features of an organism are determined by genes. However, external features depend upon a complex interaction between genes and their environment, which is known as gene-environment interaction. Gene-environment action is important, as gene action may be different in different conditions. It has been observed from the effects of diet restriction on mice and other species that gene-environment interactions can modify lifespan and the rate of ageing to a great extent. The proper understanding of interaction between genes and restricted diet is very important, as calorie restriction is known to be the most effective way for cultivating longevity as well as delaying age-related diseases in mammals.
- Different genes can be activated (switched on) or inactivated (switched off) depending on cell differentiation, which is known as differential gene expression. Gene expression may change overtime, within a given cell, during development and ageing. Differential gene expression is important for differentiation in early childhood, but it may persist further and become a key force in the ageing process.
- Some genes are beneficial and enhance longevity. For example, a gene that assists a person in metabolising cholesterol would reduce the risk of heart disease. However, some genes are harmful, like those that increase the risk of cancer. Some gene mutations are also inheritable and may shorten lifespan. Mutations may also occur after birth due to exposure to toxins, free radicals and radiations that may cause gene changes.
- It has been estimated that genes can explain a maximum of 35 per cent of lifespan, while the other larger percentage can be explained in relation to environmental and hazardous factors.
- People with long-lived parents are more likely to live long and identical twins (have exact the same genes) have a closer lifespan than siblings. Both these lend support to genetic basis of ageing.
- It has been reported that *Caenorhabditis elegan* (a microscopic nematode) with a specific single-gene mutation lived twice as long as members of the species that lacked this mutation (Kenyon et al. 1993). This finding was groundbreaking for a number of reasons. First, it challenged the prevailing concept that ageing occurs as the body deteriorates over time. Second, it led to a shift in thinking even among researchers, who already believed that ageing was subject to some sort of genetic control. Prior to this point, most scientists figured that ageing, age-related illnesses, and death were consequences of multiple cellular and physiological processes and, therefore, under the regulation of a wide and diverse set of genes. However, this finding suggested that a single gene could dramatically regulate how long an organism lived, thus opening the door to new hypothesis about modifying lifespan through genetic manipulation.
- The responsible gene is called *daf-2*. The protein coded for by this gene (designated DAF-2) looked much like the receptor protein within humans that responds to the hormone insulin. In other words, the worm protein is simply a primitive form of our own insulin receptors.
- The gene *daf-2* is now known to regulate a number of factors in addition to ageing (like stress resistance, metabolism and development). This gene and its hormone signaling protein have been conserved evolutionarily and they are found in other animals (from fruit flies to mice). In worms and flies, the gene codes for a receptor protein that is activated by an 'insulin-like' growth factor. This signaling pathway is analogous to the mammalian insulin pathway.

- Scientists have reported gene p^{63} (a sister gene to the cancer-suppressing gene p^{53} that accelerated ageing in adult mice, when it was switched off).
- Degradation of telomeres (pieces of DNA that cap the ends of chromosomes) is one of the possible causes of ageing. As cells divide repeatedly, their telomeres become increasingly shorter. Gradually, they become so short that the cells can no longer divide. Telomeric length is now known as the marker of a given cell's age. As they shorten with each division, they cause the chromosome to fold differently, exposing a different portion of the genome. This is how a person looks different, for example, at 62 years or 2 years of age.

Ageing and the Immune System

- There are two types of immune cells, viz., B cells which produce antibodies and T cells which regulate the activities of other white blood cells.
- As we age, the immune functions, particularly that of T cells tends to decline.
- The ratio of various T cells is altered and the decrease in anti-inflammatory cells results in less regulation of B cells and other white blood cells.
- As a result, the tendency for self-destruction of one's tissues, known as autoimmunity, increases.
- It is estimated that 10 to 15 per cent healthy older individuals have increased levels of antibodies against their own tissues.

The Thymus Gland

- The thymus gland is responsible for producing generic stem cells that can be programmed into different kinds of immune cells.
- The thymus is the most active during childhood and shrinks in size with decreased production of stem cells with ageing.
- There is now a widely accepted connection between ageing and immune dysfunction.
- Besides, there is also clear evidence of information exchange between the central nervous system and the immune system.
- Oxidative stress and its subsequent consequences caused by free radicals are the primary causes of ageing.
- DNA damage and reduced DNA repair capability are some critical signs in considering how to live a longer healthier life.

GENETIC ENGINEERING

Genetic engineering is a highly specialised technique by which a DNA molecule (prokaryotic or eukaryotic) is broken at two desired places to isolate a specific DNA segment and then insert it in another DNA molecule at a desired position. The resultant DNA is called 'recombinant DNA'. Using this technique, isolation and cloning of a single copy of a gene or a DNA segment into an indefinite number of identical copies has become possible. This process has become possible because bacteria, phages and plasmids reproduce in their usual manner even after the insertion of foreign DNA. The inserted DNA also replicates faithfully with the parent DNA. This technique is called 'gene cloning'. Another technique for making millions of copies of a DNA segment of choice has been developed called 'Polymerase Chain Reaction' (PCR). In PCR, a thermo-stable DNA polymerase (*Taq polymerase*) is used. The *Taq polymerase* is a DNA polymerase enzyme that is isolated from the bacterium *Thermus aquaticus* growing in the hot water springs. This enzyme acts best at 72°C and is not denatured at 90°C.

The process of isolation, cloning and transfer of desired gene or genes in the DNA of a desired organism is called 'recombinant DNA technology'. Replication or cloning is possible only when the alien gene becomes part of the chromosome (genome) or extrachromosomal DNA (extragenomic DNA) which has its own origin of replication or '*ori*'. The '*ori*' is present in the plasmid. The plasmids are the extra-chromosomal, self-replicating circular DNA segment present in the bacterial cytoplasm. Cohen and Boyer (1972) first produced a recombinant DNA by attaching an 'antibiotic resistance gene' with the plasmid of the bacterium *Salmonella typhimurium* with the help of *restriction endonuclease* enzyme. The *restriction endonuclease* enzyme cuts the two strands of the DNA double helix at a specific region that has complementary sticky ends. The cut ends of the DNA double helix are joined together by *ligase* enzyme. The recombinant DNA is now transferred into *E. coli* bacterium where it replicates by using the host nucleotides and *DNA polymerase* and forms multiple copies of the DNA. The organism with the recombinant DNA is called 'transgenic organism' or 'Genetically Modified Organism' (GMO).

Tools of Recombinant Technology

1. Source of the donor DNA
2. Restriction enzymes
3. DNA polymerase enzyme
4. Ligase enzyme
5. Vectors
6. Host cell/Organisms

E. coli can protect itself from the bacteriophage attack with the help of a defence mechanism called 'restriction-modification system'. The 'restriction-modification system' has two functional components, viz., (a) restriction enzyme and (b) methylase.

(a) **Restriction Enzyme** – It is a nuclease enzyme which recognises a specific pallindromic DNA sequence and degrades it. It is called restriction enzyme because it restricts the multiplication of bacteriophage (foreign DNA) in the bacterial cell.

(b) **Methylase** – The methylase enzyme adds methyl groups in the pallindromic region present in the bacterial DNA. The methylated bacterial DNA is not damaged by the action of the restriction enzyme.

The restriction enzyme is also called restriction endonuclease (REN). The first discovered restriction endonuclease is *Hind II* (from the bacterium *Haemophilus influenzae* by Smith, 1968) and the other is Ban HI (from the bacterium *Bacillus amyloliquefaciens*). The restriction endonuclease enzyme is of two types, viz., (a) exonucleases and (b) endonucleases. The exonucleases cut the DNA/RNA molecules from the ends, whereas the endonucleases cut the DNA molecule at intercalary positions. The cut made by restriction endonucleases at specific places are called recognition sequences. The restriction endonuclease first scans the DNA, finds out the recognition sequences, attaches itself with it and then cuts both the strands of the DNA double helix at specific places through the sugar-phosphate backbone in the DNA double helix. The recognition sequence is always a pallindromic sequence in the DNA double helix. The restriction endonuclease enzyme gets attached with the pallindromic sequence and cuts the hydrogen bonds between the two polynucleotide chains of the DNA double helix. Hence, it is also called molecular scissor or chemical scalpel. The cut ends of the DNA double helix are called staggered ends, sticky ends or cohesive ends. When the DNA from different sources are cut by the same restriction endonuclease, complementary sticky ends are formed which can be joined with the DNA ligase enzyme to form a recombinant DNA. The other group of endonuclease cuts the two strands of the DNA double helix at the same place in such a way that the single-stranded DNA polynucleotide chain is not formed at the end. The ends of the DNA double helix without single-stranded polynucleotide sequence are called 'blunt ends'. The endonuclease of this class is Sma I (from *Serratia marcescens*) and Sca I.

Seperation and Isolation of DNA Fragments

The DNA is treated with restriction endonuclease to cut into small DNA fragments of different lengths. These DNA fragments are separated by Gel electrophoresis. Gel electrophoresis is a technique of separating biomolecules under an electric field by keeping it on 'agarose' (a natural polymer extracted from sea weeds *Gracillaria* and *Gellidium*) gel matrix. The DNA fragments are placed at the cathode end in the wells. When the electric field is activated, the negatively charged DNA fragments move towards the positively charged anode. The DNA fragments separate (resolve) according to size due to sieving effect of the agarose gel. The larger DNA fragments move slowly and the smaller fragments quickly towards the anode end. The separated DNA fragments are stained with ethidium bromide and then exposed to UV radiation; bright orange-coloured bands of DNA fragments are seen. The separated DNA bands are cut out and extracted from the agarose gel. This step is called 'elution'. The purified DNA fragments are used in constructing recombinant DNAs by attaching them to the cloning vectors.

Cloning Vectors

Cloning vectors are the carriers or vehicles of desired DNA fragments (passenger DNA) which have the ability of independent DNA replication, to increase the number of desired genes, (e.g., plasmids, bacteriophages). Plasmids and bacteriophages have the ability to replicate inside the bacterial cells, independent of any bacterial genomic DNA control. The bacteriophages are present in a very high number in the bacterial cell; therefore, their genomic copy number in these bacterial cells is very high. The number of plasmid in a bacterial cell is generally 1 or 2 but it may be 15 or 100 or even higher sometimes. When an alien piece of DNA is combined with the bacteriophage or plasmid DNA, the alien DNA multiplies its number equal to the copy number of the bacteriophage or plasmid. Cloning vectors are artificially designed and synthesised in such a way that they become easily linked with the alien DNA and selection of recombinants from nonrecombinants.

Features in Cloning Vectors

A cloning vector must have the following features:

1. Origin of replication (*ori*)
2. Selectable markers (for identifying recombinants from nonrecombinants)
3. A single recognition site for cloning

Types of Cloning Vectors

Cloning vectors are of the following four types:

1. Plasmids
2. Viruses
3. Cosmids
4. Artificial chromosomes
   - Plasmid vectors for plants are different from others.
   - The best-known and commercially available vectors are pBR-322 and pUC-18. Both of these plasmids are modified natural plasmids of *E. coli*. The pUC series of plasmid vectors are named so because they were initially developed in the University of California. The plasmid pUC are derived from pBR-322. The plasmids are about 2,700 bp (base pair) long.

Amplification of Gene of Interest using PCR (Polymerase Chain Reaction)

Mullis (1985) developed the technique of Polymerase Chain Reaction (PCR). PCR is the invitro synthesis of multiple copies of a gene or desired DNA segment. In PCR, two sets of small, chemically synthesised oligonucleotide primers, which are complimentary to the DNA part and a thermo-stable DNA polymerase (Taq polymerase), are required. *Taq polymerase* is obtained from the hot spring bacterium *Thermus aquaticus*. PCR has the following three steps:

1. Denaturation
2. Annealing
3. Extension

 1. **Denaturation** – The target DNA double helix is heated at about 94°C, which causes separation of the two polynucleotide chains of the DNA double helix. The separated DNA helices act as templates.
 2. **Annealing** – The pairing or hybridisation of oligonucleotide primers with each of the single-stranded DNA templates at its 3′ end is called annealing. Annealing is done at low temperatures (at 72°C) with the addition of *Taq polymerase*, deoxynucleoside triphosphates and Mg^{++} ions.
 3. **Extension** – The synthesis of new DNA polynucleotide chains, complementary to the template DNA strands, with the help of *Taq polymerase* is called extension. This process is repeated and after every time the amount of DNA is doubled. The amplified DNA segment is ligated with the vector to form recombinant DNA (rDNA).

Bioreactors

Bioreactors are large cylindrical metal containers, of 100 to 1,000 litres capacity, with a curved base for better mixing of contents. In bioreactors, the raw materials are biologically converted into specific products, individual enzymes, etc., using microbial plant, animal and human cells. In bioreactors, the optimum growth and development conditions (temperature, pH, substrate concentration, salts, vitamins, water content and oxygen) are provided to obtain the desired product.

Downstream Processing

The process of separation and purification of useful product from the fully grown genetically modified cells is called downstream processing. The product is formulated with suitable preservatives. The formulation undergoes clinical trials to know its uses and any immediate or long-term adverse effect like drugs. The formulated product also undergoes strict quality control testing.

Biotechnology and Its Application

Plants, animals, bacteria and fungi with genetically modified genes or DNA are called genetically modified organisms.

In Agriculture

Bt Cotton

Some strains of the bacterium *Bacillus thuringiensis* produce protein crystals during a particular stage of their growth, which is toxic to certain insects. This Bt toxin protein is inactive protoxin. The inactive protoxin is converted into an active form of toxin in the alkaline pH of the cotton bollworm (caterpillar) gut. The active toxin binds to the surface of midgut epithelium and forms pores, which leads to cell swelling and lysis and finally death of the insect. From *Bacillus thuringiensis* two *cry* genes, *cryIAc* and *cryIIAb*, have been incorporated in the cotton. This genetically modified cotton plant is named Bt cotton. Bt cotton plants are resistant to the cotton bollworm due to the presence of two *cry* genes that produce toxic cry proteins. The *cry* gene *cryIAb* has been incorporated into the Bt corn plant to protect it from corn borer.

Pest Resistant Plants

Nematodes are common parasites, which infect a number of plants, animals and human beings. A nematode *Meloidegyne incognitia* causes root-knot disease in brinjals, tomatoes and tobacco. The parasitic nematode drain a lot of nutrients from the host plant, and cracks develop in the roots which lead to secondary infections. Water and mineral absorption are impaired which leads to great reduction in crop yield. The process of RNA interference (RNAi) introduced resistance in these plants. RNA interference (RNAi) is a technique of inhibiting the activity of a gene through production of sense and antisense RNAs. The two RNAs produce a double-stranded RNA (dsRNA) which undergoes fragmentation and forms short segments called short interference RNA (siRNA). The short interference RNA (siRNA) binds with the mRNA, and forms silencing complex, which cleaves the mRNA having complementary base sequence and makes the mRNA silent by stopping translation. The source of this complementary RNA may be an infection by viruses having RNA genome or mobile genetic elements (transposons) which replicate through an RNA intermediate.

Hirudin

Hirudin is a protein that prevents blood clotting. Natural source of Hirudin protein is cattle leech (*Hirudinaria)*. Gene coding for Hirudin protein was artificially synthesised and introduced in the *Brassica napus* (Rapeseed) plant. The transgenic *Brassica napus* (Rapeseed) plant synthesises the Hirudin protein in its cells and the Hirudin accumulate in the seeds in larger quantities. The Hirudin from the seed is extracted, purified and used as medicine (as an anticoagulant in surgery and dialysis). The transgenic product Hirudin is of great interest.

Genetically Engineered Insulin or Humulin

Insulin is a hormone secreted by pancreas. It is essential for maintaining the glucose–glycogen balance in the human blood. The reduced secretion of insulin leads to an increase in the glucose level in the blood because the cells fail to take up glucose from the blood. The resultant disease is called diabetes mellitus-type II. The disease can be treated by regular and small injection dose of insulin. The insulin used for diabetes was extracted from the pancreas of slaughtered cattle and pigs. This animal source insulin (protein) caused allergy or some other type of reactions in some diabetic patients.

Structurally the insulin, a protein hormone, is made up of two short polypeptide chains, viz., chain A and chain B, which are linked by disulphide bonds. In mammals, including human beings, the insulin hormone is synthesised as a pro-hormone, which is processed to become fully mature and functional hormone after releasing the extra polypeptide chain called C peptide. The C peptide is not present in the mature insulin. In 1983, a US company Eli Lilly, first prepared the two separate DNA sequences corresponding to polypeptide chain-A and chain-B of human insulin by recombinant DNA (rDNA) technology and introduced it in the plasmids of *E. coli* to synthesise insulin. Chain A and chain B were synthesised separately, extracted and joined by creating disulphide bond to form human insulin.

Biopatent

Biopatent is an official right granted by a government to an inventor or agency or an organisation. It is the sole right for commercial production, marketing and selling a particular biological material for a certain period. This right prevents the others from commercial use of the invention or product prepared by a patent holder. Biopatent is done for the commercial exploitation of the following:

(a) Strains of microorganisms
(b) Cell lines
(c) Genetically modified plants and animals
(d) DNA sequences
(e) Protein encoded by DNA sequences
(f) Biotechnological methods
(g) Products and their applications

Biopiracy

Patenting and exploitation of biological resources by other countries, multinational companies and other organisations without proper authorisation, benefit sharing agreement or compensatory payment to the countries and people concerned.

GENETICS OF BACTERIOPHAGES

- Bacteriophages are viruses that infect bacteria.
- They are also known as bacterial viruses or phages.
- The phages possess dsDNA, ssDNA, dsRNA or ssRNA as genetic material.
- Bacteriophages were discovered by F W Twort (1915) and de Herelle (1917).
- The term 'bacteriophage' was given by de Herelle.
- Bradley (1967) categorised Bacteriophages on the basis of their morphological structure.
- Each kind of bacteriophage infects a particular species of bacteria.
- Bacteriophages are the most thoroughly and well-understood viruses.
- They are the most commonly found organisms in the biosphere and they carry out similar biological functions as other viruses but they do not target human cells for infection.
- Bacteriophages have proven to be a valuable scientific research tool for a variety of applications such as models for study of viral infections mechanisms, as tools of biotechnology that introduce new genes into bacterial cells, etc.
- Three common forms of bacteriophages are known; these are tailed, cubic and filamentous.
- The tailed bacteriophages form the largest group and have been studied in detail.

Structure of Bacteriophage

- A typical bacteriophage (T_4) consists of three distinct parts, viz., head, neck and tail.
  1. **Head**
     - Head is polyhedral in shape. In many cases, it is icosahedral.
     - Head is covered with a protein covering called capsid.

- The capsid contains the nucleic acid.
- The nucleic acid may be single stranded or double stranded, linear or circular in most phages.
- Some phages may have linear RNA with one or more proteins.
- φ6 bacteriophage contains three double-stranded linear RNA molecules with variable base sequences.
- The protein coat protects nucleic acid from nucleases during phage infection.
- In the simplest phages, nucleic acid code for 3–5 gene products while in complex phage it may code for 100 gene products.

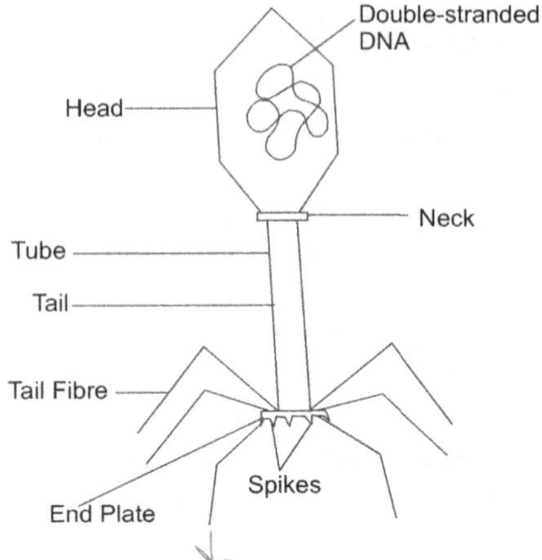

Fig. 59 A bacteriophage

2. **Neck**
   - Neck of bacteriophages is very short and has a collar.
   - Neck connects head with tail.
3. **Tail**
   - Tail is a complex multicomponent structure.
   - Tail contains a central hollow core through which nucleic acid passes into the host cell during infection.
   - Tail is surrounded by a contractile sheath.
   - The sheath has the ability to contract longitudinally.
   - The tail end of a bacteriophage is attached to the end plate, which is hexagonal.
   - The endplate terminates into tail fibres.
   - The tail fibres are involved in the attachment of the phages to the bacterial cell.
   - The process by which the tail attaches to the surface of bacteria is called adsorption.
   - Some phages lack endplate and tail fibres.
   - On the basis of their interaction with the host bacterium, the phages are divided into two groups, viz., virulent and temperate.
   - **Virulent phages** are those that normally lyse (destroy) the attached host/bacterial cells; there is probably no alternative strategy for their multiplication. This characteristic growth of virulent bacteriophages is called **'lytic cycle'** (e.g., T-even phages).
   - **Temperate phages** adopt two alternative modes for their multiplication: (a) They may enter a lytic cycle and behave like a virulent phage or (b) they may integrate themselves into the bacterial chromosome, thus resulting in an **iysogenic cycle;** in the integrated (Iysogenic) state the phage is described as a **prophage** (e.g., lambda phage).

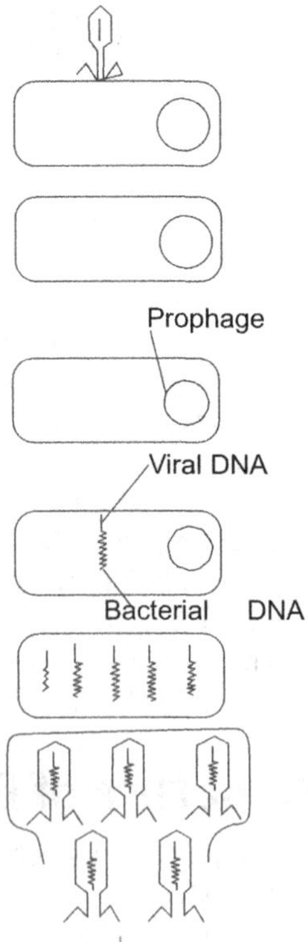

Fig. 60 Lytic cycle

Life Cycle

Bacteriophages exhibit two types of life cycles, viz., lytic cycle and lysogenic cycle.

- All the T-series bacteriophages exhibit lytic cycle. These bacteriophages infect bacteria and after completion of the life cycle, they cause rupture of the bacterial cell. Therefore, life cycle is known as lytic cycle. Lytic cycle involves the following steps:

- When a bacteriophage infects a bacterium, it gets attached with the bacterial cell by its tail.
- The tail secretes proteolytic enzymes that make a hole in the bacterial cell.
- The DNA of the head is injected into the bacterial cell.
- The protein coat remains outside the cell.
- The phage DNA inside the bacterial cell start replicating by utilising precursor substances from the host cytoplasm to form its own progeny (such a phage is called virulent phage) or becomes integrated into the host chromosome as prophage and replicate in synchrony with the bacterial chromosome (such a phage is called nonvirulent or temperate phage).
- Each copy of the DNA is enclosed by a protein coat, producing a new bacteriophage.
- The host cell containing the virulent phage soon bursts, releasing hundreds of new viral particles.
- The new viral particles may repeat the same cycle.
- The virulent phage thus destroys the host cell.
- The temperate phage does not destroy the bacterial cell.
- Rarely, a bacterium containing a temperate phage will lyse liberating many infectious viral progeny.
- As the host cells are potentially subject to lysis, they are called lysogenic.
- A virulent phage may remain in quiescent for a long time but may convert into reproductive virulent phage by some radiation sources (such as ultra violet or X-rays) and disrupt the bacterial cell.
- Virulent phages cannot become prophages. They are always lytic.
- Temperate phages can integrate into bacterial chromosome as prophages, allowing the host cells to survive in the lysogenic state.
- Prophages occasionally exit the bacterial chromosome and enter the lytic cycle.

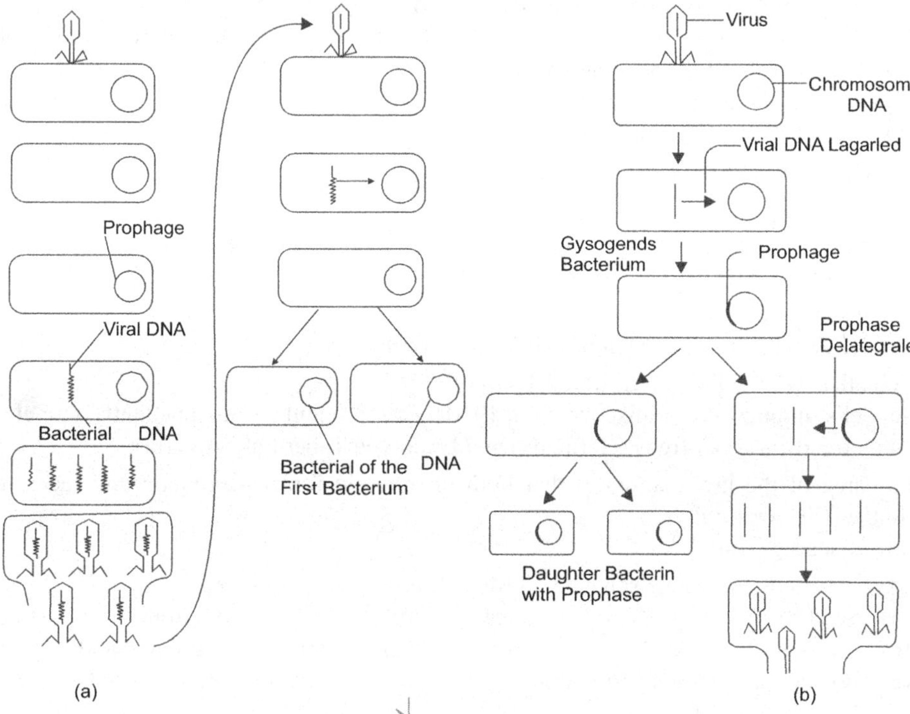

Fig. 61 Lysogenic cycle

GENETICS

MENDELISM

Short-Answer Questions

1. What is genetics?
   Answer: Genetics is the study of heredity and variations.
2. Who coined the term 'genetics'?
   Answer: W Bateson (1905).
3. What are the applications of genetic studies?
   Answer: Genetic studies have many applications in agriculture and medicine. For example, gene therapy is used to replace a defective gene with a normal gene in treating some diseases. Besides, knowledge of genetics is also helpful in knowing the mechanism of evolution.
4. Who first obtained fruitful results on the inheritance of biological traits?
   Answer: Gregor Johann Mendel was the first to obtain fruitful results on the inheritance of biological traits. He is regarded as the 'Father of Genetics'.
5. Who is called the 'Father of Modern Genetics'?
   Answer: W Bateson
6. Who is known as the 'Father of Experimental Genetics'?
   Answer: T H Morgan
7. Name the scientist associated with the paper, *Experiments in Plant Hybridisation*.
   Answer: Gregor Johann Mendel
8. On which plant did Mendel work?
   Answer: Mendel worked on the garden pea (*Pisum sativum*) plant.
9. Name the scientists who rediscovered Mendel's findings.
   Answer: Mendel's findings were rediscovered in 1900 by three scientists independently, namely, Hugo de Vries (Netherland), Karl Correns (Germany) and Erich von Tshermak (Austria).
10. What is the name of the law that states that both phenotypic forms (dominant and recessive) are expressed in the F_2 generation?
    Answer: Law of segregation
11. Name the seven pairs of characters selected by Mendel for his experiments.
    Answer:
    (a) Seed shape – Smooth and wrinkled
    (b) Seed colour – Yellow and green
    (c) Flower colour – Violet, red and green
    (d) Pod shape – Inflated and constricted
    (e) Pod colour – Green and yellow
    (f) Flower position – Axial and terminal
    (g) Stem height – Tall and dwarf
12. What were the reasons for Mendel's success?

Answer: (a) He confined his experiments to one or a few pairs of contrasting traits in each experiment.
(b) He kept accurate quantitative data, which is essential for genetic experiments.
(c) He kept track of each character separately.
(d) The unit factors in pairs for the seven characters were located on different homologous pairs in the pea plant used by him.

13. Which law of Mendel is also known as purity of gametes?
Answer: Law of segregation

14. What are genotypic and phenotypic ratios of monohybrid and dihybrid crosses?
Answer: The genotypic and phenotypic ratios of monohybrid crosses are 1:2:1 and 3:1 and those of dihybrid cross are 1:2:1:2:4:2:1:2:1 and 9:3:3:1, respectively.

15. In which type of cross, would the progeny show 64 combinations?
Answer: Trihybrid cross

16. What is incomplete dominance?
Answer: Incomplete dominance is the expression of traits in an intermediate form of two pure parents.

17. Define co-dominance.
Answer: When both alleles express themselves together, it is called co-dominance.

18. What is pure line?
Answer: A population that breeds true for a particular trait is known as pure line.

19. Who coined the term 'pure lines'?
Answer: Johannson (1903).

20. For what purpose are pure lines used?
Answer: Pure lines are used for cross-breeding in order to get the desired improvements in crops.

21. Who coined the terms F_1 and F_2 generations?
Answer: Bateson and Saunders (1905) coined these terms.

22. Who developed the Punnet square?
Answer: R C Punnet

23. How many gametes are produced in F_1 generation of a trihybrid cross?
Answer: Eight

24. In which condition is the Mendelian dihybrid ratio of 9:3:3:1 obtained?
Answer: Mendelian dihybrid ratio of 9:3:3:1 is obtained only when the alleles at both loci display dominant and recessive relationship.

25. In what circumstances does the Mendelian dihybrid ratio of 9:3:3:1 become modified?
Answer: Mendelian dihybrid ratio of 9:3:3:1 becomes modified if one or both gene loci have incompletely dominant allele or co-dominant alleles or lethal alleles.

26. Mendel could not recognise linkage in his experiments.
Answer: Because characters selected by him were located on different chromosomes.

27. Define genotype and phenotype.
Answer: The genetic make-up of an individual is called genotype and the physical appearance of a trait is called phenotype.

28. What are contrasting traits?
Answer: The alternative forms of a character are called contrasting traits.

29. Define alleles.
Answer: The two alternative forms of a gene are called alleles.

30. What is gene locus?
Answer: The location of a particular gene, on a given chromosome is known as gene locus.

31. What is syntenic?
Answer: Genes located on the same chromosomes are called syntenic.

32. Name the genes in which Mendel's law of independent assortment is not applicable.
Answer: Linked genes

33. What is atavism (reversion)?
Answer: Sudden appearance of ancestral character is known as atavism.

34. Define pleiotropism.
Answer: A gene having multiple effects is known as a pleiotropic gene and this phenomenon is known as pleiotropism.

35. Define polygenic traits.
Answer: Polygenic traits are those traits which are dominated by two or more than two genes and the inheritance of polygenic traits is called quantitative inheritance.

36. Give two examples of polygenic inheritance.
Answer: Kernel colour in wheat and human skin colour are two important examples of polygenic inheritance.

37. Name the gene responsible for albinism in maize.
Answer: Lethal gene

38. What do the following symbols mean: XX, Xx, xx?
Answer: (a) XX – Homozygous dominant (b) Xx – Heterozygous dominant
(c) xx – Recessive

39. What is test cross?
Answer: When F_1 individuals are crossed with a recessive parent, it is called test cross. For monohybrid cross, the test cross ratio is 1:1 and for dihybrid cross, it is 1:1:1:1.

40. Name the genes that produce the following ratios:
Answer: (a) 9:7 – Complementary genes (b) 9:3:4 – Supplementary genes
(c) 12:3:1 – Dominant epistasis genes (d) 9:3:4 – Recessive epistasis
(e) 15:1 – Duplicate gene

41. What is lethal factor?
Answer: The gene that causes the death of the individual carrying it is known as lethal factor.

42. Give two examples of lethal genes.
Answer: Yellow fur in mice (absolute lethality) and sickle cell anaemia (sublethality).

43. Define epistasis.
Answer: The phenomenon in which one gene masks the effect of the other gene, which are non-allelic, is known as epistasis. The gene that masks the effect of the other gene is known as epistatic gene.

44. Differentiate between dominance and epistasis.
Answer: In dominance, genes are allelic, whereas in epistasis, genes are non-allelic.

45. What is overdominance?
Answer: In certain cases, the phenotype of F_1 offspring is more pronounced than either of the parents; it is termed as overdominance.

46. What are modifying genes?
Answer: Genes which modify the major phenotype of a gene are called modifying genes.

47. What is the nature of lethal genes?
Answer: Lethal genes may be dominant, incompletely dominant or recessive.

48. In which circumstances, the Mendelian dihybrid ratio of 9:3:3:1 becomes modified to 3:6:3:1:2:1?
Answer: It occurs when the dihybrid parents have dominant and recessive alleles at one gene locus and co-dominant alleles at second gene locus.

49. What is hybridisation?
Answer: Crossing of two genetically different individuals is called hybridisation.

Long-Answer Questions

1. Describe Mendel's law of inheritance with the help of suitable examples.
2. Why did Mendel select the pea plant for his experiment? Mention the characters of the pea plant chosen by him for his experiment.
3. Distinguish between:
   (a) Genotype and phenotype
   (b) Homozygous and heterozygous
   (c) Dominant and recessive
   (d) Backcross and test cross
   (e) Monohybrid cross and dihybrid cross
   (f) Incomplete dominance and co-dominance
4. Write short notes on:
   (a) Reasons for Mendel's success
   (b) Law of independent assortment
   (c) Complementary genes
   (d) Epistasis
   (e) Lethal genes
   (f) Polygenic traits
5. Write an essay on Mendelism.

MULTIPLE ALLELES

Short-Answer Questions

1. Define multiple alleles.
Answer: When a set of more than two alleles, occupying the same locus in the homologous chromosomes, affects the expression of the phenotypic character, they are called multiple alleles and the phenomenon is called multiple allelism.

2. In which organism was the first case of multiple allele discovered?
Answer: Drosophila

3. Give two examples of multiple alleles in humans.
Answer: ABO blood groups and Human Leucocyte-associated Antigen (HLA) genes

4. How many alleles control ABO system in humans?
Answer: ABO system in humans is controlled by three alleles.

5. How many genotypes are possible in ABO blood groups?
Answer: Six
6. Which allele in ABO system does not code for any enzyme?
Answer: O allele
7. Individuals of which blood group have predominantly IgG antibody?
Answer: Blood group O
8. Which blood group is an example of co-dominance?
Answer: AB blood group
9. Ram is of blood group B and his mother is of blood group O. What is the blood group of his father?
Answer: Probable blood group of the father is either B or AB.

Long-Answer Question

1. What are multiple alleles? Describe multiple alleles with the help of ABO blood groups.

LINKAGE

Short-Answer Questions

1. What is linkage?
Answer: Linkage is the tendency of two or more genes which are located on the same chromosome to remain together from generation to generation.
2. Name the two kinds of linkages.
Answer: Complete linkage and incomplete linkage
3. Define cis arrangement and trans arrangement.
Answer: In cis arrangement, the dominant linked genes are located on one chromosome and their alleles on its homologue. But in trans arrangement, the dominant and their recessive genes are located on the same chromosome and their alleles on the homologous chromosome.
4. What are linked genes?
Answer: Genes located on the same chromosome and are fairly close to each other are called linked genes.
5. Who gave the coupling and repulsion hypothesis?
Answer: Bateson and Punnet (1905)
6. The test cross of offsprings of a dihybrid cross manifests a ratio of 1:1:1:1. State whether the genes are linked or not linked?
Answer: The genes are not linked.
7. Corn (*Zea mays*) has 10 pairs of chromosomes. How many linkages groups are in it?

Answer: 10 linkage groups

8. Define coupling and repulsion.

   Answer: Alleles (dominant or recessive) which come from the same parents have a tendency to enter the same gamete. This is known as coupling. Likewise, alleles (dominant or recessive) which come from different parents have a tendency to enter different gametes, known as repulsion.

9. Name the scientists who observed complete linkage in *Drosophila.*

   Answer: T H Morgan (1919).

10. Why Mendel missed linkage?

    Answer: Because the seven pairs of characters of the pea plant selected by him were located on different (nonhomologous) chromosomes or were far apart on the same chromosome due to which they were easily separated by crossing over.

11. Who described incomplete linkage in maize?

    Answer: Hutchinson

12. Differentiate between linkage and independent assortment.

    Answer: Linkage occurs in the genes which are located on the same chromosome (linked genes), whereas independent assortment occurs in genes located on different chromosomes.

13. Give three differences between linkage and crossing over.

    Answer:

	Linkage	**Crossing over**
(a)	Linkage keeps the genes of a chromosome together.	Crossing over separates genes of a chromosome forming new combinations.
(b)	It keeps parental characters together in offsprings.	Crossing over causes change in the parental characters in offsprings.
(c)	The strength of linkage decreases with increase in distance between the genes.	The frequency of crossing over increases with increase in the distance between genes.
(d)	Linkage prevents formation of new species.	Crossing over results in the formation of new characters and thus helps in the formation of new species.

14. Name the factors that affect the strength of linkage.

    Answer: (a) Age (b) Temperature (c) X-rays

Long-Answer Questions

1. What is linkage? Describe different types of linkage with suitable examples.
2. Write short notes on:
   (a) Complete linkage
   (b) Coupling and repulsion hypothesis
   (c) Linkage groups
   (d) Chromosomal theory of linkage

CROSSING OVER

Short-Answer Questions

1. Define crossing over.
   Answer: The process of exchange of segments between homologous chromosomes is known as crossing over.
2. In which stage of meiosis does crossing over takes place?
   Answer: Pachytene
3. Who gave the first cytological proof of genetic crossing over?
   Answer: The first cytological proof of genetic crossing over was given by Stern in *Drosophila* and Creighton and Mc Clintock in maize (1931).
4. What is the use of genetic map?
   Answer: (a) Genetic map helps in determining location, arrangement and linkage of genes on a chromosome.
   (b) It helps in predicting the results of dihybrid and trihybrid crosses.
5. Give one example of an organism in which crossing over does not occur.
   Answer: Male *Drosophila*
6. What is the name given to sites where crossing over takes place?
   Answer: Chiasmata
7. What is crossover value?
   Answer: The frequency of recombination is termed crossover value.
8. What is interference?
   Answer: The phenomenon in which one crossover reduces the chances of crossing over near it is termed interference.
9. Define single crossing over, double crossing over and multiple crossing over.
   Answer: (a) Single crossing over – When crossing over takes place at only one point (single chiasma)
   (b) Double crossing over – When crossing over takes place at two points (two chiasmata)
   (c) Multiple crossing over – When crossing over takes place at three or more sites (multiple chiasmata)
10. On which factor does the crossover percentage depend?
    Answer: Crossover percentage depends on the relative distance between genes on the chromosomes. Greater the distance, higher is the amount of crossing over.
11. Name the factors that influence the frequency of crossing over.
    Answer: (a) Distance (b) Age (c) Temperature
    (d) X-rays (e) Location of genes
12. Give the maximum frequency of a recombination at two loci.
    Answer: 50 per cent
13. Define coefficient of coincidence.

Answer: The ratio between observed and expected double crossovers is known as the coefficient of coincidence.

14. What is the unit of measurement for genetic linkage?
Answer: Map unit or centiMorgan (cM)

15. State whether the location of genes near a centromere will enhance or reduce the frequency of crossing over?
Answer: Location of genes near a centromere reduces the frequency of crossing over.

16. What is a genetic map?
Answer: A genetic map is a diagrammatic representation of the location and arrangements of genes as well as the relative distance between linked genes on a chromosome.

17. What is the significance of crossing over?
Answer: (a) It confirms the linear arrangement of genes in chromosomes.
(b) It produces a new combination of genes.
(c) It causes appearance of new characters.
(d) It causes variation among offsprings of the same parents and among the same species.
(e) The process is applied for improvement of economically important varieties of plants and animals.

Long-Answer Questions

1. Define crossing over. Describe the process of crossing over.
2. What is crossing over? Describe relationship between crossing over and chiasma formation. Add a note on significance of crossing over.
3. Write short notes on:
(a) Genetic map
(b) Factors affecting crossing over
(c) Three-point test cross
(d) Interference and coincidence
(e) Somatic crossing over

SEX DETERMINATION

Short-Answer Questions

1. Who discovered the role of chromosomes in sex determination?
Answer: Clarence Erwin Mc Clung

2. Explain sex determination in human beings.
Answer: (a) In human beings, there are 23 pairs of chromosomes.
(b) Of these 23 pairs, 22 pairs are autosomes and the remaining one pair forms sex chromosomes.
(c) The sex chromosomes are X and Y chromosomes that determine the sex.

(d) Males have one X and one Y chromosome, whereas females have two X chromosomes.
(e) Males are heterogametic and produce two types of sperm in equal ratio (50 per cent X and 50 per cent Y), whereas females are homogametic and produce only one type of gamete (X).
(f) If the sperm carrying X chromosome fertilises the ovum, the resulting foetus will be XX that is female.
(g) If the sperm carrying Y chromosome fertilises the ovum, the resulting foetus will be XY, which is male.
(h) Thus in human beings, the male is responsible for determination of sex of the child.

3. What will be the sex of zygotes of *Drosophila* having the following chromosomal complements: 2A+XXX, 3A+ X, 3A+ XX, 2A+ XY and 2A+XX?
Answer: Superfemale, supermale, intersex, diploid male and diploid female

4. What is a freemartin?
Answer: In cattle, whenever twins of the opposite sex are born, the male is perfectly normal while the female becomes sterile. Such a sterile female is called a freemartin and this phenomenon is known as freemartinism.

5. Which mammal has 10 sex chromosomes present but no mammalian sex-determining gene SRY?
Answer: Duck-billed platypus

6. Give examples of animals in which the sex is determined by temperature at which the egg is incubated.
Answer: Alligators, some turtles and tuatara (*Sphenodon*)

7. Give examples of animals in which XX–XO type sex determination is seen.
Answer: Bugs, grasshoppers and cockroaches

8. Which gene codes for the testes-determining factor?
Answer: SRY gene, which is located on the Y chromosome

9. What is Lyon hypothesis?
Answer: Lyon hypothesis states that one X chromosome in each cell becomes inactivated during early development. The inactivation is random. This results in genetic mosaics.

10. What is Barr body?
Answer: Barr body is a darkly staining body present in the nucleus of cells of females. The number of Barr body is one less than the number of X chromosomes.

11. What is the number of Barr body in Klinefelter syndrome?
Answer: One

12. Give some examples of fish that are confirmed to engage in sex changes throughout their life cycle.
Answer: Blue-headed wrasse, blue-banded gobies, Japanese gobies, angelfish, parrotfish, etc.

13. Give one example in which both sexes are chromosomally same and sex is determined by the temperature at which embryos are allowed to develop.
Answer: Alligators

14. Who coined the term 'dosage compensation'?
Answer: H J Muller (1932)

15. How is dosage compensation of genes done?
Answer: The dosage compensation of genes is done either by hypo-production due to inactivation of one X chromosome in homogametic female (mammals) or by hyperproduction due to hyperactivity of X chromosome in heterogametic male sex (*Drosophila*).

16. Name the animal in which the male with one sex chromosome (XO) and a pair of chromosomes (XX) is hermaphrodite.
Answer: *C. elegans*

Long-Answer Questions

1. Give an account of various mechanisms of sex determination in animals.
2. Describe sex-determining mechanism in:
(a) Humans (b) Grasshoppers (c) Honeybees
3. Write short notes on:
(a) Genic balance theory (b) Freemartins (c) Gynandromorphs
(d) Human sex abnormalities (e) Barr body

SEX-LINKED INHERITANCE

Short-Answer Questions

1. Who discovered sex-linked inheritance?
Answer: T H Morgan (1910)
2. What is sex-linked inheritance?
Answer: Genes located on the sex chromosomes are called sex-linked genes and their mode of inheritance is called sex-linked inheritance.
3. Give two examples of sex-linked diseases.
Answer: Haemophilia and colour-blindness
4. What are hologynic and holandric genes?
Answer: Characters passing directly from the mother to the daughter are known as hologynic, whereas characters passing directly from the father to the son are called holandric.
5. Give one example of X-linked dominant inheritance.
Answer: Xg blood group
6. Can a son with normal vision have a colour-blind mother?
Answer: No
7. What are digenic genes?
Answer: Digenic genes are located on the X chromosomes and are inherited from male to male through a female.
8. What are sex-limited genes?
Answer: Those autosomal genes whose phenotypic expressions depend on the presence or absence of sex hormones are called sex-limited genes (e. g., beard development in men).

9. Which type of haemophilia is caused by the deficiency of thromboplastin in blood plasma?
Answer: Haemophilia B

10. What is criss-cross inheritance?
Answer: The inheritance in which a father transmits his sex-linked characters to his grandson through his daughter is known as criss-cross inheritance.

Long-Answer Questions

1. What is sex-linked inheritance? Explain the phenomenon with suitable examples.
2. What is criss-cross inheritance? Explain the phenomenon of criss-cross inheritance with suitable examples.
3. Write short notes on:
(a) Colour-blindness
(b) Sex-limited genes
(c) Sex-influenced genes
(d) Inheritance of Y-linked genes

CHROMOSOMAL ABERRATIONS

Short-Answer Questions

1. What is chromosomal aberration?
Answer: Any abnormality in the structure or number of chromosomes is known as chromosomal aberration.

2. What are the four structural abnormalities in chromosomes?
Answer: (a) Deletion (b) Duplication (c) Inversion (d) Translocation

3. How are duplications recognised genetically?
Answer: Duplications are recognised genetically by
(a) Lack of revertability (b) Pseudodominance (c) Recessive lethality
Cytologically, duplications are recognised by deletion loop.

4. Which is the most common type of translocation?
Answer: Reciprocal translocation

5. Name the structural abnormalities in the chromosome that generate semisterility.
Answer: (a) Deletions (b) Inversions (c) Translocations

6. Name the syndrome which is caused due to deletion arm of short chromosome 5 in humans.
Answer: Cri-du-Chat syndrome (Cat-cry syndrome)

7. How can structural changes in chromosome be identified?
Answer: By chromosome banding and in situ hybridisation (ISH)

8. Name the abnormal chromosomes that form a bridge at anaphase.
Answer: Chromosomes having two chromosomes (dicentric)
9. Which disorder involves 9–22 translocation in a chromosome?
Answer: Chronic mylegenous leukaemia (CML)
10. What is the medical significance of inversion?
Answer: It results in the increased risk of generating unbalanced gametes.
11. How does a ring chromosome arise?
Answer: A ring chromosome arises from a break in both arms of a chromosome; the terminal ends are lost and the two sticky ends reunite to form a ring.
12. Who first observed translocation?
Answer: Hugo de Vries in the evening primrose (*Oenothera*)
13. What is unbalanced translocation?
Answer: Translocation that involves loss of DNA from a cell is known as unbalanced translocation.
14. What are autopolyploids?
Answer: Polyploids having multiplications of basic set of chromosomes are known as autopolyploids.
15. What is allopolyploidy?
Answer: The doubling of chromosomes in F_1 hybrid which is derived from two different species is known as allopolyploidy.
16. What is aneuploidy?
Answer: Aneuploidy is the loss or gain of a chromosome in the somatic chromosome.
17. How does aneuploidy arise?
Answer: Generally, aneuploidy arises due to the failure of paired chromosomes to separate (nondisjunction) at anaphase.
18. Give three examples of trisomy in human beings.
Answer: (a) Down syndrome (Trisomy of chromosome 21)
(b) Patau syndrome (Trisomy of chromosome 13)
(c) Edward syndrome (Trisomy of chromosome 18)

Long-Answer Questions

1. What is chromosomal aberration? Describe structural abnormalities in chromosomes.
2. Give an account of numerical abnormalities in chromosomes.
Write short notes on:
(a) Allopolyploidy
(b) Isochromosomes
(c) Importance of chromosomal aberrations
(d) Aneuploidy
3. Distinguish between:
(a) Paracentric and pericentric inversion
(b) Translocation homozygote and heterozygote
(c) Deletion and duplication
(d) Monosomy and nullisomy
(e) Diploidy and polyploidy

MUTATION

Short-Answer Questions

1. Who coined the term 'mutation'?
Answer: Hugo de Vries (1891).
2. What is gene mutation?
Answer: A permanent change in the DNA is known as gene mutation.
3. What is mutagen?
Answer: Any substance or agent that can cause mutation is called mutagen.
4. What is reverse mutation?
Answer: The sudden change of mutant phenotype to wild type is called reverse mutation.
5. What are sex-linked mutations?
Answer: Those mutations which occur in sex chromosomes (X, Y) are called sex-linked mutations (e.g., sex-linked diseases like haemophilia and colour-blindness).
6. What is tautomerism?
Answer: When a molecule exists in more than one chemical form, it is called tautomeric and the phenomenon is called tautomerism.
7. Give a classical example of germinal mutation.
Answer: The occurrence of ancon breed of ship (short-legged variety); it was first discovered by Seth Wright.
8. What is transversion?
Answer: Transversion is the replacement of a purine base by pyrimidine base or vice versa.
9. What are base analogues?
Answer: Base analogues are chemicals having molecular structure similar to nitrogenous bases of DNA.
10. Name the radiations that cause mutations by creating positive and negative ions.
Answer: Ionising radiations
11. Name the radiations that cause the distortion of DNA helix due to dimerisation.
Answer: Ultraviolet (UV) rays
12. How can sex-linked mutations be detected in *Drosophila*?
Answer: By CIB method or Muller-5 method
13. Which test is performed to detect mutations in bacteria?
Answer: Fluctuation test
14. How do X-rays bring about mutation?
Answer: X-rays bring about mutation by breaking the phosphate ester linkages in DNA.
15. Who developed Ames test?
Answer: Bruce Ames
16. What is silent mutation?
Answer: When there is change in codon but not in amino acid, it is called silent mutation.

17. How do spontaneous mutations occur?
Answer: Spontaneous mutations occur due to errors in:
(a) DNA replication
(b) DNA repair
(c) DNA recombination

18. Give one example of intercalating agent.
Answer: Ethidium bromide

19. Name one method to determine how mutagenic an agent is?
Answer: Ames test

20. Who discovered that X-rays cause mutations in *Drosophila*?
Answer: H J Muller (1920)

Long-Answer Questions

1. What is gene mutation? Explain the molecular mechanism of gene mutation.
2. What are mutagens? Describe how mutagens cause mutations.
3. Write short notes on:
(a) Mutagens
(b) Transition
(c) Transversion
(d) Frameshift mutation
(e) Silent mutation
(f) Role of mutation in evolution
(g) Ames test

MODERN CONCEPT OF GENE

Short-Answer Questions

1. Who coined the term 'gene'?
Answer: Johannson (1909)

2. Who proposed fine structure of gene?
Answer: Benzer

3. Who gave one-gene–one-enzyme hypothesis?
Answer: Beadle and Tatum (1940)

4. What is nested gene?
Answer: A gene that produces two proteins simultaneously from a long transcript by changing the end point of protein synthesis is known as nested gene.

5. What are pseudogenes?
Answer: The genes that have lost their ability to perform a function due to mutation are called pseudogenes.

6. Define overlapping genes.
   Answer: The genes that can be read or translated in two different ways to produce different proteins are termed as overlapping genes.
7. What causes a gene to change its wild form?
   Answer: Mutation
8. How may genes in a particular chromosome shift to their homologous pair?
   Answer: By crossing over
9. What is the unit of recombination and mutation?
   Answer: Recon and muton
10. Give one example which exhibits that one gene controls the synthesis of one polypeptide chain.
    Answer: Sickle cell anaemia, which is produced by a single change in any of the two β-polypeptide chains
11. How do genes direct the synthesis of proteins?
    Answer:
    $$\text{Genes (DNA)} \xrightarrow{\text{Transcription}} \text{mRNA} \xrightarrow{\text{Translation}} \text{Protein}$$
12. What are jumping genes?
    Answer: There are certain genes which move from one location to another on a chromosome and are known as jumping genes.
13. What is multigene family?
    Answer: A multigene family is a group of more or less similar genes of the same organism that encode proteins of the same sequence, formed by duplication of a single original gene.
14. What are housekeeping genes?
    Answer: Housekeeping genes are those genes that are required for the maintenance of basic cellular function and are present in all cells of an organism.
15. What is a split gene?
    Answer: A strand of DNA containing exons and introns is known as split gene.
16. Who discovered split gene?
    Answer: Roberts and Sharp (1977)
17. Write two differences between eukaryotic and prokaryotic genes.
    Answer: (a) Eukaryotic genes contain introns which are lacking in prokaryotic genes.
    (b) 3S box factor sequence is absent in eukaryotic genes but is present in prokaryotic genes.

Long-Answer Questions

1. Give an account of modern concept of gene.
2. Write short notes on:
   (a) Split gene
   (b) Overlapping genes
   (c) Homeotic genes
   (d) One-gene–one-polypeptide concept
   (e) Cistron

HUMAN GENETICS

Short-Answer Questions

1. What is the use of studying human genetics?
Answer: The study of human genetics is very important as it helps:
(a) In understanding the nature of humans
(b) In understanding the diseases
(c) In developing effective treatments

2. Human karyotype contains how many groups?
Answer: Seven

3. Who gave the correct number (46) of chromosomes in humans?
Answer: Tijio and Levan (1956)

4. How can human chromosomes be identified?
Answer: Human chromosomes can be identified on the basis of morphological differences and by the banding pattern.

5. What is a chromosome band?
Answer: The segment of chromosomes having characteristic lighter and darker appearance as a result of staining is termed as band.

6. Name the telocentric chromosomes in humans.
Answer: In humans, there is no telocentric chromosome.

7. Name the chemical that stimulates white blood cells or lymphocytes to undergo mitosis.
Answer: Phytohaemmagglutinin (PHA) which is extracted from red kidney bean.

8. Name the longest and smallest chromosomes of humans.
Answer: The largest pair is chromosome 1 and the smallest pair is chromosome number 22.

9. Give the accepted designation of long arm and short arm of the chromosome.
Answer: Long arm – q and short arm – p

10. Which banding is useful in staining centromeres?
Answer: C banding

11. What is the use of R banding?
Answer: R banding is used to detect:
(a) Minor deletions (b) Inversions (c) Chromosome polymorphisms

12. Give the use of fluorescent in situ hybridisation (FISH).
Answer: Fluorescent in situ hybridisation can be used to study:
(a) Deletions (b) Insertions (c) Translocations

13. Name the human chromosome pairs with which Nucleolar Organising Regions (NORs) are associated.
Answer: Chromosome pairs 13, 14, 15, 21 and 22

14. How are chromosome bands classified?
Answer: Chromosome bands are classified according to their relative location on the short arm (p) or long arm (q) of specific chromosomes (e.g., 12q1 means band 1 of the long arm of chromosome 12).

15. Name a substance that inhibits DNA synthesis but does not interfere with other cell functions.
Answer: Amethopterin

16. Which banding visualises telomere?
Answer: T banding

17. Is NOR banding similar to all persons?
Answer: No, it varies from person to person.

18. Name the abnormality that results from partial deletion of short arm of chromosome 4 in humans.
Answer: Wolf–Hirschhorn syndrome

19. In a normal karyotype, how can aneuploidy be detected?
Answer: By the presence of an extra chromosome or a missing chromosome

20. What is the name of the event that causes an increase or decrease in the number of chromosomes in an individual pair of chromosomes?
Answer: Nondisjunction

21. Name one dominant X-linked trait in humans.
Answer: Defective tooth enamel

22. In which arm of Y chromosome are the genes for maleness located?
Answer: Short arm

23. Why dominant X-linked traits are more frequent in females than males?
Answer: Because of the presence of two X chromosomes in females and only one X chromosome in males

24. What is Barr body?
Answer: Barr body is an inactivated X chromosome.

25. What is the chromosomal complement of Klinefelter syndrome?
Answer: 44+XXY

26. Name the syndrome which is due to monosomy of X chromosome.
Answer: Turner syndrome (44+XO)

27. Name the syndrome which is associated with the disorder of connective tissue.
Answer: Marfan syndrome

28. Name the syndrome which is the most common form of mental retardation.
Answer: Fragile X syndrome

29. Who described the first human trait and what is its name?
Answer: Farabee (1905), brachydactly

30. Name three autosomal recessive diseases in humans.
Answer: (a) Albinism (b) Phenylketonuria (c) Cystic fibrosis

31. Name one genetic disorder in humans which is determined by dominant allele.
Answer: Achondroplasia

32. Name the most commonly occuring type of chromosome abnormality in humans.
Answer: Aneuploidy

33. Name five diseases associated with interstitial chromosomes deletions.
Answer: (a) Di George's syndrome del (22) (pter-q11)
(b) Duchenne's muscular dystrophy del (X) (p-21)
(c) Prader–Willi syndrome del (15) (q11-13)

(d) Retinoblastoma del (13) (q14)
(e) Wilms' tumour del (11) (p13)

34. Is tongue rolling a dominant or recessive trait?
Answer: Dominant

35. Name the syndrome which is characterised by the overproduction and accumulation of uric acid.
Answer: Lesch–Nyhan syndrome

36. If both parents belong to blood group O, what is the chance of blood group O in their children?
Answer: 100 per cent

37. How is diagnosis of chromosome aneuploidy of an unborn child done?
Answer: By a combination of amniocentesis, cell culture and karyotyping

38. Define pedigree analysis. Give various symbols used in pedigree analysis.
Answer: The method of studying inheritance of characters by acknowledging the ancestral history of an individual is known as pedigree analysis. The various symbols used in pedigree analysis are shown below:

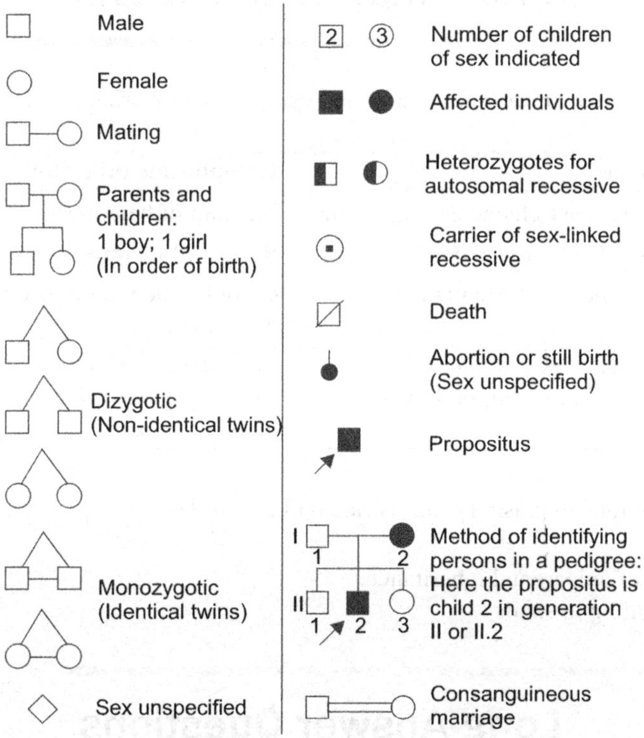

39. A child has blood group O and his father belongs to blood group A and mother to blood group B. What are the possible genotypes of the mother and the father of the child?
Answer: Father – I^AI^O
Mother – I^BI^O

Long-Answer Questions

1. Describe autosomal and sex abnormalities in humans.
2. Give an account of genetic diseases in humans.
3. Write short notes on:
   (a) Erythroblastosis foetalis (b) Sickle cell anaemia (c) Human karyotype
   (d) Blood groups (e) Muscular dystrophy (f) Pedigree analysis

CYTOPLASMIC INHERITANCE

Short-Answer Questions

1. Differentiate between, nuclear inheritqance and cytoplasmic inheritance (extranuclear inheritance).
   Answer:

Nuclear inheritance	Cytoplasmic inheritance
(a) The unit of hereditary character is the gene.	The unit of hereditary character is plasma gene.
(b) Results of reciprocal crosses do not differ.	Results of reciprocal crosses differ.
(c) Nuclear inheritance is of Mendalian behaviour.	Cytoplasmic inheritance is of non-Mendelian behaviour.
(d) The offspring receives cytoplasm from both gametes and thus from both parents.	The offspring receives cytoplasm only from the female gametes. Hence, plasma genes of only the mother are contributed to the offspring.

2. What is plasmon?
   Answer: The sum total of plasma genes is called plasmon. The term 'plasmon' was given by Wettstein (1924).
3. Give an example of uniparental inheritance.
   Answer: Shell coiling in snails

Long-Answer Questions

1. What is cytoplasmic inheritance? Explain with suitable examples.
2. Write short notes on:
   (a) Kappa particles (b) Shell coiling in snails
   (c) Plastid inheritance in *Mirabilis jalapa*

NUCLEIC ACIDS

Short-Answer Questions

1. Who discovered nucleic acid?
Answer: Friedrich Meischer (1869) first discovered nucleic acid in a pus cell in the discarded surgical bandages and he called it nuclein.

2. What allows an organism to transfer genetic information from one generation to another?
Answer: Nucleic acid

3. What are the components of nucleic acids?
Answer:

	Sugar	Phosphate	Bases	
			Purine	**Pyrimidine**
DNA	Deoxyribose	Present	Adenine Guanine	Cytosine Thymine
RNA	Ribose	Present	Adenine Guanine	Cytosine Uracil

4. What is the name of monomer subunits that form nucleic acids?
Answer: Nucleotides

5. Which type of sugar is found in DNA?
Answer: Pentose and deoxyribose type

6. What is nucleoside and nucleotide?
Answer: Nucleoside is a combination of sugar and nitrogenous base, whereas nucleotide is a combination of sugar, nitrogenous base and phosphoric acid.

7. Which scientists first gave the experimental proof that DNA is genetic material?
Answer: Avery Mac Leod and Mc Carty

8. Write one basic difference between pyrimidine and purine.
Answer: Pyrimidine (cytosine and thymine) is a single-ring structure, whereas purine (adenine and guanine) is a double-ring structure.

9. Write four differences between DNA and RNA.
Answer:

	DNA	RNA
(a)	DNA is a double helical structure	RNA is a single helical structure
(b)	In DNA, sugar is deoxyribose type	In RNA, sugar is of ribose type
(c)	In DNA, nitrogenous bases are of four types: adenine, guanine, cytosine and thymine	In RNA also nitrogenous bases are of four types but uracil is present in place of thymine
(d)	Generally, DNA is the genetic material	Generally, RNA is not the genetic material

10. Which form of DNA was proposed by Watson and Crick?
Answer: B form of DNA

11. Give the base pairs per turn of the helix in the following forms of DNA: A-DNA, B-DNA, D-DNA and Z-DNA

Answer: Base pairs per turn of the helix:

A-DNA	B-DNA	D-DNA	Z-DNA
11	10	8	12

12. Name the form of DNA having left-handed helical form with a zigzag phosphate backbone in antiparallel organisation.
Answer: Z form (Z-DNA)

13. In which bacteriophage is the DNA single stranded and remains coiled?
Answer: φX174

14. State whether phosphate makes nucleotide positively or negatively charged?
Answer: Negatively charged

15. If the percentage of guanine is 23.5 in a double-stranded DNA, what is the percentage of other three nitrogenous bases in the DNA?
Answer: The amount of guanine is equal to cytosine, so the percentage of cytosine is also 23.5. The other two bases, adenine and thymine, are present in an equal amount and in combination they represent the rest of the bases (100 – 47 = 53). Thus, A=T = 53/2 =26.5%.

16. Give an example where RNA is the genetic material.
Answer: Tobacco Mosaic Virus (TMV)

17. What are guide RNAs (gRNAs)?
Answer: gRNAs are RNA genes that function in RNA editing.

18. Name the RNA which is used for the formation of spliceosome.
Answer: Small nuclear RNA (SnRNA)

19. It is a bit difficult to break cytosine–guanine pairing in comparison to adenine–thymine pairing.
Answer: Because adenine–thymine base pair is held together by two hydrogen bonds, whereas cytosine–guanine pairing is held by three hydrogen bonds. Thus, cytosine–guanine pairing requires more energy in comparison to adenine–thymine pairing.

20. What is RNA world theory?
Answer: RNA world theory states that primitive RNA molecules evolved before protein and DNA.

21. Name the longest RNA.
Answer: mRNA

Long-Answer Questions

1. Describe the structure and function of DNA.
2. Describe different types of RNA and mention their roles in protein synthesis.
3. Differentiate between:
   (a) DNA and RNA
   (b) B-DNA and Z-DNA
   (c) Nucleoside and nucleotide
   (d) mRNA and tRNA
   (e) Palindromic DNA and B-DNA

REPLICATION OF DNA

Short-Answer Questions

1. Which type of replication occurs in DNA?
Answer: Semiconservative
2. Who proved that replication of DNA is semiconservative?
Answer: Meselson and Stahl (1958)
3. Name the first event that occurs during replication of DNA.
Answer: Unwinding of parental strand
4. During which phase of interphase does synthesis of DNA occur?
Answer: S (synthetic) phase
5. Give one example of unidirectional DNA replication.
Answer: Replication of mitochondrial DNA
6. In which direction does synthesis of DNA occur on the DNA strand?
Answer: $5' \longrightarrow 3'$ direction
7. In Meseleson and Stahl experiment, what percentage of DNA was composed of light and heavy chains after one generation?
Answer: 100 per cent
8. What are leading and lagging strands?
Answer: The strand of DNA in which synthesis occurs continuously and synthesises on $3' \longrightarrow 5'$ strand of parental DNA is termed as leading strand, whereas the strand of DNA in which synthesis occurs discontinuously and synthesises on $5' \longrightarrow 3'$ strand of parental DNA is called lagging strand.
9. Which strand of DNA, i.e., leading or lagging strand needs DNA ligase for its synthesis?
Answer: DNA ligase is needed by lagging strand to join Okazaki fragments.
10. What is DNA ligase?
Answer: DNA ligase is an enzyme which joins the ends of two DNA chains by catalysing the synthesis of the phosphor diester bond between 3′-OH at the end of one chain, and 5′ end of the other.
11. What is replication fork?
Answer: During DNA replication, the two strands uncoil at a point forming a Y-shaped structure, which is termed as replication fork.
12. What are Okazaki fragments?
Answer: DNA is synthesised in a series of small fragments (1,000–2,000 nucleotides) called Okazaki fragments. It was discovered by Reiji Okazaki in 1968.
13. What are replicons?
Answer: Replicons are the replicating regions of DNA.
14. Name the DNA virus that does not replicate in the cell nucleus using the host enzyme.
Answer: Parvovirus
15. Why is the rate of DNA replication in eukaryotic chromosomes much slower?
Answer: Due to the complex nature of chromatin

16. What are basic enzymes involved in DNA replication called?
Answer: DNA polymerases
17. RNA primer is considered essential by DNA polymerase during the synthesis of DNA?
Answer: Because DNA polymerase needs a free 3′-OH group
18. Which enzyme is responsible for removal of RNA primer?
Answer: The RNA primer is removed by DNA polymerase I.
19. What is the function of RNA primase?
Answer: RNA primase synthesises short RNA sequences on the DNA strand that serve as primer for DNA polymerase III.
20. Discovery of Okazaki fragments is the indication of which fact?
Answer: The mode of replication in DNA is discontinuous.
21. Who discovered DNA polymerase?
Answer: DNA polymerase I (DNA pol I) was discovered by Arthur Kornberg (1955) in *E. coli*.
22. What is the function of DNA polymerase I?
Answer: DNA polymerase I polymerises reaction in $5' \longrightarrow 3'$ direction. Besides, it also has proof-reading activity.
23. What is the function of DNA polymerase III?
Answer: DNA polymerase enzyme is responsible for DNA replication and synthesises DNA in $5' \longrightarrow 3'$ direction.
24. Can DNA polymerases initiate the chain *de novo*?
Answer: No
25. What is the function of the enzyme helicase?
Answer: It provides unwinding of DNA through hydrolysis of ATP.
26. Which type of replication occurs in eukaryotic DNA?
Answer: Semiconservative and bidirectional
27. How many types of DNA polymerases are found in eukaryotes?
Answer: Five types DNA polymerases are found in eukaryotes:
    (a) DNA polymerase α (alpha)
    (b) DNA polymerase β (beta)
    (c) DNA polymerase γ (gamma)
    (d) DNA polymerase δ (delta)
    (e) DNA polymerase ε (epsilon)
28. In which direction, mRNA is synthesised from DNA?
Answer: $5' \longrightarrow 3'$

Long-Answer Questions

1. Describe the experimental evidence in support of semiconservative mode of DNA replication.
2. Give an account of DNA replication in *E. coli*. How does it differ from DNA replication in eukaryotes?
3. Write short notes on:
    (a) Meselson and Stahl experiment
    (b) Okazaki fragments
    (c) DNA polymerases
    (d) Unwinding proteins
    (e) Unidirectional DNA replication

PROTEIN SYNTHESIS

Short-Answer Questions

1. What are the two main steps of protein synthesis?
   Answer: 1. Transcription
   2. Translation
2. What is central dogma and who proposed it?
   Answer: Central dogma in molecular biology states that the flow of biological information is unidirectional. Crick (1958) proposed the central dogma.

   $$DNA \xrightarrow{\text{Transcription}} RNA \xrightarrow{\text{Translation}} Protein$$
3. Who discovered reverse transcriptase?
   Answer: Temin and Baltimore (1970) independently discovered reverse transcriptase in the RNA containing Rous sarcoma virus and they were awarded the Nobel Prize in 1975 for this discovery.
4. What is transcription?
   Answer: Synthesis of mRNA on DNA strand is known as transcription.
5. What is the name of the specific area on DNA where RNA polymerase binds?
   Answer: Promoter
6. What is the difference between the first product of transcription in prokaryotes and eukaryotes?
   Answer: In prokaryotes, the first product of transcription is the mRNA, which needs no post-transcriptional modifications. However, the first transcriptional product in eukaryotes is called primary transcript, which needs post-transcriptional modifications to form hnRNA that undergoes splicing of introns, to form the final product called mRNA.
7. What is translation?
   Answer: Translation is the process during which ribosomes synthesise proteins using mature mRNA transcript, synthesised during the process of transcription.
8. In eukaryotes, where does synthesis of mRNA molecules occur and where do they migrate?
   Answer: In eukaryotes, mRNA molecules are synthesised in the nucleus and after synthesis they come to cytoplasm through nuclear pores, where they become associated with ribosomes.
9. If there are 21 nucleotides in the coding region of DNA, how many amino acids will be in the polypeptide chain which is synthesised?
   Answer: Seven
10. An mRNA molecule codifies only one polypeptide chain in eukaryotes. Why?
    Answer: In eukaryotes, mRNA is monocistronic, so it codifies only one polypeptide chain, whereas in prokaryotes mRNA is polycistronic and it may codify more than one polypeptide chain.
11. During translocation, ribosomes move along mRNA. Why?
    Answer: The ribosomes move along mRNA to expose new codons to be translated. This movement is known as ribosomal translocation.
12. Name the cellular structure to which mRNA binds, to start the process of translation.
    Answer: Ribosomes

13. Name the structure responsible for positioning and exposing codons of mRNA.
Answer: Ribosomes

14. In prokaryotes, transcription and translation can occur simultaneously but it is not possible in eukaryotes. Why?
Answer: In eukaryotes, synthesis of mRNA occurs in the nucleus, whereas proteins are synthesised in the cytoplasm.

15. What is the function of initiation factors—IF1, IF2 and IF3?
Answer: IF1 and IF2 are involved in binding of F-Met-tRNA complex to 30S ribosome mRNA, whereas IF2 is involved in recognition of correct codon of mRNA by formyl methionine tRNA. IF3 is essential for binding of m-RNA to 30S sub-unit of ribosome. It also prevents reassociation of dissociated 30S and 50S sub-units of ribosomes.

16. Name the enzymes involved in the following reactions:
Answer: (a) Attachment of activated amino acid to tRNA – Amino acyl tRNA synthetase
(b) Synthesis of rRNA, mRNA and tRNA – RNA polymerase I, II and III
(c) Peptide bond formation – Peptidyl transferase
(d) Movement of ribosomes along mRNA (translocation) – Translocase

17. Differentiate between replication and transcription.
Answer:

	Replication	Transcription
(a)	Replication occurs in the S phase.	Transcription occurs during the G_1 or G_2 phase.
(b)	It is catalysed by the enzyme DNA polymerases.	It is catalysed by the enzyme RNA polymerases.
(c)	It occurs on both strands.	Transcription occurs on only one strand.
(d)	Replication involves copying of the entire genome.	It involves copying of a few genes.
(e)	Its product remains in the nucleus	Major part of the products pass into the cytoplasm.

18. Name three inhibitors of protein synthesis.
Answer: Streptomycin, tetracycline and abrin

19. What is the function of the enzyme tRNA deacylase?
Answer: It helps in the separation of methionine and tRNA after the formation of peptide bond.

20. Enzyme peptidyl transferase is present in which ribosomal subunit: 30S or 50S?
Answer: 50S ribosomal subunit

21. Write three differences between prokaryotic and eukaryotic translation process.
Answer:

	Prokaryotes	Eukaryotes
(a)	The initiating amino acid methionine is formylated.	The initiating amino acid methionine is not is is formylated.
(b)	Ribosomes enter the mRNA at AUG codon or near Shine Delagarmo site.	Ribosomes enter at the capped 5′ end of the mRNA and then move to AUG codon
(c)	The smaller 30S ribosomal sub-unit can engage mRNA before binding of F-Met $tRNA^{met}$.	The smaller 40S ribosomal sub-unit binds firmly to mRNA only after initiating Met-$tRNA^{met}$ initiator attached to it.

22. A polypeptide found in the cytoplasm of a cell contains 15 amino acids. How many nucleotides will be required in the mRNA for this polypeptide to be translated?
Answer: 45

23. Which parts of RNA are not translated?
Answer: In case of polycistronic message, the sequences between the two coding regions are called intercistronic region which do not code for protein.

24. What are enhancers?
Answer: Enhancers are sequences which greatly enhance the transcription of a gene. They were first discovered in a mammalian virus, SV40.

25. How do enhancers differ from promoters?
Answer: (a) The position of enhancers is not fixed.
(b) They can function in either orientation.
(c) The modules of enhancers are continuous rather than spaced apart.

26. How many high-energy phosphate bonds are required for the formation of one peptide bond?
Answer: Four high-energy phosphate bonds; two for initial activation, one for EF-1 step and one for EF-2 step

Long-Answer Questions

1. Give an account of protein synthesis.
2. Write short notes on:
(a) Central dogma
(b) Steps involved in the activation of amino acids
(c) Initiation complex
(d) Elongation factors
(e) RNA polymerase
(f) Peptidyl transferase

GENETIC CODE

Short-Answer Questions

1. Who coined the term 'genetic code'?
Answer: George Gamow

2. What is the name of RNA sequence that codifies one amino acid?
Answer: Codon, which is a sequence of three nitrogenous bases

3. What are the characteristics of genetic code?
Answer: (a) Genetic code is triplet.
(b) Genetic code is commaless.
(c) Genetic code is non-overlapping.
(d) Genetic code is universal.
(e) Genetic code is degenerate.

4. Who first deciphered the genetic code using trinucleotide synthesis?
Answer: Marshall Nirenberg
5. Which is the first codon deciphered by Nirenberg and Mathaei?
Answer: UUU
6. Name the amino acids that are coded by only one codon.
Answer: Methionine and tryptophan
7. Name the amino acids having six different codons.
Answer: Leucine, serine and arginine
8. Name the codons that do not code for any amino acid.
Answer: UAA, UAG and UGA
9. Name the initiating codon.
Answer: AUG is the initiating codon (rarely GUG) which codes for methionine.
10. Who proposed wobble hypothesis?
Answer: Crick (1965)
11. What is the significance of wobble hypothesis?
Answer: It provides economy of tRNA molecules.
12. Genetic code is read in which direction?
Answer: $5' \longrightarrow 3'$ direction
13. Give examples where genetic codes overlap.
Answer: Φ X174 and SV40 viruses
14. Name the scientists who used the filter binding technique for deciphering genetic code.
Answer: Nirenberg and Lederberg
15. Name the terminating codons of human mitochondria.
Answer: AGG and AGA
16. Name the molecules that bear codons and anticodons.
Answer: mRNA bears codons, whereas tRNA contains anticodons.
17. Which nitrogenous base varies in multiple codons of an amino acid?
Answer: In multiple codons for an amino acid, the third nitrogenous base varies, whereas the first two nitrogenous bases remain constant.
18. Name the organism which translates codon UGA as tryptophan.
Answer: Mycoplasma

Long-Answer Questions

1. What is genetic code? Describe the salient features of genetic code.
2. Give an account of different processes for deciphering genetic code.
3. Write short notes on:
   (a) Gamow hypothesis
   (b) Wobble hypothesis
   (c) Chain terminating codons
   (d) Triplet-binding technique
   (e) Exception of genetic code
   (f) Codon dictionary

GENE REGULATION

Short-Answer Questions

1. What is gene expression?
Answer: The process by which cells translate their genetic information contained in DNA into proteins is known as gene expression.
2. What is gene regulation?
Answer: The activation and suppression of genes at different times is called gene regulation.
3. Name the first-discovered example of gene regulation.
Answer: Lac operon model
4. Who discovered the lac operon model?
Answer: Jacob and Monod (1961)
5. What are the components of a lac operon?
Answer: The lac operon model consists of:
(a) Structural genes (b) Promoter genes (c) Operator genes
(d) Regulator genes (e) Effector or inducer
6. State the function of a regulator gene.
Answer: A regulator gene controls the operator gene with the help of some chemical compounds known as inducers and co-repressors. Regulator genes produce a substance called repressor, which may bind with the operator gene and switch off the structural genes.
7. Define promoter gene?
Answer: Promoter gene is the site where the RNA polymerase binds. It lies left to the operator and is the actual site of transcription.
8. State whether catabolite repression is an example of positive or negative control?
Answer: Positive control
9. Which type of regulation (positive or negative) is involved in the removal of an inhibitor?
Answer: Negative control
10. Give three main features of lac operon regulation.
Answer: (a) The lac operon is under negative control. When it reacts with the lac repressor, it causes inhibition of lac transcription. The repressor is produced by the regulatory gene (i).
(b) When the lac repressor binds with the operator, it prevents transcription. Whenever an inducer (lactose or lactose analogue) is added, the repressor dissociates from the operator and as a result, lac transcription takes place.
(c) The lac operon is also under positive control. Operon transcription is enhanced when operon interacts with a regulator protein.
11. How many structural genes does lac operon involve?
Answer: Three structural genes
12. Write three differences between repression and induction.
Answer:

	Repression	Induction
(a)	It operates in anabolic pathway.	It operates in catabolic pathway.
(b)	It causes operon to turn off.	It causes operon to turn on.
(c)	It stops transcription and translation.	It initiates transcription and translation.

13. What would be the result if a mutation occurs in the CAP gene of the lac operon?
Answer: There will be no expression of lac operon as RNA polymerase would not recognise the promoter.
14. At what level does repression or depression of trp operon occur?
Answer: Repression or depression of trp operon occurs at the level of transcription.
15. What is gene battery model and who proposed it?
Answer: Gene battery model is a model for gene regulation in eukaryotes, proposed by Britten and Davidson. According to this model, a set of structural genes are controlled by one sensor site called battery.

Long-Answer Questions

1. What is an operon? Describe the positive and negative control of lac operon.
2. Describe the role of operator gene, regulator gene, structural genes, promoter and inducer.
3. Write short notes on:
(a) Attenuation (b) Hogness box (c) TATA box
(d) Enhancer and silencer (e) ara C operon (f) Gene battery model
4. Give an account of regulation of gene expression in eukaryotes.

HUMAN GENOME PROJECT

Short-Answer Questions

1. What is the Human Genome Project (HGP)?
Answer: In 1990, a project was started to recognise genome of the human race and to collect data on it. It was a mega project and termed as Human Genome Project.
2. In which year was the Human Genome Project started and when was it completed?
Answer: The Human Genome Project was started in October 1990 and completed in April 2003. However, sequence of individual chromosomes was published in 2006.
3. Who was the first director of the Human Genome Project?
Answer: James Watson. He was replaced by Francis Collins in 1993.

4. What were the two main aims of the Human Genome Project?
Answer: (a) To determine the sequence of all 3.3 billion nucleotides on the human genome.
(b) To map the location of every gene on each chromosome.

5. Why is the Human Genome Project (HGP) called 'mega project'?
Answer: (a) Human genome contains 3.3×10^9 base pairs and if the cost of sequencing is US \$3 per base pair, then the approximate cost will be US \$10 billion.
(b) If the sequence obtained were to be stored in a typed form in books and if each page contained 1,000 letters and each book contained 1,000 pages, then 3,300 such books would be needed to store the complete information.
(c) The enormous quantity of data expected to be generated also necessitates the use of high-speed computer hard-drives for data storage and supercomputers were used for retrieval and analysis.
(d) Therefore, HGP is called a mega project due to the investment of a lot of money, use of advanced techniques and involvement of many scientists in the work.

6. Name the chromosome number of humans whose complete sequence was first published.
Answer: Chromosome number 22

7. Why it is easier to prepare DNA cleanly from sperms than other cell types?
Answer: Because it has high ratio of DNA to protein and it contains all the chromosomes necessary for study.

8. Name some model organisms studied in the Human Genome Project.
Answer: Escherischia coli, Saccharomyces cerevisiae, Drosophila melanogaster and *Caenorhabditis elegans*

9. Which chromosomes of humans contain maximum and minimum number of genes?
Answer: Chromosome 1 contains maximum number of genes and Y chromosome has minimum number of genes.

10. Name the largest human gene?
Answer: Dystrophin having 2.4 million base pairs

11. What percentage of nucleotide bases is exactly similar in all people?
Answer: 99.9 per cent

12. Name some organisms whose genomes have been sequenced completely (in addition to humans).
Answer: (a) *H. influenzae* (First free-living organism to be desequenced)
(b) *Drosophila melanogaster*
(c) *Caenorhabditis elegans* (First multicellular eukaryote sequenced)
(d) *Escherichia coli*
(e) Yeast (*Saccharomyces cerevisiae*)
(f) *Arabidopsis thaliana* (First plant sequenced)
(g) Mouse (*Mus musculus*)

13. Name the hosts used in the Human Genome Project for cloning.
Answer: Bacterial artificial chromosome (BAC) and yeast artificial chromosome (YAC)

Long-Answer Question

1. Give an account of the Human Genome Project (HGP) mentioning its objectives and achievements.

DNA FINGERPRINTING

Short-Answer Questions

1. What is DNA fingerprinting?
Answer: DNA fingerprinting is a technique to identify a person on the basis of his/her DNA.
2. Can DNA fingerprinting change by any type of treatment?
Answer: No
3. Who invented the technique of DNA fingerprinting?
Answer: Sir Alec Jeffrey (1984) invented the technique of DNA fingerprinting.
4. What is called the specific type of DNA sequences used in DNA fingerprinting?
Answer: Microsatellite
5. What are microsatellites?
Answer: Microsatellites are short pieces of DNA which repeat many times in a given person's DNA.

Long-Answer Question

1. What is DNA fingerprinting? Describe the technique of DNA fingerprinting and write a note on its significance.

MOLECULAR GENETICS OF CANCER

Short-Answer Questions

1. What is cancer?
Answer: Cancer may be defined as the uncontrolled growth of the cells (mitosis) that ultimately kills the organisms.
2. Name the cells that are never in G_0 state.
Answer: Cancer cells
3. Distinguish between benign tumour and malignant tumour.
Answer:

Benign tumour	Malignant tumour
(a) They are well differentiated.	Malignant tumour lacks differentiation.
(b) They are not cancerous.	They are cancerous.
(c) They lack invasiveness.	They have the ability of invasiveness.
(d) The rate of growth is slow.	The rate of growth is rapid.

4. What is metastasis?
Answer: The ability of cells to invade other cells is known as metastasis. Metastasis is a characteristic of cancer cells. A cancerous tumour may shed cells in the blood and lymphatic vessels and disperse to distant sites where they form secondary tumours.

5. Why is it impossible to remove cancer cells surgically?
Answer: Due to their ability to invade other cells (metastasis) of the body

6. Name three DNA-containing oncoviruses that cause human cancer.
Answer: (a) Epstein-Barr virus – Burkitts lymphoma and nasopharyngeal carcinoma
(b) Hepatitis B virus – Liver cancer
(c) Papillomaviruses – Cervical and other anogenital cancers

7. In which virus were viral oncogenes first reported?
Answer: Viral oncogenes were first reported by Peyton Rous (1911) in Rous sarcoma virus.

8. Name the cancers that arise from mesodermal connective tissue.
Answer: Sarcomas

9. Name the virus that causes non-Hodgkin's lymphoma.
Answer: Simian virus 40 (SV40)

10. When was the first oncogene discovered and what is its name?
Answer: The first oncogene was discovered in 1970 in chicken retro virus and was named Src.

11. Name three tumour-suppressor genes.
Answer: (a) p^{53} located on chromosome 17 (b) RB gene located on chromosome 13
(c) BRA1 gene located on chromosome 17

12. What is oncogenesis?
Answer: Oncogenesis is the initiation of cancer.

13. Define proto-oncogenes.
Answer: Proto-oncogenes are normal cell genes from which the retroviral oncogenes originate.

14. What is carcinogen?
Answer: A carcinogen is an agent that causes or promotes tumour formation.

15. Name two carcinogens which are not mutagens.
Answer: Alcohol and estrogen

16. How the proto-oncogenes change into oncogenes?
Answer: Proto-oncogenes can change to oncogenes by
(a) Point mutation (b) Chromosomal translocation
(c) Insertional activation (d) Gene amplification

17. Name the disease in which the c-myc containing segment of chromosome 8 translocates to chromosome 14.
Answer: Burkitt's lymphoma

18. Write the difference between v-oncs and c-oncs.
Answer: The cancer-inducing genes in a virus are called v-oncs, whereas cellular genes having the ability to become oncogenes are termed as c-oncs.

19. Which types of cancers are caused by BRAC1 and BRAC2 genes?
Answer: Breast, ovarian and pancreatic cancers

20. What are tumour-suppressor genes?
Answer: Genes that suppress division of cells and survival are known as tumour-suppressor genes.

21. Name a parasite that induces cancerous growth.
Answer: Schistosoma haematobium, which induces cancer of the urinary bladder

22. Name three treatments of cancer.
Answer: (a) Surgery (b) Radiation therapy (c) Chemotherapy

IMMUNOGENETICS

Short-Answer Questions

1. Define immunogenetics.
Answer: The branch of medical science that deals with the relationship between the immune system and genetics is known as immunogenetics.

2. What are immunoglobulins?
Answer: Immunoglobulins are glycoproteins that function as antibodies which are found attached with the cell membrane, secretions or circulating in blood.

3. Antibodies are negatively or positively charged immunoglobulins?
Answer: Negatively charged immunoglobulins

4. What are antigens?
Answer: The foreign bodies that cause the formation of antibodies are termed as antigens.

5. How many types of antibodies (immunoglobulins) are found in mammals?
Answer: Five classes of antibodies are found in mammals. These are IgA, IgD, IgE, IgG and IgM with its own class of heavy chain—α, δ, ε, γ and μ, respectively.

6. Name the class of antibody which develops first.
Answer: IgM

7. Which immunoglobulin plays an important role in allergic reaction and its concentration increases during worm infection?
Answer: IgE

8. Which immunoglobulin has the ability to cross the placenta?
Answer: IgG

9. How is the antigen-binding site of an IgG molecule formed?
Answer: The variable domains of a light and heavy chain together form the antigen-binding site of an IgG molecule.

10. Write two advantages and disadvantages of IgG.
Answer:

Advantages	Disadvantages
(a) It has the ability to cross the placenta and thus protect the developing foetus.	(a) It has lower capacity to agglutinate antigens.
(b) Due to its smaller size, it diffuses more readily into intracellular fluids.	(b) It activates the complement system.

11. How is an antibody class determined?
Answer: An antibody class is determined by the constant region sequence of the heavy chain.

12. Which antibody is responsible for Rhisoimmunisation?
Answer: IgG

13. How many binding sites are found in an IgG?
Answer: Two

14. Give the number and names of domains for an antigen-binding site.
Answer: Two V_1 and C_1

15. How many variable and constant domains are present in each heavy chain of IgG?
Answer: One variable domain and three constant domains

16. What are complement factors?
Answer: Complement factors are a set of plasma proteins present in the serum. The lysis of cells by antibody is mediated by the complement system. There are nine components in the complement system, designated as C1 to C9.

17. What are isotypes, allotypes and idiotypes?
Answer: **Isotypes** – These are variants in molecules present in all normal persons (e.g., classes and subclasses of antibodies).
Allotypes – Allotype is the genetically determined difference in molecules between individuals of the same species.
Idiotype – Idiotype is the unique difference between antibodies of different antigen-binding specificities. They are determined by the conformation of the heavy and light chain variable regions of the antibodies.

18. Define opsonisation.
Answer: The process by which microorganisms or other foreign particles are coated by a substance called opsonin (antibody and/or complement) leading to recognition and ingestion by phagocytic cells is known as opsonisation.

19. How is V domain formed?
Answer: V domain is generated by joining the variable region of the light chain and that of the heavy chain.

20. What is junctional diversification?
Answer: The random loss and gain of nucleotides at joining sites is known as junctional diversification.

21. VH segment cannot join directly with a JH segment in a heavy chain gene arrangement. Give reasons.
Answer: Because both VH and JH segments are flanked by recombinant single sequences having 23 bp (2 turns) spacer, and as per one turn/two turn joining rule, signal sequences containing two turn spacer can join only with signal sequences having a one turn spacer.

22. Though B cell is diploid, it expresses the rearranged heavy chain genes from only one chromosome and rearranged light chain genes from only one other chromosome. What is the name of the process through which it is done?
Answer: Allelic exclusion

23. How is V domain generated?
Answer: V domain is generated by joining of the variable region of the light chain and that of the heavy chain

24. Draw a well-labelled diagram of an immunoglobulin molecule.

Answer:

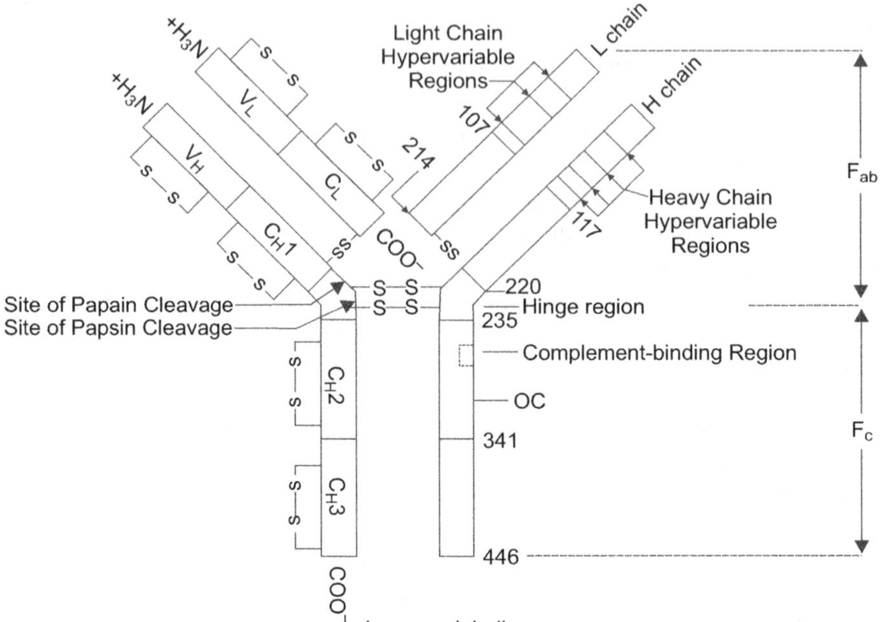

Immunoglobulin

25. Define combinatorial association.

Answer: During the synthesis of an immunoglobulin molecule, any set of light chain molecules may combine with any set of heavy chains resulting in the formation of an antibody. This process is known as combinatorial association. However, there is only one set of heavy and one type of light chain within a given immunoglobulin. Different antibody proteins are possible only by combinatorial association.

Long-Answer Questions

1. What are immunoglobulins? Describe the structure and function of immunoglobulins.
2. Give an account of organisation and expression of immunoglobulin genes.
3. Write short notes on:
   (a) Allelic exclusion
   (b) Junctional and insertional diversification
   (c) Productive and nonproductive rearrangements of Ig genes
   (d) Lambda (λ) light chain gene
   (e) Kappa (κ) light chain gene
   (f) Opsonisation
   (g) Allotypes and idiotype

PRIONS

Short-Answer Questions

1. Name the smallest form on the earth.
Answer: Prions
2. Who coined the term 'prions'?
Answer: Stanley B Prusiner (1982)
3. What are prions?
Answer: Prions are infectious agents which are made up of proteins, lack nucleic acids (DNA and RNA) and cause fatal brain diseases.
4. Do prions obey the central dogma rule?
Answer: No
5. Prions diseases are also called transmissible spongiform encephalopathies. Why?
Answer: Because they are transmissible from one host to another of the same species and sometimes even from one species to another, and they destroy brain tissue, giving it a spongy appearance.
6. Which body part is affected by prion disease?
Answer: Structure of the brain or neural tube
7. Name some prions diseases.
Answer: (a) Scrapie in sheep
(b) Mad cow disease (Bovine spongiform encephalopathy or BSE) in cattle
(c) Kuru and Creutzfeldt–Jakob disease in humans
8. How are prions detected?
Answer: Prions can be detected with high sensitivity by Protein Misfolding Cyclic Amplification (PMCA) technology.
9. Name the prion gene and on which chromosome is it found?
Answer: PrP gene and it is located on chromosome number 20

Long-Answer Question

1. Give an account of prions. Do prions obey the central dogma rule?

TRANSPOSONS

Short-Answer Questions

1. What are transposons?

Answer: Transposons are segments of DNA that move around different locations in the genome of a single cell.

2. Who discovered first transposons?
Answer: First transposons were discovered by Mc Clintock (1940) in maize.
3. Which enzyme is needed by transposons for their mobile ability?
Answer: Transposase
4. Transposons are sometimes referred to as jumping genes. Why?
Answer: Because they are the segments of DNA that move around different locations in the genome of a cell
5. What is transposition?
Answer: The movement of transposons is known as transposition.
6. What are Insertion Sequences (IS)?
Answer: Insertion Sequences are specific genetic units whose genes are only associated with transposition and self-propagation.
7. Can transposons replicate DNA in eukaryotes?
Answer: No
8. How retrotransposons move in the genome?
Answer: Retrotransposons (Class I) move in the genome by being transcribed to RNA and then back to DNA by reverse transcriptase.
9. Name some important eukaryotic transposons.
Answer: Some important eukaryotic transposons are P elements in *Drosophila melanogaster*, Ty elements in yeast and IAP and VL elements in rodents. The most common form of transposons in humans is the AlU sequence which is approximately 300 bases long.
10. What is the significance of transposons?
Answer: Transposons:
    (a) Cause mutation and chromosomal break
    (b) May be used as genetic markers
    (c) Regulate gene expression
    (d) Form large part of constitutive heterochromatin region of the chromosomes that are permanently condensed and for most part generally inactive in every cell
    (e) Some have been found to cause genetic diseases such as haemophilia, severe combined immune deficiency, Duchenne's muscular dystrophy, etc.
    (f) May be used as transformation vectors
    (g) Are used for tagging desirable genes

Long-Answer Questions

1. What are transposons? Describe prokaryotic and eukaryotic transposons. Give evolutionary and genetical significances of transposons.
2. Write short notes on:
    (a) Mobile genetic elements
    (b) IS elements
    (c) Ac–Ds system
    (d) Retrotransposons

APOPTOSIS

Short-Answer Questions

1. What is apoptosis?
Answer: Apoptosis is the programmed cell death, which occurs in the life cycle of all multicellular organisms.
2. Who coined the term 'apoptosis'?
Answer: Kerr et al. (1972)
3. What are the chief features of apoptosis?
Answer: (a) Cell shrinkage
(b) Blebbing
(c) Chromatin condensation
(d) Preservation of organelles and cell membrane
(e) DNA fragmentation
(f) Rapid engulfment by neighbouring cell, preventing inflammation
4. What is the effect of too much and too less apoptosis?
Answer: Excessive apoptosis may cause atrophy, such as in neurodegenerative diseases, whereas too less apoptosis results in uncontrolled cell proliferation as in cancer.
5. Why are caspases so named?
Answer: Because of the presence of a key cysteine residue in the catalytic site and they selectively cleave proteins at sites just terminal to aspartate residues.
6. Name the initiator and effector caspases.
Answer: Initiator caspases – Caspases 2, 8, 9 and 10
Effector caspases – Caspases 3, 6 and 7
7. What is the difference between necrosis and apoptosis?
Answer: In necrosis, cellular debris is discharged, which can damage the cell, but in apoptosis, the contents of cells are not discharged into the extracellular mileu.
8. Name some proapoptic and antiapoptic proteins.
Answer: Proapoptic proteins – Bax, Bad and Bik
Antiapoptic proteins – Bcl2 and Bcl-X1 (inhibit apoptosis)

Long-Answer Questions

1. What is apoptosis? Describe the mechanism of apoptosis. How does it differ from necrosis?
2. Write short notes on:
(a) Significance of apoptosis
(b) Cellular events of apoptosis
(c) Caspases
(d) Role of mitochondria in apoptosis
(e) Apoptosis and diseases
(f) Regulation of apoptosis

MOLECULAR BIOLOGY OF AGEING

Short-Answer Questions

1. Define ageing.
Answer: The collection of changes that lead human beings progressively more likely to die is termed as ageing.

2. According to Leonard Hayflick, cells can divide a maximum of about __________ times and as we age, our cells become increasingly less capable of dividing.
Answer: 75–80 times

3. What is the other name for biological ageing?
Answer: Senescence

4. Name the part of an ear in which structural changes is most associated with hearing loss.
Answer: Cochlea

5. What is Apoplipoprotein D gene?
Answer: Apoplipoprotein D gene is involved in the mechanism that protects cells from ageing.

6. What is gene p63?
Answer: Gene p63 gene is a sister gene of p53. It has been observed by scientists that gene p63, when switched off, induces ageing in adult mice.

7. What is the name of the single greatest factor that may accelerate the process of ageing?
Answer: Excess levels of insulin

8. Which sex hormone declines significantly with ageing?
Answer: Testosterone

9. What is TA-65?
Answer: TA-65 is an anti-ageing medicine that is extracted from a Chinese herb. It prevents telomeres from shortening, though in some cases it actually lengthens telomeres.

10. How do telomeres affect ageing?
Answer: It has been observed that telomeres get shorten with successive cell division. As telomeres become shorter, cells are unable to divide and ultimately die leading to ageing.

11. Who predicted that length of telomeres is the 'Molecular Clock'?
Answer: Leonard Hayflick

12. Which theory of ageing states that humans age because normal metabolism of the body produces unstable oxygen that accumulates around the cells damaging DNA and other cellular structures?
Answer: Free radical theory

13. According to which theory, can the process of ageing be slowed down by the ingestion of β (beta) carotene?
Answer: Free radical theory

14. What is Hutchinson–Gilford Progeria Syndrome (HGPS)?
Answer: It is a rare genetic condition in which symptoms resembling ageing appear at an early age. It is caused by point mutation in 1824 position of LMNA gene.

Long-Answer Questions

1. Define ageing. Describe various structural and functional changes that occur at cellular and tissue levels.
2. Give an account of mechanism of ageing.

GENETIC ENGINEERING

Short-Answer Questions

1. What is genetic engineering?
   Answer: The process in which recombinant DNA (rDNA) technology is used to introduce new traits into organisms is called genetic engineering.
2. What is horizontal gene transfer?
   Answer: Pieces of genetic material may be picked up by cells directly from the environment, which may be inserted into the genome. This illicit trafficking of genes is termed as horizontal gene transfer.
3. Where does vertical gene transfer takes place?
   Answer: Vertical gene transfer takes place during reproduction.
4. *Taq polymerase* is obtained from which organism?
   Answer: The *Taq polymerase* is a DNA polymerase enzyme which is isolated from the bacterium *Thermus aquaticus* growing in the hotwater springs. This enzyme acts best at 72°C and is not denatured even at 90°C.
5. What is plasmid?
   Answer: The plasmids are the extrachromosomal, self-replicating circular DNA segments present in the bacterial cytoplasm.
6. What are tools of recombinant technology?
   Answer: (a) Source of the donor DNA (b) Restriction enzymes
   (c) DNA polymerase enzyme (d) Ligase enzyme
   (e) Vectors (f) Host cell/organisms
7. What are cloning vectors?
   Answer: The DNA molecule with which the foreign DNA is joined and inserted into the host cell is termed as vector.
8. What are the features of cloning vectors?
   Answer: A cloning vector must have the following features:
   (a) Origin of replication (*ori*)
   (b) Selectable markers (for identifying recombinants from nonrecombinants)
   (c) A single recognition site for cloning
9. What are the different types of cloning vectors?
   Answer: There are four types of cloning vectors:

(a) Plasmids (b) Viruses
(c) Cosmids (d) Artificial chromosomes

10. Name the best-known and commercially available plasmid vectors.
Answer: The best-known and commercially available vectors are pBR-322 and pUC-18. Both of these plasmids are modified natural plasmids of *E. coli*.

11. Why are the plasmids of pUC series so named?
Answer: The pUC series of plasmid vectors are named so because they were initially developed in the University of California. The plasmid pUC are derived from pBR-322.

12. What is Polymerase Chain Reaction (PCR)?
Answer: Polymerase Chain Reaction is the *in-vitro* synthesis of multiple copies of a gene or desired DNA segment. Mullis (1985) developed the technique of Polymerase Chain Reaction (PCR).

13. Give the name of bacteria known as the 'natural engineer of plants'.
Answer: Agrobacterium tumefaciens

14. What is the name of the substance that is added to culture medium to induce callus induction?
Answer: Auxins

15. Name two products produced by genetic engineering.
Answer: (a) Human insulin (b) Hepatitis B vaccine

16. What is the name of the first intergenetic somatic hybrid between tomatoes and potatoes?
Answer: Pomatoes or Topatoes

17. Which is the most commonly used culture medium for plant cells and tissues?
Answer: MS medium (developed by Murashige and Skoog, 1962)

18. Name the basic technique used in plant tissue culture.
Answer: Totipotency

19. In which plant was the gene coding for hirudin protein introduced?
Answer: Brassica napus (Rapeseed) plant

20. Name the genetically engineered plants that produce polyhydroxy butyrate globules in their chloroplasts.
Answer: Arabidopisis

21. What is the use of callus culture?
Answer: Callus culture is used for:
(a) Plant regeneration (b) Preparation of single suspensions and protoplasts
(c) Genetic transformation studies

22. What is explants culture?
Answer: The culture of plant parts is known as explants culture.

23. What is the name of first gene available for genetic engineering of crop plants for pest resistance?
Answer: Cry genes (popularly known as Bt genes) from the bacterium *Bacillus thuringiensis.*

24. What is Golden rice?
Answer: Golden rice is the genetically engineered rice rich in pro vitamin A. It was developed by Prof. Ingo Potrykus and Dr Peter Beyer by introducing three genes rich in carotenoid (precursor of pro vitamin A).

25. Humans are not affected by glyphophate in plant crops?
Answer: Because humans do not produce aromatic amino acids

26. Bacterial DNA are not cleaved by their own restriction enzymes?
Answer: Because bacteria add methyl groups to their own DNA

27. For what purpose is DNA ligase used in genetic engineering?
Answer: As a sealing enzyme
28. What is the purpose of sticky ends on fragment DNA?
Answer: To hold the pieces of DNA together
29. What is the function of restriction enzymes?
Answer: To cut DNA at specific sites
30. What is the name given to the organism generated through the introduction of DNA by recombinant DNA technology?
Answer: Genetically Modified Organism (GMO)
31. Name the first genetically engineered organism.
Answer: Bacteria (1973) and then mice (1974)
32. Name the most important enzyme used in DNA recombination.
Answer: Restriction endonuclease
33. Name the first discovered restriction endonuclease.
Answer: The first discovered restriction endonuclease is *Hind II* (from the bacterium *Haemophilus influenzae* by Smith, 1968)
34. What are the items for which biopatent is done for commercial exploitation?
Answer: (a) Strains of microorganisms
(b) Cell lines
(c) Genetically modified plants and animals
(d) DNA sequences
(e) Protein encoded by DNA sequences
(f) Biotechnological method
(g) Product and its application

Long-Answer Questions

1. Define genetic engineering. Give an account of genetic engineering and mention its applications.
2. What are restriction endonucleases? What is their source? Describe the role of restriction endonucleases in genetic engineering.
3. Give an account of applications of genetic engineering.
4. Describe enzymes and vectors used in DNA recombinant technology.
5. Write short notes on:
(a) cDNA
(b) Restriction endonuclease
(c) Cosmid vectors
(d) Shotgun technique
(e) Gene splicing
(f) Application of rDNA technology

GENETICS OF BACTERIOPHAGES

Short-Answer Questions

1. What are bacteriophages?
Answer: Bacteriophages are viruses that infect bacteria.
2. What is the genetic material of bacteriophage?
Answer: The genetic material of bacteriophages may be ssRNA, dsRNA, ssDNA or dsDNA
3. Name the largest group of bacteriophages.
Answer: Tailed bacteriophages
4. In how many different shapes are bacteriophages found?
Answer: Bacteriophages are found in three different forms:
(a) Tailed (b) Cubic (c) Filamentous
5. Name the different ways by which transfer of genetic material takes place from one bacterium to another.
Answer: (a) Transformation (b) Transduction (c) Conjugation
6. What are virulent and temperate phages?
Answer: The phages that undergo only lytic cycles are called virulent phages, whereas those phages that undergo lysogenic cycles are called temperate phages.
7. Bacterial viruses have only lytic cycle?
Answer: Because they cause death and destruction of the host bacterium
8. Give an example of virulent phages.
Answer: T-even phages
9. What is lysogeny?
Answer: The integration of phage DNA with bacterial chromosome is known as lysogeny.
10. Give an example of lysogenic virus.
Answer: Bacteriophage of *E. coli* called λ (lambda)
11. What is a lysogen?
Answer: A cell that contains a prophage is termed as lysogen.
12. How is lysogeny maintained?
Answer: Lysogeny is maintained by a repressor protein encoded by the DNA of λ phage.
13. Name the viruses having the following types of nucleic acids: dsDNA, ssDNA, dsRNA, (+)ssRNA and (–)ssRNA
Answer: (a) ds DNA – Herpes simplex virus
(b) ssDNA – Parvovirus
(c) dsRNA – Rotavirus
(d) (+) ssRNA – Retrovirus
(e) (–) ssRNA – Orthomyxovirus
14. Differentiate between bacteriophages and eukaryotic viruses.
Answer:

	Bacteriophages	Eukaryotic viruses
(a)	Generally, viral genome is injected into cell without capsid.	Generally, virion is taken up inside cell (viral genome is not injected into host cell).
(b)	There is no 5′ capping.	Viral mRNA is capped.
(c)	Poly-A tails are lacking.	There is poly-A tail at 3′ end of ssRNA viruses.
(d)	No segmented genomes are present.	Some ssRNA viruses have segmented genome.

15. Name the bacteriophages whose proteins penetrate the host cell.
Answer: Phage M13
16. Name the viruses whose genomes replicate in the cytoplasm.
Answer: Poxviruses
17. Name the RNA viruses that do not replicate in the cytoplasm.
Answer: (a) Orthomyxoviruses (influenza A and B)
(b) Retroviruses
(c) Hepatitis delta virus
18. What are virusoids?
Answer: Virusoids are satellite nucleic acids which may be single-stranded RNA, DNA or double-stranded RNA.
19. Which group of viruses does not encode RNA-dependent RNA polymerase?
Answer: Retroviruses
20. Name the enzyme having both DNA polymerase and RNAase activities.
Answer: Reverse transcriptase
21. Name the virus having segmented genome.
Answer: Influenza virus
22. Name the various genes found in retroviruses.
Answer: (a) *gag*, encoding structural proteins
(b) *pol*, encoding reverse transcriptase and integrase
(c) *env*, encoding envelope proteins

Long-Answer Questions

1. What are bacteriophages? Describe the life cycle of a lytic phage.
2. Write short notes on:
(a) Bacteriophages
(b) Hfr bacteria
(c) Replication of φX174 DNA
(d) Lysogenic cycle of λ phage
(e) Adsorption of phage on a bacterial cell
(f) Plaque assay

MENDELISM

Multiple-Choice Questions

1. Study of inheritance of characters from one generation to another is known as:
   (a) Gerontology (b) Genetics (c) Evolution (d) Cytology
2. Who is regarded as the 'Father of Genetics'?
   (a) T H Morgan (b) H J Muller
   (c) Gregor Johann Mendel (d) Archibald Garrod
3. Mendel was born in:
   (a) 1802 (b) 1822 (c) 1852 (d) 1874
4. Mendel published his first paper in:
   (a) 1860 (b) 1866 (c) 1872 (d) 1876
5. The paper entitled-experiments on plant hybridisation is associated with:
   (a) Hugo De Vries (b) Carl Correns
   (c) Erich Von Tschermak (d) Gregor Johann Mendel
6. Mendel published his work in the journal:
   (a) Proceeding of the Brunn Natural Science Society (b) Systema Naturae
   (c) Die Naturlischen Pflanzen Familien (d) Scala Naturae
7. Mendel's original paper was republished in:
   (a) 1901 (b) 1905 (c) 1910 (d) 1915
8. Mendel's work was rediscovered independently by:
   (a) Hugo De Vries, Morgan and Muller (b) Lari Correns, Erich Von Tschermak and Bateson
   (c) Carl Correns, Batson and Punnet (d) Hugo Devries, Carl Correns and Erich Von Tschermak
9. Mendel performed his work on:
   (a) *Pisum sativum* (b) *Oenothera lamarckiana*
   (c) *Drosophila melanogaster* (d) *Datura*
10. Mendal was born in:
    (a) Heinzendorf (b) Peru (c) Canada (d) Sydney
11. The various forms of given genes are called:
    (a) Genome (b) Genotype (c) Alleles (d) Null alleles
12. An allele is:
    (a) Dominant (b) Recessive (c) Null, wild type or mutant (d) All
13. Mendel studied ________ traits in pea plants:
    (a) 3 (b) 5 (c) 7 (d) 9

14. Which one of the following is a monohybrid ratio?
(a) 2:1 (b) 3:1 (c) 2:1:2 (d) 5:1
15. The genotypic ratio of monohybrid cross in F_2 is:
(a) 1:2:1 (b) 1:1:1 (c) 1:1:2 (d) 3:1
16. The character not chosen by Mendel for his work:
(a) Location of flower (b) Location of pod
(c) Colour of flower (d) Colour of pod
17. Like begets like is applicable to:
(a) Cytology (b) Genetics (c) Evolution (d) Bioinformatics
18. Mendel's law of independent assortment is based on the ratio:
(a) 9:7 (b) 9:3:4 (c) 9:3:3:1 (d) 1:1:1:1
19. Which one of the following is a test cross ratio?
(a) 1:1:1:1 (b) 9:3:4 (c) 1:2:1 (d) 9:7
20. 9:3:3:1 ration is due to:
(a) Complementary genes (b) Monohybrid cross
(c) Dihybrid cross (d) Trihybrid cross
21. F_2 ratio of 1:2:1 in monohybrid cross is applicable to:
(a) Linked genes (b) Segregation
(c) Dominance and recessive (d) Pleiotropic genes
22. Which one of the following terms has not been coined by Bateson?
(a) Homozygous (b) Heterozygous (c) Allele (d) Genotype
23. The scientist associated with the law of inheritance:
(a) Darwin (b) Lamarck (c) Mendel (d) Morgan
24. Mendel is not associated with:
(a) Segregation (b) Independent assortment
(c) Dominance (d) Incomplete dominance
25. Genotype is influenced by:
(a) Ageing (b) Developmental stage (c) Environment (d) None
26. Consider the following statements:
(A) Linkage is an exception to Mendel's principle
(B) Punnet square is used to know the probable result of a cross
(C) Repeated selfing produces heterozygosity
(D) Proteome interacts with environment, generating phenome

The correct statements are:
(a) All (b) A, B and D (c) C and D (d) A and D
27. Mendel could not observe linkage in his experiment due to:
(a) Segregation (b) Dominance
(c) Independent assortment (d) Crossing over
28. The parental genotype of a plant is XxYy. Its F_2 ratio will be:
(a) 9:3:4 (b) 9:7 (c) 1:1:1:1 (d) 9:3:3:1
29. Which one of the following is responsible for independent assortment?
(a) Mutation (b) Genetic recombination
(c) Linkage (d) Co-dominance

30. Which one of the following is not associated with Mendel?
(a) Austria (b) Pea plant (c) Dominance (d) Phenome
31. Homozygosity or heterozygosity of a genotype can be confirmed by:
(a) Complementation test (b) Reciprocal cross
(c) Backcross (d) Test cross
32. What are the chances of a homozygous individual in Tt × Tt cross?
(a) 0 per cent (b) 75 per cent (c) 50 per cent (d) 25 per cent
33. Mendel was lucky that the character selected by him for monohybrid, dihybrid and trihybrid crosses were:
(a) Located on homologous chromosomes
(b) Located on different and nonhomologous chromosomes
(c) Located on homologous chromosomes and were far away from each other
(d) Unable to undergo recombination
34. The maternal and paternal characters that are transmitted to offspring:
(a) Segregate (b) Assort independently (c) Do not mix (d) All
35. The trihybrid ratio is:
(a) 27:9:9:9:3:3:3:1 (b) 27:7:7:7:5:5:5:1 (c) 27:5:5:5:9:9:9:1 (d) 27:9:9:7:5:3:3:1
36. The genotype-phenotype concept was given by:
(a) Carl Correns (1900) (b) Johannsen (1909)
(c) Sutton and Boveri (1902) (d) C B Bridges (1933)
37. The character, which does not appear in F_1 generation, is called:
(a) Mutant (b) Dominant (c) Recessive (d) Co-dominant
38. Cross between F_1 individuals with homozygous recessive parents is known as:
(a) Reciprocal cross (b) Backcross (c) Test cross (d) All
39. Mendel proposed:
(a) Blending inheritance (b) Linkage
(c) Crossing over (d) None
40. What is incorrect about phenocopies?
(a) Have same genotypes (b) Have different genotype
(c) Have same phenotypes (d) Are nonheritable
41. The genotypic ratio of a dihybrid cross in F_2 generation is:
(a) 9:3:3:1 (b) 1:2:2:4:1:2:1:1:1:21
(c) 2:1:2:3:4:3:1 (d) 3:1:1:4:3:3:1
42. Genetic recombination is due to:
(a) Independent assortment (b) Linkage
(c) Mutation (d) Incomplete dominance
43. Which one of the following is applicable to *Mirabilis jalapa?*
(a) Complete dominance (b) Incomplete dominance
(c) Pleiotropy (d) Co-dominance
44. Which of the following cross is used to confirm the law of independent assortment?
(a) Dihybrid cross (b) Backcross (c) Test cross (d) Reciprocal cross
45. In incomplete dominance, phenotypic ratio in F_2 progeny is:
(a) 3:1 (b) 1:2:1 (c) 1:1:1:1 (d) 15:1

46. Match column I with column II and select the correct answer using answer codes:

Column I	Column II
(A) Reciprocal cross	1. M-N blood group
(B) Test cross	2. Andalusian fowl
(C) Incomplete dominance	3. Mule
(D) Co-dominance	4. 1:1:1:1

Answer codes:

	A	B	C	D
(a)	4	3	2	1
(b)	3	2	1	4
(c)	3	4	2	1
(d)	2	4	1	3

47. How many types of gametes can be produced by an organism having XxYyZz?
(a) 5 (b) 9 (c) 27 (d) None

48. Which one of the following is an example of co-dominance?
(a) Sickle cell anaemia (b) Blood group
(c) Roan coat colour of cattle (d) All

49. When an Andalusian fowl having black feather is crossed with homozygous white feather fowl in F_1 generation, all fowls have blue feathers. It is an example of:
(a) Co-dominance (b) Incomplete dominance
(c) Pseudodominance (d) Epistasis

50. In the above question, if F_1 hybrids are crossed in F_2 generation, then black, blue and white feather fowls appears in the ratio of:
(a) 1:1:1 (b) 9:3:4 (c) 1:2:1 (d) 2:1:2

51. Chromosomes theory of heredity was proposed by:
(a) Sutton and Boveri (1902) (b) Bateson and Punnet (1905)
(c) Shull and East (1908) (d) Fisher (1930)

52. A brown-eyed couple has a blue-eyed daughter, the probable genotype of the parents is:
(a) BB and Bb (b) Bb and Bb (c) bb and BB (d) BB and BB

53. Which one of the following scientists performed hybridisation experiment before Mendel?
(a) Winsor (b) Kolreuter (c) Lamprecht (d) Ferguson

54. The term factor was used by:
(a) Mendel (b) Lamarck (c) Darwin (d) Lampercht

55. Which one of the following is an incorrect match?
(a) Polygenic inheritance – Nilsson Ehle
(b) Nucleus is responsible for heredity – Earnst Haeckel
(c) Incomplete dominance – Correns
(d) Chromosomes are some how – Sutton

56. The percentage of tt in a cross between TT × tt is:
(a) O (b) 25 (c) 50 (d) 75

57. Consider the following statements:
(A) Mendelian inheritance is not applicable to haploid organisms
(B) Mendel's law of inheritance can be explained with the help of meiosis
(C) Mendel's law of segregation is applicable to genes having multiple effects
(D) The characters studied by Mendel were located on chromosome 9

The correct statements are:
(a) All (b) A, B and C (c) A and B (d) B

58. A PTC taster couple have a nontaster child, the genotype of the child is:
(a) TT (b) Tt (c) tt (d) Both (a) and (b)

59. What is the frequency of XxYy in a cross between XxYy × XxYy?
(a) 8/16 (b) 6/16 (c) 4/16 (d) 1/16

60. Genetic variability decreases in:
(a) Inbreeding (b) Incomplete dominance (c) Test cross (d) All

61. Mendel died in the year:
(a) 1896 (b) 1890 (c) 1884 (d) 1880

62. The scientist associated with the discovery of nonblending inheritance is:
(a) Morgan (b) Sutton (c) Bateson (d) Mendel

63. In a cross between two AaBbCc individuals, the frequency of AaBbCc individuals will be:
(a) 1/4 (b) 1/16 (c) 1/32 (d) 1/64

64. Consider the following statements about co-dominance:
(A) In co-dominance, alleles are tightly linked on the same chromosome
(B) Both alleles are independently expressed in heterozygotes
(C) Both alleles are recessive to each other
(D) The alleles are expressed in different developmental stages

The correct statements are:
(a) B and C (b) B and D (c) A, C and D (d) All

65. Which one of the following is applicable to a mule?
(a) Outbreeding (b) Inbreeding (c) Co-dominance (d) Heterozygosity

66. The F_2 phenotype number in trihybrid cross is:
(a) 6 (b) 8 (c) 9 (d) 12

67. Match column I with column II and select the correct answer using answer codes:

Column I	Column II
(A) Emasculation	1. Sickle cell anaemia
(B) YyRr × YyRr	2. Incomplete penetrance
(C) Pleiotropioc gene	3. 9:3:3:1
(D) Polydactyly	4. Hybridisation

Answer codes:

	A	B	C	D
(a)	3	2	4	1
(b)	2	3	1	4
(c)	4	3	1	2
(d)	4	2	1	3

68. Besides pea, Mendel also worked on:
(a) Beans (b) *Hieracium*
(c) *Drosophila* (d) Beans and *Hieracium*

69. Mendel started his experiments on inheritance of characters in:
(a) 1852 (b) 1855 (c) 1857 (d) 1859

70. In which one of the following is the Mendelian rule of dominance and recessiveness not true?

(a) Flower colour of *Mirabilis jalapa* (b) ABO blood group
(c) MN blood group (d) All

71. Which one of the following is an essential feature of Mendel's particulate theory of inheritance?
(a) Variability is conserved (b) Variability is not destroyed in cross-breeding
(c) There is nonmixing of maternal and paternal factors in hybrids (d) All

72. In which one of the following crosses are the F_1 hybrids not of uniform type?
(a) Monohybrid cross (b) Dihybrid cross (c) Trihybrid cross (d) None

73. Mendel cultivated tested (between 1856 and 1863) some__________ pea (*Pisum sativum*) plants:
(a) 5,000 (b) 29,000 (c) 35,000 (d) 2,900

74. Mendel died on 6 January 1884 from:
(a) Chronic nephritis (b) Tuberculosis (c) Liver cirrhosis (d) Typhoid

75. Who is associated with modern synthesis of evolutionary biology?
(a) Karl Pearson (b) Carl Correns (c) William Bateson (d) R A Fisher

76. Bateson coined the term:
(a) Genetics (b) Gene (c) Allele (d) All

77. Which one the following is not a dominant trait in humans?
(a) Hitchhiker's thumb (b) Widow's peak
(c) Wet Carwax (d) Face freckles

78. Test cross is used to determine whether an individual dominant phenotype is:
(a) Homozygous (b) Heterozygous
(c) Homozygous or heterozygous (d) None

79. Which one of the following is used for genotyping of an individual?
(a) ASO prove (b) DNA sequencing and PCR
(c) Hybridisation to DNA microarrays (d) All

80. What is incorrect about polygenic inheritance?
(a) Also known as quantitative or multifactorial inheritance
(b) It is the inheritance of phenotypic characteristic that varies in degree
(c) May be due to the interaction between two or more genes and their environment
(d) It follows the pattern of Mendelian inheritance

81. Lethal genes are:
(a) Recessive (b) Dominant (c) Conditional/Semilethal (d) All

82. First recessive lethal gene was discovered by:
(a) Cuenot (b) Castle (c) Little (d) Davenport

83. The lethal gene ratio is:
(a) 1:1 (b) 1:2:1 (c) 3:1 (d) 2:1

84. When Baur (1907) crossed two aurea Snapdragon (*Anthrrhinum*) plants, the ratio observed by him was:
(a) 3:1 (b) 2:1 (c) 1:2:1 (d) 1:1

85. Inheritance of traits is non-Mendelian in:
(a) Viruses (b) Bacteria (c) Fungi (d) All

86. Non-Mendelian pattern of inheritance is not shown by:
(a) Trinucleotide repeat disorders (b) Genomic imprinting
(c) Gene conversion (d) All

87. Which one of the following is a polygenic trait in humans?
(a) IQ (b) Personality (c) Weight and height (d) All

88. What is incorrect about a Mendelian trait?
(a) Controlled by a single gene
(b) There are only two alleles possible in the population for the individual
(c) Out of two alleles, one is dominant and the other is recessive
(d) May be affected by the environment

89. In which one of the following do both alleles contribute equally to phenotype?
(a) Complete dominance (b) Incomplete dominance
(c) Co-dominance (d) Pseudodominance

90. Two F_1 hybrid plants having yellow-coloured seeds were crossed. The percentage of green colour in F_2 generation will be:
(a) 0 (b) 25 (c) 50 (d) 75

91. Which one of the following helps in understating Mendel's law of segregation?
(a) Mitosis (b) Meiosis (c) Crossing over (d) Linked genes

92. 3:1 phenotypic ratio can be obtained when:
(a) The alleles are identical (b) The alleles do not segregate during meiosis
(c) The alleles segregate during meiosis (d) The alleles are incompletely dominant

93. If F_1 hybrids having yellow seeds are crossed, the percentage of yellow seeds in F_2 is:
(a) 0 per cent (b) 25 per cent (c) 75 per cent (d) 100 per cent

94. A phenotypic ratio of 9:3:3:1 in the progenies for a dihybrid cross occurs when:
(a) Genes pairs are linked (b) Gene pairs assort independently
(c) Gene undergo crossing over (d) All

95. In a dihybrid cross between AaBb × AaBb, the fraction of homozygous aabb will be:
(a) 1/16 (b) 4/16 (c) 1/4 (d) 0/16

96. A man heterozygous for cystic fibrosis marries a homozygous normal woman. The chances of cystic fibrosis in their children are:
(a) 0 per cent (b) 25 per cent (c) 50 per cent (d) 100 per cent

97. A brown-eyed man whose father was blue-eyed married a woman whose mother was brown-eyed. The genotype of the man and woman is:
(a) BB (b) Bb (c) bb (d) BB or Bb

98. Which one of the following is a recessive trait in pea plants?
(a) Round seeds (b) Wrinkle seeds (c) Yellow cotyledon (d) Axillary flower

99. Generally a gamete contains:
(a) One allele of a gene (b) Two alleles of a gene
(c) Three alleles of a gene (d) All alleles of a gene

100. Which one of the following is not an interaction of genes?
(a) Pleiotropy (b) Recessiveness (c) Dominance (d) All

101. Match column I with column II and select the correct answer using answer codes:

Column I		Column II
(A) Blending inheritance	1.	2:1
(B) Complementary genes	2.	1:2:1
(C) Lethal genes	3.	15:1
(D) Duplicate genes	4.	9:7

Answer codes:

	A	B	C	D
(a)	2	4	1	3
(b)	4	3	1	2
(c)	3	4	2	1
(d)	2	4	3	2

102. Mendel was unable to observe linkage due to:
(a) Dominance (b) Multiple effect
(c) Crossing over (d) Independent assortment

103. Which one of the following is an incorrect match?
(a) Pleiotropy – Sickle cell anaemia (b) Co-dominance – Roan coat colour in cattle
(c) Incomplete dominance – Anthrrhinum (d) Polygenes – Same genotypes and phenotypes

104. Isoalleles can be distinguished by a change in:
(a) pH and humidity (b) Temperature (c) Biochemicals (d) All

105. In which one of the following do genes not occur in a pair?
(a) Gametes (b) Zygote
(c) Dividing cells (d) Spermatocytes and oocytes

106. The genotype of an organism is WwXxYyZz, the number (types) of gametes produced by this organism:
(a) 64 (b) 32 (c) 16 (d) 8

107. Which one of the following genes controlled fruit shape of Shepherd's purse plant?
(a) Epistatic gene (b) Duplicate genes
(c) Supplementary genes (d) Complementary genes

108. Which one of the following was not proposed by Mendel?
(a) Blending inheritance (b) Polygenic inheritance
(c) Linkage (d) All

109. Mendel's law a purity of gametes is based on:
(a) Monohybrid cross (b) Dihybrid cross (c) Trihybrid cross (d) Backcross

110. Which one of the following is an incorrect match?
(a) Darwin – Theory of pangenesis (b) Weismann – Theory of germplasm
(c) Malpighi – Preformation theory (d) Empedocles – Gemmule theory

111. The ratio of dominant epistasis is:
(a) 13:3 (b) 12:3:1 (c) 9:3:4 (d) 15:1

112. Which one of the following is not applicable to Mendel?
(a) Austrian Monk (b) Experiments in plant hybridisation
(c) Emasculation (d) Inbreeding

113. Monohybrid test cross ratio is:
(a) 1:2:1 (b) 9:7 (c) 2:1 (d) 1:1

114. Red:Pink:White is applicable to:
(a) Monohybrid cross (b) Dihybrid cross
(c) Incomplete dominance (d) Co-dominance

115. 9:3:3:1 is the ratio shown by:
(a) Complementary genes (b) Collaborative genes
(c) Supplementary genes (d) None

116. Which one of the following is lethal in human beings during homozygous condition?
(a) Thalassemia
(b) Congenital icthyosis
(c) Infantile amaurotic idiocy
(d) All

117. Which one of the following increases homozygosity?
(a) Mutation
(b) Crossing over
(c) Outbreeding
(d) Inbreeding

118. A hybrid having genotype RrTt is crossed with a plant having genotype rrtt, the phenotypic ratio of the offspring will be:
(a) 3:1
(b) 13:3
(c) 1:1:1:1
(d) 9:3:4

119. A cross between individuals heterozygous for two gene pairs will produce _______ genotypes:
(a) 4
(b) 5
(c) 9
(d) 11

120. In the above cross number of homozygous individuals will be:
(a) 1
(b) 3
(c) 4
(d) 6

121. Haploid organisms are:
(a) Homozygous
(b) Heterozygous
(c) Both homozygous or heterozygous
(d) Neither homozygous or heterozygous

122. Epiloia genes in human beings is:
(a) Dominant lethal
(b) Conditional lethal
(c) Recessive lethal
(d) Semilethal

123. Which one of the following statements is incorrect?
(a) Mating between a wide variety of genotypes results in continued production of new combinations in each generation.
(b) The offsprings of a heterozygous individual always segregate for a variety of gene differences.
(c) Semilethal genes do not cause death of all individuals.
(d) In cats, black skin colour is a dominant trait.

124. In a cross between YYRR and yyrr, all F_1 individuals are YyRr. When F_1 hybrids are crossed in F_2 generation, the number of yyRR will be:
(a) 0
(b) 1
(c) 2
(d) 3

125. Sudden appearance of ancestral character is known as:
(a) Atavism
(b) Pleiotropism
(c) Freemartinism
(d) Gene conversion

126. Which one of the following is not an allelic interaction?
(a) Multiple alleles
(b) Lethal factor
(c) Duplicate gene
(d) Incomplete dominance

127. Which one of the following is an example of a complementary gene?
(a) Flower colour is sweet pea
(b) Hair direction in guinea pig
(c) Pigment glands in cotton
(d) Kernel colour in wheat

128. Which one of the following ratios is applicable to multiple alleles?
(a) 13:3
(b) 7:6:3
(c) 9:3:4
(d) None

129. Which one of the following statements is incorrect about modifier genes?
(a) Modifier genes modify the effect of other genes
(b) Reduce the effect of other gene
(c) Enhance the effect of other gene
(d) Reduce or enhance the effect of other gene in qualitative manners

130. What is incorrect about outbreeding?
(a) Increases heterozygosity
(b) Enhances the vigour of progeny

(c) Always leads to better breeding value than the parental populations
(d) Involves crossing individuals of different families or crossing different breeds of livestock

131. Match column I with column II and select the correct answer using answer codes:

Column I		Column II
(A) Inbreeding	1.	Self-sterility in tobacco
(B) Outbreeding	2.	Male *Drosophila*
(C) Complete linkage	3.	Mule
(D) Multiple alleles	4.	Decreased fertility and mortality

Answer codes:

	A	B	C	D
(a)	3	1	4	3
(b)	4	3	2	1
(c)	4	4	1	2
(d)	3	1	2	3

132. Hybrids plantings are not so common due to the:
(a) Lack of proper soil
(b) Lack of reproductive fitness
(c) Difficulty of ensuring cross-pollination
(d) None

133. Who is known as the 'Father of Experimental Genetics'?
(a) Gregor Mendel (b) H J Muller (c) C E McClung (d) T H Morgan

134. Blood group of an individual is determined by:
(a) Combination of RBC and WBC
(b) Shape of RBC and amount of haemoglobin
(c) Presence or absence of antigens and antibodies
(d) All

135. Which one of the following is not applicable to ABO blood groups?
(a) Multiple alleles
(b) Co-dominance
(c) Incomplete dominance
(d) Genetic marker

136. A person while driving his/her car gets injured and loses a lot of blood. There is no time to analyse his/her blood group. It is safe to transfer blood of group:
(a) A^+ (b) B^+ (c) O^+ (d) O^-

137. If the blood group of a child is O, the possible blood group of his/her parents is:
(a) A and O (b) B and O (c) AB and O (d) O and O

138. Antigen is lacking in the blood group:
(a) A (b) B (c) AB (d) O

139. What is incorrect about the blood group AB?
(a) Lack antibody AB
(b) Universal recipient
(c) Oldest blood group
(d) More frequent in India

140. The frequency of O blood group in children of parents belonging to the blood group AB:
(a) 0 per cent (b) 25 per cent (c) 50 per cent (d) 75 per cent

141. Which one of the following is an incorrect match?

Blood group of parents		Blood group of children
(a) A	1.	50 per cent A and 50 per cent O
(b) B	2.	50 per cent B and 50 per cent O
(c) O	3.	00 per cent A or B
(d) B × O	4.	100 per cent B

142. A man having blood group A wishes to have the possibility of all four types of blood groups in his children. He should marry a woman having the blood group:
(a) A (b) B (c) AB (d) O

143. Suppose a particular population X of Delhi migrated from Jammu and Kashmir, and has been residing in Delhi for about 150 years. This migrant population shows changed gene frequencies when compared with their counterparts still residing in Jammu and Kashmir. These changed frequencies in the blood group might be due to:
(a) Temporal variation (b) Natural selection
(c) Genetic drift and geographical isolation (d) All

144. For a person of blood group B, the blood group acceptable for transfusion is:
(a) A and O (b) B and O (c) AB and O (d) A and B

145. Blood group is generally not affected by:
(a) Genetic drift (b) Natural selection (c) Diet (d) Temporal variation

146. Which one of the following is a dominant trait?
(a) Blood group O (b) Rh^+ blood group (c) Rh^- blood group (d) AB blood group

147. In humans, multiple allelism at a locus was first demonstrated by the:
(a) ABO blood group genes (b) Rh blood group genes
(c) A MN blood group genes (d) None

148. A boy of blood group O has a sister of AB group. The blood groups of his parents are:
(a) O and AB (b) A and B (c) O and A (d) O and B

149. The possible number of genotypes in the ABO blood group is:
(a) 2 (b) 4 (c) 6 (d) 8

150. A person with AB blood group could not be possibly the parent of an offspring with blood group:
(a) O (b) A (c) B (d) AB

151. The most prevalent blood group in India is the:
(a) A (b) B (c) AB (d) O

152. Which one of the following is an incorrect match?

Genotype of parents	Possible blood group of children
(a) $I^AI^A \times I^OI^O$	1. A
(b) $I^AI^B \times I^AI^B$	2. A, B or AB
(c) $I^OI^O \times I^OI^O$	3. O
(d) $I^AI^O \times I^AI^B$	4. A, B or O

153. Which one of the following is an exception to the Mendelian rule?
(a) Linkage (b) Lethal gene (c) Multiple gene (d) All

154. Which one of the following statements is incorrect?
(a) Meiotic division is the basis of segregation.
(b) Independent assortment is valid to the genes which are located on different chromosomes.
(c) There are some exceptions to the law of segregation.
(d) The Mendelian rule is not applicable to interaction of genes.

155. What is incorrect about multiple alleles?
(a) Always occupy the same locus (b) Always control the same character
(c) Exhibit crossing over (d) Related as dominant and recessive

156. The frequencies of genetic disorder may increase:

(a) Inbreeding (b) Outbreeding (c) Linkage (d) Mutation

157. Match column I with column II and select the correct answer using answer codes:

Column I	Column II
(A) Polydactly	1. Do not show independent assortment
(B) Sickle cell anaemia	2. Genetic recombination
(C) Genes present on the same chromosomes	3. Point mutation
(D) Independent assortment	4. Variable expressivity

Answer codes:

	A	B	C	D
(a)	4	3	1	2
(b)	4	1	3	2
(c)	3	4	1	2
(d)	4	1	2	3

158. Which one of the following is not an autosomal co-dominant trait?
(a) Human leucocyte antigen system (b) Haptoglobin
(c) *Acid phosphatase* and *adenylate kinase* (d) None

159. What is incorrect about autosomal dominant inheritance?
(a) Variable expressivity (b) Horizontal pedigree pattern
(c) Disease expressed in heterozygote (d) On average, half of the offsprings are affected

160. Which one of the following is not an example of multiple allelism?
(a) ABO blood groups (b) Eye colour of *Drosophila*
(c) Coat colour of *Drosophila* (d) Wing colour of *Drosophila*

161. Rh blood group derives it name from:
(a) *Rhesus macaca* (b) *Home* macaca (c) *Gorilla gorilla* (d) Baboon

162. Match column I with column II and select the correct answer using answer codes:

Column I	Column II
(A) Epistasis	1. Roan coat colour in cattles
(B) Phenocopy	2. Two genes interact to produce a particular trait
(C) Complementary genes	3. Environmentally induced phenotypes
(D) Incomplete dominance	4. Masking of effect of one gene by another which are non-allelic

Answer codes:

	A	B	C	D
(a)	4	1	3	2
(b)	3	2	4	1
(c)	4	3	2	1
(d)	4	3	1	2

163. When a white colour cow was mated with a red colour cow, all F_1 progenies were roan coat colour. On crossing F_1 individuals, red-, roan- and white-coloured calves were found in the ratio of 1:2:1. It presents an example of:
(a) Co-dominance (b) Epistastis
(c) Incomplete dominance (d) Gene interaction

164. The modes of Mendenlian traits are:
(a) Autosomal dominant (b) X-linked dominant and recessive
(c) Autosomal recessive (d) All

165. The first one to work on *Drosophila* was:
(a) H J Muller (b) T H Morgan (c) W Bateson (d) E O Dodson

166. In rodents, skin colour is under the control of multiple alleles and these are agouti (CC), chinchilla ($C^{ch}C^{ch}$), himalayans (C^hC^h) and albino (cc). The gradation of dominance of this colour:
(a) $CC>C^{ch}>C>C^h$ (b) $CC>C^h>C^{ch}>c$ (c) $C^{ch}>CC>C^h>c$ (d) $CC>C^{ch}>C^h>c$

167. Which one of the following statements is incorrect with reference to Mendel?
(a) Mendel established the laws that are the foundation of classical genetics.
(b) Charles Darwin very carefully read the copy of Mendel's study paper received by him.
(c) Nageli misinterpreted the results of Mendel's.
(d) Mendel performed his first set of hybridisation experiments with pea plants.

168. Mendel studied ________________ at the University of Vienna, but he failed.
(a) Physics (b) Mathematics (c) Biology (d) All

169. Mendel was born on ____________ 1822:
(a) 6 January (b) 12 June (c) 6 July (d) 6 October

170. Mendel worked on:
(a) *Pisum sativum* (b) *Hieracium* (c) Heredity (d) All

171. Three scientists who rediscovered Mendel's work do not belong to:
(a) Germany (b) Poland (c) Austria (d) Holland

172. A man of blood group B (whose mother was O blood group) married a woman having blood group A. The blood group of their child is O. What is the genotype of the man and woman?
(a) I^BI^o and I^AI^o (b) I^BI^o and I^AI^B (c) $I^AI^B \times I^oI^o$ (d) $I^oI^o \times I^oI^o$

173. Which one of the following statements is incorrect?
(a) In polygenic inheritance, the phenotype is contributed by two or more pairs of genes.
(b) Polygenic traits are quantified by measurement.
(c) The phenotypic expression in polygenic inheritance does not vary.
(d) In humans, the skin colour presents an example of polygenic inheritance.

174. Inbreeding may lead to:
(a) Increased genetic disorders (b) Loss of immune system function
(c) Increased homozygosity (d) All

175. The strength of linkage is affected by:
(a) Distance between genes (b) Increasing age
(c) Increase in temperature (d) All

176. In multiple alleles, the gamete contains:
(a) More than three alleles (b) Three alleles
(c) Two alleles (d) One allele

177. In multiple allelic system, an individual possesses only:
(a) One allele (b) Two alleles
(c) Three alleles (d) More than three alleles

178. Which one of the following is an interaction of genes?
(a) Formation of combs in poultry (b) Skin colour in humans
(c) Purple colour in sweet pea plant (d) All

179. In humans, the number of linkage group is:
(a) 46 (b) 23 (c) 12 (d) 6

180. The diploid number of chromosomes in *Pisum satium* is:
(a) 10 (b) 12 (c) 14 (d) 16

181. Which of the following characters pass directly from the mother to the daughter?
(a) Diagenic (b) Hologenic (c) Diandric (d) Holandric

182. In *Harmonia axyridis*, the pigmentation pattern is under the control of:
(a) Dominant allele (b) Recessive allele (c) Codomiant allele (d) Pleiotropic gene

183. Which one of the following statements is incorrect?
(a) Mendel started his work with 34 varieties. (b) Mendel reduced the number of 34 varieties to 22.
(c) Mendel worked on only seven varieties. (d) None

184. The scientist who worked on *Pisum sativum* before Mendel was:
(a) Kolreuter (b) Goss (c) Dodson (d) Naudin

185. An individual has genotype XXYYZZ, the number (type) of gametes produced by this individual is:
(a) 8 (b) 6 (c) 4 (d) 1

186. Which one of the following is not applicable to the law of independent assortment?
(a) Parental types (b) Recombinant types (c) Reciprocal cross (d) Co-dominance

187. In a particular genotype, seven out of ten individuals express the diseases phenotype, the penetrance is:
(a) 70 (b) 1.42 (c) 0.7 (d) None

188. In *Mirabilis jalapa*, incomplete dominance was observed by:
(a) Hugo de Vries (b) Correns (c) Goss (d) Bateson

189. Which one of the following is not an example of incomplete dominance?
(a) Inheritance of kappa particles *P. aurelia* (b) Coat colour in rodents
(c) Self-sterility in tobacco (d) All

190. The expression of gene is affected by:
(a) Even its own allele (b) Specific gene at the other loci involved in the same pathway
(c) The remainder of the genome (d) All

191. Who first showed that there is no blending of an individual's characters in a hybrid?
(a) Knight (b) Goss (c) Correns (d) Mendel

192. In Mendel's experiment, alleles were:
(a) Dominant (b) Co-dominance (c) Incomplete dominance (d) All

193. What is correct about independent assortment?
(a) Different genes assort independently. (b) Different genes located on homologous chromosomes assort independently.
(c) Centromeres assort independently. (d) Different chromosomes assort independently.

194. What is true about progeny of a hybrid plant?
(a) Becomes sterile (b) Breeds true
(c) Segregate (d) Resembles parental plant

195. Colour-blindness was discovered by:
(a) Schimper (1883) (b) Wilson, Yule (1906) (c) Wilson (1911) (d) Carothers (1913)

196. A cross between yellow wrinkle and yellow wrinkle gives 115 yellow wrinkle and 32 green wrinkle plants, the genotype of yellow wrinkle is:
(a) YYRR and yyrr (b) YyRr and YyRr (c) yyRR and yyrr (d) Yyrr and Yyrr

197. Which one of the following could have caused the failure of Mendel's results if the plant selected for study were:
    (a) Heterozygote
    (b) Linked genes
    (c) Occurrence of crossing over
    (d) All
198. Which one of the following is a dominant trait?
    (a) Syndactylism (b) Brachydactylism (c) Polydactylism (d) All
199. Which one of the following is an incorrect match?
    (a) Kernel colour in maize – Multiple alleles
    (b) AB blood group – Co-dominant
    (c) Congenital icthyosis – Recessive lethal gene
    (d) Incomplete penetrance – Diabetes mellitus
200. Mendel's original paper was republished in:
    (a) Flora in 1901
    (b) Flore in 1905
    (c) Proceedings of Brunn Natural Science Society 1901
    (d) Proceedings of National Science Academy
201. Which one of the following is an incorrect match?
    (a) *Pisum sativum* – Mendel
    (b) *Drosophila melanogaster* – Morgan
    (c) Induced mutation by radiation – Muller
    (d) Cis and trans arrangement – Lamprecht
202. Lamprecht (1961) demonstrated that seven genes selected by Mendel belonged to four linkage groups located on chromosomes:
    (a) 1, 3, 5 and 7 (b) 1, 5, 7 and 9 (c) 3, 7, 9 and 10 (d) 1, 4, 5 and 7
203. What is incorrect about Rh factor?
    (a) Dominant trait
    (b) Human beings have 10 different types of Rh antigens
    (c) The production of Rh antigen depends on three closely set autosomal genes
    (d) Rh^- father and Rh^+ mother causes *erythroblastosis foetalis* in the child
204. Coupling and repulsion hypothesis was given by:
    (a) Sutton and Boveri
    (b) Bateson and Punnet
    (c) Morgan
    (d) Lewis
205. What is wrong about T H Morgan?
    (a) First to work on *Drosophila*
    (b) Worked on *Drosophila* to show mechanism of segregation
    (c) Discovered pseudoallelism in *Drosophila*
    (d) Discovered sex linkage
206. Which one of the following is an X-linked trait?
    (a) Cock feathering in male fowl
    (b) Horned character in sheep
    (c) Barred plumage in fowl
    (d) Crown baldness in humans
207. Mendel got success in his experiments because:
    (a) He selected pea plant for his experiment having a variety of characters
    (b) The factors responsible for seven pairs of characters were linked and crosses were made between parents that differed in only one trait
    (c) The factors responsible for seven pairs of characters were located on different chromosomes
    (d) Mendel made proper interpretations of the data
208. A gene that suppresses the effect of other genes, which are non-allelic, is called:
    (a) Epistatic gene (b) Hypostatic gene (c) Pleiotropic gene (d) Dominant gene
209. In the above question, the gene whose effect is being suppressed is called:
    (a) Recessive gene (b) Dominant gene (c) Epistatic gene (d) Hypostatic gene

210. The number of genotypes of the coat colour of rabbit is:
(a) 6 (b) 8 (c) 10 (d) 12

211. Balanced lethal system:
(a) Only dominant homozygotes die
(b) Only recessive homozygous die
(c) Is found in *Oenothera* and *Drosophila*
(d) All

212. The number of possible gametes formed by an organism having genotype AaBbCcDdEe:
(a) 8 (b) 16 (c) 32 (d) 64

213. Which one of the following is not a polygenic trait?
(a) Intelligence in humans
(b) Eye colour in humans
(c) Frizzle trait in chicken
(d) Skin colour in humans

214. Which one of the following statements is incorrect?
(a) Homoploid hybrid speciation is rare.
(b) Homoploid hybrid does not involve change in chromosome number.
(c) *Heliconius* butterflies and sunflowers are homoploid hybrids.
(d) None

215. A phenotype is ____________ detectable characteristic of an organism:
(a) Structural
(b) Biochemical and physiological
(c) Behavioural
(d) All

216. Which one of the following contribute to the phenotype of an individual?
(a) Inherited genotype
(b) Environmental variations
(c) Transmitted epigenetic factors
(d) All

217. Zebronky is a hybrid of:
(a) Zebra and mare (b) Zebra and donkey (c) Mare and donkey (d) Donkey and mare

218. Consider the following statements:
(A) There is no experimental evidence as yet to indicate that alleles mix with each other in hybrid
(B) Independent segregation and assortment of the alleles during gamete formation in F_1 hybrids were the sources of genotypic and phenotypic variations among the F_2 progeny
(C) In co-dominance, the phenotype of heterozygote may or may not be similar to homozygous genotypes
(D) In fowls, walnut comb is the interaction of the product of two genes, viz., R and P

The correct statements are:
(a) All (b) A and C (c) C and D (d) A, B and C

219. What is correct about Mendelian disorders?
(a) Caused by single gene mutation
(b) Causes abnormality that is confined to an organ system
(c) Huntington disease is a Mendelian genetic disorder, which affects the central nervous system
(d) All

220. Variable expressivity is a common feature in:
(a) Autosomal dominant traits
(b) Autosomal recessive traits
(c) X-linked recessive traits
(d) All

221. Which one of the following is a good example of variable expressivity?
(a) Retinoblastoma
(b) Duchenne muscular dystrophy
(c) Neurofibromatosis
(d) Baldness

222. Which one of the following will not create a problem in analysing Mendelial inheritance?
(a) Lethal gene (b) Blending (c) Epistasis (d) Barr body

223. Consider the following statements:
(A) Polygenic inheritance is the most common cause of genetic disorders
(B) Expression of polygenic traits follows a normal or bell-shaped curve
(C) Polygenic inheritance tends to run in families
(D) In polygenic inheritance, sex of the affected child as well as nongenetic factors have no role

The incorrect statement is:
(a) A (b) B (c) C (d) D

224. Which one of the following statements is incorrect?
(a) Mendel was a monk.
(b) Mendelism is applicable to both plants and animals.
(c) According to the law of segregation, genes occur in pairs and during the formation of gametes, each pair separates. So then, each gamete receives only one member of the pair.
(d) According to Mendel, the genetic traits of parents generally blend in their children.

225. Which one of the following is a modified Mendelian ratio?
(a) 9:3:4 (b) 9:7 (c) 15:1 (d) All

226. Which one of the following is responsible for the formulation of coupling and repulsion by Bateson and Punnet?
(a) Failure of segregation (b) Failure of independent assortment
(c) Blending nature of genes (d) Co-dominance

227. In male *Drosophila*, crossing over does not take place except:
(a) *D. melanogaster* (b) *D. ananassae* (c) *D. malerkotliana* (d) *D. bipectinata*

228. Which one of the following may reduce the frequency of heterozygous parents?
(a) Mutation (b) Independent assortment
(c) Linkage (d) Crossing over

229. What is wrong about Correns?
(a) Is one of three rediscoverer's of Mendel's principles (b) Proposed law of filial regression
(c) Discovered incomplete dominance (d) Discovered cytoplasmic inheritance

230. Linkage in a pea plant was described by:
(a) Stern (1931) (b) Sutton (1902)
(c) Morgan (1933) (d) Bateson and Punnet (1906)

231. Which one of the following is an incorrect match?
(a) Crossing over – Produces variations
(b) Phencopy – Goldschmidt
(c) Discontinuous variations – Acquired characters
(d) Polygenic inheritance – Genes together affect a particular trait

232. Match column I with column II and select the correct answer using answer codes:

Column I		Column II
(A) Incomplete dominance	1.	Galton
(B) Hyperchromism	2.	Crossing over
(C) Law of filial regression	3.	Red:Pink:White
(D) Continued variation	4.	Same types of chromosome more than one

Answer codes:

	A	B	C	D
(a)	3	1	2	4
(b)	3	4	1	2
(c)	4	2	1	3
(d)	2	4	1	2

233. Advantages of cross-pollination in plants were noted by:
    (a) Kolreuter (1763) (b) Sprengel (1793) (c) Darwin (1874) (d) All
234. Which one of the following is a correct match?
    (a) Linked genes – Increased heterozygosity
    (b) Inbreeding – Increased frequency of recessive genetic abnormalities
    (c) Origin of alleles – Crossing over
    (d) Overdominance hypothesis – Crow (1944)
235. Which one of the following statements is correct about the offsprings of mating between Ss × ss?
    (a) All offsprings will show dominant trait.
    (b) 75 per cent offsprings will show dominant trait.
    (c) 50 per cent offsprings will show recessive trait.
    (d) 25 per cent offsprings will show recessive trait.
236. How many different types of gametic chromosomal combination may result if $2n = 14$?
    (a) 28 (b) 56 (c) 64 (d) 128
237. Which one of the following is not applicable to segregation?
    (a) Separation of maternal chromosome from paternal chromosome during meiosis
    (b) Consequent separation of their alleles
    (c) The phenotypic difference is shown by the offspring
    (d) None
238. The shape of RBCs becomes semicircular in:
    (a) Haemophilia (b) Anaemia (c) Sickle cell anaemia (d) None
239. Independent assortment is applicable to:
    (a) Plasmagenes (b) Linked genes (c) Sex-linked genes (d) None
240. Genes that have similar phenotypic effects when present separately, but when together, interact to produce a different trait, are known as:
    (a) Multiple genes (b) Duplicate genes (c) Complementary genes (d) Co-dominant genes
241. The first to translate Mendel's paper into English was:
    (a) Bateson (b) Punnet (c) Morgan (d) Boveri
242. Mendel studied:
    (a) Two genes in chromosome 4, three genes in chromosome 1 and one gene in each chromosome 5 and 7
    (b) Three genes in chromosome 4, two genes in chromosome 1 and one gene each in chromosome 5 and 7
    (c) One gene each in chromosome 4 and 1, three genes in chromosome 5 and two in chromosome 7
    (d) One gene in chromosome 4, two genes in chromosome 4, three genes in chromosome 5 and one in chromosome 7
243. The impact of gene on the phenotype depends on:
    (a) Its dominance relations
    (b) Condition of the rest of genome
    (c) Condition of the environment
    (d) All

244. Consider the following statements:
   (A) Mendelian inheritance occurs in organisms with meiosis as part of its life cycle
   (B) Gene interaction occurs in haploid as well as in diploid organisms
   (C) Mendel himself did not postulate any genetical law
   (D) It was Correns who thought that Mendel's discovery could be represented by the fundamental principle of heredity

   The correct statements are:
   (a) All (b) B, C and D (c) A and B (d) A, C and D

245. Which one of the following statements is incorrect about Hb^A allele?
   (a) Hb^A alleles is dominant with regard to anemia
   (b) Co-dominant in regard to haemoglobin
   (c) Incomplete dominance with regard to blood cell shape
   (d) None

Answers to Multiple-Choice Questions

1. (b)	2. (c)	3. (b)	4. (b)	5. (d)	6. (a)	7. (a)	8. (d)
9. (a)	10. (a)	11. (c)	12. (d)	13. (c)	14. (b)	15. (a)	16 (b)
17. (b)	18. (c)	19. (a)	20. (c)	21. (b)	22. (d)	23. (c)	24. (d)
25. (d)	26. (b)	27. (c)	28. (d)	29. (b)	30. (d)	31. (d)	32. (d)
33. (b)	34. (d)	35. (a)	36. (b)	37. (c)	38. (c)	39. (d)	40. (a)
41. (b)	42. (a)	43. (b)	44. (a)	45. (b)	46. (c)	47. (d)	48. (d)
49. (b)	50. (c)	51. (a)	52. (b)	53. (b)	54. (a)	55. (d)	56. (a)
57. (d)	58. (c)	59. (c)	60. (a)	61. (c)	62. (d)	63. (d)	64. (c)
65. (a)	66. (b)	67. (c)	68. (d)	69. (c)	70. (a)	71. (d)	72. (d)
73. (b)	74. (a)	75. (d)	76. (d)	77. (a)	78. (c)	79. (d)	80. (d)
81. (d)	82. (a)	83. (d)	84. (b)	85. (d)	86. (d)	87. (d)	88. (d)
89. (c)	90. (b)	91. (b)	92. (c)	93. (c)	94. (b)	95. (a)	96. (a)
97. (b)	98. (b)	99. (a)	100. (d)	101. (a)	102. (d)	103. (d)	104. (d)
105. (a)	106. (c)	107. (b)	108. (d)	109. (a)	110. (d)	111. (b)	112. (d)
113. (d)	114. (c)	115. (b)	116. (d)	117. (d)	118. (c)	119. (c)	120. (d)
121. (d)	122. (a)	123. (d)	124. (b)	125. (a)	126. (c)	127. (a)	128. (d)
129. (d)	130. (c)	131. (b)	132. (c)	133. (d)	134. (c)	135. (c)	136. (d)
137. (d)	138. (d)	139. (b)	140. (a)	141. (d)	142. (b)	143. (d)	144. (b)
145. (c)	146. (b)	147. (a)	148. (b)	149. (c)	150. (a)	151. (b)	152. (d)
153. (d)	154. (c)	155. (c)	156. (a)	157. (a)	158. (d)	159. (b)	160. (d)
161. (a)	162. (c)	163. (c)	164. (d)	165. (b)	166. (d)	167. (b)	168. (d)
169. (c)	170. (d)	171. (b)	172. (a)	173. (c)	174. (d)	175. (d)	176. (d)
177. (b)	178. (d)	179. (b)	180. (c)	181. (b)	182. (c)	183. (d)	184. (b)
185. (d)	186. (d)	187. (c)	188. (b)	189. (d)	190. (d)	191. (d)	192. (a)
193. (b)	194. (c)	195. (c)	196. (d)	197. (d)	198. (d)	199. (a)	200. (a)
201. (d)	202. (d)	203. (b)	204. (b)	205. (c)	206. (c)	207. (c)	208. (a)
209. (d)	210. (c)	211. (d)	212. (c)	213. (c)	214. (d)	215. (d)	216. (d)
217. (b)	218. (a)	219. (d)	220. (a)	221. (c)	222. (d)	223. (d)	224. (d)
225. (d)	226. (b)	227. (b)	228. (c)	229. (b)	230. (d)	231. (c)	232. (b)
233. (d)	234. (b)	235. (c)	236. (d)	237. (d)	238. (c)	239. (d)	240. (c)
241. (a)	242. (b)	243. (d)	244. (a)	245. (d)			

Fill in the Blanks

1. The alleles which express themselves both in homozygous and heterozygous conditions are known as _______________ alleles.
2. The gene or the genetic material of an organism, which is somewhat altered is known as ________ type.
3. An individual who is not true for its character is known as _______________.
4. A cross in which F_1 individuals are crossed with one parent, from which they are derived, is known as _______________.
5. The term 'heterosis' was coined by _______________.
6. When both allele express themselves together, it is known as _______________.
7. The removal of anther is called _______________.
8. The process of determining the genotype of an individual by the use of biological assays is known as _______________.
9. _______________ in humans is an example of polygenic inheritance.
10. The law of independent assortment is also known as _______________.
11. Independent assortment occurs during _______________ in eukaryotes.
12. Individuals having two similar alleles are called _______________.
13. Individuals having two different alleles are called _______________.
14. A gene that affects more than one phenotype is known as _______________ gene.
15. Lethal genes with less than 100 per cent penetrance and expressing are called _______________ genes.
16. The ability of a gene to be expressed to any degree is known as _______________.
17. The principle of segregation is limited to _______________ chromosomes while the principle of independent assortment is applicable to nonhomologous chromosomes.
18. Variants of a gene controlling the same characteristic are called _______________.
19. Particular position of a gene on a chromosome is called _______________.
20. RrYy forms _______________ types of gametes.
21. Independent assortment can be proved by _______________ cross.
22. A cross between F_1 hybrid and recessive parent gives _______________ ratio.
23. Pure line organisms are _______________.
24. Mendelian factors represent _______________.
25. Pre-Mendelian theories are called _______________ theories.
26. Theory of epigenesis was given by _______________
27. Vapour theory was given by _______________
28. Environmentally induced phenotype is known as _______________.
29. The term 'pure line' has been given by _______________.
30. _______________ is the measure of the number of inherited _______________ traits in a population.
31. The character which remains hidden in F_1 generation is known as _______________.
32. The relationship which exists between genetically dissimilar members of the same species is known as _______________.

33. _______________ is the relationship which exists between members of genetically different species.
34. _______________ is an exception to the Mendelian principle.
35. All seven traits in a pea plant studied by Mendel were located on ____________ different chromosomes.
36. _______________ carried out several dihybrid crosses in *Drosophila*.
37. A square which is constructed to show phenotypes and genotypes of the offspring is known as _______________ square.
38. The term _______________ used by Mendel, nowadays is called genes.
39. _______________ is used as an example of monohybrid inheritance.
40. The genotype of blood group A is _______________ or _______________.
41. _______________ first associated specific gene with a specific chromosome.
42. Punnet square is used to predict the probability of _______________ and _______________ taking place in a particular cross or mating.
43. _______________ disease is characterised by incomplete dominance.
44. Albinism is a _______________ trait.
45. Transfer of genes from one population to another is known as _______________.
46. _______________ is the opposite of inbreeding depression.
47. Linkage was first discovered by _______________ and _______________.
48. The first linkage map was developed by _______________.
49. _______________ are offsprings of mating between two pure strains.
50. Gene action is of two types, viz., _______________ and _______________.
51. All linked genes present on a pair of chromosomes constitute a _______________.
52. Coupling and repulsion are the two aspects of the same phenomenon called _______________.
53. Coat colour in mice shows _______________ Epistasis.
54. Punnet square is of three types, viz., _______________, _______________ and _______________.
55. Albinism is corn is due to _______________ gene.
56. An organism having normal allele is called _______________ type.
57. The number of phenotypic classes will be _______________ if two genes affect the same characteristics.
58. _______________ is the sum total of genotypes of all individuals in a population.
59. Sickle cell anaemia is a very good example of _______________ pleiotropic gene.
60. The seeds taken by Mendel in his dihybrid cross were _______________ and _______________.
61. The expression of more than one character by a single gene is called _______________.
62. The pattern of inheritance of white eye colour in *Drosophila* manifested that the genes are located on the _______________ chromosome.
63. Mendel's law of independent assortment is not true for _______________ genes.
64. Gene maps are constructed by using the frequency of _______________ between genes.
65. _______________ gene masks the effect of other non-allelic genes.
66. The number of linkage groups is equal to the _______________ chromosome number of that species.
67. Complete linkage is shown by _______________ and _______________.
68. Mendel carried out his experiment on _______________ and _______________ generations.
69. Mendel's law of independent assortment is not applicable in case of _______________.
70. There are _______________ linkage groups in *Drosophila*.

71. If dominant allele of either gene pair, separately or together, produce the similar phenotype, such a gene is called ______________.
72. According to Mendel's hypothesis, pairs of factors are separated during ______________.
73. Linkage was discovered by ______________ in ______________.
74. Albinism in corn is due to ______________ gene.
75. In F_1 generation of a trihybrid cross, ______________ types of gametes are produced.
76. The number of chromosomes in a mule is ______________.
77. Alleles for different genes on different chromosomes are generally not linked due to ______________ during meiosis.
78. An individual contains ______________ alleles for each trait.
79. Morgan's experimental organism was ______________.
80. Height in humans is a ______________ trait.
81. The random fluctuation in the frequency of an allele is called ______________.
82. Hybridisation without change in chromosome number is called ______________ hybrid.
83. Undesirable gene flow in wild populations is known as ______________.
84. ______________. is a good example of age-related penetrance.
85. A person with one A-blood-type allele and one B-blood-type allele would have the blood type of ______________.
86. Linkage is of two types, viz., ______________ and ______________.
87. Overdominance hypothesis was given by Shull and ______________.
88. ______________ was the first to show that Mendel's theory is also applicable to animals.
89. The type of dominance in sickle cell anemia depends on the phenotypic level at which the observations are being made- ______________, ______________ or ______________.
90. In ______________, the heterozygote shows the phenotypes of both homozygotes.
91. Mendelian inheritance occurs in any organism with ______________ as part of its life cycle.
92. When the phenotype of heterozygote is not similar to either of homozygote, it is called ______________.

Answers to Fill in the Blanks

1. Dominant	2. Mutant	3. Heterozygous
4. Backcross	5. Shull	6. Co-dominance
7. Emasculation	8. Genotyping	9. Skin colour
10. Inheritance law	11. Meiosis I	12. Homozygous
13. Heterozygous	14. Pleiotropic	15. Semilethal
16. Penetrance	17. Homologous	18. Allele
19. Locus	20. Four	21. Dihybrid
22. 1:1	23. Homozygous	24. Gene
25. Blending	26. Wolf	27. Pythagoras
28. Phenocopy	29. Johannson	30. Geneticaload, deleterious
31. Recessive	32. Allogenic	33. Xenogenic
34. Linkage	35. Four	36. T H Morgan
37. Punnet	38. Factor	39. Albinism
40. I^AI^A, I^AI^O	41. T H Morgan	42. Genotypes, phenotypes
43. Tay–Sachs	44. Pleiotropic	45. Gene flow

46. Heterosis
47. Bateson, Punnet
48. Sturtevant
49. Hybrids
50. Intragenic, intergenic
51. Linkage group
52. Linkage
53. Recessive
54. Genotypic, phenotypic, gametic
55. Lethal
56. Wild
57. Less
58. Gene pool
59. Lethal
60. Yellow round, green wrinkled
61. Pleiotropism
62. X
63. Linked
64. Crossing over
65. Epistatic
66. Haploid
67. *Drosophila, Bombyx mori*
68. F_2, F_3
69. Linked gene
70. Four
71. Duplicate gene
72. Gametogenesis
73. T H Morgan, *Drosophila melanogaster*
74. Lethal
75. Eight
76. 63
77. Independent assortment
78. Two
79. *Drosophila melanogaster*
80. Polygenic
81. Genetic drift
82. Homoploid
83. Genetic pollution
84. Multiple Endocrine Neoplasia1(MEN-1)
85. AB, A, B or O
86. Complete, incomplete
87. East (1908)
88. Bateson
89. Organismal, cellular or molecular
90. Co-dominance
91. Meiosis
92. Overdominance

True or False

1. Mendel was the first to carry out the hybridisation experiment.
2. In 1854, Mendel became a teacher.
3. Phenotype is the physical appearance of an individual.
4. Self-fertilisation of a heterozygote does not allow a homozygous recessive to appear in the next generation.
5. Mendel used discontinuous characters in his experiment.
6. Haploid organisms are homozygotes.
7. Bateson confirmed Mendel's work by a series of hybridisation experiments.
8. The Hbs allele in homozygous condition acts as a lethal gene.
9. Individuals having identical phenotypes also have identical genotypes.
10. Inflated pod shape in pea plant is a recessive trait.
11. In cattle, hornless condition is dominant over horned condition.
12. For monohybrid, the test cross ratio is 1:1.
13. Mule is a hybrid between a male horse and a female donkey.
14. The gene of phenylketonuria is pleiotropic.
15. It is difficult to get pure line in a pea plant.
16. Lethal gene causes deviation from the expected ratio.
17. Blending inheritance is a common phenomenon.
18. All characters are true dominant.
19. Morgan was the first to rediscover Mendel's law.
20. Inbreeding is also known as cross-fertilisation.
21. Gametes are never hybrid.

22. *Anthrrhinum* show incomplete dominance.
23. All alleles assort independently.
24. Linked genes are present on the same chromosome.
25. Linkage was first worked out in maize.
26. The strength of linkage depends on the number of genes.
27. Linked genes assort independently.
28. Linkage helps in locating genes on chromosomes.
29. In garden pea, yellow pod and long stem are dominant traits.
30. Mendel was made priest at the age of 50 years.
31. Mendel pointed out that inheritance is based on the segregation of alleles.
32. A test cross differentiates two heterozygous forms.
33. A completely penetrant gene is always expressed.
34. Linkage increases the possibilities of variations.
35. Sex-linked inheritance was discovered by Bateson.
36. Environmental effects may modify the expression of genes.
37. Hereditary characters are dependent on each other.
38. Mendel obtained pure strain by selfing.
39. Polygenes are linked genes.
40. Inbreeding helps in conserving useful characters.
41. A dominant allele is expressed both in homozygous and heterozygous conditions.
42. An organism having genotype AaBbCc can produce eight types of gametes.
43. Lethal alleles may be pleiotropic.
44. The phenotypic ratio obtained from Aa×Aa crossing, will be 1:1.
45. Ttrr represents a true dihybrid condition.
46. Two allelic traits are located on two homologous chromosomes.
47. If the test cross ratio of a dihybrid cross is 1:1:1:1, it indicates that the genes are not linked.
48. Gene mutation is an exception to the Mendelian rule.
49. Dihybrid ratio is the best evidence for law of independent assortment.
50. Linked genes obey the law of independent assortment.
51. Haemophilic genes do not obey Mendelian rules.
52, Linkage is shown by only those genes which are located on the same chromosome.
53. Mendelian recombination is due to crossing over.
54. In balanced lethal system, surviving phenotype is the same and consists of only heterozygotes.
55. A gene showing co-dominance has alleles expressed at the same time in development.
56. Atvism is a lethal phenomenon.
57. A genetic map is a gene map.
58. The distance between gene is called a centimorgan or map unit.
59. Mendel died on 6 January 1984, due to chronic nephritis.
60. Polygenic traits are expressed as absolute or discrete characters.
61 Mendelism is not sufficient to forecast experiment observations in all cases.
61. Gene interaction may involve two or more genes.

62. Gene coding for a widow's peak would be marked by a gene causing baldness.
63. It has been reported that gene-for-gene interactions help bacteria in killing their host.
64. Duffy antigen provides some resistance against malaria.
65. When a phenotype is expressed to a different degree among individuals having the same genotype, it is called variable expressivity.
66. Genotype and phenotype are always directly correlated.
67. A single nucleotide polymorphism typically has three genotypes.
68. Rh^- individuals may be sensitive to Rh^+ blood.
69. Epistatic and hypostatic genes are present on the same chromosome.
70. On self-fertilisation of heterozygotes, homozygous recessive appear in the next generation.
71. Diseases caused by pleiotropic genes are reversible by gene therapy.
72. In any cross, the number/type of genotype and phenotype in F_1 generation is always one.
73. In a cross between PpQq × PpQq, the frequency of PpQq is 6/16.
74. Curly hair is dominant over straight hair.
75. Mendel's law of segregation is not applicable to genes having multiple effects.

Answers to True or False

1. False	2. True	3. True	4. False	5. True	6. False	7. True	8. True
9. False	10. False	11. True	12. True	13. False	14. True	15. False	16. True
17. False	18. False	19. False	20. False	21. True	22. True	23. False	24. True
25. False	26. False	27. False	28. True	29. False	30. False	31. True	32. False
33. True	34. False	35. False	36. True	37. False	38. True	39. False	40. True
41. True	42. True	43. True	44. False	45. True	46. True	47. True	48. True
49. True	50. False	51. True	52. True	53. False	54. True	55. False	56. False
57. False	58. True	59. True	60. False	61. True	62. True	63. True	64. True
65. True	66. False	67. True	68. True	69. False	70. True	71. False	72. True
73. False	74. True	75. True					

Give Reasons

1. Mendel pointed out that F_1 tall plants are not similar to parental tall plants.
   – Because F_1 tall plants also had alleles for dwarf plants.
2. In the cross made by Mendel between pure tall and dwarf plants, the F_1 tall plants were heterozygous.
   – Because F_1 tall plants had alleles for both tallness (T) and dwarfness (t), so they were heterozygous and were tall as allele for tallness (T) is dominant over dwarfness (t).
3. Monohybrid ratio of 3:1 is not universal.
   – Because sometimes the dominant allele is unable to mask completely the phenotypic expression of recessive allele, as a result an intermediate type appears.
4. In diploid organisms, a large number of genetic variations are hidden.

– Because they are present in recessive forms.

5. To determine the genotype of an organism, it is necessary to perform genetic crosses for several generations.
   – Because individuals having similar phenotypes may have different genotypes.
6. Test cross is so called.
   – Because it is used to test whether an individual is homozygous or heterozygous.
7. Mendelian phenotypic ratios are different in some cases.
   – Because often a particular allele may be incompletely or equally dominant over the other due to the presence of more than two alleles or due to lethal alleles.
8. *Drosophila* is an ideal organism for genetics.
   –Because it has:
   (a) Small size
   (b) Short generation time
   (c) Easy to handle
   (d) Can be cultured in laboratory
   (e) Susceptibility to mutations
9. The first generation of hybrid shows greatly increased vigour and better yield.
   – Mainly, because many genes for recessive (sometimes deleterious) traits from one parent is marked by the other.
10. In humans, the skin, hair and eye colour are polygenic traits.
   – Because they are influenced by more than one allele at different loci.
11. Alleles on the same chromosome can be segregated.
   – Due to a little crossing over of DNA.
12. Two or more pairs of genes do not always assort independently to one another.
   – Because often genes are linked and so they do not assort independently to one another.
13. Mendel could not find linkage.
   – Because the genes (factors) of the characters selected by him were not linked.
14. *Neurospora* is suitable by for genetical experiments.
   – Because:
   (a) It can be grown on a minimal medium.
   (b) It has a short life cycle.
   (c) It is a haploid organism, so recessive genes can express themselves.
15. Incomplete dominance does not favour blending theory.
   – Because 25 per cent of F_1 progenies show parental characters.
16. In *Drosophila*, genes for eye colour and wing length are inherited together.
   – Because they are located on the same chromosome.
17. The shape of RBCs becomes semicircular in sickle cell anaemia.
   – Because of reduced oxygen tension.
18. Bateson is considered by many workers as the real founder of genetics.
   – Because he coined the term genetics and was the first to translate Mendel's work into English as well as the first to show that Mendel's theory is also applicable to animals.
19. Mule is sterile.
   – Because of difference in chromosome number of horse (64) and donkey (62).

SEX DETERMINATION

Multiple-Choice Questions

1. Who first showed that sex is determined by the presence or absence of X?
   (a) Henking (1891) (b) Stevens (1906) (c) McClung (1902) (d) Correns (1906)
2. Which one of the following is an incorrect match?
   (a) X/A ratio – Bridges (b) Theory of heterogamesis – Correns (1906)
   (c) TDF – Longest gene (d) Freemartin – Cattle
3. What is incorrect about intersex with reference to *Drosophila*?
   (a) They are interfile. (b) Gonads are absent.
   (c) They are modified females. (d) Intersex characters are dominant.
4. No sex chromosome is known to carry structural genes with sex-determining properties except:
   (a) Y chromosome of mammals (b) X chromosomes of *C. elegans*
   (c) W chromosome of birds (d) All
5. In vertebrates, sex is determined by:
   (a) Chromosomal factors (b) Environmental influences
   (c) Both (a) and (b) depending on the species group (d) Nutritional status and chromosomal factors
6. XX♀ – XY ♂ sex-determination system is:
   (a) Human (b) *Drosophila* and humans
   (c) *Caenorhabditis elegans* (d) *Ginko*
7. In which one of the following animals are females heterogametic and males homogametic?
   (a) Some amphibians (b) Reptiles (c) Birds (d) All
8. The XY–XY system of sex determination is found in:
   (a) Organisms having alternation of generations (b) Vascular plants
   (c) Liverworts (d) All
9. In the XY–XY system:
   (a) Male gametophytes have X chromosomes (b) Female gametophytes have X chromosomes
   (c) Sporophytes have XY chromosomes (d) All
10. In grasshoppers, sex-determining mechanism is:
    (a) XX♀ –XY ♂ (b) XX♀ –XY ♂ (c) XX♀ –XO ♂ (d) $^{X}/_{A}$ ratio
11. Which one of the following statements is incorrect?
    (a) In alligators, sex is determined by the temperature at which the embryos are allowed to develop.
    (b) In alligators, both males and females are chromosomally the same.
    (c) In *Drosophila* and *Caenorhabditis elegans*, sex is determined by the relative ratio of sex chromosomes X and autosomes.
    (d) None

12. Dosage compensation is unaffected by:
    (a) X chromosome (b) Y chromosome
    (c) Both X and Y chromosomes and sexual physiology (d) Sexual physiology
13. Single gene controlling sex is found in:
    (a) *Neurospora* (b) *Chlamydomonas* (c) *Asparagus* (d) All
14. The existence of sex chromosomes X and Y was cytologically demonstrated by:
    (a) Painter (1923) (b) McClung (1902)
    (c) Morgan and Bridges (1917) (d) Goldschmidt (1934)
15. Multiple set of sex chromosomes are present in:
    (a) Marsupials (b) Monotremes
    (c) Eutherians (d) Monotremes and marsupials
16. Which one of the following statements is incorrect with reference to marsupials?
    (a) XO – Leads to the development of empty scrotum
    (b) XXY – Scrotum fails to develop
    (c) XX – Leads to the development of pouch and mammary glands
    (d) None
17. Consider the following statements:
    (A) X and Y chromosomes share common genes
    (B) All genes on a given chromosome belong to the same linkage group
    (C) Deletion on Y chromosome may cause male infertility
    (D) The Y chromosome is one-third the size of the chromosome

    The incorrect statement is:
    (a) None (b) A (c) B (d) C
18. In which one of the following is sex determined by the temperature at which the eggs are incubated?
    (a) Lizards (b) Alligators (c) Turtles (d) All
19. In which one of the following are sex chromosomes indistinguishable or slightly differentiated?
    (a) Ostriches (b) Emus
    (c) Ostriches, emus and penguins (d) Australian crossowaries, ostriches and emus
20. Which one of the following statements is incorrect?
    (a) Some species of fishes are hermaphrodite.
    (b) Hermaphroditic fishes have no sex chromosomes.
    (c) In lizards, high temperature can convert a genotypic male (ZZ) into a female.
    (d) None
21. In which one of the following does an XO karyotype result in a female?
    (a) Mice (b) Sheep (c) Monkeys (d) All
22. In which one of the following if the male dies or disappears, the top-ranking female gradually changes into a male and starts producing sperms?
    (a) *Petromyzon* (b) Coral reef fish (c) Dogfish (d) *Crepidula*
23. Y chromosome is larger than X chromosome in:
    (a) *Lychnis* (b) *D. melanogaster* (c) *Romalea* (d) *C. elegans*
24. Match column I with column II and select the correct answer using answer codes:

Column I (X/A ratio)	Column II (Sex)
(A) 0.5	1. Superfemale

(B) Less than 0.5 — 2. Normal male
(C) 0.67 — 3. Supermale
(D) More than 1 — 4. Intersex

Answer codes:

	A	B	C	D
(a)	4	2	1	3
(b)	2	3	4	1
(c)	2	4	1	3
(d)	3	1	2	4

25. Which one of the following is lacking in a Turner syndrome individual?
(a) Barr body (b) Drumstick (c) Y chromosome (d) All

26. Sex-determining gene SRY is lacking in:
(a) Duck-billed platypuses and opossums
(b) Kangaroos and walruses
(c) Nonmammalian vertebrates
(d) Duck-billed platypuses and nonmammalian vertebrates

27. In which one of the following is sex determination progamic?
(a) Queen honeybees
(b) Drone of honeybees
(c) *Crepidula*
(d) Duck-billed platypuses and nonmammalian vertebrates

28. Consider the following statements:
(A) In the absence of *Tdx* gene, the embryo develops into a female
(B) If *Tdx* gene is defective, the embryo develops into a female
(C) If *Tdx* gene is present in small quantity, the embryo develops into a female
(D) Foetal ovaries play a key role in sex differentiation

The incorrect statements are:
(a) A and D (b) B and C (c) B, C and D (d) D

29. Consider the following statements:
(A) *C. elegans* has two sexes, viz., hermaphrodite and male
(B) In *C. elegans*, males are generated through spontaneous X chromosome loss
(C) Males of *C. elegans* produce equal numbers of X haplo X and null 0X sperm
(D) In *C. elegans*, hermaphrodites have five pairs of sex chromosomes and one X chromosome

The incorrect statements are:
(a) B, C and D (b) A and B (c) C and D (d) D

30. Which one of the following is applicable to organisms that develop parthenogenetically?
(a) Meiosis and crossing over
(b) Reduction in number of chromosomes
(c) Segregation
(d) None

31. Consider the following statements:
(A) The first satellite DNA from snakes was isolated from *Elaphe radiata* to know the evolution of their Z and W chromosomes
(B) This satellite DNA was found to be present in males
(C) Thus, it indicates that this satellite DNA is derived from the Z chromosome
(D) BKm-related DNA sequences have also been reported in the genome of other families of snakes, namely, Boidae, Riperidae, Colubridae, Elapidae, etc.

The correct statements are:
(a) All (b) A, B and D (c) A and D (d) None

32. Which one of the following genes do not play a role in determination of sex in mammals?
(a) *WT1* (b) *SF1* (c) *DAX1* (d) *RAF1*

33. Haplodiploidy is a common feature of:
(a) Bees (b) Ants (c) Mites and ticks (d) All

34. Dosage compensation occurs in:
(a) *Drosophila* and mammals (b) Mealy bugs
(c) *Gryllus* and *Gryllotalpa* (d) All

35. Freemartin is due to:
(a) Gene mutation (b) Chromosomal deletion
(c) Hormonal control (d) Environmental factor

36. Gynandromorphs are:
(a) Normal male and normal female individuals (b) Half-male and half-female individuals
(c) Male with phenotypic female characters (d) Male or female with abnormal sex chromosomes

37. Genic balance theory was proposed by:
(a) T H Morgan (b) C B Bridges (c) H J Mullar (d) Wilson and Stevens

38. Environmental sex determination is found in:
(a) *Bonellia* (b) *Crepidula* (c) *Daphnia* (d) All

39. What is incorrect about honeybees?
(a) Drones develop from unfertilised eggs. (b) Queen and workers develop from fertilised eggs.
(c) Queen lays two types of egg, viz., unfertilised and fertilised. (d) Drone produces two types of sperms.

40. Match column I with column II and select the correct answer using answer codes:

Column I	Column II
(A) XO male	1. Queen and workers
(B) ZW female	2. Alligators
(C) Sex is determined by temperature	3. Bugs
(D) Royal jelly	4. Birds

Answer codes:

	A	B	C	D
(a)	3	4	2	1
(b)	4	2	3	3
(c)	2	3	4	1
(d)	4	1	3	2

41. What will be the sex of the *Drosophila* zygote having 2A+XXXY chromosomal complement?
(a) Intersex (b) Supermale (c) Superfemale (d) Normal female

42. A gynandromorph may have:
(a) Bilateral symmetry (b) Sexual dimorphism
(c) Mosaicism (d) All

43. Consider the following statements:
(A) In marsupials, X and Y chromosomes are shorter than eutherian X and Y chromosomes
(B) In marsupials, Y chromosome is male determining
(C) In marsupials, TDF gene is located on the autosome
(D) Genes *OTC*, *SYN1*, *MDM* and *STC* are located on the X chromosome in both marsupials and eutherians

The correct statements are:
(a) All (b) A, B and C (c) A, B and D (d) A, C and D

44. In cattle, whenever twins of the opposite sex are born, the males are normal but the females become sterile. Such sterile females are called:
(a) Gynander (b) Freemartin (c) Superfemale (d) Gypsy female

45. Genetic father is lacking in males of:
(a) Ants (b) Wasps (c) Bees (d) All

46. Sequential hermaphrodite is applicable to:
(a) *Genicanthus melanospilos* (b) *Drosophila melanogaster*
(c) *Petromyzon* (d) Kangaroo rat

47. Which one of the following statements is incorrect?
(a) Recombination between X and Y chromosomes is useful.
(b) Such recombination results in males without necessary genes previously present on the X chromosome.
(c) Such recombination results in females having genes previously located on the Y chromosome.
(d) In Dunnart (a marsupial), the *SRY* gene is the smallest known mammalian Y chromosome.

48. What is incorrect about freemartins?
(a) Genetically female (b) Phenotypically female
(c) Infertile (d) Fertile

49. Which one of the following is an incorrect match?
(a) Quantitative theory of sex determination – Goldschmidt
(b) Genic balance theory – Bridges
(c) Theory of heterogamesis – Wilson
(d) Haplo-diploid mechanism – Honeybees

50. Parasitic castration is shown by:
(a) *Nosema* (b) *Physalia* (c) *Sacculina* (d) *Myxine*

51. Consider the following statements:
(A) *Bonellia* is a marine worm
(B) In *Bonellia*, males are parasites in the uterus of females
(C) Males are smaller than females
(D) Larvae are juvenile and if a larva becomes attached with the proboscis of a female, it becomes intersex

The incorrect statements are:
(a) All (b) A, B and C (c) C and D (d) D

52. In which one of the following annelids is sex governed by the number of body segments?
(a) *Ophryotrocha* (b) *Bonellia* (c) *Polygordius* (d) *Sabella*

53. Progamic sex determination is found in:
(a) Drone of queenbee (b) *Drosophila*
(c) Males of *C. elegans* (d) *Melandrium*

54. Which one of the following is fertile?
(a) Intersex (b) Gynandromorph (c) Freemartin (d) None

55. Match column I with column II and select the correct answer using answer codes:

Column I	Column II
(A) Dipoidisation of male pronucleus	1. Beetles
(B) Sex mosaic	2. Sex hormones
(C) Compound X's and compound Y's chromosomes	3 Gynandromorphs
(D) Freemartins	4. Hydatiform moles

Answer codes:

	A	B	C	D
(a)	3	1	4	2
(b)	4	2	3	1
(c)	4	3	1	2
(d)	1	4	3	2

56. In which one of the following is the female heterogametic?
(a) *Fragaria elatior* (b) *Xenopus laeveis* (c) *Lymantria dispar* (d) All

57. XX-female and XY-male type of sex-determination mechanism is found in:
(a) Fumea (b) *C. elegans* (c) *Blaps polych*resta (d) *Ambystoma*

58. Which one of the following is a sex reversal gene?
(a) *Tfm* (b) *Tdy* (c) Tda^{-1} (d) *Tas*

59. Which one of the following is an anti-testis gene?
(a) *WT-1* (b) *DAX-1* (c) *SF-1* (d) *SOX9*

60. What is incorrect about *Ascaris incurve?*
(a) There are 35 chromosomes in male and 42 in female, of which 26 are autosomes.
(b) In males, sex chromosome is 8X+Y.
(c) In females, sex chromosome is 16X.
(d) Its Y chromosome bears a gene, which is comparable to mammalian testis-determining gene *Sry.*

61. The Y borne gene *SRY* is lacking in:
(a) Birds (b) Reptiles (c) Duck-billed platypuses (d) All

62. Match column I with column II and select the correct answer using answer codes:

Column I	Column II
(A) XXXXX and YYYYY sperm	1. Arthropods
(B) Dominant individual in the group becomes female	2. Clownfish
(C) Sex is determined by infection	3. Alligators
(D) Genes are not the primary factor in the determination of sex	4. Duck-billed platypuses

Answer codes:

	A	B	C	D
(a)	3	2	4	1
(b)	4	2	1	3
(c)	2	4	3	1
(d)	4	1	3	2

63. In which one of the following does the sex-determining gene have multiple allele?
(a) Habrobracon (b) *Agama agama* (c) *Chrysema picta* (d) *Lepisma*

64. In which one of the following is the female responsible for the determination of sex?
(a) Moths (b) Butterflies (c) Birds (d) All

65. Which one of the following statements is incorrect with reference to honeybees?
(a) Queen of honeybees contains 32 chromosomes.
(b) Drone of honeybees has 16 chromosomes.

(c) There are 8 chromosomes in a sperm. (d) There are 16 chromosomes in an egg.

66. In honeybees, biparental inheritance is not applicable to:
(a) Queen (b) Workers (c) Drones (d) All

67. Temperature-dependent sex determination is prevalent in reptiles except:
(a) Snakes (b) Turtles (c) *Sphenodon* (d) *Gavialis*

68. Consider the following statements:
(A) *DAX1* gene is located on the short arm of the X chromosome
(B) Duplication of *DAX1* gene causes male to female sex reversal
(C) Deletion of *DAX1* gene causes adrenal hypoplasia congenital (AHC)
(D) *DAX1* is initially expressed in genital ridges of both the sexes
The correct statements are:
(a) All (b) A, B and C (c) B and C (d) A and D

69. *WN4* is a:
(a) Ovary-determining gene (b) Testis-differentiation-suppressor gene
(c) Scrotum-differentiation gene (d) Sex-reversal gene

70. Which one of the following is a sex chromosomal gene?
(a) *DMRT1* (b) *SOX9* (c) *SF1* (d) None

71. XX–XO mechanism of sex determination is not found in:
(a) Grasshoppers (b) Crickets (c) Wasps (d) Beetles

72. Match column I with column II and select the correct answer using answer codes:

Column I	Column II
(A) Female heterogamety	1. *Agama agama*
(B) Male heterogamety	2. Lizards
(C) Both male and female heterogamety	3. Turtles
(D) Incubation of eggs at different temperature changes the sex ratios of hatching	4. Snakes

Answer codes:

	A	B	C	D
(a)	4	1	3	2
(b)	2	1	4	3
(c)	3	4	1	2
(d)	4	3	2	1

73. Absence of *SOX9* causes:
(a) Campomelic dysplasia (b) XY sex reversal
(c) Both (a) and (b) (d) Secretion of anti-Mullerian hormone

74. XX–XY mechanism of sex determination is found in:
(a) *Crepidula* (b) *Ophryotrocha* (c) *Tubifex* (d) None

75. Environmental factors play a key role in the determination of sex in:
(a) Kiwis (b) Kraits (c) Parasitic wasps (d) None

76. Which one of the following is the most conserved gene in sex determination?
(a) *DMRT-1* (b) *WT-1* (c) *SF1* (d) *DAX1*

77. Klinefelter syndrome is characterised by:
(a) Low concentration of gonadotropins (b) High concentration of testosterone
(c) High level of virilisation (d) None

78. Sex anomalies alteration is:
(a) Phenotypical (b) Chromosomal (c) Gonadal (d) All
79. Consider the following statements:
(A) Every zygote of *Crepidula* contains genes necessary for the development of male and female reproductive systems
(B) Young *Crepidula* remains a functional female and produces only ova
(C) As a young *Crepidula* matures, it gradually becomes transformed into a functional male and starts producing sperms
(D) This functional male may again become transformed into a functional female
The incorrect statements are:
(a) A, B and C (b) B, C and D (c) B and C (d) C and D
80. In *Drosophila,* sex is determined by the ratio of:
(a) Sex chromosomes X and Y (b) Sex chromosome X and autosome
(c) Sex chromosome Y and autosome (d) Sex chromosomes X, Y and autosomes
81. Which one of the following presents a good example of X chromosome inactivation?
(a) Freemartins (b) Tortoise shell cat
(c) Masculine pseudohermaphroditism (d) None
82. Lyon (1972) confirmed the existence of Barr body in:
(a) Normal females (b) Superfemales (c) Klinefelter males (d) All
83. Sex chromosomes are lacking in mammalian:
(a) RBCs (b) Kidneys (c) Liver (d) Neurons

Answers to Multiple-Choice Questions

1. (b)	2. (c)	3. (b)	4. (d)	5. (c)	6. (c)	7. (d)	8. (d)
9. (d)	10. (c)	11. (d)	12. (c)	13. (d)	14. (a)	15. (b)	16. (d)
17. (a)	18. (d)	19. (d)	20. (d)	21. (d)	22. (b)	23. (a)	24. (b)
25. (d)	26. (d)	27. (b)	28. (d)	29. (d)	30. (d)	31. (c)	32. (d)
33. (d)	34. (d)	35. (c)	36. (b)	37. (b)	38. (d)	39. (d)	40. (a)
41. (a)	42. (d)	43. (c)	44. (b)	45. (d)	46. (a)	47. (a)	48. (d)
49. (c)	50. (c)	51. (d)	52. (a)	53. (a)	54. (d)	55. (c)	56. (d)
57. (a)	58. (d)	59. (b)	60. (d)	61. (d)	62. (b)	63. (a)	64. (d)
65. (c)	66. (c)	67. (a)	68. (a)	69. (a)	70. (d)	71. (c)	72. (d)
73. (c)	74. (d)	75. (d)	76. (a)	77. (d)	78. (d)	79. (b)	80. (b)
81. (b)	82. (d)	83. (a)					

Fill in the Blanks

1. The term 'dosage compensation' was coined by ________________.
2. In *C. elegans*, X/A ratio greater than ________________ leads to the development of hermaphrodites.
3. Inactivation of X chromosome by hyperproduction occurs in________________.

4. ____________________ discovered the role of sex chromosomes in sex determination.
5. *C. elegans* has____________________ pairs of autosomes and ____________________ of sex chromosomes.
6. In XY females, gene *SRY* is____________________.
7. In the absence of ____________________ antigen, primordial gonads become ovaries.
8. Sexual differentiation of mammalian brain is dependent on the presence or absence of ____________ which is derived from testosterone.
9. Mutations in gene ____________________ are associated with male to female sex reversal in humans.
10. Heterosomes are ____________________ chromosomes.
11. In human beings, unfertilised ovum contains ____________________ chromosome.
12. In plants, sex chromosomes were first discovered in ____________________.
13. The term 'gynandromorphy' was coined by ____________________.
14. The Y-linked zinc protein gene is termed as ____________________.
15. In *D. melanogaster,* no structural gene is present on X chromosome except the ________________ gene.
16. The first evidence of common evolution on the basis of sex determination in animals has been provided by the discovery of the homology of the key sex-determining genes __________________ in *Drosophila* and ____________________ in *C. elegans*.
17. In habrobracon, the haploid embryos develop into ____________________, while fertilised eggs into____________________.
18. In *Drosophila*, ____________________ gene converts a female into a sterile male, when it is present in the homozygous state.
19. Sequential hermaphroditism is a type of hermaphroditism where an individual can change its ____________________ at some point in its life.
20. XX male syndrome is formed when __________________ occurs during the formation of male gametes.
21. The male *Protenor* forms two types of sperms, viz., one _______________ chromosome and the other ____________________ chromosome.
22. Freemartin was discovered by ____________________.
23. The ____________________ X chromosome is not transcribed.
24. In hydatiform mole, chromosome is solely derived from the ____________________.
25. Heterogametic females have ____________________ eggs.
26. The process by which the total gene expression by one X chromosome in one sex is made equal to the gene expression by XX chromosome in another sex is known as ____________________.
27. ____________________ is the most common cause of male pseudohermaphroditism.
28. *SRY* gene was isolated in the year ____________________.
29. In duck-billed platypuses, there are ____________________ chromosomes and 5Y chromosomes.
30. In *C. elegans*, the male bears one (XO) sex chromosome while with a pair of XX chromosomes, it is a ____________________.
31. Chromosomes in ZW region of birds are ____________________ in mammals and vice versa.
32. Sexual differences between individuals most probably first originated in their ____________________.
33. In homozygous condition, __________________ allele of *Drosophila* transform a normal diploid female into a sterile male.
34. *SRY* gene is located near the tip of the ____________________ of Y chromosome.
35. Gene *SOX9* was discovered in an investigation of ____________________ disease.

36. *WT1* gene was discovered in a patient with ____________________.
37. One extra X chromosome in males due to nondisjunction of X chromosome during oogenesis leads to ____________________ syndrome.
38. Sex chromosomal abnormalities occur due to ____________________ changes in chromosomes.
39. In majority of animals, sex is established at the time of ____________________.
40. XO *Drosophila* flies are ____________________ but sterile.
41. In bacteria, sex determination is controlled by ____________________.
42. The humans, X chromosome does not combine with Y chromosome except for small pieces of ____________________ regions of the telomeres.

Answers to Fill in the Blanks

1. Muller (1932)
2. 0.75
3. *Drosophila*
4. McClung
5. Five, one pair
6. Deleted or mutated
7. H-Y
8. Estradiol
9. *SOX9*
10. Sex determining
11. X
12. *Sphaerocarpus*
13. R B Goldschmidt
14. *ZFY*
15. *Sxl*
16. *Doublesex, mab3*
17. Males, females
18. Transformer (tra)
19. Sex
20. Recombination
21. With X, without X
22. Lillie
23. Inactive
24. Father
25. Megalecithal
26. Dosage compensation
27. Testicular feminisation
28. 1990
29. 5X
30. Hermaphrodite
31. Autosomal
32. Gametes
33. *tra*
34. Short arm
35. Campomelic dysplasia
36. Wilm's tumour
37. Klinefelter
38. Numeric
39. Fertilisation
40. Females
41. Sex factor
42. Pseudoautosomal

True or False

1. In honeybees, the ratio of X chromosomes to autosomes remains the same in both the sexes.
2. Errors in meiosis can produce sperms or egg cells that have abnormal number of sex chromosomes.
3. Birds lack clear homolog of *SRY.*
4. Sex lethal (SXI) gene plays a key role in the development of sexual dimorphism in *Drosophila.*
5. All animals have sex chromosomes.
6. Males are not found in whiptail lizards of Northern Mexico and Southwestern United States.
7. Genes are enough to make a male or female.
8. In *Drosophila,* dosage compensation occurs in all tissues except polytene salivary glands and imaginal disc.
9. Y chromosomes play a key role in compensatory activity.
10. All larvae of *Bonellia* are cytologically and genetically similar.
11. Drone of queenbee is infertile.
12. A strongly BKm positive region is present in the X chromosome of *D. melanogaster.*

13 In *Drosophila*, the Y chromosome does not have any sex-determining factor.
14. Many egg-laying reptiles do not exhibit genotypic sex determination.
15. Fish like barramundi can change sex at some part in their life (from male to female and female to male).
16. In whiptail lizards, eggs remain unfertilised and develop into females.
17. The XO and XY males in *Drosophila* have similar levels of X-linked products.
18. In *Drosophila*, sex determination and dosage compensation are controlled by two different gene products.
19. In *Dinophilus*, the size of the egg determines the sex.
20. The Bkm DNA quantitatively remain the same in all taxa.
21. In whiptail lizards, only females are born.
22. Chimera individuals are derived from different zygotes.
23. Dosage compensation occurs in Z-linked genes in birds.
24. Gynandromorphs are fertile.
25. Many ectothermic vertebrates have no sex chromosomes.
26. The human Y chromosome is able to recombine with itself, using palindrome base pair sequence.
27. Goby fishes have the ability to transform from one sex to another depending upon the local environmental condition.
28. A calico tomcat is generally fertile.
29. Haplodiploidy is also known as the thelytoky.
30. The inactivation of X chromosome is a random phenomenon.
31. Male pseudohermaphrodites are sex chromatin positive.
32. A gene isolated from the Y chromosome of teleost fish medaka is functionally comparable to the mammalian testis-determining gene *Sry*.
33. Mammalian X chromosome is highly conserved in placental mammals.
34. The transformer *(tra)* gene of *Drosophila* has no effect in heterozygous condition.
35. Environmental factors may cause sex reversal in amphibians.
36. Z and W chromosomes share homology with the mammalian X and Y chromosomes.
37. The differential region of Y chromosome is mostly euchromatic.
38. Male tortoise shell cats have only one X chromosome.
39. Among mammals, the marsupial Y chromosome is the longest one.
40. In humans, sex determination is syngamic.
41. There is no sex chromosome in many ectothermic vertebrates.
42. *DYS* is used to designate a segment of DNA on the Y chromosome.

Answers to True or False

1. True	2. True	3. True	4. True	5. False	6. True	7. False	8. False
9. False	10. True	11. False	12. True	13. True	14. True	15. True	16. True
17. True	18. False	19. True	20. False	21. True	22. True	23. False	24. False
25. True	26. True	27. True	28. False	29. False	30. True	31. False	32. True
33. True	34. True	35. True	36. False	37. False	38. True	39. False	40. True
41. True	42. True						

Give Reasons

1. Sex lethal (*SXL*) gene is of special interest.
   - Because its developmental function is controlled by both transcriptional and post-transcriptional mechanisms.
2. The YO condition is fatal.
   - Because Y chromosome carries many genes which are essential for survival.
3. Y chromosome is also called androsome.
   - Because it controls male sex.
4. Gynandromorphs are infertile.
   - Because they are not internally developed as male or female.
5. XXY *D. melanogaster* are female.
   - Because in *D. melanogaster*, Y chromosomes do not contain necessary genes for the development of maleness.
6. Sex which develops at high temperature is larger in body size.
   - Because high temperature causes faster growth, larger juvenile size and shorter incubation periods.
7. In honeybees, sperms produced by an individual drone are genetically identical.
   - Because drones are haploid and do not undergo meiosis for the production of sperms.
8. Monotremes are unusual with regard to sex determination.
   - Because they have 10 sex (5X and 5Y) chromosomes.
9. Duck-billed platypuses are an important link between chromosomal sex determination in birds and mammals.
   - Because genes located on X chromosomes (situated at one end of sex chromosome) are orthologous with those on human X chromosome. However, genes on chromosome, situated at the other end of the chain, are homologous with those on the bird's Z chromosome.
10. Marsupials are different from other mammals in differentiation of accessory organs.
    - Because in marsupials formation of scrotum, pouch and mammary glands are dependent on genes located on the X chromosome and not on gonadal hormones as in eutherians.
11. X and Y chromosomes differ from each other.
    - Because of accumulation of sex-determining genes on the respective sex chromosome.
12. *DMRT-1* is one of the most conserved genes in sex determination.
    - Because it is present in all invertebrate to vertebrate phyla.
13. Male tortoise shellcats are very rare.
    - Because they have only one X chromosome.
14. Many species of invertebrates have no sex-determination mechanism.
    - Because in such species, each individual has both sets of reproductive organs.
15. Vital genes lackY chromosomes.
    - Because 50 per cent of humans (females) lack Y chromosomes.

CHROMOSOMAL ABERRATIONS

Multiple-Choice Questions

1. Chromosomal abnormalities usually include:
   (a) Numerical abnormalities
   (b) Structural abnormalities
   (c) Both (a) and (b)
   (d) Crossing over
2. Chromosomal aberrations are extremely common, as they affect at least_________ of all conceptions:
   (a) 2 per cent
   (b) 5 per cent
   (c) 7.5 per cent
   (d) 10 per cent
3. Chromosomal abnormalities:
   (a) May involve either sex chromosomes or autosomes
   (b) May occur as a result of germ cell mutation in the parent or more remote ancestor
   (c) May occur as a result of somatic mutation in which only a part of the cell is affected
   (d) All
4. An isochromosome is an abnormal chromosome which has:
   (a) Deletion of one arm
   (b) Deletion of one arm with duplication of the other
   (c) Deletion of one arm with inversion of the other
   (d) Deletion of one arm with translocation of the other
5. Aneuploidy at a mitotic cell division may result in:
   (a) An individual with cell lines of two or more different chromosomal complements derived from a single zygote
   (b) An individual with cell lines of two or more different chromosomal complements derived from two different zygotes
   (c) Hydatiform mole
   (d) Turner syndrome
6. Aneuploidy:
   (a) Can arise during mitotic nondisjunction
   (b) Can arise during meiotic nondisjunction
   (c) Both (a) and (b)
   (d) None of the above
7. The frequency of meiotic nondisjunction possibly increases with:
   (a) Maternal age
   (b) Maternal hypothyroidism
   (c) After irradiation or viral infection or as a familiar tendency
   (d) All
8. Which one of the following statements is incorrect?
   (a) If the nondisjunction is at the first meiotic division, then the gamete with the extra chromosome will be either maternal or paternal in origin.
   (b) If the nondisjunction is at the second meiotic division, then both normal and the extra copy of that chromosome will be either maternal or paternal in origin.
   (c) If the nondisjunction is at the first meiotic division, then the chromosome will be maternal in origin.

(d) If the nondisjunction is at the second meiotic division, then the gamete with the extra chromosome will contain both parental homologous of that chromosome.

9. Mutation that causes visible change in the structure of chromosomes is known as:
   (a) Chromosomal aberration (b) Spontaneous mutation
   (c) Transposon (d) Aneuploidy
10. Translocation is:
   (a) Exchange of chromosomal segments between homologous chromosomes
   (b) Exchange of chromosomal segments between nonhomologous chromosomes
   (c) Change in order of gene on the chromosome
   (d) Loss of a segment from a chromosome
11. The rearrangement of genes on the chromosome in a reverse order is caused due to:
   (a) Deletion (b) Duplication (c) Inversion (d) Translocation
12. A duplication can result from meiotic events in a parent with:
   (a) Isochromosome (b) Translocation (c) Inversion (d) All
13. Which one of the following have played a part in human evolution?
   (a) Point mutations (b) Chromosomal rearrangements
   (c) Gene duplication and subsequent diversification of function (d) All
14. In paracentric inversion, if crossover occurs within the loop, then it will result in:
   (a) A dicentric chromatid and an acentric fragment
   (b) Both of these are unstable and rarely result in abnormal offsprings
   (c) Two dicentric chromatids, which are stable and result in abnormal offsprings
   (d) Both (a) and (b)
15. The chromosomal abnormality, which is most damaging in the homozygous condition?
   (a) Deletion (b) Duplication (c) Inversion (d) Translocation
16. Edward syndrome is an example of:
   (a) Translocation (b) Deletion (c) Aneuploidy (d) Monoploidy
17. If inversion is present in one of the chromosome, it suppresses:
   (a) Translocation (b) Crossing over (c) Mutation (d) Deletion
18. The gene sequence of a chromosome is changed from ABC. DEF. GH to ABCDFEDGH. It is due to:
   (a) Deletion (b) Inversion
   (c) Translocation (d) Spontaneous mutation
19. Deletion occurs more frequently in:
   (a) Polytene chromosome (b) Lampbrush chromosome
   (c) Metacentric chromosome (d) Chromosome without centromere
20. Which one of the following is applicable to inversion and translocation?
   (a) All segments are present. (b) All segments are distributed in a new fashion.
   (c) Qualitative alterations (d) Quantitative alterations
21. Which one of the following is responsible for deletion?
   (a) Unequal crossing over (b) Breaking without rejoining
   (c) Losses from translocations (d) All
22. 11q deletion disorder is associated with:
   (a) Jacobsen syndrome (b) Sezary syndrome
   (c) Retinoblastoma (d) Li–Fraumeni syndrome

23. Which one of the following is not applicable to the origin of inversion?
    (a) Parts of a chromosome become detached (b) Turning through 180º
    (c) Linear arrangement of genes (d) Reversed order of genes
24. Quantitative alterations occur in the chromosome complement during:
    (a) Deficiencies (b) Duplications
    (c) Both (a) and (b) (d) Inversions and translocations
25. In which one of the following chromosomal abnormalities does the total mass of a chromosome remain the same?
    (a) Deletion (b) Deletion and duplication
    (c) Inversion (d) Inversion and translocation
26. Chromosome aberrations generally occur when there is an error in:
    (a) Mitosis (b) Meiosis (c) Both (a) and (b) (d) None
27. Which one of the following is the best criterion for identifying deletion?
    (a) Failure of the chromosome to survive as a homozygote.
    (b) Chromosomes with deletions can never revert to a normal condition.
    (c) Lack of phenotypic expression of a recessive gene on a normal chromosome.
    (d) Failure of crossing over in the region having deletion.
28. Duplication of the gene encoding peripheral myelin protein 22 (PMP22) on chromosome 17 is associated with:
    (a) Loffler's syndrome (b) Charcot–Marie–Tooth disease
    (c) Nezelof syndrome (d) Chediak–Higashi syndrome
29. Gene duplication may be produced by:
    (a) Unequal crossing over (b) Clustering of genes
    (c) Nondisjunction (d) Inversion
30. Centric fusion of chromosomes____________ is the single-most frequent type of translocation in man:
    (a) 8th and 10th (b) 9th and 10th (c) 13th and 14th (d) 14th and 21st
31. Chromosomal abnormality in which one or more genes are added to a chromosome by the attachment of a chromosomal segment is known as:
    (a) Translocation (b) Inversion (c) Duplication (d) Deletion
32. Chromosomal mutation that causes removal of genes from genome is termed as:
    (a) Deletion (b) Duplication (c) Inversion (d) Translocation
33. Replacement of a gene from one linkage group to another is caused by:
    (a) Nondisjunction (b) Crossing over (c) Translocation (d) Inversion
34. Match column I with II and select the correct answer using answer codes:

Column I		Column II
(A) Deletion	1.	Suppression of crossing over
(B) Duplication	2.	Centric fusion of chromosome 13 and 14
(C) Inversion	3.	Prader–Willi syndrome
(D) Translocation	4.	Unequal crossing over

Answer codes:

	A	B	C	D
(a)	2	1	3	4
(b)	3	4	1	2
(c)	3	1	2	4
(d)	2	4	1	3

35. What is wrong about duplication?
    (a) Duplication may originate by unequal crossing over.
    (b) Duplications are more common than deletions.
    (c) Duplications may play an important role in permitting gene diversification during evolution.
    (d) Duplication is more harmful than deletion.
36. Duplication:
    (a) Increases the number of genes in the genotypes
    (b) Protects the organism against harmful mutations
    (c) Meiotic events in a parent with translocation, inversion or isochromosome may result in duplication
    (d) All of the above
37. Notched wing margin in *Drosophila* is formed by:
    (a) Deletion of a segment in X chromosome (b) Duplication of a segment in X chromosome
    (c) Inversion (d) Translocation
38. Chromosome aberrations take place due to:
    (a) Physical effects (b) Polyploidy (c) Aneuploidy (d) All
39. The four types of chromosomal aberrations can be detected by looking through a microscope during:
    (a) Interphase (b) Prophase I of meiosis
    (c) Prophase II of meiosis (d) Anaphase II of meiosis
40. Translocation may result in:
    (a) Duplications (b) Deficiencies (c) Sterility (d) All
41. Which one of the following pairs cause problem during meiosis?
    (a) Deletion and duplication (b) Duplication and translocation
    (c) Inversion and translocation (d) Translocation and duplication
42. What is incorrect about duplication?
    (a) Persons with heterozygous small deletions are viable while those with homozygous deletions do not survive.
    (b) The smallest visible chromosomal loss is about 4 million base pairs.
    (c) Deletion may arise when one parent has a pericentric inversion.
    (d) None
43. Which one of the following statements is incorrect?
    (a) A hydatiform mole is always 46 + XY.
    (b) Centric fusion translocation results in a trivalent at meiosis.
    (c) Chromosomal aberrations arise mainly due to nondisjunction and chromosomal breakage.
    (d) Mosaicism may be confined to the gonad.
44. The genetic map in humans and chimpanzees is very similar if pericentric inversion of chromosomes ____________ and ____________ is accounted for:
    (a) 10th and 12th (b) 12th and 14th (c) 12th and 17th (d) 13th and 19th
45. Sometimes, the original gene is found in one species and the duplicated in a totally unrelated organism. This phenomenon is known as:
    (a) Horizontal transfer (b) Vertical transfer
    (c) Descendant transfer (d) Parallel transfer
46. Match column I with column II and choose the correct answer using answer codes:

Column I	Column II
(A) Deletion	1. Reduced lifespan

(B) Mosaicism
(C) Carriers of balanced translocation
(D) Nondisjunction

2. $n + 1$ or $n - 1$
3. Congenital abnormalities
4. Always arises after fertilisation

Answer codes:

	A	B	C	D
(a)	3	1	2	4
(b)	3	4	1	2
(c)	1	3	2	4
(d)	4	1	3	2

47. During translocation:
(a) Loss of genes occurs
(b) Addition of genes occurs
(c) Rearrangement of gene occurs
(d) None

48. When translocation involves more than two nonhomologous pairs of chromosomes, the number of chromosome-containing meiotic rings is:
(a) 6 (b) 8 (c) More (d) All

49. Ring chromosomes in heterozygotes are manifested by:
(a) Translocations (b) Inversions (c) Duplications (d) Deletions

50. Which one of the following produces semisterility?
(a) Deletion (b) Inversion (c) Translocation (d) All

51. Which one of the following is responsible for relocation of a chromosomal segment to another position in the genome?
(a) Deletion (b) Duplication (c) Inversion (d) Translocation

52. An isochromosome of human autosomal chromosomes, usually results in an early spontaneous abortion with rare exceptions of isochromosomes of the short arms of chromosomes:
(a) 9 and 12 (b) 9 and 21 (c) 6 and 12 (d) 4 and 8

53. Endopolyploidy is common in:
(a) Tumour cells (b) Plant cells (c) Animal cells (d) All

54. Polyploid animals do exist in:
(a) Brine shrimps (b) Flatworms (c) Leeches (d) All

55. What is incorrect about autopolyploids?
(a) Longer vegetative period in comparison to corresponding diploids
(b) Often have lower osmotic pressure
(c) Retarded rate of cell division
(d) Increased rate of cell division

56. Which one of the following leads to gene amplification?
(a) Deletion (b) Duplication (c) Inversion (d) Translation

57. Deletions and deficiencies:
(a) Cause removal or addition of block of genes
(b) Change the genetic balance of the individual
(c) Change the morphology and physiology of their possessors
(d) All of the above

58. Waltzing disease in mice is associated with:
(a) Deletion (b) Duplication (c) Inversion (d) Translocation

59. The first case of translocation was studied in:
(a) *Drosophila* (b) *Oenothera* (c) *Triticum* (d) Maize

60. In humans, the deleted part of chromosome 22 is translocated to chromosome:
(a) 5 (b) 9 (c) 15 (d) 21

61. In humans the chromosome number ___________ is formed by the centric fusion of two chromosomes of primate ancestor:
(a) 2 (b) 4 (c) 10 (d) 20

62. Rearrangement of genes takes place due to:
(a) Deletion and duplication (b) Inversion and duplication
(c) Inversion and translocation (d) Translocation and duplication

63. Translocation involves:
(a) Addition of genes (b) Loss of genes
(c) Rearrangement of genes only (d) All

64. The bulging of an unpaired loop is characteristic of:
(a) Deletion (b) Duplication (c) Inversion (d) Translocation

65. If during deletion, the dominant gene 'A' is lost, the recessive gene 'a' becomes expressed. This is known as:
(a) Dominance (b) Pseudodominance
(c) Incomplete dominance (d) Co-dominance

66. The term 'polyploidy' was given by:
(a) Strasburger (1910) (b) T H Morgan (1910)
(c) Karpechenko (1928) (d) L J Stadler (1938)

67. Which one of the following is applicable to common chromosomes in the somatic cells of organisms?
(a) Haploid (b) Diploid (c) Triploid (d) Tetraploid

68. Which one of the following is monosomic?
(a) n (b) n+1 (c) 2n–1 (d) 2n+1

69. A genome having chromosome number more than the diploid level is known as:
(a) Aneuploidy (b) Polyploidy (c) Hyperploidy (d) Nullisomy

70. $2N + 1A - 1B$ represents:
(a) Mixed aneuploidy (b) Nullisomy
(c) Monosomy (d) None

71. Which one of the following is a nullisomic condition?
(a) n (b) $2n + 1$ (c) $2n - 2$ (d) $2n + 2$

72. Aneuploidy is due to:
(a) Disjunction (b) Nondisjunction (c) Chromosomal breakage (d) Deletion

73. If haploid chromosome is 15, the tetrasomic number is:
(a) 32 (b) 60 (c) 62 (d) 120

74. *Raphnobrassica* is:
(a) An allotetraploid (b) An autotetraploid (c) A mutant variety (d) A hexaploid

75. Tetraploidy is usually due to the failure to complete the______ zygotic division:
(a) First (b) Second (c) Third (d) Fourth

76. A double monosomic is:
(a) 2n+1+1 (b) 2n–1–1 (c) 2n–2 (d) 2n+2

77. Among chromosomal variations, the easiest to observe are that involve changes in:
(a) Structure (b) Number (c) Both (d) None

78. The animal species that appear to be allopolyploid are generally:
(a) Sexually reproducing (b) Parthenogenetic
(c) Short lived (d) All

79. Formation of polyploids directly through increase in lower number is known as:
(a) Monoploids (b) Autopolyploids (c) Allopolyploids (d) Aneuploids

80. Which one of the following represent autoalloploid?
(a) AAAABB (b) AABBBB (c) AAAABBBB (d) All

81. $4n$, $8n$ or $16n$ chromosome sets are often found in the vertebrate _________ cell nuclei:
(a) Liver (b) Kidney (c) Testis (d) Brain

82. What is true about segmental polyploidy?
(a) The term segmental polyploidy was coined by Stebbins.
(b) In this type of polyploidy there are two pairs of genomes.
(c) This is an intermediate polyploidy in between autopolyploidy and allopolyploidy.
(d) All

83. Monosomic condition may be:
(a) $2n–1$ (b) $2n–1–1$ (c) $2n–1–1–1$ (d) All

84. The genomic formula of double tetrasomic is:
(a) $2n+2$ (b) $2n+2+2$ (c) $2n+4$ (d) $2n+4+4$

85. Which one of the following statements is incorrect?
(a) Poyploid organisms have more than two genomes
(b) Polyploidy in humans always results in spontaneous abortion
(c) In animals, polyploidy is rare because of sex balance
(d) Autotriploids are fertile

86. Monoploidy is common in:
(a) Rotifers (b) Drones of honeybees (c) Wasps (d) All

87. Allopolyploidy among animals is limited due to:
(a) Rarity of interspecific cross-fertilisation (b) Less frequency of asexual reproduction
(c) Chromosomal mechanism of sex determination (d) None

88. The chromosome number can be altered by:
(a) Polyploidy, crossing over and aneuploidy (b) Polyploidy, aneuploidy and translocation
(c) Aneuploidy and centric fusion (d) Centric fusion, polyploidy and aneuploidy

89. Blakeslee and Belling (1924) reported _______ possible trisomics in *Datura stramonium*:
(a) 6 (b) 8 (c) 10 (d) 12

90. Double trisomic can by represented by:
(a) $2n+1$ (b) $2n+1+1$ (c) $2n+2$ (d) $2n+3$

91. Polyploidy can be induced by:
(a) Nitrous acid (b) Ethyl-Methyl-Sulphonate (EMS)
(c) Phospho-Ethylene Glycol (PEG) (d) Colchicine

92. Edward syndrome is characterised by trisomy of:
(a) 13th (b) 18th (c) 21st (d) 22nd

93. Endopolyploidy is applicable to:
(a) Polytene chromosome (b) Tapetum and endosperm
(c) Suspensor of many plants (d) All

94. Translocations in heterozygous condition are frequently found in nature in case of:
(a) *Tradescantia* (b) *Oenothera* (c) *Rhoeo* (d) All

95. In humans, polyploidy have been found in:
(a) Nerve cells and cancer cells (b) Liver cells and nerve cells
(c) Liver cells and cancer cells (d) None

96. Which one of the following is not a consequence of chromosomal mutation?
(a) Dosage and position effects (b) Fusion
(c) Disruption (d) None

97. Consider the following statements:
(A) Duplications of an original globin gene such as the myoglobin gene found in whales have served as a source of new genetic material
(B) Duplications play a positive role in evolution by providing the raw material for new genes
(C) Chromosomal rearrangements are the only cause of sterility in hybrids
(D) Chromosome pairing in an individual homozygous for the deletion will appear normal although chromosomes are longer

The correct statements are:
(a) All (b) A and B (c) C and D (d) A, B and D

98. The first association between cancer and chromosomal aberration was reported by:
(a) Pauling and Heartman (1955) (b) Nowell and Hungerford (1960)
(c) Walker and Jones (1964) (d) Walter and Roberts (1964)

99. The first consistent chromosomal abnormality found in human cancer was:
(a) Deletion of chromosome 13 (b) Deletion of chromosome 21
(c) Translocation between chromosome 8 and 22 (d) Translocation between chromosome 9 and 22

100. Which one of the following chromosomal rearrangements have been identified in neoplastic cells?
(a) Deletion and duplication (b) Inversion
(c) Insertion and translocation (d) All

101. Ring chromosome is applicable to:
(a) Lipoma (b) Hibernoma (c) Atypical lipoma (d) Fibromatosis

102. Which one of the following is not applicable to ring chromosome?
(a) Aberrant chromosome (b) With no ends
(c) Circular (d) With well-mark ends

103. $2n+2$ condition represents:
(a) Tetrasomic (b) Tetraploid (c) Trisomic (d) Monosomic

104. 2N+1A–1B represents:
(a) Nullisomic (b) Aneuploidy (c) Mixed aneuploidy (d) Tetrasomic

105. If haploid number of chromosome is 18, what will be a monosomic number?
(a) 36 (b) 37 (c) 35 (d) 38

106. If haploid number is 12, what will be tetrasomic number?
(a) 48 (b) 24 (c) 23 (d) 26

107. Which one of the following chromosomal complement represents monosomic condition?
(a) AB, AB (b) ABC, ABC (c) ABC, AB (d) ABC, A

108. Who suggested induction of polyploidy by cochicine?
(a) Blakeslee (1937) (b) Stebbins (1950) (c) Roux (1902) (d) Muller (1927)

109. Double monosomic condition is represented by the figure:

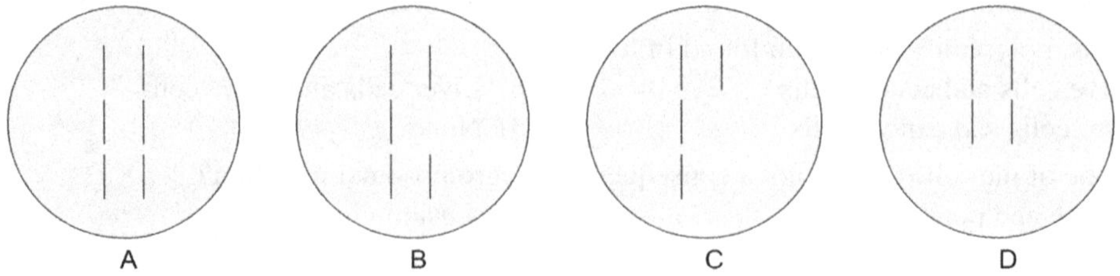

Answer codes:
(a) A (b) B (c) C (d) D

110. Trisomic condition is shown by the figure:

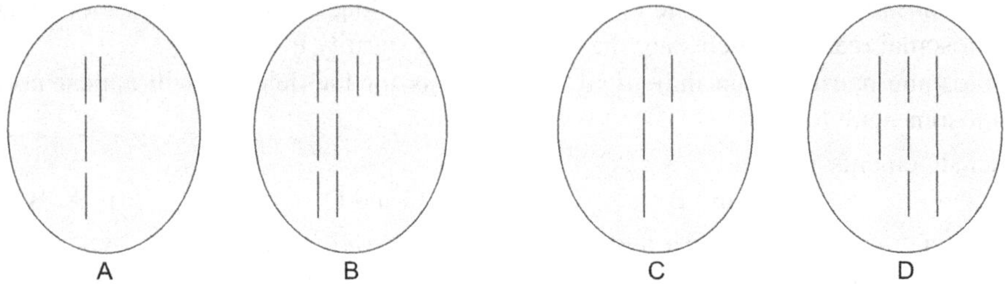

Answer codes:
(a) A (b) B (c) C (d) D

111. Exact multiple of haploid is:
(a) Aneuploid (b) Euploid (c) Hypoloid (d) Alloploid

112. Cell division is arrested by colchicine at:
(a) Prophase (b) Metaphase (c) Anaphase (d) Telophase

113. The phenotypic changes which are the result of differences in the position of genes are called:
(a) Position effects (b) Trans effects (c) Reverse effects (d) None

114. The banding techniques are very helpful in detecting cases of:
(a) Paracentric inversions (b) Pericentric inversions
(c) Balanced translocations (d) Deletions

115. The best-known balanced translocation is the:
(a) Philadelphia chromosome (b) Isochromosome
(c) Patau syndrome (d) Cri-du-chat syndrome

116. In the paracentric inversion________ chromatids are formed:
(a) Telocentric and acentric (b) Dicentric
(c) Acentric and dicentric (d) Acentric and telocentric

117. Gene mutation differs from chromosomal mutation in that:
(a) In gene mutation, the changes are produced at the molecular level
(b) In gene mutation, the alteration resides in the genetic code
(c) Gene mutation can be detected only by its molecular expression
(d) All

118. In humans, which one of the following causes chromosomal aberrations?
(a) Heavy medication radiation (b) Nuclear accidents
(c) Radioactive substances (d) All

119. The evolution of chromosomes in primates consists of modifications of the morphology in the form of:
    (a) Inversions (b) Reciprocal translocation
    (c) Fusion and fission (d) All of the above
120. Buckling effect is related with:
    (a) Deletion (b) Duplication (c) Ring chromosome (d) Isochromosome
121. Cytologically, deleted chromosome sections can be detected in:
    (a) Metaphase of mitosis (b) Anaphase of meiosis I
    (c) Prophase of meiosis I (d) Anaphase of meiosis I
122. Deletion may be caused by ________ due to error in recombination:
    (a) Viruses (b) Chemicals (c) Heat or radiation (d) All of the above
123. Chromosome having a gene sequence LMNOPQRS represents LMNOPOPQRS aberration. It is a case of:
    (a) Deletion (b) Duplication (c) Inversion (d) Translocation
124. Generally inversion results in:
    (a) Loss of genetic material (b) Gain of genetic material
    (c) Both loss or gain of genetic material (d) Neither loss nor gain of genetic material
125. Heterozygous inversions can be recognised by the presence of loops during ________ stage of meiosis:
    (a) Zygotene (b) Pachytene (c) Diplotene (d) Diakinesis
126. Which one of the following statements is incorrect?
    (a) Translocation heterozygotes are semisterile.
    (b) Animal gametes with deleted or duplicated chromosome parts are nonfunctional.
    (c) Deletion is of two types, viz., terminal and interstitial.
    (d) The zygotes formed by the gametes having deleted or duplicated chromosome parts, generally die.
127. Which one of the following is an incorrect match?
    (a) $2n-1$ – Monosomic (b) $2n+1$ – Triploid
    (c) n – Haploid (d) $2n+2$ – Tetrasomic
128. Long term survival occurs in trisomy ____________ patients.
    (a) 13th (b) 18th (c) 21st (d) None of these
129. The presence of varying numbers of quadrivalents during meiosis suggests:
    (a) Allopolyploidy (b) Autopolyploidy (c) Amphiployploidy (d) Aneuploidy
130. Polyploidy arises due to errors during:
    (a) Meiosis (b) Fertilisation
    (c) Both meiosis and fertilisation (d) None of these
131. Which one of the following is applicable to deletion?
    (a) Deficiency (b) Deletion (c) Gene deletion (d) All
132. Robertsonian translocation occurs with chromosomes:
    (a) 12, 13, 15, 17 and 22 (b) 13, 14, 15, 21 and 22
    (c) 2, 8, 12, 16 and 18 (d) 3, 7, 9, 13 and 21
133. Deletion is recognised genetically by:
    (a) Pseudodominance (b) Recessive lethality (c) Lack of reversibility (d) All
134. Translocation may:
    (a) Alter shape of the chromosome (b) Alter size of the chromosome
    (c) Alter position of the centromere (d) Alters size of the chromosome as well as position of its centromere

135. A wide variety of structural changes are possible in the chromosome depending upon the:
(a) Number of breaks (b) Location of breaks
(c) The pattern in which broken ends join together (d) All

136. The first cytological demonstration of chromosomal rearrangement in plants was made in:
(a) Wheat (b) Maize (c) Bean (d) Pea

137. The scientist who first demonstrated chromosomal rearrangement in plants:
(a) T H Morgan (b) H J Muller (c) T Dobzhansky (d) B McClintock

138. Natural selection maintains the polymorphism for gross chromosomal rearrangement in:
(a) *Datura* (b) *Campanula* and *Rhoeo*
(c) *Oenothera* (d) All

139. Chromosomal aberrations play a key role in:
(a) Genetic analysis (b) Evolution (c) Breeding (d) All

140. Structural changes in chromosomes can be identified by:
(a) Chromosome banding (b) ISH
(c) Both (a) and (b) (d) None

Answers to Multiple-Choice Questions

1. (c)	2. (c)	3. (d)	4. (b)	5. (a)	6. (c)	7. (d)	8. (b)
9. (a)	10. (b)	11. (c)	12. (d)	13. (d)	14. (d)	15. (a)	16. (c)
17. (b)	18. (b)	19. (d)	20. (d)	21. (d)	22. (a)	23. (c)	24. (c)
25. (d)	26. (c)	27. (b)	28. (b)	29. (a)	30. (c)	31. (c)	32. (a)
33. (c)	34. (b)	35. (d)	36. (d)	37. (a)	38. (d)	39. (b)	40. (d)
41. (d)	42. (d)	43. (a)	44. (c)	45. (a)	46. (b)	47. (c)	48. (d)
49. (a)	50. (d)	51. (d)	52. (a)	53. (d)	54. (d)	55. (d)	56. (b)
57. (d)	58. (a)	59. (b)	60. (b)	61. (a)	62. (c)	63. (c)	64. (a)
65. (b)	66. (a)	67. (b)	68. (c)	69. (b)	70. (a)	71. (c)	72. (b)
73. (a)	74. (a)	75 (a)	76. (b)	77. (b)	78. (b)	79. (b)	80. (d)
81. (a)	82. (d)	83. (d)	84. (b)	85. (d)	86. (d)	87. (a)	88. (d)
89. (d)	90. (b)	91. (d)	92. (a)	93. (d)	94. (d)	95. (c)	96. (d)
97. (b)	98. (b)	99. (d)	100. (d)	101. (c)	102. (d)	103. (a)	104. (c)
105. (c)	106. (d)	107. (c)	108. (a)	109. (c)	110. (b)	111. (b)	112. (b)
113. (a)	114. (c)	115. (a)	116. (c)	117. (d)	118. (d)	119. (d)	120. (a)
121. (c)	122. (d)	123. (b)	124. (d)	125. (b)	126. (b)	127. (b)	128. (c)
129. (b)	130. (c)	131. (d)	132. (b)	133. (d)	134. (d)	135. (d)	136. (b)
137. (d)	138. (d)	139. (d)	140. (c)				

Fill in the Blanks

1. Cytologically, deletions are recognised by ____________________.
2. In deletion, a single break may cause a ____________________ deletion, while two breaks can produce a ____________________ deletion.

3. In the heterozygote condition, translocations produce ____________________ meiotic products.
4. Inversions are caused by ____________________ degree turn of a segment of a chromosome.
5. In humans, the gene-determining structure of ____________________ is one of the best evidence for the unequal crossover origin of tanden duplication.
6. Jacob and Strong first observed ___________ anomalies in patients with sexual development disorders.
7. For any structural changes in the chromosome, ____________________ in the chromosome are essential.
8. Organisms are either ____________________ or ____________________ for an inversion.
9. In translocation, homozygotes break occurs in ____________________ chromosomes of two different chromosome pairs.
10. Generally, ____________________ deletion is lethal.
11. The first case of translocation was noticed in ____________________.
12. Inverted segment does not include centromere in ____________________ inversion.
13. When a translocation is present in one of the two sets of chromosomes, it is known as __________ translocation.
14. Aneuploidy occurs during ____________________.
15. ____________________ involves a loss of any part of a chromosome.
16. Numerical abnormalities of chromosomes may be ____________________ and ____________________
17. Nondisjunction may occur during ____________________ or ____________________.
18. Pericentric inversion include ____________________ within the inverted region.
19. Monosomic is formed by the union of ____________________ with normal gamete (n).
20. ____________________ in *Drosophila* is a classical example of duplication.
21. A reverse tandem duplication involves a reversal of the ____________________
22. Direct fission of centromere of metacentric chromosome results in two ______________ chromosomes.
23. For homologous chromosomes to pair, one member must form a __________ in the region of inversion.
24. Isochromosome of the ____________________ is the most common isochromosome in live births.
25. Insertional translocation involves ____________________ breaks in one or two chromosomes.
26. The four chromosomes involved in the translocation heterozygote may form _____________ chiasmata.
27. The ____________________ techniques are useful in detecting cases of balanced translocation.
28. Aneuploidy is produced by ____________________.
29. In fungi, ______________ analysis may be very useful in detecting chromosome aberrations in general.

Answers to Fill in the Blanks

1. Deletion loop
2. Terminal, interstitial
3. Duplication-deletion
4. 180°
5. Haemoglobin
6. Sex chromosome
7. Breaks
8. Homozygous, heterozygous
9. Both
10. Homozygous
11. *Oenothera*
12. Paracentric
13. Heterozygote
14. Gametogenesis
15. Deletion
16. Aneuploidy, polyploidy
17. Mitosis, meiosis
18. Centromere
19. Deficient gamete (n–1),
20. Bar eye
21. Duplicated segment
22. Telocentric
23. Loop
24. Long arm of X
25. Three
26. Two
27. Banding
28. Nondisjunction
29. Tetrad

True or False

1. Generally, translocation heterozygotes are more fertile than the homozygotes.
2. Gene amplification is a gene rearrangement that alters gene expression.
3. Translocation is of two types, viz., reciprocal and nonreciprocal.
4. Repeated duplications may lead to gene amplification.
5. The hyperploids may involve loss of one or more chromosomes.
6. Inversion involves reverse order of the genes in part of a chromosome.
7. In the presence of deficiency, a recessive allele behaves like a dominant allele.
8. Inversions are often described as crossover suppressors.
9. Duplications are easy to detect and are common.
10. Bar eye in *Drosophila* is a classical example of duplication.
11. Duplications are more common than deletions.
12. Aneuploidy is a common event in germ cells.
13. Aneuploidy may be considered as a means of speciation.
14. Gene order changed by inversion leads to the formation of unbalanced gametes.
15. Duplication is more harmful than deletion.
16. Dicentric bridges are formed in pericentric inversion.
17. Chromosomal aberrations can be diagnosed even before a child is born.
18. In humans, polyploids always abort.
19. In terminal duplication, lost segment does not include telomere.
20. If nondisjunctions occurs in meiosis I, all four gametes are abnormal.
21. A translocation always results in an abnormal phenotype.
22. Animals having homozygous deficiency, usually do not survive.
23. Banana is autotriploid.
24. Colchicine inhibits replication of chromosome.
25. Trisomy 21 patients have long-term survival.
26. Heterozygous deficiencies generally increase the general viability.
27. Crossing over within and outside inversion results in various kinds of deletions and duplications.
28. Polyploid plants are less vigorous than diploids.
29. Monoploid organisms are highly sterile.
30. Tetrasomics and nullisomics are less viable.
31. The gene dosage is not affected by euploidy and aneuploidy.
32. Centric fusion translocation is also known as Robertsonian translocation.
33. Autopolyploids are sterile to varying degrees.
34. Deletion loops are present in the polytene chromosome.

Answers to True or False

1. False	2. False	3. True	4. True	5. False	6. True	7. True	8. True
9. False	10. True	11. False	12. True	13. False	14. True	15. False	16. False
17. True	18. True	19. False	20. True	21. True	22. True	23. True	24. True
25. False	26. False	27. True	28. False	29. True	30. True	31. False	32. True
33. True	34. True						

Give Reasons

1. In agriculture, the occurrence of translocations in certain crops may reduce yield.
   – Because of the number of unbalanced zygotes formed.
2. Sometimes chromosome studies are performed, if a child is found to have chromosome anomly.
   – Because chromosome anomalies can be inherited from a parent.
3. In pericentric inversion, disjunction of crossover chromosome is normal.
   – Because the centromeres are present in the inverted regions.
4. Chromosomal aberrations play a key role in evolution.
   – Because they create variations in natural populations.
5. Homozygosity for a large deletion is often lethal.
   – Because many genes are vital to normal development.
6. Deletion of genes involved in regulation of cell division (tumour-suppressor genes) leads to an increased risk for cancer in heterozygotes.
   – Because loss of function of the one remaining copy in many cells may lead to unchecked cell division.
7. Aneuploidy is not considered as a means of speciation.
   – Because aneuploidy is common in the somatic tissue and not in the germplasm.
8. Polyploidy is supposed to inhibit evolution of animals.
   – Because of duplication of so many homologous chromosomes, the spread of new advantageous genes becomes quite impossible.
9. Allopolyploids are comparatively larger in size than the diploids from which they are derived.
   – Because increased cell size results in larger size of the plants.
10. Individuals with odd sets of chromosomes ($3n$, $5n$, $7n$) are sterile.
    – Due to abnormal meiosis which leads to the formation of abnormal gametes.
11. In animals, polyploidy is greatly restricted.
    – Because in animals the sex-determining mechanism is XX–XY type, which can be easily upset when chromosome sets are duplicated.

MUTATION

Multiple-Choice Questions

1. Who first suggested the possibility that new types of inherited characters may appear suddenly?
(a) Lamarck (b) Mendel (c) Hugo De Vries (d) Seth Wright
2. The short-legged flock of sheep raised by Seth Wright was named as:
(a) Brown swiss (b) Ancon (c) Sahiwal (d) Surti
3. Consider the following statements:
(A) Sudden, heritable change in genotype is known as mutation
(B) Mutations are deviations from normal genotypic and phenotypic conditions
(C) Mutations act as tool to understand the structure and functioning of a gene
(D) Mutations are an important source of genetic variability in living organisms
The correct statements are:
(a) All (b) A, B and C (c) C and D (d) A and D
4. The mutation frequencies (both spontaneous and induced) differ from locus to locus was observed by:
(a) T H Morgan in *Drosophila* (b) H J Muller in *Drosophila*
(c) L J Stadler in maize (d) Barbara McClintock in corn
5. The scientific study on mutation was started by:
(a) T H Morgan (1910) (b) H J Mullar (1927)
(c) Barbara McClintock (1940) (d) W R Singleton (1962)
6. Mutation rarely occurs in:
(a) TATA sequence of the promoter (b) The Shine–Dalgarno sequence of the leader
(c) The coding region (d) All
7. Point mutation in this region may turn an active gene into an inactive state:
(a) The coding region (b) TATA sequence
(c) The Shine–Dalgarno sequence (d) All
8. CIB test system was developed by:
(a) Morgan (b) Muller (c) Auerbach (d) Gustafsson
9. The mutagenic effect of X-rays on *Drosophila* was first demonstrated by:
(a) Gustafsson (b) Morgan (c) Stadler (d) Muller
10. A mutant phenotype may revert to normal phenotype:
(a) Due to reverse mutation at the same locus (b) Due to suppressor mutation at another locus
(c) Both (a) and (b) (d) Cannot revert to normal phenotype
11. The term 'mutation' was coined by:
(a) Hugo de Vries (b) Seth Wright (c) Singleton (d) Tessman
12. The scientist credited to make differentiation between heritable and environmental variations:

(a) Dulbecco (b) Tschermak (c) Ide Graaf (d) Hugo de Vries

13. The gigas mutant in *Oenothera lamarckiana* was due to:
(a) Gene mutation (b) Polyploidy (c) Deletion (d) Aneuploidy

14. 5-bromouracil is a base analogue of:
(a) Thymine (b) Guanine (c) Cytosine (d) Adenine

15. Which one of the following is a transition?
(a) $A \rightleftharpoons C$ (b) $G \rightleftharpoons C$ (c) $A \rightleftharpoons T$ (d) $A \rightleftharpoons G$

16. A change from codon AGG to AAG is a:
(a) Silent mutation (b) Neutral mutation (c) Missense mutation (d) Nonsense mutation

17. Mutations are caused by:
(a) Exposure to ultraviolet ionising radiation (b) Chemical mutagen or viruses
(c) Errors in genetic material during replication (d) All

18. *Epidermolysis bullosa* is an example of:
(a) Neutral mutation (b) Nonsense mutation (c) Missense mutation (d) Frameshift mutation

19. Marfan syndrome is an example of:
(a) Lethal mutation (b) Reverse mutation
(c) Missense mutation (d) Dominant negative mutation

20. Albino mutations resulting from deficiency of chlorophyll are:
(a) Biochemical mutations (b) Lethal mutations
(c) Resistant mutations (d) Conditional mutations

21. Match column I with column II and select the correct answer using answer codes:

Column I	Column II
(A) L J Stadler	1. Mutations can be induced by X-rays
(B) Kelner	2. Paramutation
(C) H J Muller	3. Detection of mutations in plants
(D) Brink	4. Photoreactivation

Answer codes:

	A	B	C	D
(a)	4	1	2	3
(b)	3	4	1	2
(c)	2	4	1	3
(d)	3	1	4	2

22. Replacement of purine by pyrimidine or pyrimidine by purine is:
(a) Transition (b) Transversion (c) Reversion (d) Deletion

23. Which one of the following tests is used for mutagenecity?
(a) Ames test (b) Hay's test (b) Jaffe's test (d) Rothera's test

24. Methyl cytosine residues on deamination produce:
(a) Thymine (b) Adenine (c) Guanine (d) Uracil

25. Mutations are called new when they occur in:
(a) Eggs (b) Sperms (c) Just after fertilisation (d) All

26. Which one of the following is incorrect about ultraviolet radiation?
(a) Ultraviolet radiation is a mutagen. (b) It is ionising radiation.
(c) It can react with DNA. (d) It can react with other biological molecules.

27. Mutation is transmitted to future generations, if it occurs in:
(a) Germ cells (b) Somatic cells (c) Body cells (d) Tissues

28. Which one of the following is incorrect about mutation?
(a) Rare (b) Random (c) Harmful or fatal (d) Directed

29. The frequency of spontaneous mutation is ______________ per organism:
(a) $10^2 - 10^5$ (b) $10^{-7} - 10^{-12}$ (c) $10^{-15} - 10^{-25}$ (d) $10^7 - 10^7$

30. The rate of detectable mutations in an average gene is:
(a) 1 in 10^6 (b) 5 in 10^{10} (c) 0 in 10^5 (d) 1 in 10^{10}

31. Which one of the following mutations results in death of the cells or the organism?
(a) Suppressor mutation (b) Lethal mutation
(c) Supervital mutation (d) Subvital mutation

32. The correct order of frequency of deamination of cytosine, adenine and guanine is:
(a) Guanine, adenine and cytosine (b) Adenine, guanine and cytosine
(c) Cytosine, guanine and adenine (d) Adenine, cytosine and guanine

33. Match column I with column II and select the correct answer using answer codes:

Column I	Column II
(A) Hydroxylamine	1. Forms single bond with cytosine
(B) Acridine orange	2. Breaks the phosphodiester linkages in DNA
(C) X-rays	3. Transition and mispairing
(D) 2-aminopurine	4. Distortation in DNA

Answer codes:

	A	B	C	D
(a)	3	4	2	1
(b)	1	2	4	3
(c)	2	3	4	1
(d)	3	2	4	1

34. The degree of phenotypic effect produced by a mutation depends on:
(a) Cell type (b) Nature of mutant gene (dominant or recessive)
(c) Stage of the life cycle of an organism (d) All

35. Which one of the following is an example of a dominant germinal mutation?
(a) Cancer (b) Delicious apple (c) *Ancon* (d) Naval orange

36. The deamination of cytosine at the 6-position results in the formation of:
(a) Thymine (b) Uracil (c) Guanine (d) Adenine

37. A missense mutation can be caused due to:
(a) Insertion (b) Substitution (c) Deletion (d) All

38. Most of the variations observed by Hugo de Vries in *Oenothera lamarckiana* were due to:
(a) Gene mutations (b) Chromosomal aberrations
(c) Suppressor mutations (d) Conditional mutations

39. The first chemical mutagen to be discovered was:
(a) Nitrous acid (b) 5-bromouracil (c) Dimethyl sulphate (d) Mustard gas

40. Tautomeric forms may result in the pairing of:
(a) C–A (b) G–T (c) T–G and A–C (d) All

41. The pairing behaviour of hypoxanthine is like that of:

(a) Guanine (b) Cytosine (c) Adenine (d) Thymine

42. In which one of the following is 5-hydroxymethyl cystosine present as the normal constituent of DNA instead of cytosine?
(a) T_2 bacteriophage (b) ϕX174 bacteriophage (c) SV_{40} (d) None

43. Gene mutation occurs by:
(a) Due to loss or gain of a single nucleotide base (b) Mistake/chance
(c) Repair and recombination (d) All

44. Recessive mutations are expressed in:
(a) First generation (b) Second generation
(c) Homozygous condition (d) Heterozygous condition

45. Which one of the following is not a base analogue?
(a) 5-bromouracil (b) 2-aminopurine
(c) Colchicine (d) Ethyl ethane sulphonate

46. Ultraviolet radiation results in:
(a) Deletion of base pairs (b) Addition of base pairs
(c) Formation of thymine dimers (d) Formation of cytosine dimers

47. Which one of the following genes controls the rate of mutation of the other gene?
(a) Epistatic gene (b) Regulator gene (c) Mutator gene (d) C. gene

48. Which one of the following mutations involves addition or deletion of nucleotide?
(a) Silent mutation (b) Frameshift mutation (c) Nonsense mutation (d) Missense mutation

49. Deamination of which one of the following bases does not cause heritable mutation?
(a) Adenine (b) Guanine (c) Cytosine (d) All

50. Which one of the following is an incorrect match?
(a) Morgan – Mutant in *Drosophila*
(b) Hugo de Vries – Mutants in *Oenothera lamarckiana*
(c) New Combe – Mutants in humans
(d) Beadle and Tatum – Biochemical mutants in *Neurospora*

51. Presence of which one of the following during DNA replication causes frameshift mutation?
(a) Hydroxylamine (b) 2-aminopurine (c) Nitrous acid (d) Acridine

52. Which one of the following is not a mutagenic agent?
(a) Mycoplasma (b) Viruses (c) Temperature change (d) None

53. A mutation in the germ cell can be detected:
(a) In the same generation (b) Next generation
(c) Homozygous (d) None

54. In dominant autosomal mutation, what percentages of affected heterozygous individuals are expected to have this trait?
(a) 25 per cent (b) 50 per cent (c) 75 per cent (d) 100 per cent

55. The method of detection of mutation depends whether it is located on:
(a) Autosome (b) Sex chromosome
(c) Autosome or sex chromosome (d) None

56. A mutation that inserts or deletes a single base:
(a) Results in frameshift mutation
(b) Will change the reading frame for the entire subsequent sequence

(c) The entire amino acids of protein is altered beyond the site of mutation
(d) All

57. Consider the following statements:
(A) Base mispairing and direct chemical conversion of one base into another are sources of transitions
(B) The occurrence of changes caused by transposable elements is indifferent to mutagens
(C) The deamination of cytosine generates uracil
(D) Tautomerism involves changes in chemical forms that occurs through rearrangement of electrons and protons in the molecule

The incorrect statements are:
(a) None (b) A and D (c) B and C (d) B, C and D

58. The most common type of dimer formed by ultraviolet ray is the:
(a) Cytosine–cytosine (b) Cytosinc–thymine
(c) Thymine–thymine (d) Adenine–adenine

59. Thymine in its tautomeric state pairs with:
(a) Cytosine (b) Guanine (c) Adenine (d) Cannot pair

60. Which one of the following is applicable to sickle cell anemia?
(a) Transition (b) Tautomerism (c) Transversion (d) Reversion

61. Which one of the following is incorrect with reference to frameshift mutation?
(a) Insertion or deletion of nucleotides (b) Production of useless protein
(c) May lead to thalassemia (d) Never results in premature chain termination

62. Mutations are:
(a) Harmless or beneficial (b) Deleterious
(c) Fatal (d) All

63. Which one of the following mutagens does not produce ions?
(a) X-rays (b) Alpha and beta rays (c) Gamma rays (d) Ultraviolet rays

64. The enzyme that helps in the removal of damaged nitrogenous bases:
(a) Resolvase (b) Transpoase (c) Glycosylase (d) Glycosyl transferase

65. In which one of the following diseases is DNA replication inhibited by ultraviolet rays and has defective ability to excise pyrimidine dimers?
(a) Bloom syndrome (b) Fanconi's anemia
(c) Xeroderma pigmentosum (d) Cockayne syndrome

66. Deamination of this base is unable to pair with cytosine or thymine, leading to inactivation of the DNA and often there is a lethal effect:
(a) Guanine (b) Cytosine (c) Thymine (d) Adenine

67. Which one of the following cannot be detected?
(a) Autosomal dominant mutants (b) Autosomal recessive mutants
(c) Sex-linked recessive mutants (d) None

68. Which one of the following is applicable to base analogues?
(a) Keto or enol form (b) Amino and imino form
(c) Substitution mutation and tautomeric shift (d) All

69. Back mutation can produce phenotype reversion in:
(a) Bloom syndrome (b) Kartagener syndrome
(c) Klippel–Feil syndrome (d) Angelman syndrome

70. PCR mutagenesis is a simple method for generating:
(a) Transposon mutagenesis (b) Insertional mutagenesis
(c) Directed mutagenesis (d) Site-directed mutagenesis

71. The nitrogenous bases of DNA that are most vulnerable to radiation are:
(a) Adenine and cytosine (b) Cytosine and thymine
(c) Cytosine and guanine (d) Thymine and guanine

72. It has been estimated that in humans, gene for muscular dystrophy mutate once in every:
(a) 10^{-5} people (b) 10^{5} people (c) 10^{15} people (d) 10^{20} people

73. Mutations are responsible for:
(a) Haemophilia (b) Albinism
(c) Certain kinds of cancer and brain malfunction (d) All

74. Which one of the following mutation codes for the same amino acid?
(a) Silent mutation (b) Missense mutation (c) Nonsense mutation (d) Neutral mutation

75. Which one of the following is the ultimate source of genetic variability?
(a) Genetic drift (b) Mutation (c) Nonrandom mating (d) Migration

76. Which one of the following is a disorder due to mutation lacking or reducing DNA repair mechanism?
(a) Albinism (b) Cystic fibrosis
(c) Osteosarcoma (d) Xeroderma pigmentosum

77. Mutation may occur in a single cell within an early embryo. As all cells divide during growth and development, the individual will have some cells with mutations and some cells without any genetic change. This situation is known as:
(a) Freemartinism (b) Gynandromorphism (c) Mosaicism (d) None

78. Point mutation does not occur due to:
(a) Mistakes in DNA replication (b) Insertion of mobile element
(c) DNA damage by chemical mutagens and misrepair (d) Unequal crossing over and insertion of mobile element

79. Site-directed mutagenesis was first described by:
(a) H J Muller (1935) (b) Michael Smith (1978)
(c) Kary B Mullis (1993) (d) T H Morgan (1940)

80. A triplet sequence is repeated many times in:
(a) Fragile X syndrome (b) Huntington disease
(c) Myotonic dystrophy (d) All

81. In the figure given below, transversion is not shown by the letter:
(a) I (b) III (c) II (d) IV

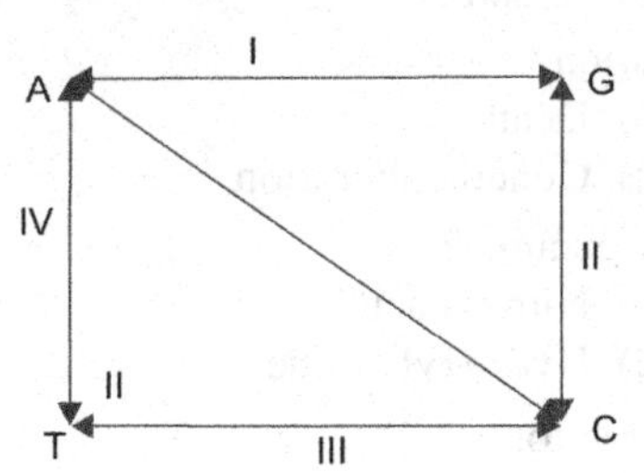

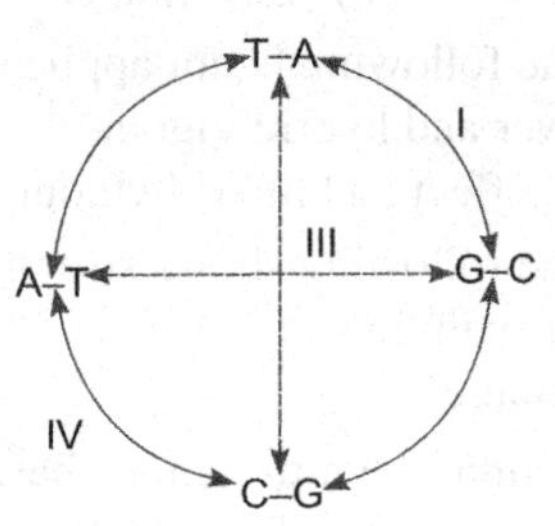

82. Presence of gene mutations in the human body appears to:
(a) Lower levels of bad cholesterol (b) Increase levels of good cholesterol
(c) Prevents heart disease (d) All

83. The first biochemical mutation in *Neurospora* was found in:
(a) 1930 (b) 1941 (c) 1952 (d) 1960

84. Which one of the following mutagens acts preferentially at the DNA-replicating forks?
(a) Ethyl methane sulfonate (b) Proflavin
(c) Nitrous acid (d) Nitrosoguanidine

85. Pyrimidine dimers are a target of the enzyme:
(a) Photolyase (b) Glycosylase (c) Transpoase (d) Resolvase

86. Which one of the following is an incorrect match?

	Normal pairing	Bases formed by deamination	New pairing
(a)	C–G	Uracil	U–A
(b)	A–T	Hypoxanthine	G–C
(c)	G–C	Xanthine	X–C
(d)	T–A	5-bromouracil	5BU–G

87. Tautomeric shift refers to:
(a) Isomerisation of nitrogen base, altering its hydrogen-bonding condition
(b) Isomerisation of deoxyribose sugar
(c) Isomerisation phosphate, altering its hydrogen-bonding capacity
(d) All

88. The mutagen being used extensively to induce mutations in microorganisms, plants and animal is:
(a) Ethyl methane sulphonate (b) Hydrazine
(c) Diepoxybutane (d) Dimethyl nitrosamine

89. X-rays bring about mutations by:
(a) Addition of water molecules to pyrimidines (b) Breaking the phosphate ester linkages in DNA
(c) Hydrogen bonds between nucleotides (d) Alkylation of phosphate groups of DNA

90. Mutation produces variations, which are:
(a) Useful (b) Useless (c) Harmful (d) All

91. Consider the following statements:
(A) Generally, transition mutations code for chemically similar amino acids
(B) Transition and transversion may cause nonsense mutations
(C) Normal purine nucleosides have an antimutagenic effect
(D) Ultraviolet rays cause excitation through energy transfer

The incorrect statements are:
(a) A, B and D (b) B, C and D (c) C and D (d) None

92. Which one of the following is not applicable to mutation?
(a) Crossing over and hybrid vigour (b) Death
(c) Deleterious effect and hybrid vigour (d) Genetic alteration

93. Which one of the following shows strong affinity for guanine?
(a) Ethyl ethane sulphonate (b) Nitrous acid
(c) Acridine orange (d) Hydroxyl amine

94. Addition or deletion of a nitrogenous base is applicable to:
(a) Transcription (b) Transduction (c) Frameshift mutation (d) Missense mutation

95. Alkylating agents cause mutations by:
(a) Frame shifts (b) Transitions (c) Transversions (d) All

96. Which one of the following is wrong about Cockayne syndrome?
(a) Autosomal dominant (b) DNA replication is inhibited by ultraviolet rays
(c) Dwarfism (d) Premature ageing

97. Consider the following statements:
(A) Mutations can be inherited (B) Mutations can be acquired
(C) Forwarded mutation and reverse mutation do not occur on the same locus
(D) Frameshift mutation often results in the formation of a new termination codon

The correct statements are:
(a) All (b) A, B and C (c) A, B and D (d) B and C

98. Which one of the following is not applicable to ethyl ethane sulfonate mutagen?
(a) Deletion and insertion of bases (b) $G = C \longrightarrow T = A$
(c) $A = T \longrightarrow C \equiv G$ (d) $G \equiv C \longrightarrow C \equiv G$

99. Match column I with column II and select the correct answer using answer codes:

Column I (Base analogues)	Column II (Organisms)
(A) 5-glucosyl hydroxymethyl cytosine	1. *E. coli*
(B) 5-hydroxymethyl uracil	2. Some bacteria
(C) 5- methyl cytosine	3. Wheat and grasses
(D) 6-methyl purine	4. Certain viruses

Answer codes:

	A	B	C	D
(a)	1	4	3	2
(b)	4	3	1	2
(c)	2	1	3	4
(d)	3	4	1	2

100. Which one of the following is applicable to mutation?
(a) Sports (b) Saltatory variations
(c) Discontinuous variations (d) All

101. Which one of the following enzymes is not involved in correcting distortion caused by thymine dimers?
(a) Exonuclease (b) DNA polymerase (c) DNA ligase (d) Helicase

102. Mutations are immediately expressed in:
(a) Haploid (b) Diploid (c) Aneuploid (d) Polyploid

103. Mutations are errors in DNA that:
(a) Occur spontaneously at a very fast rate (b) Cause diseases or death of the cells
(c) Occur only on the X and Y chromosome (d) Occur spontaneously at low rate

104. Which one of the following is responsible for the origin of alleles?
(a) Mutations (b) Polyploidy (c) Crossing over (d) Nondisjunction

105. Which one of the following is correct about recessive mutants?
(a) Expressed only in homozygous condition (b) Expressed only in heterozygous condition
(c) Always useful and rarely expressed (d) Expression depends on the physiological condition of the cell

106. Mutations are:
(a) Controllable (b) Directed (c) Purposive (d) None
107. The scientists involved in the study of mutation:
(a) Hugo de Vries (b) T H Morgan (c) Calvin Bridges (d) All
108. Spontaneous mutations occur due to:
(a) Errors in DNA replication (b) Repair of DNA
(c) Recombination (d) All
109. Mutations can be induced by:
(a) Radiations (b) Chemicals
(c) Transposable insertions (d) All
110. A DNA segment has the following base sequence CCA ACC ATG AGG AUG. The first base undergoes silent mutation. Its effect on the coding will result in:
(a) Change in amino acid sequence (b) A change in codon but not in amino acid
(c) A change both in codon as well as amino acid (d) There will be no coding
111. Generally mutations are:
(a) Dominant (b) Co-dominant (c) Incomplete dominance (d) Recessive
112. In Uganda around 15 per cent of elephants are now born without tusks. This is the result of:
(a) An inherited defect, a mutation which prevents tusks from developing (b) Environmental degradation
(c) Genetic erosion (d) Adaptation
113. Paramutation was first discovered in ______ and studied by:
(a) *Neurospora* – Beadle and Tautum (1950) (b) *Zea mays* – R Brink (1950)
(c) *Drosophila* – T H Morgan (1910) (d) *E. coli* – Jacob and Monad (1961)
114. Paramutation violates Mendel's:
(a) First law (b) Second law (c) Third law (d) All
115. Which one of the following is an antimutagen?
(a) Curcumin (b) Vitamin A (c) Vitamin C and E (d) All
116. Which one of the following is the most mutable virus?
(a) HIV (b) ϕX174
(c) TMV (d) Alfaalfa mosaic virus
117. Point mutation is:
(a) Acceptable (Hb Bristol) (b) Partially acceptable (HbIS)
(c) Unacceptable (HbM) (d) All
118. Site-directed mutagenesis technique has been described by:
(a) Bruce N Ames (b) Michael Smith (c) Andrew V Schally (d) Alexander R Todd

Answers to Multiple-Choice Questions

1. (c)	2. (b)	3. (a)	4. (c)	5. (a)	6. (d)	7. (d)	8. (b)
9. (d)	10. (c)	11. (a)	12. (d)	13. (b)	14. (a)	15. (d)	16. (b)
17. (d)	18. (c)	19. (d)	20. (b)	21. (b)	22. (b)	23. (a)	24. (a)
25. (d)	26. (b)	27. (a)	28. (d)	29. (b)	30. (a)	31. (b)	32. (d)
33. (a)	34. (d)	35. (c)	36. (b)	37. (d)	38. (b)	39. (d)	40. (d)

41. (a)	42. (a)	43. (d)	44. (c)	45. (c)	46. (c)	47. (c)	48. (b)
49. (b)	50. (c)	51. (d)	52. (d)	53. (b)	54. (b)	55. (c)	56. (d)
57. (a)	58. (c)	59. (b)	60. (c)	61. (d)	62. (d)	63. (d)	64. (c)
65. (c)	66. (a)	67. (b)	68. (d)	69. (a)	70. (d)	71. (b)	72. (b)
73. (d)	74. (a)	75. (b)	76. (d)	77. (c)	78. (d)	79. (b)	80. (d)
81. (c)	82. (d)	83. (b)	84. (d)	85. (a)	86. (d)	87. (a)	88. (a)
89. (b)	90. (d)	91. (d)	92. (a)	93. (a)	94. (c)	95. (d)	96. (a)
97. (c)	98. (a)	99. (a)	100. (d)	101. (d)	102. (a)	103. (d)	104. (a)
105. (d)	106. (d)	107. (d)	108. (d)	109. (d)	110. (b)	111. (d)	112. (a)
113. (b)	114. (d)	115. (a)	116. (a)	117. (d)	118. (b)		

Fill in the Blanks

1. A mutation arises mainly by __________ copy errors.
2. The term 'mutation' was coined by__________.
3. Substitution of a purine base by a pyrimidine base or vice versa is known as __________.
4. __________ involves replacement of a purine base by another purine base or a pyrimidine base by another pyrimidine base.
5. The ability of a molecule to exist in more than one form is called __________.
6. A chemical resembling a base in its structure is called__________.
7. Guanine is deaminated to __________.
8. Mutations are the primary source of __________.
9. __________ mutations are not transmitted to future generations.
10. CIB method of detecting mutations in *Drosophila* was discovered by __________.
11. Nitrous acid causes changes in the DNA bases by replacing the __________ with a __________.
12. The short-legged lamb discovered by __________ is an example of dominant germinal mutation.
13. The gene of sickle cell anaemia is a __________ gene.
14. Tautomers are produced by the rear rearrangements of __________ and __________ in molecules.
15. The mutagenic effect of ultraviolet rays was discovered by__________.
16. It was __________ who first reported that mutations can be induced by chemicals.
17. 2-aminopurine can pair with thymine by two hydrogen bonds and with cytosine by __________ hydrogen bond.
18. Transitions may be produced by__________.
19. Ochre mutation is the mutation of __________ codon.
20. A single mutation affecting more than one character is called__________.
21. __________ are the favourable sites for mutations.
22. H J Muller received the Nobel Prize for discovering __________.

23. Enzyme ____________________ catalyses removal of the alkyl group from the base.
24. Attached X chromosome technique for detection of mutation is simple as mutation can be detected in ____________________ generation.
25. In *Salmonella typhimurium*, glycine tRNA contains ____________________ instead of the triplet CCC in the anticodon position.
26. The effect of a nonsense mutation can be suppressed by ____________________ suppression.
27. Mutations occurring under natural conditions are called ____________________ mutations.
28. Deamination of adenine causes formation of hypoxanthine, which pairs with ____________________.
29. Mutations that inactivate genes are called ____________________ mutations.
30. Sites containing ____________________ provide hot spots for spontaneous mutations.
31. The frequency of mutations can be increased by____________________.
32. Mutation theory of evolution was proposed by ____________________.
33. Enol state of thymine pairs with____________________.
34. The smallest segment of gene capable of undergoing mutation is called____________________.
35. Mutation involving the removal of a gene is known as____________________.
36. Mustard gas as a chemical mutagen was first of all used by ____________________.
37. During the course of development, the phenotype changes while the genotype remains relatively constant, except the rare changes called ____________________.
38. Homoeosis is the replacement of one body part by a different one. The mutants causing such effects are called ____________________.
39. A large number of cells become affected, if mutation occurs during ________________ development.
40. Single base substitution occurs when DNA replication takes place during ____________ of the cell cycle.
41. Sometimes point mutations are reversed by____________________.
42. Suppressor mutations can be distinguished from reverse mutation by ________________________the reversed stock to wild type.
43. In dominant autosomal mutation of an affected heterozygous individual, around____________________ offsprings are expected to show this trait.
44. Photoreactivation is catalysed by the enzyme ____________________.
45. A dietary source of mutagen is ____________________ and is a possible cause of cancer.
46. A single individual will carry mutation, if it occurs in ____________________ or ____________________.
47. Genetic mutations are the result of________________,________________ and, ____________________ in the DNA.
48. The first individual showing a mutation is called____________________.
49. A new mutation that was not inherited from either parent is called a ____________________.
50. The effects of ____________________ do not influence the fitness of an individual.
51. An induced mutation variety of wheat is the ____________________.
52. The primary mutagenic effect of ultraviolet rays is due to the formation of ____________________.
53. Mutation in ____________________ gene is responsible for majority of human cancers.
54. ____________________ destroys mutagens.
55. ____________________ mutations are useful to us.

56. Mutations that result in missing DNA are called ___________________.
57. All polymorphisms are the result of a ___________________ in the gene.
58. These days ___________________ elements are being used to create new mutations.

Answers to Fill in the Blanks

1. DNA
2. Hugo de Vries
3. Transversion
4. Transition
5. Tautomerism
6. Base analogue
7. Xanthine
8. Genetic variations
9. Somatic
10. H J Muller
11. Aminogroup, hydroxyl group
11. Seth Wright
13. Mutant recessive
14. Electrons, protons
15. Altenburg in 1930
16. Auerbach
17. Single
18. Tautomeric shift
19. UAA
20. Pleiotropic
21. Hot spots
22. Induced mutations by X-rays
23. Alkyl transferase
24. First
25. Quadruplet CCCC
26. Intergenic
27. Spontaneous
28. Cytosine
29. Forward
30. 5-methyl cytosine
31. Mutagens
32. Hugo de Vries (1901)
33. Guanine
34. Muton
35. Deletion
36. Auerbach and Robinson
37. Mutations
38. Homeotic
39. Early embryonic
40. S phase
41. Suppressor mutations
42. Back crossing
43. 50 per cent
44. Photolyase
45. Lipids
46. Gamete, zygote
47. Base substitution, nucleotide insertion and nucleotide deletion
48. Mutant
49. De novo mutation
50. Neutral mutations
51. Sharbati sonara
52. Thymine dimers
53. p^{53}
54. Superoxidase mutase
55. 0.1 per cent
56. Deletions
57. Mutation
58. Transposable

True or False

1. Mutation is the sudden heritable change.
2. Mutations provide raw materials for evolution.
3. Missense mutation changes a codon into a terminator codon.
4. Mutation is the change in gene frequency.
5,. Gene mutation always involves only one region of a gene structure.
6. The frequency of forward mutation is generally higher than reverse mutation.
7. Nutritional mutants are called auxotrophs.
8. Pedigree analysis is helpful in analysing mutations in humans.
9. Caffeine is a base analogue.
10. Most mutations are submicroscopic.
11. Cytosine after deamination changes to thymine.
12. Azaserine is a strong mutagen.

13. Gigas mutant is *Oenothera lamarckiana*, which is due to gene mutation.
14. Auerbach first reported that mutations can be induced by chemicals.
15. A nonsense mutation leads to changes in the protein structure.
16. Mutator gene increases the rate of mutations.
17. Transversion involves replacement of purine bases by pyrimidine and vice versa.
18. Mutations are always harmful.
19. Crossing over may result in mutation.
20. Harmful deleterious genes are unable to affect the organism in its heterozygous condition with a normal gene.
21. CIB technique is able to detect mutations in the F_1 generation.
22. Mutations are essential for evolution.
23. Mutations are dominant, if they are expressed in heterozygous condition.
24. Gametic mutations occur only on sex chromosomes.
25. The probability of a gene to undergo mutation in single generation is known as mutation rate.
26. Suppressor mutation reverses the original mutation.
27. The changes caused by mutagens are called induced mutations.
28. Any base pair of a DNA may be mutated.
29. Hot spots are universal for all types of mutations.
30. Sites having 5-methyl cytosine provide hot spots for spontaneous mutations.
31. Peroxides are mutagenic.
32. Germinal mutations are the cause of most genetically determined diseases.
33. Majority of spontaneous mutations are conditional.
34. Mutations tend to have little effect on allele frequency.
35. Acquired mutations are changes in DNA that develop throughout the life of an individual.
36. Dividing cells are more susceptible to mutations when exposed to radiations.
37. The frequency of forward mutation is more than the frequency of reverse mutations.
38. Mutations are more common in males than females.
39. Mutation rate varies across species.
40. All genes are not mutable.
41. Transversions are by far the most common type of mutation.
42. Acridine mutations cause transitions like base analogues.
43. Glycosylase enzymes are lacking in mammals.
44. From a recent study it has been found that children having mutation in a gene called the *melanocortin* 3 receptor consume more calories than those with normal *MC3R* genes.
45. Most single nucleotide mutations are reversible.
46. Suppressor genes act by changing the nucleotide sequence of a mutant gene.
47. The mutagens, which cause chromosomal mutations, are unable to produce extrachromosomal mutants.
48. Electrophoretic mobility of mutant sickle cell haemoglobin is similar to that of normal haemoglobin.
49. Parasitic fungi of field of crops are one of the major sources of natural mutagenic agents.
50. Mutations have led to the evolution of new genes.

51. It is possible to forecast when a gene will mutate.
52. Most single nucleotide mutations are reversible.
53. The first disease caused by mutation in mitochondrial DNA was reported in 1988.
54. Muller reported white-eyed mutation in *Drosophila.*
55. Recessive mutations are expressed in the same generation.
56. Hormonal imbalances cause mutation.
57. Kepone is a mutagen.
58. Antigenic mutations, which occur sometimes in microorganisms, are useful to human beings.
59. Silent mutations have metabolic effect.

Answers to True or False

1. True	2. True	3. False	4. False	5 False	6. True	7. True	8. True
9. False	10. True	11. False	12. True	13. False	14. True	15. False	16. True
17. True	18. False	19. False	20. True	21. False	22. True	23. True	24. False
25. True	26. False	27. True	28. True	29. False	30. True	31. True	32. True
33. True	34. True	35. True	36. True	37. True	38. True	39. True	40. False
41. False	42. False	43. False	44. True	45. True	46. False	47. True	48. False
49. False	50. True	51. False	52. True	53. True	54. False	55. False	56. False
57. False	58. False	59. True					

Give Reasons

1. Addition or deletion of nucleotide disrupts the reading frame.
   – Because of triplet nature of gene expression.
2. Mutations can be considered as one of the raw materials of evolution.
   – Because they are the source of new variations in organisms.
3. Frameshift mutation is one of the most severe types of mutation.
   – Because of this mutation more than one amino acid or entire proteins are affected.
4. There is no basic difference between gene mutations and chromosomal mutations.
   – Because both result in sudden changes in the genetic material of an organism.
5. Bromodeoxyuridine is a more effective mutagen.
   – Because it can be easily converted to deoxyribonucleoside triphosphate.
6. Single base substitution mutations are critical for evolution.
   – Because such mutations make changes in ways that are to enough to form useful variants.
7. Some mutations which are harmful do not get eliminated from the gene pool.
   – Because they are recessive and are present in heterozygous condition.
8. Sex-linked recessive mutations can be detected.
   – Because a female heterozygous with the mutant gene produces 50 per cent of male progeny with the mutant gene.

9. It is easy to detect dominant mutation.
   – Because they exert their effect immediately.
10. Base analogues are responsible for mutations.
   – Because they can be easily incorporated into the DNA.
11 It is easy to detect dominant mutations.
   – Because of their immediate expression.
12. Detection of mutations in microorganisms is easy.
   – Because they contain haploid set of chromosomes. So, the mutations if generated (dominant or recessive), will express themselves in the next generation.
13. Deamination of guanine is not mutagenic.
   – Because guanine is deaminated to xanthine which behaves like cytosine and thus instead of G–C pairing, X–C pairing is formed which is not mutagenic.
14. Sometimes mutations in DNA may cause changes in the way a cell behaves.
   – Because a gene contains the instructions required for a cell to work. If some of the instructions are wrong, then the cell will not work in the manner it is supposed to.
15. Nonsense and frameshift mutations are generally unnoticeable.
   – Because these mutations result in the formation of shorter or abnormal proteins, which are generally inactive.
16. Mutations that pass from the parents to the child are called germline mutations.
   – Because they are present in egg cells or sperm cells.
17. Mutation in a single cell of a tissue may not impair the organism, even if the mutation is detrimental.
   – Because in the tissues of an adult organism, many cells perform the same function.
18. Mutations as a source of variations were rejected by Darwinians.
   – Because they were mostly considered deleterious by them.
19. Frameshift mutations are so named.
   – Because the whole frame of triplet reading gets altered, resulting in the formation of a changed protein.
20. Mutation in the junk region of the DNA usually has no effect on the individual.
   – Because the junk region does not code for proteins.
21. Deamination of guanine does not cause a heritable variation.
   – Because deamination of guanine results in the formation of xanthine pairing behaviour, which is like that of guanine.
22. Generally, insertion or deletion mutations cause more damage than single pair substitutions.
   – Because they may drastically change the sequences of codons, downstream from the mutation.
23. When a transposable element moves to a new location, the result is a new mutant.
   – The mutant arises because the presence of a piece of DNA in a wild-type gene disrupts the normal functioning of that gene.
24. In animals, mutations are more likely to occur in the sperm than the ova.
   – Because a large number of cell divisions are required in the production of sperms.

MODERN CONCEPT OF GENE

Multiple-Choice Questions

1. Which one of the following statements is incorrect?
   (a) Gene is the basis of life.
   (b) The term 'gene' was coined by Johannsen (1903). It was referred to as factor by Mendel (1866).
   (c) The physical and chemical nature of a gene forms the founding principles of genetics.
   (d) The mRNA, rRNA and tRNA are transcribed by the same gene.
2. Match column I with column II and select the correct answer using answer codes:

Column I		Column II
(A) T H Morgan	1.	Mutations in genes result in errors in steps in metabolic pathways
(B) F Griffith	2.	Genes are located on specific chromosomes
(C) Beadle and Tatum	3.	Genes can be transferred
(D) W Fiers	4.	The first man to determine sequence of a gene

Answer codes:

	A	B	C	D
(a)	2	1	3	4
(b)	2	3	1	4
(c)	3	2	4	1
(d)	4	3	2	1

3. Genes are made up of:
   (a) Sugar (b) Phosphate
   (c) Adenine, thymine, cytosine and guanine (d) All
4. Consider the following statements:
   (A) Genes control inheritance
   (B) Genes are arranged in a linear order on a chromosome
   (C) The linear arrangement of genes on the chromosome may be changed by inversion and translocation
   (D) The characters of an individual are controlled by paired genes
   The correct statements are:
   (a) All (b) A, C and D (c) A and D (d) A, B and D
5. Normally a gene exists in:
   (a) One form (b) Two forms
   (c) Three forms (d) More than three forms
6. Most eukaryotic genes are interrupted. This means:
   (a) They lack cistrons (b) Cistrons contain both exons and introns
   (c) Cistrons contain only exons (d) Cistrons, recons and mutons are lacking

7. Which one of the following is applicable to genes?
(a) Autocatalysis (b) Heterocatalysis (c) Mutation (d) All
8. In which one of the following genes are there two introns?
(a) Albumin (b) Collagen (c) Ova albumin (d) Globin
9. Introns are lacking in:
(a) Histone gene (b) Interferon genes
(c) Histone and interferon genes (d) Interferon and vitellogenin genes
10. Introns code endonuleases in:
(a) Vertebrate mitochondria (b) Fungal mitochondria
(c) Amylase gene in mouse (d) Immunoglobulin genes
11. Overlapping sequences have been reported in __________ of *E. coli*:
(a) Leucine mRNA (b) Methionine mRNA
(c) Tryptophan mRNA (d) Phenylalanine mRNA
12. The defective copy of a functional gene is called:
(a) Split gene (b) Selfish gene (c) Pseudogene (d) Jumping gene
13. Split genes have been reported in:
(a) Ribosomal genes of *Drosophila* (b) tRNA genes of yeast
(c) Beta (β) globin genes of mice (d) All
14. Genes having exons and introns are termed as:
(a) Pseudogenes (b) Split genes (c) Overlapping genes (d) Homeotic genes
15. Split genes were discovered by:
(a) Mitchell and Tjian (1989) (b) Sharp and Roberts (1993)
(c) Zouros et al. (1992) (d) Baker and Wickner (1992)
16. The terms 'exons' and 'introns' were used by:
(a) N Davidson (b) P Chambon (c) W Gilbert (d) R A Flavell
17. In which one of the following does the amylase gene contains two promoters?
(a) Humans (b) Mouse (c) Krait (d) Pigeon
18. Exon theory of genes was given by:
(a) Bera (1991) (b) Noller (1991) (c) Gilbert (1987) (d) Wang (1991)
19. The genome of ϕX174 contains ____________cistrons:
(a) 3 (b) 5 (c) 7 (d) 9
20. Consider the following statements:
(A) The yeast mitochondrial genome is smaller than the mammalian mitochondrial genome
(B) The yeast mitochondrial genome has five introns
(C) The yeast and mammalian mitochondria code for different proteins
(D) The introns of yeast have been derived from different/common ancestors
The incorrect statements are:
(a) A, B and C (b) B and C (c) C and D (d) All
21. Overlapping gene has been reported in:
(a) *E. coli* (b) *D. melanogaster* (c) Yeast (d) ϕX174
22. Which one of the following is a selfish gene?
(a) Mitochondrial male sterile gene in plants (b) Segregation distorter gene in *Drosophila*

(c) Replicative transposons (d) All

23. Repeat DNAs have abundant:
(a) Exons (b) Introns (c) Both exons and introns (d) Overlapping genes

24. Consider the following statements:
(A) Introns are noninformative
(B) Introns occupy much portion of the genome
(C) Introns are essential components in recombination
(D) Introns promote gene inversion between homologous genes
The incorrect statements are:
(a) A, B and C (b) A, B and D (c) C and D (d) None

25. Introns are lacking in the genome of:
(a) *E. coli* (b) Yeast
(c) Vertebrate mitochondria (d) Adenovirus

26. Genes within genes is applicable to:
(a) *E. coli* (b) ϕX174 (c) Both (a) and (b) (d) *C. elegans*

27. The segment of DNA, which is being used to produce more than one different polypeptides is called:
(a) Pseudogenes (b) Selfish genes (c) Split genes (d) Overlapping genes

28. Match column I with column II and select the correct answer using answer codes:

Column I		Column II
(A) Pseudogenes	1.	Codes for 10 proteins
(B) *Tetrahymena* rRNA	2.	Notch locus
(C) ϕX174	3.	Splicing of intron itself in the absence any of protein
(D) Unusual complementation pattern	4.	Unable to produce functional mRNA

	A	B	C	D
(a)	3	1	4	2
(b)	4	3	1	2
(c)	2	3	1	4
(d)	4	3	1	2

29. Mutant genes have been compared with wild-type genes by:
(a) T H Morgan (1910) (b) H J Mullar (1932)
(c) S Benzer (1950) (d) Stern (1936)

30. The scientist associated with complementation test:
(a) S Benzer (b) T H Morgan (c) H J Muller (d) J R Lyod

31. Which one of the following is an incorrect match?
(a) Ova albumin – 7 introns (b) Albumin – 14 intron
(c) Vitellogenin – 28 introns (d) Collagen – 20 introns

32. Which one of the following is considered as a dead byproduct of evolution?
(a) Split genes (b) Overlapping genes (c) Pseudogenes (d) Jumping genes

33. One-gene–one-polypeptide hypothesis was given by:
(a) Beadle and Tautum (1945) (b) Pauling and Itano (1949)
(c) Grahm and Morgan (1966) (d) Gurdon (1962)

34. Match column I with column II and select the correct answer using answer codes:

Column I		Column II
(A) Introns	1	Adh locus in the second chromosome of *D. malanogaster*

(B) Homeotic genes — 2. Noncoding part of the gene
(C) Exons — 3. Coding part of the gene
(D) A gene within gene — 4. Determine body plans

Answer codes:

	A	B	C	D
(a)	3	1	2	1
(b)	4	1	3	2
(c)	2	4	3	1
(d)	4	2	1	3

35. A set of multiple copies of a gene descended by duplications and variations is called:
(a) Multiple allele (b) Pleiotropic gene (c) Multigene family (d) Mosaic gene family

36. An example of multigene family, which has all genes occurring on the same locus is:
(a) Aldolase gene (b) Growth hormone gene (c) Homeobox gene (d) All

37. Split genes contain:
(a) Recons and mutons (b) Cistrons and exons
(c) Exons and introns (d) Introns and mutons

38. Which one of the following is the coding sequence of split genes?
(a) Cistrons (b) Exons (c) Introns (d) Mutons

39. The intervening sequences of genes are called:
(a) Exons (b) Introns (c) Mutons (d) Recons

40. Which one of the following is an example of multigene families?
(a) Genes coding for the various forms of actins (b) Genes coding for tubulins
(c) Genes coding of the globins (d) All

41. Consider the following statements with reference to globin gene:
(A) Globin gene consists of three exons and two introns
(B) In humans, the total length of three exons is 50 bp
(C) Globin genes show age-dependent changes in the expression
(D) The exons of each type of globin genes vary in length
The incorrect statement is:
(a) D (b) B (c) A (d) C

42. The exons and introns of a globin molecule have been derived from a single ancestral gene by duplication followed by:
(a) Mutations (b) Insertions (c) Translocations (d) All

43. Pseudogenes have been reported in:
(a) Human beings (b) Mice (c) *Drosophila* (d) All

44. Exons are those parts of genes that are represented in their final:
(a) mRNA product (b) rRNA product (c) tRNA product (d) All

45. The term 'selfish gene' was coined by:
(a) David Stove (1975) (b) Richard Dawkins (1976)
(c) George C Williams (1980) (d) Maki (1981)

46. Class-I-type intron is characterised by:
(a) Lacking consensus sequences (b) Their restricted distribution
(c) Having a short conserved sequence element at the junction of the exon and the intron (d) Its peculiar splicing mechanism

47. Which one of the following is a classical example of class-I-type intron?
(a) rRNA of *Tetrahymena*
(b) Cytochrome C gene of yeast mtDNA
(c) tRNA gene of chloroplast
(d) tRNA gene of *E. coli*

48. Which one of the following is an example of selfish genes?
(a) Homing endonuclease gene
(b) Transposons
(c) Supernumerary B chromosome
(d) All

49. Twintrons were discovered by:
(a) Copertinol and Hallck (1991)
(b) Wong (2004)
(c) Clayton et al. (2007).
(d) Craig Venter (2007)

50. Introns:
(a) Are DNA bases, which are found between exons
(b) Are not transcribed
(c) Possess genes called interrupted genes
(d) All

51. Almost all eukaryotic nuclear introns start with________ nucleotide sequence and end with
(a) AG, GC (b) AG, GT (c) GT, AG (d) GC, GT

52. In prokaryotes, introns are only found in:
(a) mRNA (b) tRNA (c) rRNA (d) None

53. Twintrons are:
(a) Exons within exons
(b) Exons within introns
(c) Introns with introns
(d) Introns with introns

54. Exon shuffing theory has been given by:
(a) W Gilbert (1987)
(b) E Neher (1976)
(c) Cooper and Martin (1982)
(d) J M Gilbert (1972)

55. In which type of introns, the RNA itself functions as an enzyme called ribozyme?
(a) Type I (b) Type II (c) Type III (d) All

56. Which one of the following is associated with RNA splicing?
(a) Spliceosome (b) Ribozyme (c) Ribonuclease (d) All

57. In human genome,__________ are more GC rich than others:
(a) First exon and intron
(b) Internal exons and introns
(c) Last exons and introns
(d) First exon

58. Which one of the following provide a record that the genomic DNA has been changed?
(a) Pseudogenes
(b) Overlapping genes
(c) Housekeeping genes
(d) Selfish genes

59. Which one of the following is correct about pseudogenes?
(a) Genomic sequence similar to normal genes
(b) Not expressed into functional proteins
(c) Defunct relatives of functional genes
(d) All

60. Which one of the following is applicable as a function of pseudogenes?
(a) Longevity of genome
(b) Endogenous siRNA derived pseudogenes
(c) Enhancement of the gametes
(d) All

61. Consider the following statements:
(A) Pseudogenes are DNA sequences similar to normal genes
(B) The first pseudogene was reported by Jacq et al. (1977)

(C) *Eta* globin pseudogene is found both in humans and chimpanzees
(D) *Alu* sequences are involved in gene regulation

The incorrect statements are:
(a) A and B (b) C and D (c) D (d) None

62. Pseudogenes are involved in:
(a) Gene conversion with functional genes (b) Recombination with functional genes
(c) Both (a) and (b) (d) None

63. Pseudogenes:
(a) Exhibit evolutionary consideration of genes sequences
(b) Reduced nucleotide variability
(c) Excess synonymous over nonsynonymous nucleotide polymorphism
(d) All

64. Which one of the following is an example of functional pseudogenes?
(a) *Makorin 1 p1* (b) *Anti NOS-1* (c) *Anti NOS-2* (d) All

65. Molecular fossils are applicable to:
(a) Overlapping genes (b) Pseudogenes
(c) Housekeeping genes (d) Split genes

66. Which one of the following statements is correct?
(A) Recently, it has been reported that a snail's pseudogenes can direct synthesis of a useful shortened peptide.
(B) Gene members of multigene family are closely linked and have overlapping functions.
(C) Pseudogenes have been identified in bacteria.
(D) All

67. Which one of the following is a set of junk DNA?
(a) Pseudogenes and retro pseudogenes (b) Minisatellites and microsatellites
(c) Satellites, transposons and retro transposons (d) All

68. Selfish DNA:
(a) Have the habit of jumping from one location to another
(b) Have the habit of sending copies to new locations
(c) Are apt to be shifted in the middle of working genes
(d) All

69. Which one of the following is not an example of selfish DNA?
(a) Microsatellite (b) Nucleosome
(c) Transposon (d) Homing endonuclease gene

70. The term 'selfish DNA' was coined by:
(a) Orgel and Crick (1980) (b) Shapeio (1983)
(c) Schule and Evans (1991) (d) Murray and Krischner (1989)

71. Which one of the following is not an example of multigene families?
(a) Histocompatibility antigens (b) Heat shock proteins
(c) Zein protein (d) Yolk proteins

72. The gene as the unit of gene function is based on:
(a) Complementation test (b) Test cross
(c) Linear arrangement (d) One-gene–one-polypeptide hypothesis

73. Overlapping genes have been reported in:
(a) *D. melanogaster* (b) Rats and mice (c) Bacteriophage G4 (d) All

74. Overlapping genes are located on:
(a) The same DNA strand
(b) Opposite DNA strands
(c) May be on the same DNA strand or on the opposite DNA strand
(d) In bacteriophage on the same DNA strand and in other on the opposite DNA strand.

75. In the SV40 virus, genes overlapped for as many as:
(a) 10 nucleotides (b) 50 nucleotides (c) 100 nucleotides (d) 122 nucleotides

76. Nested genes were first demonstrated in:
(a) *Adh* locus of *Drosophila*
(b) *Gart* locus of *Drosophila*
(c) *Ty* element in yeast
(d) *Bx-c* locus of *Drosophila*

77. Nested genes were first reported by:
(a) Henikoff et al. (1986)
(b) Lyons et al. (1988)
(c) Singer and Berg (1991)
(d) Portin (1993)

78. In humans, nested genes were reported by:
(a) Dillon (1987)
(b) Carlson (1990)
(c) Barbara Levinson et al. (1990)
(d) Portin (1993)

79. Which one of the following is a nested gene in humans?
(a) Neurofibromatosis gene
(b) 5SRNA gene
(c) Human leukocyte antigen genes
(d) Duchenne muscular dystrophy gene

80. The first enhancers in nucleated cells were demonstrated in the:
(a) Ribosomal RNA gene
(b) Immunoglobulin heavy chain gene
(c) Gene for kappa light chain
(d) Haptoglobin

81. Which one of the following is correct about enhancers?
(a) Their regulative effect may be positive or negative.
(b) They are located at 5´ and 3´ end of the gene.
(c) Enhancers can also be located within the introns.
(d) All

82. Which one of the following is an incorrect match?
(a) Mono intron genes – tRNA *tyr.* gene
(b) The first observation of repeated structure of genes – Amphibian ribosomal RNA
(c) The first observation of overlapping genes – ϕX174
(d) Alternative splicing first observed –TMV

83. Which one of the following is a polyprotein gene?
(a) Neuropeptide gene of humans
(b) Immunoglobulin heavy chain gene of cluster of humans
(c) Interferon gene of humans
(d) Duchenne muscular dystrophy gene of humans

84. The condition in which one gene resides within an intron of the other gene is known as:
(a) Overlapping gene
(b) Nested gene
(c) Split gene
(d) Pleiotropic gene

85. Processed pseudogenes:
(a) Lack certain regular genes
(b) Lack introns
(c) Sometimes terminate in adenine series
(d) All

86. Processed pseudogenes are:
    (a) Only a partial copy of the corresponding mRNA
    (b) Copy of the mRNA that contain sequences in addition to those expected to be present in the mRNA
    (c) Are a complete copy of the mRNA
    (d) All
87. Which one of the following is a recently formed processed pseudogene in humans?
    (a) *HBB* (b) *DHFR psi1* (c) *GALK* (d) *APOE*
88. Which one of the following is incorrect about pseudogenes?
    (a) Are regions of DNA (b) Are different from normal genes
    (c) Have no defined functions (d) They can provide information about genomic history and evolution
89. As per the recent concept, the rapid evolution of pseudogenes is almost entirely due to:
    (a) Genetic drift (b) Temporal variation (c) Natural selection (d) Mutation
90. Defects in pseudogenes include:
    (a) Lack of a start codon (b) Presence of extra stop signals
    (c) Abnormal or absent flanking regulatory elements (d) All
91. Processed pseudogenes were first found only in:
    (a) Bacteria (b) Viruses (c) Amphibians (d) Mammals
92. Which one of the following is not a proto-oncogene?
    (a) *VRT* (b) *ERK* (c) *TRK* (d) *RAS*
93. It has been estimated that ____ of the 25,000 or so genes in the human genome are proto-oncogenes:
    (a) 0.25 per cent (b) 0.50 per cent (c) 1 per cent (d) 5 per cent
94. Which one of the following is incorrect about homeotic genes?
    (a) They are located on a particular segment of chromosomes.
    (b) They may be as multigene or as cluster of genes.
    (c) A mutation in one homeotic gene has no influence on the other homeotic gene.
    (d) They are involved in the regulation of development.
95. In descending order, the correct sequence is:
    (a) Recon-cistron-muton (b) Cistron-recon-muton
    (c) Cistron-recon-muton (d) Muton-cistron-recon
96. Alterations in which of the following genes, result in change in patterns of body parts:
    (a) Homeotic genes (b) Split genes (c) Housekeeping genes (d) None
97. Homeobox genes were discovered by:
    (a) Rosenberg and Court (b) Edward Lewis
    (c) Hartwell and Weinert (d) Clarke and Carbon
98. Overlapping genes have been reported in the genome of:
    (a) Viruses (b) Prokaryotes and Eukaryotes
    (c) Mitochondria (d) All
99. Consider the following statements:
    (A) A gene family is a set of genes which have descended from different ancestral genes by duplication and variation
    (B) The members of gene family are always clustered on the same chromosome
    (C) The members of gene family perform related or identical functions

(D) The members of gene family may be expressed at different times or in different cell types.

The incorrect statements are:

(a) None (b) A, B and C (c) A and B (d) C and D

100. An interrupted gene retains the same structure in all tissues, except:

(a) Germline tissue (b) Somatic tissue (c) Gametic tissue (d) None

101. The consensus sequence is lacking in introns of:

(a) Mitochondria (b) Chloroplasts (c) Yeast tRNA (d) All

102. Which one of the follwoing is applicable to pseudoalleles?

(a) Closed-linked genes (b) Similar phenotypic effects
(c) Recombination (d) All

103. In which one of the following genes an intron may be retained or spliced out?

(a) Fibronectin (b) Transposase of *P* element in *Drosophila*
(c) Both (a) and (b) (d) None

104. The beta (β) gene family resembles heavy chain gene family in having:

(a) D segment (b) J segment (c) V segment (d) All

105. Which one of the following genes lack introns?

(a) Human cytochrome C gene (b) Rat cytochrome gene
(c) *Drosophila* alcohol dehydrogenase gene (d) Yeast alcohol dehydrogenase gene

106. Small nuclear RNAs (snRNAs) are typically:

(a) Transfer RNAs (b) Ribosomal RNAs (c) Both (a) and (b) (d)

107. Which one of the following small nuclear RNAs are present in the nucleolus?

(a) U_1snRNA (b) U_2 snRNP (c) U_3 snRNP (d) U_4 snRNP

108. In homeotic genes, one gene:

(a) Activates the other gene (b) Represses the other gene
(c) May activate or repress the other gene (d) Neither activates nor represses the other gene

109. Match column I with column II and select the correct answer using answer codes:

Column I		Column II
(A) Tn_3 elements	1.	Members generally function in all living cells at all times
(B) Viral oncogenes	2.	May be used as genetic markers
(C) snRNA genes	3.	Largest family of prokaryotic transposon
(D) Transposable elements	4.	Single exon

Answer codes:

	A	B	C	D
(a)	4	1	3	2
(b)	3	1	4	2
(c)	3	4	1	2
(d)	4	3	2	4

110. Insertion element at the end lacks:

(a) Tn_3 (b) Tn_5 (c) Tn_9 (d) Tn_{10}

111. Which one of the following is responsible for sexual dimorphism in mice?

(a) GACA (b) GATA (c) CATA (d) AGTA

112. The human tumour necrosis factor (TNF) was cloned in:

(a) 1980 (b) 1985 (c) 1990 (d) 1998

Answers to Multiple-Choice Questions

1. (d)	2. (b)	3. (d)	4. (a)	5. (b)	6. (b)	7. (d)	8. (d)
9. (c)	10. (b)	11. (c)	12. (c)	13. (d)	14. (b)	15. (b)	16. (c)
17. (b)	18. (c)	19. (d)	20. (d)	21. (d)	22. (d)	23. (b)	24. (d)
25. (c)	26. (b)	27. (d)	28. (d)	29. (c)	30. (a)	31. (d)	32. (c)
33. (a)	34. (c)	35. (c)	36. (b)	37. (c)	38. (b)	39. (b)	40. (d)
41. (a)	42. (d)	43. (d)	44. (d)	45. (b)	46. (c)	47. (a)	48. (d)
49. (a)	50. (d)	51. (c)	52. (b)	53. (c)	54. (a)	55. (b)	56. (d)
57. (a)	58. (a)	59. (d)	60. (d)	61. (d)	62. (c)	63. (d)	64. (d)
65. (b)	66. (d)	67. (d)	68. (d)	69. (b)	70. (a)	71. (c)	72. (a)
73. (d)	74. (c)	75. (d)	76. (b)	77. (a)	78. (c)	79. (a)	80. (b)
81. (d)	82. (d)	83. (a)	84. (b)	85. (d)	86. (d)	87. (b)	88. (b)
89. (a)	90. (d)	91. (d)	92. (a)	93. (c)	94. (c)	95. (c)	96. (a)
97. (b)	98. (d)	99. (c)	100. (d)	101. (d)	102. (d)	103. (c)	104. (d)
105. (d)	106. (d)	107. (c)	108. (c)	109. (c)	110. (b)	111. (a)	112. (b)

Fill in the Blanks

1. The alpha (α) globin cluster contains ______________ functional gene and ______________ pseudogene.
2. The beta (β) globin gene complex consists of ______________ functional gene and ______________ pseudogene.
3. Introns are removed by a process called ______________.
4. There are two types of pseudogenes called ______________ and ______________ pseudogenes.
5. The conserved nucleotide sequence involved in splicing is recognised by ______________.
6. There are six pseudogenes in the genome of______________.
7. Any DNA that is not translated to protein is called ______________.
8. Homeotic genes are also called ______________ genes.
9. Most pseudogenes are recognised by______________.
10. The DNA sequence of the ______________ exons in humans, chimpanzees and gorillas are similar.
11. Mutation in ______________ gene will affect the rate of synthesis of polypeptide chain.
12. ______________ is a sequence of few bases that lies at the junction between intron and exon.
13. ______________ is the smallest unit of gene capable of undergoing recombination.
14. ______________ is the unit of mutation.
15. Introns are also called ______________.
16. Pyknons form______________ of the human intergenic and intronic regions.
17. ______________ is a gene that is attached to another gene of interest by scientists in cell culture.
18. Genes having introns are called ______________ genes.
19. A gene cluster may expand or contract by unequal______________.
20. Two or more identical genes present on the same chromosome are known as ______________ copies.
21. The Antennapedia complex contains three homeotic genes called ______________, ______________

and ___________________.

22. Multigene families have arisen by ___________________ crossing over.
23. Some genes are constantly expressed in bacteria cells called ___________________ genes.
24. Besides the protein coding sequences of gene, exons also include the ___________________ residues present in the mRNA.
25. ___________________ different types of small nuclear RNAs are present in all vertebrate cells.
26. All small nuclear RNAs (snRNAs) have unusual trimethylated cap structure at their 5´ ends except for ___________________.
27. In higher eukaryotes, the chromosomal region adjacent to the centromere is composed of very long blocks of ___________________.
28. The gene having the ability to move from one site to another in the chromosome is called ___________.
29. Transposable elements are of two types, viz., the short insertion sequences (Is elements) and the ______________.
30. ___________________ are essential components of the recombination process.
31. ___________________ is an example of one-gene–one-polypeptide hypothesis.
32. Oncogenes arise from the normal ___________________ of a normal cell.
33. The DNA segment having the ability to insert itself at one or more sites in the genome of prokaryotic or eukaryotic organism is called ___________________ genetic element.
34. Members of multigene families *lac* occur in clusters and differ from ___________________ clusters.
35. Members of multigene families have homologies with ___________________ families.
36. Gene ___________________ may result in a multigene family.
37. ___________________, ___________________ and ________________ have the same number of beta (β) globin genes arranged on the same chromosome.
38. Pseudogenes have been originally derived from ___________________ genes.
39. The human TNF gene contains ___________________ exons.
40. A female individual cannot reproduce sexually except ___________________.

Answers to Fill in the Blanks

1. Three, three
2. Five, one
3. RNA splicing
4. Processed, unprocessed
5. UsnRNAs
6. *Mycobacterium tuberculosis*
7. Junk DNA
8. *Hox*
9. Sequence alignments
10. *Eta globin* pseudogene
11. Regulatory
12. Consensus sequence
13. Recon
14. Muton
15. Intervening sequences
16. One-sixth
17. Reporter gene
18. Interrupted
19. Crossing over
20. Non-allelic copies
21. Proboscipedia (*pb*), sex-comb reduced (*sxr*), antennapedia (*Ant.*)
22. Unequal
23. Constitutive
24. Untranslated 5´ and 3´
25. Six
26. U6
27. Highly repetitive DNA
28. Mobile gene
29. Longer transposons (Tn)
30. Introns
31. Sickle cell anaemia
32. Protocogenes
33. Transposable
34. Tandem
35. Single gene
36. Duplication
37. Humans, chimpanzees, gorillas
38. Functional
39. Four
40. Parthenogenesis

True or False

1. The proportion of interrupted genes varies with each organism.
2. In eukaryotes, intron removal and splicing is completed within the nucleus.
3. In dystrophin gene of humans, there are 75 introns.
4. The length of one intron varies from 31 nucleotides (in SV40) to over 2,10,000 in human dystrophin gene.
5. About 98 per cent of the genes that cause cystic fibrosis consist of introns.
6. Spliceosomal introns are conserved in all species.
7. In higher organisms, genes are continuous.
8. A gene may be overlapping.
9. Multigene families of genes are a group of similar and totally identical sequences; each sequence representing a gene.
10. Selfish genes are detrimental.
11. Intron of one gene may contain exon of another gene.
12. Fungal mitochodria lack split genes.
13. Pseudogenes are nearly as abundant as functional genes.
14. Bacteriophage T4 contains introns.
15. Mature mRNA contains both exons and introns.
16. All protein-coding genes have introns.
17. Pseudogenes are produced from functional genes during evolution.
18. Pseudogenes act as a switch that control gene expression.
19. Selfish DNA is detrimental to other genes.
20. All genes containing a homeobox are homeotic in nature.
21. Processed pseudogenes are found on different chromosomes from their functional counterparts.
22. All introns encode a maturase.
23. In both prokaryotes and eukaryotes, multiple gene families exist as a gene cluster.
24. All globin genes (in two clusters) are transcribed from the 5′ to 3′ direction.
25. Repeat DNAs are abundant in introns.
26. Two genes of the same size can have different number of introns.
27. Individuals having five alpha (α) globin genes show no change in haemoglobin synthesis.
28. Gene is a unit of mutation s or recombination.
29. Five members of aldolase gene family are located on the same chromosome.
30. In SV40, different genes are expressed at different times during lytic infection.
31. Presence of introns is not essential for gene expression.
32. Small nuclear RNAs (snRNAs) are relatively unstable.
33. Small nuclear RNAs are present in all vertebrate cells.
34. Small nuclear RNAs exist as naked RNA molecules.
35. A locus may contain many genes.
36. A gene, which is always expressed, has 100 per cent penetration.
37. Transposable element can be used as transformation vector.
38. The alpha (α) globin gene family genes are located on the chromosome 16.
39. One-gene–one-enzyme hypothesis was given by Beadle and Tatum.
40. Outbreeding leads to homozygosity.
41. Homoalleles are not separable by recombination.

42. Heteroalleles are structurally allelic and functionally non-allelic.
43. Gene is a nondivisible unit.

Answers to True or False

1. True	2. True	3. True	4. True	5. True	6. False	7. False	8. True
9. False	10. True	11. True	12. False	13. True	14. True	15. False	16. False
17. True	18. True	19. True	20. False	21. True	22. False	23. True	24. False
25. True	26. True	27. False	28. False	29. False	30. True	31. True	32. False
33. True	34. False	35. True	36. True	37. True	38. False	39. True	40. False
41. True	42. False	43. False					

Give Reasons

1. Sea urchins and amphibians have the most repetitious histone genes.
   – Due to their need of rapid cell divisions during early development.
2. U3SnRNP is represented as single word.
   – Because they attach polypeptides unique to each of these particles.
3. It is possible to construct a gene map.
   – Because crossovers and recombinations take place in genes.
4. Housekeeping genes are always transcribed.
   – Because the proteins coded by them are always required for the basic vital processes of a cell.
5. Insertion elements are so named.
   – Because they can insert at different sites of bacterial chromosomes and plasmids.
6. *Etat* gene is unable to synthesise functional protein.
   – Because it lacks initiation codon (AUG) and has several stop codons. So, no mRNA is made and hence no protein.
7. Natural selection cannot act directly on an individual gene in a multigene family.
   – Because they have identical and overlapping functions.
8. Selfish DNA is abundant in the genome.
   – Because it has the ability to increase the number of copies within the genome with time.
9. Most selfish DNA are harmless.
   – Because if they damage the host, their survival will be affected.
10. Some pseudogenes cannot be expressed.
    – Because they lack promoters and introns needed for transcription.
11. The unprocessed pseudogene is generally not harmful.
    – Because it originates from gene duplication, in which, one copy acquires mutation that inactivates it. But the other copy remains intact, so they are harmless.
12. In bacteriophage ϕX174, overlapping gene is necessary.
    – Because the amount of DNA present in a single circular DNA is not sufficient to encode the I11 proteins, if transcription takes place in a linear fashion.

HUMAN GENETICS

Multiple-Choice Questions

1. Karyotype is used to study:
   (a) Chromosomal aberrations (b) Taxonomic relationships
   (c) Cellular functions (d) All
2. Consider the following statements:
   (a) Karyotypes are generally prepared from culture cells
   (b) Quinacrine mustard stains G–C-rich areas
   (c) In situ hybridisation technique is used to locate definite areas on chromosomes
   (d) Flow cytometry technique is useful in the detection of deletion, duplication and aneuploidy
   The correct statements are:
   (a) All (b) B, C and D (c) A, C and D (d) C and D
3. The in situ hybridisation technique was developed initially by:
   (a) Schwarzacher (b) Pardue and Gall (c) Mitchell and Tjian (d) Wilkins
4. The first banding to be described was:
   (a) G banding (b) Q banding (c) C banding (d) R banding
5. The first Mendelian trait in humans was described by:
   (a) Farabee (1905) (b) Hardy (1908) (c) Shull (1910) (d) Rhoades (1938)
6. The first Mendelian trait in humans reported was:
   (a) Polydactly (b) Syndactly (c) Brachydactly (d) Club foot
7. The first successful attempt to count the number of human chromosomes was made by:
   (a) Wini Warter (1912) (b) Weinberg (1908)
   (c) Warthin (1925) (d) Wittmann (1960)
8. Culturing of lymphocytes from human blood for chromosomes preparation was described by:
   (a) Nowell and Hungerford (1960) (b) Moorhead et al. (1960)
   (c) Harda et al. (1964) (d) Zuckerkandl (1965)
9. Consider the following statements:
   (A) Chromosomes are present in all human cells
   (B) In male human beings, one of each pair of autosomes and X chromosome are of maternal origin
   (C) Chromosomes are lacking in mitochondria
   (D) The 28S and 18S ribosomal genes are present in the short arms of chromosomes 13, 15, 21 and 22
   The incorrect statements are:
   (a) A, C and D (b) B and D (c) A and C (d) A and D
10. In humans, which one of the following tissues is used for studying chromosomes?
    (a) Amniotic fluid or chorionic villi (b) Bone marrow and skin fibroblasts

(c) Peripheral blood lymphocytes (d) All

11. Fluorescent in situ hybridisation (FISH) can be used to study:
(a) Translocations (b) Inversions (c) Deletions (d) All

12. The characteristic chromosomal complement of a eukaryotic species is known as:
(a) Karyotype (b) Karyogram (c) Idiogram (d) Karyology

13. Who first observed that in humans, the diploid number of chromosome is 46?
(a) Winiwarter (1912) (b) Tijo and Levan (1856)
(c) Lejeune et al. (1959) (d) Ford and Hamerton (1959)

14. Which one of the following is correct about human chromosomes?
(a) Different sizes (b) Different locations of centromeres
(c) Some haves small satellites (d) All

15. Which one of the following groups of human chromosomes are medium sized, acrocentric and have satellites?
(a) 6–12 (b) 13–15 (c) 16–18 (d) 19–20

16. In which one of the following human chromosomes is the centromere approximately median?
(a) 12 (b) 14 (c) 16 (d) 21

17. The normal human karyotype has 46 chromosomes:
(A) 23 chromosomes derived from each parent (B) Males are XY and females are XX
(C) The sex of an offspring is determined by the Y chromosome (D) Sex chromosomes are carried in the sperms
The correct statements are:
(a) All (b) A, B and C (c) A and D (d) A, B and D

18. Which one of the following bandings stains centromeres?
(a) R banding (b) C banding (c) Q banding (d) T banding

19. Which one of the following banding is the reverse of G banding?
(a) R banding (b) Q banding (c) C banding (d) T banding

20. Telomeres are visualised by:
(a) T banding (b) Q banding (c) NOR banding (d) G banding

21. The G bands are richer in:
(a) Adenine–thymine (b) Cytosine–guanine (c) Cytosine–thymine (d) Adenine–guanine

22. Consider the following statements:
(A) G banding is obtained with Giemsa stain
(B) G banding gives a series of light and dark bands
(C) The dark bands tend to be hetrechromatic and are rich in a adenine–thymine
(D) The light bands tend to be euchromatin, are late replicating and are guanine–cytosine rich
The incorrect statement is:
(a) A (b) B (c) C (d) D

23. Band 1923 refers to:
(a) Band 1 on the short arm of chromosome 23 (b) Band 23 on the long arm of chromosome 1
(c) Band 23 on the short arm of chromosome 1 (d) Band 1 on the long arm of chromosome 23

24 During the cell cycle, G bands replicate in:
(a) G1 phage (b) G2 phage
(c) First half of the S phage (d) Second half of the S phage

25. None of the human chromosomes is:
(a) Telocentric (b) Acrocentric (c) Metacentric (d) Submetacentric

26. R banding is used to detect:
(a) Minor deletions (b) Inversions
(c) Chromosome polymorphism (d) All

27. Which one of the following bandings is applied to evaluate terminal bands that appear light after G banding?
(a) C banding (b) NOR banding (c) R banding (d) T banding

28. The longest submetacentric chromosomes of humans are:
(a) 1–3 (b) 4–5 (c) 6–12 (d) 16–18

29. The smallest human chromosome pair is of chromosome number:
(a) 1 (b) 10 (c) 20 (d) 22

30. The number of groups in human karyotype is:
(a) 3 (b) 5 (c) 7 (d) 9

31. The number of autosomes in human ovum is:
(a) 46 (b) 44 (c) 23 (d) 22

32. The number of chromosomes in human spermatocyte is:
(a) 46 (b) 23 (c) 22 (d) 44

33. Which one of the following is not an autosomal trisomy syndrome?
(a) Patau syndrome (b) Edward syndrome
(c) Wolf –Hirschhorn syndrome (d) Down syndrome

34. Cri-du-chat syndrome was discovered by:
(a) Le Jeune (1963) (b) Edward (1960) (c) Patau (1960) (d) Lyon (1961)

35. Deletion of short arm of chromosome 4 results in:
(a) Burkitt's lymphoma (b) Wolf –Hirschhorn syndrome
(c) Rett syndrome (d) Martin–Bell syndrome

36. Prader–Willi syndrome is caused due to a small deletion in the long arm of chromosome:
(a) 15 (b) 17 (c) 21 (d) 22

37. Which one of the following fragile sites is associated with clinical abnormality?
(a) Xq27 (b) 20 p11 (c) 6p23 (d) 12q13

38. The scientist associated with inactivation of X chromosome:
(a) Lejeune (b) Jeffreys (c) Lyon (d) Barr

39. Match column I with column II and select the correct answer using answer codes:

Column (Syndrome)	Column II (Site of the microdeletion)
(A) DiGeorge syndrome	1. Xq27.3
(B) Langer–Giedion syndrome	2. 8q23
(C) Aniridia syndrome	3. 22q11
(D) Martin–Bell syndrome	4. 11p13

Answer codes:

	A	B	C	D
(a)	3	2	4	1
(b)	3	1	4	2
(c)	4	2	3	1
(d)	2	4	1	3

40. Each type of autosomal has been seen with the exception of chromosome:
(a) 1 (b) 3 (c) 9 (d) 13

41. Which one of the following is incorrect about Down syndrome?
(a) Trisomy of chromosome 21 (b) Low IQ and congenital heart malformations
(c) Males are generally infertile (d) Pregnancies are not common in females

42. Cleft lip and palate occur in:
(a) Down syndrome (b) Edward syndrome (c) Patau syndrome (d) Cat-cry syndrome

43. Match column I with column II and select the correct answer using answer codes:

Column I	Column II
(A) Marfan syndrome	1. Autosomal co-dominant
(B) Lesch–Nytian syndrome	2. Autosomal recessive
(C) Zellweger syndrome	3. Autosomal dominant
(D) Aipha1-antitrypsin	4. X-linked recessive

Answer codes:

	A	B	C	D
(a)	4	1	3	2
(b)	2	3	4	1
(c)	3	4	2	1
(d)	3	4	1	2

44. Which one of the following is incorrect about Y chromosomes?
(a) The genes on Y chromosomes that determine maleness are located on the long arm of the Y chromosomes.
(b) The number of Y chromatin bodies is identical to the number of Y chromosomes.
(c) A large portion of Y chromosomes is heterochromatic.
(d) Y chromosomes direct the organogenesis of testis.

45. In retinoblastoma, there is a deletion in the:
(a) Short arm of chromosome 13 (b) Long arm of chromosome 13
(c) Long arm of chromosome 3 (d) Short arm of chromosome 15

46. Balanced translocation between chromosome 9 and 22 occurs in:
(a) Burkitt's lymphoma (b) Rett syndrome
(c) Apert syndrome (d) Chronic myeloid leukaemia

47. In all human beings, all monosomic are believed to be invisible except the monosomy of chromosome:
(a) 3 (b) 13 (c) 21 (d) X

48. Which one of the following is an X-linked dominant trait?
(a) Intestinal polyposis (b) Vitamin D resistant rickets
(c) Cystinuria (d) Gaucher disease

49. Consider the following statements:
(A) Xg blood group is an X-linked dominant trait
(B) The gene of Xg blood is located near the tip long arm of the X chromosome
(C) Colour-blindness is more frequent in Eskimos
(D) X-linked dominant disorder is more frequent in females than males

The incorrect statements are:
(a) B and C (b) C and D (c) A and C (d) B, C and D

50. In human beings, variations in the size of centromeric heterochromatin are relatively frequent for chromosomes:

(a) 1, 5 and 8 (b) 1, 9 and 16 (c) 5, 9 and 19 (d) 7, 11 and 21

51. Barr body and drumstick are lacking in:
(a) Klinefelter's syndrome (b) Turner syndrome
(c) Lesch–Nyhan syndrome (d) Prader–Willi syndrome

52. Match column I with column II and select the correct answer using answer codes:

Column I		Column II
(A) Lejeune (1959)	1.	Established the first British genetic counseling clinic in 1946
(B) John Fraser Roberts	2.	Sickle cell anaemia
(C) Pauling (1949)	3.	Assigned gene for colour-blindness to the X chromosome
(D) E B Wilson (1911)	4.	Discovered first chromosomal disease in humans

Answer codes:

	A	B	C	D
(a)	4	3	2	1
(b)	4	1	2	3
(c)	3	2	1	1
(d)	2	3	1	4

53. Which one of the following is incorrect about inactive X hypothesis?
(a) Proposed by Lyon
(b) Only one of the X chromosomes is genetically active
(c) The X chromosome undergoing inactivation is maternal
(d) Inactivation occurs early in embryonic life

54. In which one of the following syndromes is the number of Barr body one?
(a) Down syndrome (b) Edward syndrome (c) Patau syndrome (d) All

55. In humans, sex-determining mechanism is:
(a) XX♀ – XY♂ (b) XY♀ – XX♂ (c) XX♀ – XX♂ (d) Through X/A ratio

56. Which one of the following is due to defects in metabolism?
(a) Phenylketonuria (b) Albinism (c) Alacptonuria (d) All

57. A case of triploid human being (2n = 69) was reported in 1960 by:
(a) John and Milkos (b) Brook and Santess on (c) Green and Green (d) Peterson and Tjian

58. Human chromosome __________ has resulted from the fusion of two acrocentric ape chromosomes (numbers 12 and 13):
(a) 2 (b) 5 (c) 11 (d) 18

59. The gene for haemophilia is:
(a) 18,60,000 base pairs (b) Contains 26 exons
(c) Exons are separated by 25 introns (d) All

60. Haemophilia is due to defective factor VIII, which can be detected in foetuses:
(a) 5 weeks old (b) 10 weeks old (c) 20 week old (d) 40 weeks old

61. Which one of the following is applicable to Queen Victoria of England?
(a) Colour-blind (b) Deaf (c) Cancer (d) Haemophilic

62. Which one of the following is not a sex-linked trait in humans?
(a) Haemophilia (b) Colour-blindness (c) G-6-PD deficiency (d) Sickle cell anaemia

63. Holandric genes are located on:
(a) Autosomes (b) X chromosome (c) Y chromosome (d) All

64. The movement of *SRY* portion of the Y chromosome due to recombination during formation of gametes results in:
(a) XXYY syndrome (b) XX male syndrome
(c) XYY syndrome (d) XO syndrome

65. Which one of the following is applicable to human Y chromosome?
(a) Gene conversion (b) *SRY* gene
(c) Sex-determining chromosome (d) All

66. A translocation between chromosome 8 and 14 causes:
(a) Chokenflok syndrome (b) Kearns–Sayre syndrome
(c) Burkitt's lymphoma (d) Wernicke–Korsakoff syndrome

67. Consider the following statements:
(A) 95 per cent Down syndrome patients are trisomy 21 and 4 per cent are translocations of the 21 with a D or G group chromosome
(B) Turner syndrome is the only viable on monosomy
(C) It has been estimated that 10 per cent sperms and 50 per cent eggs contain abnormal chromosomes
(D) Noonan syndrome is an autosomal dominant trait whose phenotype overlaps with Turner syndrome

The incorrect statements are:
(a) A and D (b) B and C (c) C and D (d) None

68. In which one of the following is the paternal chromosome always activated?
(a) Humans (b) Marsupials (c) Apes (d) Echidnce

69. In human-centric translocation, transfusion involves:
(a) Any of the chromosomes (b) Sex chromosomes
(c) Acrocentric chromosomes 13–15 and 21–22 (d) 21 and 22

70. Sex chromosome abnormalities are rare amongst early abortuses, with the exception of:
(a) 45, X (b) 45, XX (c) 47, XYY (d) Trisomy 13

71. Which one of the following is a deletion syndrome?
(a) Glycerol kinase deficiency (b) Duchenne muscular dystrophy
(c) Adrenal hypoplasia (d) All

72. The usual age of onset of alkaptonuria is:
(a) Prenatal (b) Birth (c) 16–19 years (d) 20–30 years

73. The usual age of onset of blood group antigens is:
(a) Birth to four months (b) Birth
(c) Prenatal (d) One year

74. Fanconi's anaemia is:
(a) Autosomal dominant (b) Autosomal recessive
(c) X-linked dominant (d) X-linked recessive

75. Which one of the following statements is incorrect?
(a) The maximum number of Barr bodies is one less than the number of X chromosomes.
(b) XYY combination does not have any obvious effect on sexual phenotype.
(c) Klinefelter syndrome individuals are chromatin negative.
(d) X-linked recessive traits may occur in females with X-autosome translocations.

76. Which one of the following syndrome is usually caused by deletion or inactivation of maternally inherited chromosome 15?
(a) Patau syndrome (b) Turner syndrome
(c) Angelman syndrome (d) Cri-du-chat syndrome

77. Phytohaemagglutinin (PHA) is used to:
(a) Arrest cell division (b) Stimulate cell division
(c) Spread chromosomes (d) Stain chromosomes

78. Who is regarded as the 'Father of Human Genetics'?
(a) Galton (b) Garrod (c) Sturli (d) Muller

79. The karyotype of Edward syndrome is:
(a) 45 + XO (b) 44 + XXY
(c) 45 + XX or 45 + XY (d) 46 + XO or 46 + XY

80. Which one of the following syndrome is a classic example of genetic imprinting?
(a) Angelman syndrome (b) Apert syndrome
(c) Bloom syndrome (d) Klinefelter syndrome

81. Which one of the following is a set of SAT chromosomes in humans?
(a) 1, 10, 14, 16 and Y (b) 1, 10, 12, 16 and X
(c) 13, 14, 15, 16, 21 and 22 (d) 2, 6, 10, 13 and Y

82. Trisomy of chromosome 21 results in Down syndrome which is due to:
(a) Nondisjunction (b) Mutation (c) Linkage (d) Recombination

83. Hypertrichoses is associated with:
(a) Autosome (b) X–Y chromosome (c) X chromosome (d) Y chromosome

84. Which one of the following is incorrect about haemophilia?
(a) Royal's disease (b) Bleeder's disease
(c) More common in males (d) Y-linked disorder

85 Which one of the following syndromes mostly affects females?
(a) Beckwith–Wiedemann syndrome (b) Rett syndrome
(c) Menkes kinky hair syndrome (d) Apert syndrome

86. Criss cross inheritance is shown by:
(a) Sickle cell anaemia (b) Blood groups
(c) Hypertrichosis (d) Colour-blindness

87. Mongoloid looks are due to trisomy of chromosome:
(a) 13 (b) 15 (c) 18 (d) 21

88. In humans, if the number of chromosome 47 is due to trisomy of chromosome 13. It is a case of:
(a) Rett syndrome (b) Patau syndrome (c) Edward syndrome (d) Down syndrome

89. Which one of the following has the longest survival?
(a) Trisomy 13 (b) Trisomy 18 (c) Trisomy 21 (d) All

90. Which one of the following is not applicable to Barr body?
(a) X chromatin (b) Drumstick (c) Sex chromation (d) Euchromatin

91. Which one of the following is applicable to Group G (21–22) of chromosome?
(a) Medium to large, submetacentric, SAT (b) Medium, acrocentric, SAT
(c) Smallest, metacentric, SAT (d) Smallest, acrocentric, SAT

92. Which one of the following is a correct match?

(a) Leprosy – Hereditary disease
(b) Webbed neck – 44 + XO
(c) Mongolism – Nullisomy
(d) Klinefelter syndrome – No Barr body

93. Christmas disease is applicable to:
(a) Colour-blindness (b) Hypertrichosis (c) Hepatitis (d) Haemophilia B

94. Holandric genes are transmitted through:
(a) X chromosome
(b) Y chromosome
(c) Both X and Y chromosomes
(d) Autosomes

95. Which one of the following is incorrect about inheritance pattern of traits with an autosomal recessive?
(a) Show a vertical pattern if the condition is rare siblings may be affected but rarely are their parents.
(b) Both sexes can be affected.
(c) If the trait is rare, consanguinity may be suspected.
(d) The probability of a normal sibling being a carrier is 2/3.

96. Which one of the following is an exception to dosage compensation, which is more in females in comparison to males?
(a) Level of factor VIII in blood
(b) Level of factor IX in blood
(c) Level of steroid sulphatase in blood
(d) None

97. Which one of the following is an incorrect match?
(a) Aicardi syndrome – X-linked recessive
(b) Tay–Sachs disease – Autosomal recessive
(c) Marfan syndrome – Autosomal dominant
(d) Androgenetic alopecia – X-linked recessive

98. Which one of the following is due to deletion, extra point mutation, uniparental disomy or translocation?
(a) Angelman syndrome
(b) Apert syndrome
(c) Marfan syndrome
(d) Andriogen insensitivity syndrome

99. Match column I with column II and select the correct answer using answer codes I:

Column I	Column II
(A) Phenylketonuria	1. Disorder of mitochondrial function
(B) Lesch–Nyhan syndrome	2. Disorder of amino acid metabolism
(C) Niemann Pick disease	3. Disorder purine or pyrimidine metabolism
(D) Sayre syndrome	4. Lysomal storage disorder

Answer codes:

	A	B	C	D
(a)	3	2	4	1
(b)	4	3	1	2
(c)	2	3	4	1
(d)	2	1	3	4

100. Which one of the following is a multifactorial disorder?
(a) Stroke (b) Cancer (c) Hypertension (d) All

101. Which one of the following is incorrect about sickle cell anaemia?
(a) It is a disorder of the blood caused by inherited abnormal haemoglobin.
(b) The abnormal haemoglobin causes distorted red blood cells.
(c) In sickle cell, blood cells are rigid and not prone to rupture.
(d) Sickle cell anaemia is the most common genetic disease in England.

102. Consider the following statements about G-6-PD deficiency:
(A) An inborn error of metabolism
(B) Located on autosome
(C) More common in men than in women
(D) Deficiency persons are healthy

The correct statements are:
(a) All (b) A, B and D (c) B and C (d) A, C and D

103. Defect in thalassaemia is due to:
(a) Structural defect in alpha chain (b) Structural defect in beta chain
(c) Both (a) and (b) (d) A reduced rate of synthesis of beta chain

104. Which one of the following is a dominant trait?
(a) Epiloia (b) Albinism
(c) Microcephaly (d) Infantile amaurotic idiocy

105. Taste sensitivity to phenyithiocarbamide (PT(C) is an:
(a) Autosomal dominant trait (b) Autosomal recessive trait
(c) X-linked recessive trait (d) X–Y linked trait

106. Which one of the following is an X–Y linked trait?
(a) Xeroderma pigmentosum (b) Spastic paraplegia
(c) Retinitis pigmentosa (d) All

107. The existence of human blood group system was recognised by:
(a) Landsteiner (1900) (b) Boyd (1939)
(c) Dungern and Hirsifeld (1911) (d) Bernstein (1924)

108. Which one of the following blood groups was not recognised by Landsteiner?
(a) A (b) B (c) O (d) AB

109. The mode of inheritance of ABO blood group was described by:
(a) Mourant (1954) (b) Bhende et al. (1952)
(c) Bernstein (1924) (d) Thomsen et al. (1930)

110. The H substance remains unchanged in the presence blood group allele:
(a) A (b) B (c) O (d) All

111. Null allele is applicable to blood group:
(a) A (b) B (c) AB (d) O

112. Which one of the following is an incorrect statement?
(a) Criss cross inheritance is shown by the genes present on X and Y chromosomes.
(b) Inheritance of ABO blood groups shows multiple allelism and co-dominance.
(c) Antibodies are lacking in AB blood group.
(d) H antigen is a glycolipid produced by most of the individuals.

113. Usually, an individual has the same blood group for life. However, it may rarely change through addition or suppression of an antigen during:
(a) Malignancy (b) Autoimmune disease (c) Infection (d) All

114. The A and B alleles of blood group differ from each other by __________ nucleotide substitutions:
(a) 3 (b) 5 (c) 7 (d) 9

115. The first successful experiment on blood transfusion was conducted on a dog by:
(a) Richard Lowenin (1666) (b) Charles Bonnet (1745)
(c) Young (1802) (d) Francesco Redi (1988)

116. Consider the following statements:
(A) Environment plays a significant role in the determination of blood groups
(B) In ABO blood group, both blood group A and blood group B are dominant over blood group O
(C) A man with blood group B and his wife with blood group AB, can produce a child of O blood group

(D) If the father is of A blood group the mother is of B group their child may be of A, B, AB or O blood group

The correct statements are:

(a) All (b) A, B and D (c) A and D (d) B and D

117. If mother is A^- and father is O^+, the blood group of their child may be:

(a) A^+ or O^+ (b) A^- or O^-

(c) A^+, A^- or O^+, O^- (d) A^+ or O^+

118. Mother–foetus ABO incompatibility most commonly occurs when the mother is of _______________ type and foetus is of A, B or AB type:

(a) A (b) B (c) AB (d) O

119. If the mother is of blood group O and father is of AB, the possible blood group of their child is:

(a) A or AB (b) B, AB or O (c) A or B (d) AB or O

120. Parents having blood group AB cannot produce a child of the blood group:

(a) A (b) B (c) AB (d) O

121. Consider the following points with reference to Rh factor:

(A) Discovered by Landstainer and Wiener (B) A dominant trait

(C) Present in all vertebrates (D) Responsible for erythroblastosis foetalis

The correct statements are:

(a) A, B and D (b) A, B and D (c) B, C and D (d) A, C and D

122. A couple has three sons belonging to blood groups A, AB and O, the possible genotypes of the couple:

(a) I^AI^A and I^OI^O (b) I^AI^O and I^BI^O (c) I^AI^A and I^BI^O (d) I^AI^B and I^OI^O

123. If the blood group of the father is O and that of the mother is AB, what are the chances of blood group O in their children?

(a) 75 per cent (b) 50 per cent (c) 25 per cent (d) 0 per cent

124. Blood group of the mother is AB and that of the father is also AB, the chances of AB blood group in their children is:

(a) 25 per cent (b) 50 per cent (c) 75 per cent (d) 0 per cent

125. Which one of the following is a case of genetic imprinting?

(a) Beckwith–Wiedemann syndrome (b) Wolf–Hirschhorn syndrome

(c) Rett syndrome (d) Apert syndrome

126. Single gene mutations may be:

(a) Autosomal (b) Sex linked (c) Largely recessive (d) All

127. The most frequent type of translocation in humans is the centric fusion of chromosomes:

(a) 8 and 12 (b) 10 and 12 (c) 13 and 14 (d) 16 and 17

128. Match column I with column II and select the correct answer using answer codes:

Column I	Column II
(A) Garrod	1. First human gene assignment
(B) Wilson	2. Chromosome analysis on blood
(C) Roberts	3. Biochemical variation
(D) Moorhead	4. United Kingdom's first genetic clinic

Answer codes:

	A	B	C	D
(a)	3	4	1	2
(b)	3	1	4	2

(c) 4 2 1 4
(d) 2 1 3 4

129. Which one of the following is responsible for 47 + XXX abnormality?
(a) Nondisjunction in female meiotic division.
(b) Nondisjunction in female meiotic division or at the male second meiotic division.
(c) Nondisjunction at the male first or second meiotic division.
(d) Nondisjunction at the second meitotic division either in male or female.

130. Which one of the following is uncommon in eskimos?
(a) Cystic fibrosis (b) Diabetes inspidus
(c) Diabetes mellitus (d) Twins

131. Which one of the following is applicable to Klinefelter syndrome?
(a) Hypogonadism (b) Gynecomastia (c) Barr body (d) All

132. A karyotype is a picture of all the chromosomes during_______ state:
(a) Prophase (b) Metaphase (c) Anaphase (d) Telophase

133. Hitchhiker's thumb is:
(a) Autosomal dominant (b) Autosomal recessive
(c) X-linked dominant (d) X-linked recessive

134. In translocation Down syndrome, the number of chromosomes is:
(a) 45 (b) 46 (c) 47 (d) More than 47

135. Which one of the following is incorrect?
(a) Down syndrome – Trisomy of chromosome 21
(b) Edward syndrome – Trisomy of chromosome 18
(c) Wolf–Hirschhorn syndrome – Delection of long arm of 4
(d) Chronic myelogenods leukaemia – Reciprocal translocation between chromosome 9 and 22

136. Match column I with column II and select the correct answer using answer codes:

Column I	Column II
(A) Cystic fibrosis	1. Dominant gene on chromosome 17
(B) Tay–Sachs disease	2. Recessive gene on chromosome 7
(C) Huntington disease	3. Dominant gene on chromosome 15
(D) Neurofibro matosis	4. Dominant gene on chromosome 4

Answer codes:

	A	B	C	D
(a)	3	4	2	1
(b)	2	3	4	1
(c)	4	2	1	3
(d)	4	3	1	2

137. Sex-linked characters are generally:
(a) Dominant (b) Recessive (c) Lethal (d) Atavistic

138. Match column I with column II and select the correct answer using answer codes:

Column I	Column II
(A) Sex limited	1. AB blood group
(B) Sex influenced	2. Baldness
(C) Sex linked	3. Milk production
(D) Co-dominance	4. Duchenne muscular dystrophy

Answer codes:

	A	B	C	D
(a)	4	2	1	3
(b)	2	4	2	3
(c)	3	4	4	2
(d)	3	2	4	1

139. Dosage compensation is not applicable to:
(a) Xg red cell antigen gene
(b) Immunoglobulin heavy chain gene
(c) Fragile X syndrome gene and Xg red cell antigen gene
(d) Steroid sulfatase gene and Xg red cell antigen gene

140. Dosage compensation applicable to all X-linked genes except:
(a) Those on the long arm
(b) Those near the pairing segment at the end of the short arm
(c) Those near the pairing segment at the end of the long arm
(d) Those near the base of the long arm

141. Noonan syndrome is applicable to:
(a) Beckwith–Wiedemann syndrome
(b) Male Turner syndrome
(c) Klinefelter syndrome
(d) Apert syndrome

142. Clinodactyly is the incurved __________ fingers:
(a) Fifth (b) Third (c) Second (d) First

143. Retinitis pigmentosa is:
(a) Autosomal dominant
(b) Autosomal recessive
(c) X-linked recessive
(d) All

144. Which one of the following is a polymorphic trait in humans?
(a) ABO blood groups
(b) MN blood groups
(c) Rh factor
(d) All

145. With respect to the given symbols used for karyotype description, which one of the following is an incorrect match?
(a) p – Short arm
(b) qter – Tip of long arm
(c) i – Inversion
(d) l – Mosaicism

146. In humans, Philadelphia chromosome is formed by reciprocal translocation between chromosomes:
(a) 2 and 12 (b) 3 and 12 (c) 9 and 21 (d) 9 and 22

147. Match column I with column II and select the correct answer using answers code:

Column I	Column II
(A) Wilson disease	1. Inactivation of one of the two X chromosomes
(B) Lesch–Nyhan disease	2. Defective copper metabolism
(C) Ehlers–Danlos syndrome	3. Found only in males
(D) Lyons hypothesis	4. Autosomal dominant but X-linked and autosomal recessive forms are known

Answer codes:

	A	B	C	D
(a)	2	4	3	1
(b)	2	3	4	1
(c)	4	3	1	2
(d)	3	1	2	4

148. Which one of the following is an inherited disorder?
(a) Parkinon's disease (b) Leprosy
(c) AIDS (d) None

149. Which one of the following is a genetically transmitted trait?
(a) Albinism (b) Colour-blindness
(c) Muscular systrophy (d) All

150. Which one of the following is not shown by Down syndrome individuals?
(a) Brusffield spots (b) Saddle nose
(c) Webbed neck (d) Epicanthal skin folds

151. Crigler–Najjar syndrome is characterised by:
(a) Increased uric acid production (b) Presence of bile in various tissue
(c) Progressive mental deterioration (d) Increased production of oxidized haemoglobin

152. Philadelphia chromosome belongs to group:
(a) G (b) F (c) D (d) A

153. Which one causes erythroblastones foetalis?
(a) Rh^+ male Rh^+ female (b) Rh^- male and Rh^- female
(c) Rh^+ male and Rh^- female (d) Rh^- male and Rh^+ female

154. The most commonly associated chromosomal abnormalities with Down syndrome is:
(a) Deletion (b) Translocation (c) Mosaicsm (d) Trisomy

155. Which one of the follwoing is an incorrect match?
(a) ABO blood groups – Multiple allelism
(b) Nondisjunction – Change in chromosome number
(c) Euthansia – Selective inbreeding in human beings
(d) AB blood group – Co-dominance

156. The chronological age of a boy is 5 years and his mental age is 15, his I.Q is:
(a) 3 (b) 45 (c) 150 (d) 100

157. All enzyme deficiencies are autosomal recessive except:
(a) Tryosinase (b) Acute intermittent porphyria
(c) Galactone phosphate uridyl transferase (d) Transacetylase

158. Retinoblastoma is due to loss of heterozygosity, which may be due to:
(a) Mitotic recombination (b) Somatic mutation
(c) Loss of a chromosome (d) All

159. What is incorrect about Huntington disease?
(a) Erinucleotide CAG repeated over and over again
(b) Gene located on chromosome 4
(c) Individuals with fewer than 30 repeats of CAG develop this disease
(d) Degenerative brain disorder

160. Match column I with column II and select the correct answer using codes:

Column I	ColumnII
(A) Enzyme defect	1. G-6-PD deficiency
(B) Alteration of protein	2. Albinism
(C) Defect in receptor and transport system	3. Hyperchlesteremia
(D) Genetically determined and adverse reaction to drugs	4. Sickle cell anaemia

Answer codes:

	A	B	C	D
(a)	2	3,	1	4
(b)	3	4	2	1
(c)	4	3	1	2
(d)	2	4	3	1

161. According to a recent finding, Down syndrome individuals by the age 35 years develop:
(a) Parkinson disease (b) Alzheimer syndrome
(c) Retinoblastoma (d) Fragile X syndrome

162. Match column I with column II and select the correct match using answer codes:

Column I		Column II
(A) Eugenics	1.	Sum total of genes in a population
(B) Euphenics	2.	Improvement of human race
(C) Gene pool	3.	Increase in homozygosity
(D) Consanguinity	4.	Phenotypic improvement of humans after birth

Answer codes:

	A	B	C	D
(a)	2	4	1	3
(b)	2	3	4	1
(c)	3	2	4	1
(d)	4	2	1	3

163. In humans, the possible of number of phenotypes for ABO blood group is:
(a) 2 (b) 4 (c) 6 (d) 8

164. Which one of the following is an incorrect statement?
(a) Campomelic dysplasia is a genetic disorder.
(b) It is due to a mutation in the gene *SOX9*, located on the chromosomes 17.
(c) *SOX9* is related with sex reversal.
(d) Gene *SOX9* is responsible for Waardenburg syndrome.

165. In male human beings, which one of the following is essential for gonadal development?
(a) Testis-determining gene *SRY* (b) X-linked genes
(c) Both (a) and (b) (d) None

166. Which one of the following regions of X and Y chromosomes are paired during synapses?
(a) Differential region (b) Pairing region
(c) Both (a) and (b) (d) None

167. The expression of testis-determining genes can be detected by:
(a) Feulgen test (b) Immunochemistry
(c) Immunochemistry and in situ hybridisation (d) All

168. Which one of the following statements is incorrect?
(a) A deletion in Y chromosome may lead to infertility.
(b) According to data, it has been suggested that X and Y chromosomes evolved from ancestral autosomes nearly 300 million years ago.
(c) The Y chromosome is one third the size of X chromosome.
(d) The X and Y chromosomes do not share common genes.

169. The sexuality of an individual is determined at __________ level:

(a) Chromosomal sex (b) Gonadal sex
(c) Somatic sex and sexual orientation (d) All

170. Which one of the following is not applicable to XY gondal dysgenesis?
(a) Swyer syndrome (b) Hypogonadism
(c) Defect in specific gene on a chromosome (d) Cockayne syndrome

171. Klinefelter syndrome can be diagnosed by:
(a) Karyotyping (b) Pedigree analysis
(c) On the basis of behaviour (d) Somatic cell hybridisation

172. Which one of the following is a key gene in human sex determination?
(a) DAX_1 (b) *SRY* (c) *SOX9* (d) All

173. Consider the following statements:
(A) 47 XYY syndrome was first reported by Sandberg (1961)
(B) 47 XYY is an aneuploidy
(C) 47 XYY is inherited
(D) The incidence of 47 XYY is not affected by advanced paternal or maternal age
The incorrect statements are:
(a) A, B and C (b) B and C (c) B, C and D (d) C

174. Which one of the following is not used in the analysis of human genetics?
(a) Test cross (b) Karyotyping (c) Pedigree analysis (d) RFLP analysis

175. Which one of the following is an X-linked dominant trait within uterolethality for a hemizygous male?
(a) Marfan syndrome (b) Mucopolysaccharidosis
(c) Incontinentia pigmenti (d) Intestinal polyposis

176. Which one of the following is an X-linked recessive trait?
(a) Hurler syndrome (b) Hunter syndrome (c) Sanfilippo syndrome (d) Morquio disease

177. Prenatal diagnosis is not possible for:
(a) Gardner syndrome (b) Multiple endocrine neoplasia
(c) Hunter syndrome (d) All

178. Which one of the following is not applicable to neurofibromatosis?
(a) Autosomal dominant (b) Full penetrance but variable expression
(c) Prenatal diagnosis possible (d) von Recklinghausen disease

179. In the given pedigree, the pattern of inheritance is:

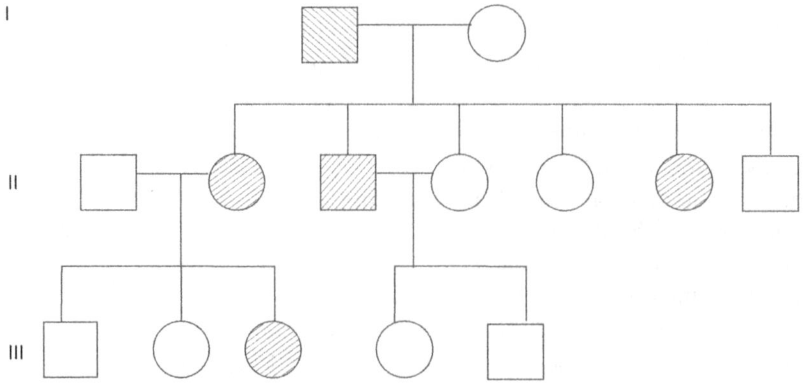

(a) Sex-linked dominant
(b) Sex-linked recessive
(c) Autosomal dominant
(d) Autosomal recessive

180. Which one of the following is an autosomal dominant trait with complete but age-dependent penetrance having locus on 4 pter?
(a) Huntington disease
(b) Wilson disease
(c) Morquio disease
(d) von Recklinghausen disease

181. The frequency of nondisjunction increases:
(a) With increasing maternal age
(b) With maternal hypothyroidism
(c) After irradiation or viral infection
(d) All

182. A consanguineous couple is at increased risk for:
(a) Congenital malformations
(b) Autosomal recessive traits
(c) Multifactorial traits
(d) All

183. Consider the following statements:
(A) Pericentric inversions do not produce clinical abnormality in the carrier
(B) In pericentric inversion, there is a risk of producing chromosomally unbalanced offsprings
(C) Diego positive blood group is found only in East Asians and North Americans
(D) The frequency of blood groups is known to be affected by genetic drift, geographical isolation, temporal variation and natural selection

The incorrect statements are:
(a) C and D (b) A and D (c) B and D (d) None

184. If the affected males of an autosomal dominant trait with sex limitation are infertile, then the pedigree pattern is identical to___________ where males do not reproduce:
(a) An X-linked dominant
(b) An X-linked recessive
(c) XY-linked
(d) Autosomal recessive

185. Restriction Fragment Length Polymorphisms (RFLPs) are important tools for:
(a) Gene mapping
(b) Prenatal diagnosis
(c) Carrier detection
(d) All

186. Which one of the following statements is incorrect ?
(a) In human beings, independent assortment of X chromosomes only occurs in females.
(b) Gene dosage cannot be assessed by the intensity of bands on autoradiographs.
(c) Tetraploidisation may well have been an early step in the origin of human karyotype.
(d) A chimaera is an individual with two cell lines, which were derived from two separate zygotes.

187. Charcot–Marie tooth disease is:
(a) Generally autosomal dominant
(b) Rarely autosomal recessive
(c) X-linked dominant
(d) All

188. Phenylketonuria is rare in:
(a) Indians (b) Negroes (c) Ashkenazi Jews (d) All

189. In which one of the following is prenatal diagnosis possible?
(a) Zellweger syndrome
(b) Waardenburg syndrome
(c) Gardner syndrome
(d) Peutz–Jeghers syndrome

190. Which one of the following is not a single gene disorder?
(a) Wilson disease (b) Haemochromatosis (c) Cleft lip (d) Galactosaemia

191. Epicanthus is applicable to which one of the following?
(a) Down syndrome (b) Turner syndrome (c) Klinefelter syndrome (d) All

192. Which one of the following is a correct match?
(a) Mongolism – Nullisomy
(b) Dosage compensation – Morgan
(c) Lyon's hypothesis – Number of Barr bodies
(d) Tongue rolling – Recessive trait

193. The parameters used to characterise a chromosome in the karyotype:
(a) Length of chromosome
(b) Shape of chromosome
(c) Centromeric index
(d) All

194. Tylosis is characterised by:
(a) Absence of some or all nails of the finger and toes
(b) Soft enamel
(c) Soft skin on palms and soles
(d) Excessive thick and hard skin on palms and soles

195. Which one of the following is a dominant trait?
(a) Large eyes (b) Schizophrenia (c) Deafness (d) Normal vision

196. Which one of the following chromosomes lack N bands?
(a) 14 (b) 21 (c) 15 (d) 5

197. Which one of the following symbols represents a monozygotic twin?

(a) 

(b) 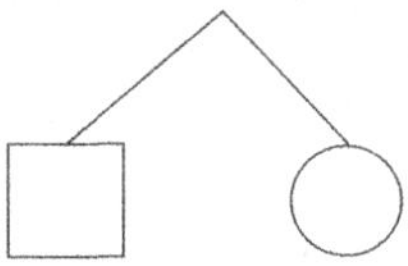

(c) 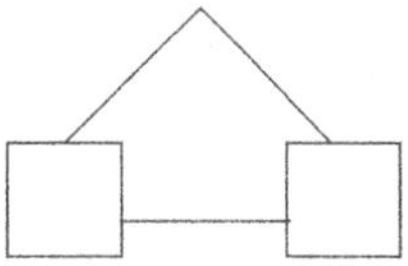

(d) 

198. Which one of the following diseases is characterised by pigmented skin spot?
(a) Porphyria
(b) Neurofibromatosis
(c) Icthyosis
(d) Tay–Sachs disease

199. Which one of the following is also known as Woody Gutherie's disease?
(a) Huntington disease
(b) Sickle cell anaemia
(c) Tay–Sachs disease
(d) Albinism

200. Which one of the following is incorrect about G bands?
(a) Stain darkly with Giemsa and Quinacrine
(b) AT rich
(c) DNAase sensitive
(d) LINE

201. Sister Chromatid Exchange (SCE) is used in the diagnosis of:
(a) Edward syndrome
(b) Down syndrome
(c) Bloom syndrome
(d) Angelman syndrome

202. Dark G bands and bright Q bands are:
(a) Late replicating
(b) Gene poor
(c) Contain LINE repeats
(d) All

203. Which one of the following is incorrect about G light bands and dull Q bands?
(a) Early replicating
(b) Gene rich
(c) Lack SINE/*Alu* sequences
(d) GC rich

204. Which one of the following symbols indicates death?

(a)    (b) 

(c) 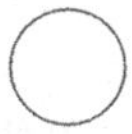   (d) 

205. Sickle cell anaemia does not form an example of:
    (a) Pleiotropism (b) Reversion (c) Lethal effect (d) Co-dominance
206. The frequency of detection of sex chromatin in females is more in:
    (a) Amniotic or chorionic epithelium (b) Oral smears
    (c) Nervous tissue (d) Constant in all
207. Multifactorial inheritance is associated with hereditary traits like:
    (a) Eye colour and skin colour (b) Finger-printing patterns
    (c) Height (d) All
208. mtDNA allows ________________ to trace their maternal lineages:
    (a) Men (b) Women
    (c) Both men and women (d) None
209. Which one of the following is a rare inherited disorder of the nervous system?
    (a) Batten disease (b) Crohn's disease (c) Tay–Sachs disease (d) Fabry disease
210. Which one of the following is known as Jewish genetic disease?
    (a) Cystic fibrosis (b) Sickle cell disease
    (c) Gaucher's disease (d) Charcot–Marie tooth disease
211. Bloom syndrome is:
    (a) Autosomal dominant (b) Autosomal recessive
    (c) X-linked dominant (d) X-Y linked
212. The most common human enzyme deficiency known is:
    (a) Phosphofructokin ase deficiency (b) Acid beta glucosidase deficiency
    (c) G-6-PD deficiency (d) Transglucosisage deficiency
213. Which one of the following is incorrect about β-thalassaemia?
    (a) Autosomal recessive inherited disorder
    (b) Caused by a mutation in of the β-haemoglobin chain
    (c) Genetic testing via blood test can defect β-thalassaemia, if a person is a carrier
    (d) Also known as Fanconi anaemia
214. Consider the following statements:
    (A) Tryosinosis causes harmful effect
    (B) Inheritance of intelligence is genetically controlled and the environment has no role in it
    (C) Asymmetric karyotype is a primitive feature
    (D) Sex-linked characters are generally lethal
    The incorrect statements are:
    (a) A and D (b) B and D (c) C and D (d) All
215. Which one of the following shows paternal disomy?
    (a) Klinefelter syndrome (b) Beckwith–Wiedemann syndrome
    (c) Rett syndrome (d) Martin–Bell syndrome

216. An isochromosome does not have:
(a) Two identical arms
(b) Two short arms
(c) Two long arms
(d) One long and one short arm

217. Which one of the following is an incorrect match?
(a) Congenital hypothyroidism – Chromosome 1
(b) Crohn disease – Chromosome 5
(c) Prader–Willi syndrome – Chromosome 15
(d) Beckwith–Wiedemann syndrome – Chromosome 19

218. Match column I with column II and select the correct answer using answer codes:

Column I (Disease)	Column II (Ethnic groups)
(A) Bloom syndrome	1. African blacks
(B) Sickle cell disease	2. Caucasians
(C) β thalassaemia	3. Italians, US blacks and Greeks
(D) Cystic fibrosis	4. Ashkenazi Jews

Answer codes:

	A	B	C	D
(a)	4	3	1	2
(b)	3	2	4	1
(c)	4	1	3	2
(d)	2	3	4	1

219. Which one of the following is an example of X-linked co-dominant inheritance?
(a) Agammaglio bulinaemia
(b) Becker muscular dystrophy
(c) Restriction fragment length polymorphisms
(d) Fragile X associated mental retardation

220. An Rh^- women cannot be sensitised by:
(a) Rh blood transfusion
(b) Abortion
(c) Amniocentesis or chorionic villus sampling
(d) AIDS

221. All human cells contain 23 pairs of chromosomes except:
(a) Sperms (b) Ova (c) Mature RBCs (d) All

222. The technique used for RFLP mapping in humans is similar to those of:
(a) *Drosophila* (b) Plants (c) Mice (d) None

223. Which one of the following is applicable to epilepsy?
(a) Chromosomal aberration
(b) Multifactorial inheritance
(c) Single gene trait
(d) All

224. Interpupillary distance above excepted is known as:
(a) Hypertelorism (b) Hypotelorism (c) Telocanthus (d) Clinodactyly

225. Match column I with column II and select the correct answer using answer codes:

Column I	Column II
(A) Turner syndrome	1. Excess of arches
(B) Trisomy 18	2. 6–10 arches
(C) Trisomy 21	3. Predominance of whorls
(D) 47, XXY	4. Generally all ulnar loops

Answer codes:

	A	B	C	D
(a)	3	2	4	1

(b)	4	3	1	2
(c)	2	4	1	1
(d)	3	4	2	1

226. Molecular pathology of β-thalassaemia does not involve:
(a) Point mutation (b) Deletion (c) Inversion (d) Insertion

227. Consider the following statements:
(A) XX males have been reported (B) XX males lack *SRY* genes
(C) These males have mutations in the *RSP 01* gene which encode growth factor called R-spondin-1
(D) According to latest research, *R-spondin-1* works with *Wnt* and *b-catenin* genes in repressing male pathway and if *RSP01* is mutated, it leads to the development of testis

The correct statements are:
(a) All (b) A, B and C (c) B and C (d) C and D

228. Generally gene *SRY* activates gene:
(a) *RSP01* (b) *SOX9* (c) *Fgf9* (d) W*nt*

229. Genes may have a strong impact on our health:
(a) Through chromosomes (b) Single gene disorders
(c) By influencing our susceptibility to diseases (d) All

230. The gene *CNTNAP2* is associated with:
(a) Deafness (b) Leprosy
(c) Specific Language Impairment (SLI) (d) Reproduction activity

231. Stainable Barr bodies can be seen in:
(a) Oogonia (b) Spermatogonia (c) Spermatids (d) None

Answers to Multiple-Choice Questions

1. (d)	2. (c)	3. (b)	4. (b)	5. (a)	6. (c)	7. (a)	8. (b)
9. (c)	10. (d)	11. (d)	12. (a)	13. (b)	14. (d)	15. (b)	16. (c)
17. (a)	18. (b)	19. (a)	20. (a)	21. (a)	22. (d)	23. (b)	24. (d)
25. (a)	26. (d)	27. (d)	28. (b)	29. (d)	30. (c)	31. (d)	32. (a)
33. (c)	34. (a)	35. (b)	36. (a)	37. (a)	38. (c)	39. (a)	40. (a)
41. (d)	42. (c)	43. (c)	44. (a)	45. (b)	46. (d)	47. (d)	48. (b)
49. (a)	50. (b)	51. (b)	52. (b)	53. (c)	54. (d)	55. (a)	56. (d)
57. (b)	58. (a)	59. (d)	60. (c)	61. (d)	62. (d)	63. (c)	64. (b)
65. (d)	66. (c)	67. (d)	68. (b)	69. (c)	70. (a)	71. (d)	72. (b)
73. (c)	74. (b)	75. (c)	76. (c)	77. (b)	78. (b)	79. (c)	80. (a)
81. (c)	82. (a)	83. (d)	84. (d)	85. (b)	86. (d)	87. (d)	88. (b)
89. (c)	90. (d)	91. (d)	92. (b)	93. (d)	94. (b)	95. (a)	96. (c)
97. (a)	98. (a)	99. (c)	100. (d)	101. (c)	102. (d)	103. (d)	104. (a)
105. (a)	106. (d)	107. (a)	108. (d)	109. (c)	110. (c)	111. (d)	112. (a)
113. (d)	114. (c)	115. (a)	116. (d)	117. (c)	118. (d)	119. (c)	120. (d)
121. (b)	122. (b)	123. (d)	124. (b)	125. (a)	126. (d)	127. (c)	128. (b)
129. (b)	130. (c)	131. (d)	132. (b)	133. (b)	134. (b)	135. (c)	136. (b)
137. (b)	138. (d)	139. (d)	140. (c)	141. (b)	142. (a)	143. (d)	144. (c)
145. (d)	146. (d)	147. (b)	148. (d)	149. (d)	150. (c)	151. (b)	152. (a)
153. (c)	154. (d)	155. (c)	156. (c)	157. (a)	158. (d)	159. (c)	160. (d)
161. (b)	162. (a)	163. (b)	164. (d)	165. (c)	166. (b)	167. (c)	168. (d)

169. (d)	170. (d)	171. (a)	172. (d)	173. (d)	174. (a)	175. (c)	176. (b)
177. (d)	178. (c)	179. (c)	180. (a)	181. (d)	182. (d)	183. (d)	184. (b)
185. (d)	186. (b)	187. (d)	188. (d)	189. (a)	190. (c)	191. (a)	192. (c)
193. (d)	194. (b)	195. (a)	196. (d)	197. (c)	198. (b)	199. (a)	200. (c)
201. (c)	202. (d)	203. (c)	204. (a)	205. (b)	206. (d)	207. (d)	208. (c)
209. (a)	210. (c)	211. (b)	212. (c)	213. (d)	214. (d)	215. (b)	216. (d)
217. (d)	218. (c)	219. (c)	220. (d)	221. (d)	222. (d)	223. (d)	224. (a)
225. (a)	226. (c)	227. (a)	228. (b)	229. (d)	230. (c)	231. (d)	

Fill in the Blanks

1. G-band patterns are essentially identical to ____________ band patterns.
2. Short arm of the chromosome is called p arm while long arm is called _______ arm.
3. The banding technique was developed by ________________ .
4. The cytogenetic technique, which can be used to detect and locate presence or absence of specific DNA sequence, is the ______________________________.
5. Three major types of chromatin, which can be recognised with G banding are ____________ and ________________.
6. The centromeric index is the ratio of the lengths of the _____________ of the chromosome.
7. Diagramatic representation of karyotype is called __________________.
8. ______________ banding has been utilised for mapping of gene sequences in human chromosomes.
9. The term 'sex chromosome' was coined by ______________________.
10. ______________________________ discovered human Y chromosome.
11. __________________________ is the act of preparing karyotype.
12. The human Y chromosome is composed of about ____________ million base pairs.
13. The human Y chromosome contains ________________working genes.
14. The human Y chromosome is able to recombine with itself using _____________ sequences.
15. X chromosome inactivation occurs if the number of X chromosome exceeds ___________.
16. Inactivation of the X chromosome takes place at ____________________________stage.
17. Sickle cell haemoglobin (Hb-S) is an abnormal variant of the ________________ chain.
18. The cause of genetic disorder is a ______________________ in every cells of an individual.
19. Prenatal deflection of chromosomal abnormalities is performed mainly by _____________.
20. Presently, more than ______________________________ human genetic traits are known.
21. Karyotype showing gradual transition from the largest to smallest chromosome is known as __________.
22. Failure of separation of sister chromatids is known as ______________________.
23. Citrate Giemsa stains regions of _______________________ heterochromatin.
24. __________banding and __________banding patterns are used for plant karyotyping.
25. ______________________________ is the diagrammatic representation of karyotype.
26. Prenatal detection of chromosomal abnormalities is mainly done by _______________.

27. ____________________ is a group of diseases in which blood does not clot normally.
28. In human beings, the largest chromosome is ______ and the smallest one is __________.
29. Q bands are identical to G bands except that Q bands ____________________________.
30. Male humans can contribute ______or _________ chromosomes to their offsprings while female humans can contribute only __________ chromosomes to their offsprings.
31. Deficiency of copper transmission is an ____________ trait in humans.
32. Colour-blindness is of three types, viz., ___________, _____________ and ____________.
33. The genes of phenylketonuria are located on chromosome _______________________.
34. Y spot was discovered by ______________.
35. The chromosomes are arranged according to __________________________ system.
36. Alkaptonurics excrete excess of _______________ acid.
37. The allele responsible for sickle cell anaemia is _______________ and is located on the short arm of chromosome 11.
38. Chromosome walking is a technique to ___________a gene from its known closet markers.
39. The first human trait proven to be due to a gene on a specific chromosome was __________________.
40. A molecular cytogenetic technique used to simultaneously visualise all the pairs of chromosomes in an organism in different colours is known as the ______________________.
41. The phenomenon showing differential pattern is an expression of genes inherited from the father and the mother is known as ________________.
42. Albinos are individuals who lack the ability to produce ___________________________.
43 Barr body is lacking in _____________________ cells of females.
44. Red and green colour detection genes are on the ___________ chromosome while the blue colour detection gene is on an ____________________.
45. Huntington disease was studied by Gusella et al., who used _______________to identify marker.
46. The cause of genetic disorder is a ___________________________.
47. Replacement of genes by corrected copies of genes to treat genetic disorders is called _________________________.
48. Marriages among blood relations are known as ____________________________ marriages.
49. Haemophilia was first studied by ____.
50. Pedigree of colour-blindness was first described by ____________________________.
51. Baldness is a sex-influenced trait in humans, which is ___________ in males and ___________ in females.
52. A marked loss of ___________________ is found in the brain of Alzheimer's patients.
53. The genotype of a person with sickle cell trait is ______________________________.
54. The term 'eugenic' was coined by ____.
55. Albinism is due to the lack of enzyme ____.
56. Deuternapoia is characterised by having no perception of ______________________ colour.
57. Agglutinogens are not found in the blood group ______________________________.
58. ABO blood group is determined by ______________________________________alleles.
59. Polydactly in humans is due to ____ gene.

60. Amniocentesis is generally performed at ________________________ weeks of gestation.
61. Phenylketonuria is autosomal recessive trait with locus on chromosome _______________.
62. Webbed toes in humans are a________________ linked trait.
63. The Rh locus is located on chromosome________________________.
64. A variation in the gene ___________ has a positive effect on the life expectancy of humans.
65. The spectral karyotyping method has been introduced by __________________________.
66. Ullrich–Turner syndrome is the most common due to ________________ chromosome defect.
67. Blood groups are formed by the molecules present on the surface of _________________.
68. International Society of Blood Transfusion (ISBT) has recognised ___________ human blood group system.
69. Defective dentine of teeth in humans is an _______________ linked trait.
70. Lesch–Nyhan syndrome lacks enzyme ____.
71. Rare X-linked dominant traits are found in_____________ females and ___________ males.
72. Y-linked inheritance is called ____ inheritance.
73. If a trait is carried by a Y chromosome without a homologue on the X chromosome, it is said to be ___ ________________________________trait.
74. In retinoblastoma, there is a deletion of the long arm of chromosome ________________.
75. All humans are heterozygous for recessive alleles, which if present in homozygous, would be lethal. This is known as ____________________________________.
76. The ______________________ is an essential precursor to the ABO blood group antigens.
77. No antibodies are formed against the H antigen except in the individuals with__________________ _________
78. Inheritance of human traits are based on_________________________ model of inheritance.
79. Erythroblastosis foetalis can be prevented by injecting a serum containing ______________ antibodies into the mother.
80. Philadelphia chromosome is found in persons suffering from ______________________.
81. Christmas disease is another name of ______________________.
82. Allele of cystic fibrosis occurs on the chromosome number_______________________.
83. The brain of a person suffering from Alzheimer's disease is unable to metabolise ________.
84. Agglutinogens are carried by ____________________.
85. The daughters of a haemophilic father and a normal mother will be ______________________.
86. The tradition of ending life in a painless manner is known as _______________________.
87. An assessment made by a medical professional about genetic disorder risk in a family is known as __________________analysis.
88. Marfan syndrome is a disorder of _______ .__________________tissue.
89. ___________________ is a red blood cell undergoing cell fragmentation.
90. In humans, ____________________ pair of chromosomes are different.
91. The famous pedigree of haemophilia in the royal family of Europe was discovered by _________________.
92. A child having an IQ of 140 or more belongs to _________________________ category.
93. If only some individual cells have chromosomal abnormality, it is known as __________________.

94. Down syndrome can occur in three different forms, viz., ________, ___________ and ____________.
95. Amyloid β-protein deposits cause damage of brain in individuals suffering from ____________ disease.
96. Antigens are lacking in the blood group______________.
97. If blood group of the father is A and that of the mother is O, the possible blood group of their children will be __________ or ____________.
98. If both parents are of blood group AB, their children's blood group will never be of blood group ______________.
99. The probability of a haemophilic son to a haemophilic father and a normal mother is ___________.
100. Haemophilia is a genetic disorder, which is caused due to the deficiency of _______________.
101. A normal female whose father was colour-blind marries a normal man, the chances of colour-blindness in their son is ___________________________.
102. Winiwarter (1912) reported____________ chromosomes in spermatogonia and _______________ chromosomes in oogonia of human beings.
103. Robertosian translocation occurs among the __________ or ___________ groups of the chromosomes.
104. The endoscopic visualisation of the foetus is called _______________.
105. In neutophil leucocytes, the X chromatin appears as a small rod-like structure called_________________.
106. The majority of hereditary metabolism disorders are inherited as_________________.
107. In Beckwith–Wiedemann syndrome, both copies of a region on the short arm of chromosome are inherited from the _______________.

Answers to Fill in the Blanks

1. Q
2. q
3. Caspersson
4. Fluorescent In Situ Hybridisation (FISH)
5. Euchromatin intercalary, heterochromatin centromeric
6. Long and short arms
7. Idiogram
8. R
9. Wilson (1906)
10. Painter (1923)
11. Karyotyping
12. 60
13. 78
14. Palindrome base pair
15. One
16. Blastocyst
17. Beta
18. Gene mutation
19. Amniocentesis
20. 3,500
21. Symmetric
22. Nondisjunction
23. Constitutive
24. C, N
25. Idiogram
26. Amniocentesis
27. Haemophilia
28. 1, 22
29. Fluorescence
30. X, Y, X
31. X linked
32. Protanopia, deuteranopia, tritanopia
33. 12
34. Zech (1970)
35. Denver
36. Homogentisic
37. Autosomal recessive
38. Clone
39. Red-green, colour-blindness
40. Spectral karyotyping
41. Genetic imprinting
42. Melanin pigments
43. Pre-meiotic
44. X, autosome
45. Restriction-fragment length polymorphisms
46. Gene mutation
47. Gene therapy
48. Consanguineous
49. John Cotto (1803)
50. Horner (1876)
51. Dominant, recessive
52. Neurons
53. Hb^S Hb^A
54. Francis Galton
55. Tyrosinase
56. Green
57. O
58. Three
59. Autosomal dominant
60. 16–18
61. 12
62. Y
63. (1) One
64. *FOX3QA*
65. Schrock et al.
66. 45, X
67. Red blood cells

68. 30
69. X
70. Hypoxanthine-guanine phosphoribosyl transterase(HGPRT)
71. Heterozygous, homozygous
72. Holandric
73. Sex limited
74. 13
75. Genetic load
76. H antigen
77. Bombay phenotype
78. Mendel's
79. Anti-Rh+
80. Leukaemia
81. Haemophilia B
82. Four
83. Amyloid β-peptide
84. Erthrocytes
85. Carrier
86. Euthanasia
87. Pedigree
88. Connective
89. Schistocyte
90. 23rd
91. Haldane
92. Genius
93. Mosaicism
94. Trisomy 21, tranlocation, mosaic
95. Alzheimer's
96. AB
97. A, O
98. O
99. 0 per cent
100. Thromboplastin
101. 50 per cent
102. 47, 48
103. D, G
104. Fetoscopy
105. Drumstick
106. Autosomal recessive
107. Father

True or False

1. Each chromosome has its individual pattern of banding.
2. Sickle cell anaemia affects karyotype.
3. Thalassaemia is an effect of regulating gene.
4. G-6-PD-deficient persons are sensitive to primaquine.
5. All species have a standard karyotype.
6. The X chromosome has a more active euchromatin than the Y chromosome.
7. The X chromosome is a gene-rich region.
8. Sickle cell trait does not cause sickle cell anaemia.
9. X chromosome lacking X Inactivation Center (XI(C) do not inactivate.
10. Autosomal crossovers are more frequent in males than females.
11. A and B antigens are also found in body fluids.
12. In some cells of male human beings, X chromosome remains inactive.
13. X-linked dominant disorder is more frequent in females.
14. Males of trisomy 21 are generally fertile.
15. Carrier detection is not possible in galactosaemia.
16. 45, X is particularly frequent in early spontaneous abortions.
17. Klinefelter syndrome is usually due to paternal nondisjunction.
18. Trisomy 21 shows marked maternal age effect.
19. Selection may result in allele fixation.
20. First prenatal chromosome analysis was done by Breg and Stec (1966).
21. In male human beings, any mutant X allele is always expressed.
21. Trisomy 21 may be complicated by hypothyroidism.
23. Barr body can be easily demonstrated in the cells of buccal epithelium.

24. Achondroplasia is the most common cause of short-trunked dwarfism.
25. Huntington disease is curable.
26. Nail–Patella syndrome is an autosomal recessive trait.
27. Riis and Fuchs (1960) performed first prenatal sexing.
28. Crown baldness in humans is a sex-influenced trait.
29. O allele does not code for any enzyme in the ABO system.
30. Spectral karyotype defect translocations, which are not recognisable by traditional banding patterns.
31. The phenotype of Ullrich–Turner syndrome is highly variable and depends mainly on the karyotype.
32. Recently congenital adrenal hyperplasia has been reported in patients with 47+XXX chromosomal complement.
33. Anti-A and anti-B antibodies have the ability to cross the placenta.
34. Bone marrow transplant is one of the most common causes of change in blood group.
35. Duchenne muscular dystrophy is common in females.
36. Retinoblastoma is a dominant trait which is due to loss of heterozygosity.
37. Females with Turner syndrome have a high risk of developing autoimmune thyroid diseases.
38. Autosomal dominant traits are sometimes complicated by variable expressivity and incomplete penetrance.
39. Marfan syndrome and homocystinuria have similar phenotypes.
40. In autosomal dominant trait, each affected person is heterozygous.
41. Achondroplasia is a rare sex-linked trait.
42. Albinism exhibits locus and allelic heterogenesity.
43. Osteogenesis imperfecta is an autosomal recessive disorder.
44. Individuals with 44+YO chromosomal complement do not survive.
45. Plants lack G bands.
46. Y chromosome bears satellite.
47. Human chromosome 5 is similar to chromosome 1 of chimpanzees, gorillas and orangutans.
48. Heart defects are common in all trisomy 18 chromosomes.
49. The distribution of blood group antigens is almost constant in different races.
50. Blood group O individuals have predominantly IgG molecules.
51. Haemolytic anaemia occurs from the mother's antibodies against foetal red blood cells.
52. The Y chromosome in humans is not essential for viability.
53. Achondroplasia in double dose is lethal.
54. Cleft lip clfet palate is more common in males.
55. H antigen is produced by the action of recessive allele.
56. The Bombay phenotype blood produces anti-O antibodies.
57. The Rh factor is found only in humans and rhesus monkey.
58. In humans, sex is established at the time of fertilisation.
59. Acquired immuno deficiency syndrome is a sex-influenced trait.
60. Sickle cell anaemia is an example of molecular mutation.
61. Rh factor is related with blood clotting.
62. Children conceived by fathers of older age have more inborn mutations.

63. A recessive trait or disease may remain hidden for several generations.
64. IQ is largely heritable.
65. Genes affect human behaviour.
66. Smooth chin is a recessive trait.
67. Autosomal dominant traits are frequent.
68. Cystic fibrosis is the most common disease in Indians.
69. 2n + 0Y foetus do not survive.
70. Sex-limited genes are located on sex chromosomes.
71. The euchromatic region of Y chromosome contains holandric genes.
72. Keratoderma dissipatum is an X–Y linked trait in humans.
73. Curly hair is a sex-linked trait.
74. Holandric genes are never found in pairs.
75. Blood group O parents will have children with blood group O and A in 1:4.
76. Spermatogenesis does not occur in Klinefelter syndrome individuals.
77. A hydatiform mole is always 46+XY.
78. Men with blood group A are not likely to get prostrate cancer as compared with men with other blood groups.
79. Presence of beard in male human beings is a Y-linked character.
80. Fabry's disease is found only in male human beings.
81. Epidermolysis is a holandric inheritance.
82. In Huntington disease, limb movement is absent.
83. X chromosome in human beings is acrocentric.
84. Cataract is a hereditary disease.
85. Leukaemia is a sex-linked disease.
86. Ishihara cards are used to test colour-blindness.
87. Peculiar dancing gait and bizarre grimacing are manifested by patients of Huntington disease.
88. The frequency of trisomy 21 increases markedly with increasing maternal age.
89. Duchenne muscular dystrophy may be allelic to Becker muscular dystrophy.
90. Duplications tend to be more harmful than deletions.
91. Klinefelter syndrome is generally due to paternal nondisjunction.
92. Chorionic villus sampling permits rapid fetal karyotyping.
93. Tay–Sachs disease in heterozygous form offers an advantage against tuberculosis.
94. Inactivation of X chromosome is a random process.
95. Defective ova and sperm usually result in inherited chromosomal abnormalities.
96. Autosomad abnormalities are less common in comparison to sex chromosome abnormalities.
97. Down syndrome patients have longer expected lifespan than average.
98. There is no relation between the age of a father and the probability of a child to acquire Down syndrome.
99. Women having Down syndrome are at a lower risk of having children with Down syndrome.
100. Leucoderma is a sex-linked character.
101. The chances of Down syndrome is about 1 in 1400 in 350 and 1 in every 100 live births to the mothers between ages 25, 35 and 40 more than 40 years, respectively.
102. Down syndrome is a genetic disorder that discriminates races.

103. Individuals of blood type O have mainly IgG molecules.
104. Haemolytic disease in newborns may be due to incompatibility of ABO and Rh blood groups.
105. *Sry* and *Sox9* are mammalian testis-determining genes.
106. Poly-X females are infertile.
107. Pseudohermaphrodites have either testicular or ovarian tissue, but not both.
108. Wolfman syndrome is X-linked dominant.
109. Down syndrome is a very good example of a characteristic pattern of abnormalities that is produced by multigene defect.
110. Balanced translocation effects health status.
111. Sickle cell disease is especially prevalent in blacks.
112. Leucoderma is a sex linked character.

Answers to True or False

1.	True	2.	False	3.	True	4.	True	5.	False	6.	False	7.	False	8.	True
9.	True	10.	False	11.	True	12.	False	13.	True	14.	False	15.	False	16.	True
17.	False	18.	True	19.	False	20.	True	21.	True	22.	True	23.	True	24.	False
25.	True	26.	False	27.	True	28.	True	29.	True	30.	True	31.	True	32.	True
33.	False	34.	True	35.	False	36.	True	37.	True	38.	True	39.	True	40.	True
41.	False	42.	True	43.	False	44.	True	45.	True	46.	False	47.	False	48.	True
49.	False	50.	True	51.	True	52.	True	53.	True	54.	True	55.	False	56.	True
57.	True	58.	True	59.	False	60.	True	61.	False	62.	True	63.	True	64.	True
65.	True	66.	False	67.	False	68.	False	69.	True	70.	False	71.	True	72.	False
73.	False	74.	True	75.	False	76.	True	77.	False	78.	True	79.	False	80.	False
81.	False	82.	False	83.	False	84.	False	85.	False	86.	True	87.	True	88.	True
89.	True	90.	False	91.	False	92.	True	93.	True	94.	True	95.	True	96.	False
97.	False	98.	False	99.	False	100.	False	101.	True	102.	False	103.	True	104.	True
105.	True	106.	False	107.	True	108.	True	109.	False	110.	False	111.	True	112.	False

Give Reasons

1. Staining with quinacrine mustard produces consistent, bright and less bright fluorescent bands in the chromosome arm.
   – Because of differences in the relative amount of cytosine–guanine (C–G) or adenine–thymine (A–T) base pairs.
2. It is difficult to examine prophase bands.
   – Because it is not possible to stop cell cycle at prophase stage.
3. Phytohaemagglutinin is used in tissue culture.
   – Because it stimulates cell division.
4. Philadelphia chromosome is so named.
   – Because it was first reported in a patient suffering from chronic myeloid leukaemia in Philadelphia.

5. Humans are not suitable for genetic studies.
   – Because of the following reasons:
   (a) Long lifespan
   (b) Very small number of offsprings
   (c) Controlled breeding is not possible
   (d) Majority of them are heterozygous for many traits, so it is not possible to get iso strain.
6. Expression of genesis highly variable.
   – Because of migration and changes in habitat.
7. In most mammals, sex is determined by whether or not Y chromosome is present.
   – Because the Y chromosome carries male-determining *SRY* gene.
8. A man can contribute X or Y chromosome to his offspring while a woman contribute only X chromosome.
   – Because a man has an XY genotype while a woman has XX genotype.
9. Sickle cell trait (carrier status) is co-dominant.
   – Because the mutant allele (Hbs) produces mutant haemoglobin and the normal heamoglobin allele (Hb(A) produces normal haemoglobin.
10. All familial traits are not hereditary.
    – Because relatives tend to share common genes as well as more or less common environments.
11. Hypertrichosis passes from father to son and from son to grandson.
    – Because hypertrichosis is a Y-linked trait and Y chromosome of the father is always inherited by the son.
12. X–Y linked traits are incompletely sex linked.
    – Because of occasional occurrence of crossing over in the homologous part of X and Y chromosome.
13. X-linked dominant traits are more variable in affected females.
    – Because of lyonisation.
14. Consangunity has no genotypic effect on the sex-linked genes in males.
    – Because males are hemizygous for each loci, which they receive from their mothers.
15. Normally, a proportion of polyploidy cells are present in the bone marrow of human beings.
    – Because mega karyocytes generally have 8–16 times the haploid number.
16. Females are more affected by sex-linked dominant traits in comparison to males.
    – Because females have two X chromosomes while males have only one.
17. Androgen insensitivity testicular feminisation is both sex linked and sex limited trait.
    – It is sex linked because its gene, the androgen receptor is located on the X chromosome and it is sex limited because it only manifests in XY foetuses.
18. Pedigree charts of royal families have a high degree of inbreeding.
    – Because marriage between royal families is a custom.
19. In human beings, it is difficult to have isostrains.
    – Because most of the human beings are heterozygous for many traits.
20. In ABO system, incompatibility is mainly limited with blood group O mothers having blood group A or B foetuses.
    – Because anti-A and anti-B antibodies present in blood group A and B individuals are mainly IgM that do not cross the placenta. However, blood group O individuals have predominantly IgG molecules that can cross the placenta.
21. Nature does not favour abnormal sex complements.
    – Because sometimes these conditions result in infertility as well as somewhat unusual physical appearance.

22. A man carrying defective Y chromosome passes it only to his sons.
    – Because sons always receive Y chromosome from the father and X chromosome from the mother, while daughters receive one X chromosome from the father and another X chromosome from the mother.
23. XXY individuals are male.
    – Because Y chromosome contains *SRY* gene which causes the development of male characteristics.
24. It is generally predicted that half of the human offsprings born will be males and half will be females.
    – Because sex is determined by the sex chromosomes in the sperm that fertilises the egg. Half of the sperm contains X chromosome and half contains Y chromosome.
25. People born with an abnormal number of chromosomes usually have genetic disorders.
    – Because their cells contain more or less genetic information.
26. Occasionally XY CIAs zygote develops into female.
    – Because in such zygote, X chromosome contains a mutant sex-linked gene which inhibits the differentiation of external genitalia as well as normal functioning of testes.
27. For karyotype preparation, cells are treated with colchicines and hypotonic solution.
    – Because to stop mitosis at metaphase, and to swell as well as scattering of the chromosome.
28. Chromosome bandings are used in clinical diagnosis.
    – Because banding patterns are characteristics and constant for each normal chromosome. So any type of chromosomal abnormalities like deletion, insertion of any chromosomal segment or whole chromosome can be easily recognised.
29. Y chromosome cannot be used to study the ancestry of paternal grandmothers.
    – Because Y chromosome is transferred only from the father to the son.
30. At birth, the infant with beta thalassemia has no symptoms.
    – Because the foetal haemoglobin still predominates.
31. All familial traits are not hereditary.
    – Because relatives are likely to share common genes and environments.
32. Mitochondrial genome disorders are inherited in non-Mendelian fashion.
    – Because mtDNA is inherited only from the mother.
33. Brothers and sisters do not look alike.
    – Because chromosome of a pair comes from different parents having a different version of genes. The pairing of chromosome and their orientation prior to separation is a matter of chance resulting in many different combinations of chromosomes in eggs and sperms. Chromosome pairing is an opportunity for crossing over between homologous chromosomes resulting in recombinant chromosomes. Further, random fertilisation of eggs by sperms takes place. Sometimes changes in genetic composition occur by mutations.

CYTOPLASMIC INHERITANCE

Multiple-Choice Questions

1. Cytoplasmic inheritance is also known as:
   (a) Nonchromosomal inheritance
   (b) Non-Mendelian inheritance
   (c) Maternal inheritance
   (d) All of the above
2. Cytoplasmic inheritance manifests:
   (a) Maternal characters
   (b) Paternal characters
   (c) Parental characters
   (d) Sometimes maternal and sometimes paternal characters
3. Cytoplasmic inheritance was first described by:
   (a) Correns (b) Sonneborn (c) Ruth Sagar (d) Conkilin
4. What is incorrect about cytoplasmic inheritance?
   (a) It is also known as nonchromosomal inheritance.
   (b) It is controlled by plasmagenes.
   (c) Results are different in reciprocal cross.
   (d) None
5. Match column I with column II and select the correct answer using answer codes:

Column I	Column II
(A) Sonneborn	1. Milk factor in mice
(B) Correns	2. Sigma particles
(C) Bittner	3. Kappa particles
(D) Heritier and Teisier	4. Plastids

Answer codes:

	A	B	C	D
(a)	3	1	2	4
(b)	3	4	1	2
(c)	2	1	3	4
(d)	1	4	2	3

6. Match column I with column II and select the correct answer using answer codes:

Column I	Column II
(A) Cytoplasmic genes first observed in	1. Plasmon
(B) Cell organelles responsible for cytoplasmic inheritance	2. Chloroplast and mitochondria
(C) Episome	3. *Mirabilis jalapa*
(D) Sum total of plasmagenes	4. Cytoplasmic inheritance

Answer codes:

	A	B	C	D
(a)	3	2	4	1

(b)	3	2	1	4
(c)	2	3	1	4
(d)	1	2	4	3

7. Mitochondrial DNA:
   (a) Usually shows maternal inheritance
   (b) Usually shows biparental inheritance
   (c) Inherited from nuclear genes
   (d) None
8. Sigma particle is a/an:
   (a) Spirochaete bacteria
   (b) Milk-producing factor
   (c) Infectious DNA virus
   (d) Paramecin producing particle
9. In snails, dextral shell coiling is dominant over sinistral coiling. If a female, having sinistral (dd) coiling is crossed with a male having dextral (DD) coiling in F_1 all progenies have:
   (a) Dextral coiling
   (b) Sinistral coiling but with dextral genotype
   (c) Dextral coiling but with sinistral genotype
   (d) Both dextral as well as sinistral coiling
10. The differences caused by extra nuclear factors:
    (a) Usually disappear after one generation
    (b) May persist as long as the extra nuclear factor can perpetuate itself
    (c) Persist throughout life
    (d) Both (a) and (b)
11. Cytoplasmic inheritance:
    (a) Enhances the chances of cytoplasmic mutation
    (b) Prevents the total loss of organelles
    (c) Suppresses cytoplasmic mutation and prevents total loss of organelles
    (d) Both (a) and (b)
12. Maternal inheritance may be due to genes present in:
    (a) Cytoplasm
    (b) Mitochondria
    (c) Lysosomes
    (d) Endoplasmic reticulum
13. The scientists associated with cytoplasmic inheritance:
    (a) Sonneborn and Morgan
    (b) Bittner and Muller
    (c) Ephrussi and Correns
    (d) Conklin and Darlington
14. Which one of the following is known as mate killer?
    (a) Kappa particle (b) Mu particle (c) Milk factor (d) Self-sterility allele
15. Match column I with column II and select the correct answer using answer codes:

Column I (Examples of cytoplasmic inheritance)	Column II (Organisms)
(A) Milk factor	1. *Paramecium*
(B) Kappa particles	2. Maize
(C) Shell coiling	3. Mice
(D) Male sterility	4. *Limnea*

Answer codes:

	A	B	C	D
(a)	3	2	4	1
(b)	3	1	2	4
(c)	3	1	4	2
(d)	3	2	4	1

16. Kappa particles are associated with:
    (a) Mitochondria
    (b) *Paramecium* – Killer strain
    (c) Nucleus
    (d) Liver fluke

Answers to Multiple-Choice Questions

1. (d)	2. (a)	3. (a)	4. (d)	5. (b)	6. (a)	7. (a)	8. (c)
9. (b)	10. (d)	11. (d)	12. (b)	13. (c)	14. (b)	15. (c)	16. (b)

Fill in the Blanks

1. Cytoplasmic inheritance is carried out by________________________________.
2. Generally, cytoplasmic inheritance is ________________________________.
3. Cytoplasmic inheritance never manifests ______________________________ ratio.
4. The genes that control maternal inheritance are located in __________ and ______________.
5. The phenotypes of reciprocal crosses are identical in __________________ inheritance.
6. Cytoplasmic inheritance was discovered by________________________________.
7. Correns (1909) gave the first conclusive example of cytoplasmic inheritance in the ______________ Colour of *Mirabilis jalapa*.
8. The sum total of plasmagenes is called____________________________________.
9. In yeast, extranuclear segregation and recombination are achieved during _______________ in diploid cytohets.
10. The cytoplasmic DNA of *Chlamydomonas* is inherited _______________.
11. Cytoplasmic inheritance is distinct from Mendelian inheritance, as shown by _______________.
12. In reciprocal crosses, poky character in *Neurospora* shows _______________ inheritance.
13. _______________ and _______________ contain the genetic determinants of maternal inheritance.

Answers to Fill in the Blanks

1. Plasmagenes	2. Uniparental	3. Mendelian
4. Mitochondria, Chloroplast	5. Cytoplasmic	6. Correns
7. Variegated leaf	8. Plasmon	9. Bud formation
10. Uniparentally	11. Reciprocal cross	12. Maternal
13. Mitochondria, chloroplast		

True or False

1. Male sterility in plants is an example of cytoplasmic inheritance.
2. Factors responsible for cytoplasmic male sterility are mainly located in mitochondrial DNA.
3. Maternal inheritance is one of the criterions for recognising organelle-based inheritance.

4. Maternal inheritance always indicates extranuclear inheritance.
5. Petotes have massively altered mitochondrial DNA.
6. In cytoplasmic petites, the mitochondrial electron transport chain is defective.
7. Cytoplasmic DNA is more complex than mitochondrial DNA.
8. The phenomenon of extranuclear inheritance does not operate in all eukaryotic organisms.
9. In cytoplasmic inheritance, male and female parents play an equal role in determining the phenotype of the progeny.
10. Pigmentation in flour moth exhibits maternal effects.
11. In *Neurospora*, mitochondria also controls inheritance patterns.
12. In extranuclear inheritance, phenotypes of reciprocal crosses are identical.
13. The regions containing cp DNA are called celluloids.
14. Egg cell does not influence the pattern of inheritance in extrachromosomal inheritance.
15. Ribulose bisphosphate carboxylase (RUBP) enzyme presents a very good example of cooperation of nuclear and chloroplast genes.

Answers to True or False

1. True	2. True	3. True	4. False	5. True	6. True	7. True	8. False
9. False	10. True	11. False	12. False	13. False	14. False	15. True	

Give Reasons

1. Cytoplasmic inheritance is of much importance.
   – Because it prevents the loss of organelles and enhances the chances of cytoplasmic mutation under favourable conditions.
2. When normal green male maize is crossed with white female, the progeny is white.
   – Because inheritance of plastid takes place through the female parent.

REPLICATION OF DNA

Multiple-Choice Questions

1. DNA can produce an exact copy of itself which is known as:
(a) Replication (b) Transcription (c) Transduction (d) Transformation
2. Which one of the following is not applicable to DNA replication?
(a) Initiation (b) Replication (c) Termination (d) Translocation
3. DNA replication takes place in:
(a) G_1 phase (b) S phase (c) G_2 phase (d) G_0 state
4. In *E. coli* replication of DNA is:
(a) Unidirectional (b) Bidirectional
(c) Multidirectional (d) May be bidirectional or multidirectional depending upon the physiological condition of the cell
5. *DNA polymerase* was discovered by:
(a) Nirenberg (b) Kornberg (c) Crick (d) Okazaki
6. Type I *topoisomerases*:
(a) Can cut both strands of DNA (b) Require ATP to work
(c) Do not require ATP to work (d) Can open the double helix at the replication fork
7. Consider the following statements:
(a) Before replication can occur, the part of DNA double helix must be unwound
(b) The two strands must be separated
(c) The separated strands must be kept apart to expose bases
(d) An RNA primer is required for the synthesis of DNA
The correct statements are:
(a) All (b) A and D (c) A, B and C (d) A, B and D
8. On the leading strand, *DNA polymerase* is able to synthesise DNA using:
(a) 3' OH group donated by a DNA strand (b) 3' OH group donated by a RNA primer
(c) 5' OH group donated by a DNA strand (d) 5' OH group donated by a RNA primer
9. The replication fork is formed by the action of enzyme:
(a) *DNA polymerase* (b) *DNA ligase* (c) *Helicase* (d) *Topoisomerse*
10. Which one of the following *DNA polymerase* helps in the replication of mitochondrial DNA?
(a) *DNA polymerase* (α) *(b) DNA polymerase* (β)
(c) DNA polymerase (γ) (d) Both (a) and (b)
11. Which one of the following statements is correct about *helicase*?
(a) It separates DNA strands with cut. (b) It separates strands of DNA without cut.

(c) It separates strands of DNA without a cut and with using energy from hydrolysis of ATP.
(d) It separates strands of DNA without a cut but using energy from hydrolysis of ATP.

12. Which one of the following is applicable to *DNA gyrase*?:
(a) Addition of nucleotides (b) Supercoiling of the DNA helix
(c) Opening of the DNA helix (d) Joining of Okazaki fragments

13. The enzyme responsible for converting linear DNA strand into a circular form in ϕX174:
(a) *DNA gyrase* (b) *DNA ligase* (c) *DNA polymerase* (d) *Helicase*

14. Which one of the following is incorrect about *DNA polymerase II*?
(a) It is a very fast enzyme. (b) It has 3′ ⟶ 5′ *exonuclease* activity.
(c) It lacks 5′ ⟶ 3′ *exonuclease* activity. (d) It was first of all isolated from *E. coli.*

15. Messelson and Stahl (1957) performed their experiment on:
(a) *Drosophila bipectinata* (b) *Neurospora*
(c) *Vivcia fava* (d) *E. coli*

16. *Topoisomerases* are involved in:
(a) Replication (b) Recombination (c) Transcription (d) All

17. Consider the following statements:
(A) In eukaryotes, DNA replication is initiated at multiple points
(B) In bacteria DNA replication has a single origin
(C) In bacteria DNA synthesis results in hemimethylated sequences
(D) In chloroplast and mitochondria, DNA replication is independent of the cell cycle
The correct statements are:
(a) A, B and C (b) B and C (c) A and D (d) All

18. The rate of the eukaryotic replication fork movement is about _______ nucleotides per second:
(a) 5 (b) 10 (c) 50 (c) 100

19. The mode of DNA replication is:
(a) Conservative (b) Semiconservative
(c) Dispersive (d) Generally semiconservative and in some cases conservative; all dispersive

20. DNA replication:
(a) Asymmetrical (b) Symmetrical (c) Zigzag (d) Circular

21. RNA primers are not required in the synthesis of DNA in:
(a) Adenoviruses (b) *Bacillus* phage ϕ 29 and adenoviruses
(c) φX174 (d) TMV

22. Replisomes are associated with:
(a) Synthesis of RNA (b) Synthesis of protein
(c) Replication of DNA (d) Transformation and translocation

23. *Gyrase* enzyme acts on:
(a) Linear DNA (b) Circular DNA (c) Fragmented DNA (d) All

24. Which one of the following is a member of the enzyme *reverse transcriptase*?
(a) Helicases (b) Topoisomerases (c) Telomerases (d) All

25. Human DNA contains repetitive sequence of TTAGGG which:
(a) Codes for RNA (b) Codes for protein
(c) Serves a definite function in replication (d) None.

26. Nicking and sealing of DNA strands occur in:
(a) DNA replication (b) DNA transcription
(c) Both (a) and (b) (d) DNA replication and transformation
27. *DNA ligase* uses________ as co-enzyme:
(a) NAD^+ (b) FAD (c) FMN (d) GDP
28. Rolling circle replication occurs mostly in:
(a) Oocytes of *Drosophila* (b) Oocytes of amphibians
(c) Mitochondria (d) All
29. Klenow fragment is applicable to:
(a) Transcription (b) Translation (c) Transduction (d) Replication of DNA
30. The mammalian *DNA polymerase* having *primase* activity.
(a) Alpha (b) Beta (c) Delta (d) Epsilon
31. Which one of the following is an inhibitor of eukaryotic *topoisomerases*?
(a) Ciprofloxacin (b) Puromycin (c) Streptomycin (d) Erythromycin
32. Doxorubicin inhibits:
(a) *Helicase* (b) *DNA polymerase* (c) *Topoisomerase II* (d) *Ligase*
33. Match column I with column II and select the correct answer using answer codes:

Column I (Enzyme)	Column II (Function)
(A) *Primase*	1. Chain elongation
(B) *DNA polymerase I*	2. Supercoiling
(C) *DNA polymerase*	3. Primer synthesis
(D) *Tpoisomerase II*	4. Gap filling and removal of RNA primer

Answer codes:

	A	B	C	D
(a)	3	4	2	1
(b)	3	4	1	2
(c)	4	2	3	1
(d)	2	1	4	3

34. During replication of DNA, Okazaki fragments are formed only in:
(a) Viruses (b) Bacteria (c) Eukaryotes (d) All
35. Which one of the following enzymes can change the topological form (shape) of DNA?
(a) *Helicases* (b) *DNA polymerases* (c) *Topoisomerases* (d) *Ligases*
36. *DNA polymerases* can:
(a) Start synthesis of DNA segment (b) Only elongates a primer molecule
(c) Join neighbouring Okazaki fragments (d) Open the double helix DNA at the replication fork
37. *DNA polymerases* can add nucleotides only to the ________ of a pre-existing primer:
(a) 3′ OH group (b) 5′ OH group
(c) Both 3′ OH and 5′ OH groups. (d) Either at 1′ OH group or 5′ OH group
38. Which one of the following statements is incorrect?
(a) In *E. coli*, replication forks proceed in directions opposite to the origin.
(b) In eukaryotes, for accurate replication of DNA, it is necessary that one initiation site is used more than once in each cell cycle.
(c) Helicase separates DNA strands without cut and without hydrolysis of ATP.

(d) DNA synthesis is continuous in the leading strand.

39. Which one of the following mammalian *DNA polymerases* lacks 3′ 5′ exonucleuose activity?
(a) Alpha (b) Beta (c) Gamma (d) Alpha and beta

40. Which one of the following enzymes produces nick in DNA?
(a) *RNA primase* (b) *Helicase* (c) *Topoisomerase* (d) *DNA polymerase*

41. The enzyme involved in the repair of DNA:
(a) *DNA polymerase I* (b) *DNA polymerase II*
(c) *DNA polymerase III* (d) *Topoisomerase* and *DNA polymerase III*

42. The presence of which atom is essential for the activity of *DNA polymerase I*?
(a) Zinc (b) Sodium (c) Cobalt (d) Selenium

43. Supercoiling of the DNA helix regulates:
(a) DNA replication (b) DNA transcription (c) Both (a) and (b) (d) None

44. During DNA synthesis, an RNA primer is required by *DNA polymerase* because the enzyme requires a:
(a) Free 3′ – PO_4 group (b) Free 5′ – PO_4 group
(c) Free 3′ – OH group (d) Free 5′ – OH group

45. The DNA strand in which synthesis is continuous is known as:
(a) Primer strand (b) Master strand (c) Leading strand (d) Lagging strand

46. In DNA, chain elongation occurs in:
(a) 5′ ⟶ 3′direction (b) 3′⟶5′ direction
(c) Minor grooves (d) Major groves

47. Okazaki fragments are built upon:
(a) Leading strand (b) Lagging strand (c) RNA primer (d) Both (a) and (b)

48. Discontinuous DNA replication was discovered by:
(a) Kornberg (b) Ochoa (c) Okazaki (d) Cairns

49. *Topoisomerases*:
(a) Produce break in the DNA strands (b) Reconnect the strands
(c) Prevent tangling and knotting during DNA replication (d) All

50. The enzyme capable of synthesising DNA in 3′⟶5′ direction is found in:
(a) *E. coli* (b) SV_{40} (c) *Neurospora* (d) None

51. The formation of Okazaki fragments suggests that replication of DNA is:
(a) Discontinuous (b) Continuous (c) Both (a) and (b) (d) Semiconservative

52. Which one of the following is known as nicking-closing enzyme?
(a) *DNA polymerase* (b) *DNAase* (c) *Topoisomerase* (d) *RNA primase*

53. The most active *DNA polymerase* is:
(a) *DNA polymerase I* (b) *DNA polymerase II*
(c) *DNA polymerase III* (d) All are equally active

54. Error during DNA replication is corrected by:
(a) *DNA ligase* (b) *DNA polymerase* (c) *Topoisomerase* (d) *DNA primase*

55. Prokaryotic and eukaryotic replication is similar:
(a) Having the same number and types of *DNA polymerases* (b) The ability to form replicating fork
(c) The rate of DNA synthesis (d) Having the same number of replicating origin

56. Match column I with column II and select the correct answer using answer codes:

Column I	Column II
(A) *DNA polymerase I*	1. Okazaki fragments
(B) DNA-binding protein	2. Major enzyme responsible for DNA replication
(C) *DNA polymerase*	3. Stabilises single-stranded DNA
(D) *DNA ligase*	4. Fills in gap between DNA fragments

Answer codes:

	A	B	C	D
(a)	4	3	2	1
(b)	3	2	4	1
(c)	4	3	1	2
(d)	2	4	3	1

57. Which one of the following *polymerases* does not show *exonuclease* $5' \longrightarrow 3'$ activity?
(a) *DNA polymerase I* (b) *DNA polymerase II*
(c) *DNA polymerase III* (d) All

58. In the Messelson and Stahl experiment, which model was eliminated by the analysis of DNA isolated from bacteria after one replication shifting from N^{15} to N^{14} medium?
(a) Dispersive (b) Conservative (c) Semiconservative (d) Both (a) and (b)

59. *DNA polymerases* are required for:
(a) DNA replication (b) DNA repair (c) Proofreading (d) All

60. Taylor et al. (1957) performed their classic experiment to show semiconservative type of replication on:
(a) *Neurospora* (b) *E. coli* (c) *Vicia fava* (d) None

61. While copying a template, the incorrect nucleotides are added at the rate of one every ____ nucleotides by *DNA polymerase*:
(a) 10^5 (d) 10^{10} (c) 10^{15} (d) 10^{50}

62. In conservative type of DNA replication, the progeny DNA consists of:
(a) All with old strands (b) All with new strands
(c) 50 per cent with parental strands and 50 per cent with new strands (d) All with patches of old and new strands

63. Okazaki fragments are precursor only to:
(a) 50 per cent of the daughter chains which grow in the $5' \longrightarrow 3'$ and $3' \longrightarrow 5'$ directions
(b) 25 per cent of the daughter chains that grow in the $3' \longrightarrow 5'$ direction
(c) 50 per cent of the daughter chains
(d) None

64. In which one of the following viruses is an RNA primer needed to convert a circular single-stranded DNA into its double helical replicating form?
(a) G_4 (b) ϕX174 (c) M13 (d) All

65. In which one of the following viruses does DNA not replicate using bidirectional replicating forks:
(a) Adenoviruses (b) T_7 virus (c) ϕ 29 virus (d) All

66. Which one of the following is a strong inhibitor of *DNA polymerase* α?
(a) Ciprofloxacin (b) Doxorubicin (c) Amphidicolin (d) Kirromycin

67. In which type of replication do the parent DNA molecules remain unchanged?
(a) Conservative (b) Semiconservative (c) Dispersive (d) None

68. In eukaryotes, DNA replication starts from:

(a) Upper end (b) Mid-point (c) A specific point (d) Many sites

69. Which one of the following is not needed in the chain formation of DNA during the course of its replication?
(a) *DNA polymerase III* (b) *Topoisomerase*
(c) ATP (d) Mg^{+++}

70. Match column I with column II and select the correct answer using answer codes:

Column I		Column II
(A) Cairns	1.	Small amount of extra DNA synthesis
(B) Primosome	2.	Can remove thymine dimmers produced due to ultra violet irradiation
(C) Pachytene	3.	Bacterial DNA replication
(D) *DNA polymerase I*	4.	Complex of proteins involved in priming reaction.

Answer codes:

	A	B	C	D
(a)	4	1	2	3
(b)	3	4	1	2
(c)	4	3	1	2
(d)	2	1	3	4

71. Which one of the following shows *helicase* activity?
(a) T antigen (b) Zidovudine (c) A antigen (d) Amantadine

72. Okazaki used tritiated thymidine in DNA of *E. coli* infected with____________, for the demonstration of replication in short fragments:
(a) T_2 virus (b) T_4 virus (c) T_7 virus (d) SV_{40}

73. Proofreading refers to:
(a) Removal of excess of amino acid in a newly synthesised polypeptide (b) Removal of errors during DNA replication
(c) Reverse transcription (d) All

74. The eukaryotic *DNA polymerase* that can initiate DNA replication in the absence of RNA primer:
(a) α (b) β (c) δ (d) None

75. Evidence of bidirectional DNA replication has been obtained from:
(a) Genetic studies (b) Autoradiography study
(c) Electron microscopy study (d) All

76. Bidirectional replication is not found in:
(a) Mitochondrial DNA (b) Yeast
(c) *Drosophila* (d) Mammals

77. Minimum number of replicons is found in:
(a) *Drosophila* (b) *E. coli* (c) Yeast (d) *Vicia fava*

78. The eukaryotic *DNA polymerase* which is tightly bound with *primase*:
(a) *DNA polymerase* (α) (b) *DNA polymerase* (β)
(c) *DNA polymerase* (γ) (d) *DNA polymerase* (δ)

79. Which one of the following is not associated with DNA replication?
(a) Messelson and Stahl's experiment (b) Cairn's autoradiography experiment
(c) Taylor and co-worker's autoradiography experiment (d) Lederberg's replica plating technique

80. Termination of DNA replication requires:
(a) *Tus* protein (TBP) (b) 'n' protein
(c) proteini (d) All

81. Repair synthesis of DNA during pachytene is carried out by:
(a) *DNA polymeras* α (b) *DNA polymerase* β
(c) *DNA polymerase* δ (d) *DNA polymerase* γ

82. The only *polymerase* whose synthesis increases during the S phase is:
(a) *DNA polymeras* α (b) *DNA polymerase* β
(c) *DNA polymerase* δ (d) *DNA polymerase* γ

83. Which one of the following is known as nicking and closing enzyme?
(a) *Helicases* (b) *Topoisomerases* (c) *DNA polymerases* (d) *DNA ligases*

84. All DNA viruses replicate in the nucleus except:
(a) Herpes viruses (b) Pox viruses (c) Adenoviruses (d) Parvoviruses

85. In Messelson and Stahl's experiment, N^{15} labelled *E. coli* were transferred to N^{14} medium for conservative mode of replication. What proportion of N^{15}, N^{14}–N^{15} and N^{14} are expected after the second generation?
(a) ½, 0, ½ (b) ½ , 0, ¼ (c) ¼ , 0, ½ (d) ¼ , 0, ¾

86. In the above experiment, for semiconservative type of replication, what proportion of N^{15}, N^{14}–N^{15} and N^{14} are expected after the fourth generation?
(a) 0, ½ , ½ (b) 0, 1/8, 7/8 (c) ¼ , 0, ¾ (d) 0, 1, 0

87. A heavy DNA (N^{15}) was transferred to N^{14} medium and allowed to replicate two times. What composition is expected in populations?
(a) All N^{15} (b) All N^{14}
(c) 50 per cent N^{14} and 50 per cent $N^{14} + N^{15}$ (d) 75 per cent N^{14} and 25 per cent $N^{14} + N^{15}$

88. In the above experiment, if the DNA is subjected to centrifugation, it would show:
(a) One band (b) Two bands (c) Three bands (d) No band

89. The DNA of an organism labelled with N^{15} was transferred to a medium containing N^{14} After one replication its 50 per cent DNA were N^{15} and 50 per cent were N^{14}. It shows:
(a) Semiconservative replication (b) Conservative replication
(c) Dispersive replication (d) May be semiconservative or conservative

90. In the above experiment, if replication of DNA is dispersive, on centrifugation it would show:
(a) One band either near to N^{14} or N^{15} (b) One band, intermediate between N^{14} and N^{15}
(c) One band close to N^{14} (d) One band close to N^{15}

91. Which one of the following is an incorrect match?

	Protein	Function
(a)	*Ligase*	Ligation
(b)	*Topoisomerase I*	Supercoiling
(c)	*DNA polymerase I*	Gap filling
(d)	DNA	Origin of replication fork

92. Telomerase synthesises only__________rich strand of telomeres:
(a) A (b) T (c) G (d) C

93. What is common between *DNA polymerase I, II* and *III?*
(a) 3′⟶5′ *exonuclease* activity (b) 5′⟶3′ *exonuclease* activity
(c) Number of polypeptide sub-units (d) Molecular weight

94. Okazaki fragments are responsible for the elongation of:
    (a) The leading strand away from the replication fork
    (b) The lagging strand towards the replication fork
    (c) The leading strand towards the replication fork
    (d) The lagging strand away from the replication fork
95. Match column I with column II and select the correct answer using answer codes

Column I	Column II
(A) *Helicase*	1. Stabilises single DNA strand
(B) SSB protein	2. Synthesises RNA primer
(C) *Primase*	3. Begins unwinding of DNA helix
(D) *DNA gyrase*	4. Assists unwinding

Answer codes:

	A	B	C	D
(a)	4	1	3	2
(b)	3	1	2	4
(c)	3	4	1	2
(d)	3	1	4	2

96. One of the major differences between DNA replication of eukaryote and prokaryote is the:
    (a) Different *DNA polymerases* (b) Difference in opening and closing of the helix
    (c) Multiple origins in eukaryotes (d) Lack of nucleus in prokaryotes
97. Which one of the following is not applicable to DNA replication?
    (a) Okazaki fragments (b) *DNAase*
    (c) *DNA polymerase* (d) *DNA ligase*
98. Which one of the following is incorrect about T_4 *DNA polymerase*?
    (a) The T_4 *DNA polymerase* functions as $5' \longrightarrow 3'$ *DNA polymerase*
    (b) It has $3' \longrightarrow 5'$ *exonuclease* activity
    (c) It lacks $5' \longrightarrow 3'$ *exonuclease* activity
    (d) None
99. Which one of the following is incorrect about *DNA polymerase I*?
    (a) $5' \longrightarrow 3'$ elongation *polymerase* activity (b) $3' \longrightarrow 5'$ *exonuclease* proofreading
    (c) $5' \longrightarrow 3'$ *exonuclease* repair activity (d) None
100. Synthesis of DNA involves ______steps:
    (a) Two (b) Three (c) Four (d) More than four steps
101. Which one of the following is applicable to replication of mtDNA?
    (a) Lack of proofreading (b) Rapid rate
    (c) Asynchronic (d) All
102. *RNA polymerase*:
    (a) Does not require RNA primer (b) Lacks proof reading capacity
    (c) Both (a) and (b) (d) None

Answers to Multiple-Choice Questions

1. (a)	2. (d)	3. (b)	4. (b)	5. (b)	6. (c)	7. (a)	8. (b)
9. (c)	10. (c)	11. (d)	12. (c)	13. (b)	14. (a)	15. (d)	16. (d)

17. (d)	18. (c)	19. (b)	20. (a)	21. (b)	22. (c)	23. (b)	24. (c)
25. (c)	26. (c)	27. (a)	28. (b)	29. (d)	30. (a)	31. (a)	32. (c)
33. (b)	34. (d)	35. (c)	36. (b)	37. (a)	38. (c)	39. (d)	40. (c)
41. (c)	42. (a)	43. (c)	44. (c)	45. (c)	46. (a)	47. (b)	48. (c)
49. (d)	50. (d)	51. (a)	52. (c)	53. (c)	54. (b)	55. (b)	56. (a)
57. (b)	58. (d)	59. (d)	60. (c)	61. (a)	62 (c)	63. (a)	64 (d)
65. (d)	66. (c)	67. (a)	68. (d)	69. (b)	70. (b)	71. (a)	72. (a)
73. (b)	74. (d)	75. (d)	76. (a)	77. (b)	78. (a)	79. (d)	80. (a)
81. (b)	82. (a)	83. (b)	84. (b)	85. (d)	86. (b)	87. (c)	88. (b)
89. (b)	90. (a)	91. (b)	92. (c)	93. (a)	94. (d)	95. (b)	96. (c)
97. (b)	98. (d)	99. (d)	100. (b)	101. (d)	102. (c)		

Fill in the Blanks

1. The process of DNA duplication is generally called_______________.
2. DNA synthesis occurs in _______________ direction.
3. _______________ is essential for initiation of synthesis of DNA strand.
4. The strand in which DNA synthesis is continuous is called_______________ strand.
5. The mode of replication in DNA is _______________ type.
6. Generally in human cells there are _______________ strands of DNA called chromosomes.
7. Replication fork may be _______________ or _______________.
8. RNA primers initiate the _______________ end of both leading strand and Okazaki fragments.
9. Short pieces of replicating DNA are called _______________.
10. Newly formed DNA fragments are joined together by the enzyme *DNA ligase* by the formation of a phosphodiester bond between _______________ end of one nucleotide chain and_______________ hydroxyl group of the other DNA fragments.
11. During DNA replication, hydrogen bonds between base pairs are disrupted by the enzyme _______________.
12. In humans, DNA is copied at about _______________ base pairs per second.
13. In yeast, multiple origins of replication are called _______________.
14. Eukaryotes require a special enzyme called _______________ to replicate their ends.
15. Proliferating cell nuclear antigen is a _______________ factor.
16. The process of _______________ replication is a good model for other DNA replication.
17. *Gyrase* acts on _______________ DNA.
18. During the replication of DNA, the wrong base is removed by _______________ activity of *DNA polymerase*.
19. The *DNA polymerase* can only work on a template that runs in the _______________ direction.
20. Retroviruses encode an unusual *DNA polymerase* by the enzyme called _______________.
21. The RNA primer provides the free _______________ needed by the *DNA polymerase III*.
22. During DNA replication, relief of supercoil needs _______________.

23. *Telomerase* was discovered by _______________.
24. In bacteria, *DNA polymerase* _______________ is the main replicating enzyme.
25. The RNA primer is synthesised by the enzyme _______________.
26. *Telomerase* recognises _______________ end of the telomere.
27. *Telomerase* synthesises _______________ repeats on the telomere sequence.
28. Polymerisation of new strands of DNA occurs in _______________ direction, meaning that template is read in the _______________ direction.
29. *DNA polymerase I* is known as _______________.
30. DNA synthesis is towards replication fork in _______________ strand and away from the fork in – strand.
31. In certain small viruses, _______________ are needed to convert the circular single-stranded DNA into their double helical replicating forms.
32. Cells contain a specific class of proteins called _______________ that help in the separation of DNA strands.
33. In newly synthesised DNA strands, RNA primers are removed by the enzyme _______________.
34. Eukaryotic *DNA polymera* β is involved in the _______________ of DNA.
35. Discontinuous DNA synthesis was discovered by _______________.
36. The replication of DNA in which half molecules are conserved is known as _______________.
37. _______________ is essential for initiation of DNA chain.
38. _______________ is the DNA replication from single origin.
39. A _______________ is formed by the unwinding of DNA and synthesis of new strands.
40. D-loop replication is found in _______________ and _______________.
41. The _______________ activity of *DNA polymerase* makes it suitable for proof reading.
42. *DNA polymerase II* lacks _______________ activity.
43. At the time of DNA replication, the separated strands are stabilised by _______________.
44. Okazaki fragments are joined by _______________.
45. *DNA polymerases* are essential for DNA replication and _______________ of DNA.
46. Okazaki fragments start with _______________.
47. Okazaki fragments are short pieces of _______________.
48. Eukaryotic DNA replication can be best studied by the technique of DNA fibre _______________.
49. _______________ is an analogue which can be incorporated in the chromosome in place thymidine.
50. Mammalian *topoisomerases* are inhibited by _______________ and _______________.
51. *DNA ligase* is able to join the _______________ end of one DNA strand to the _______________ end of another DNA strand.
52. The open area of a chromosome between replication forks is called _______________.
52. The amount of DNA replicated from a single strand is called _______________.
54. RU7 stops the replication of bacterial DNA by disrupting _______________.
55. Klenow fragment is formed when *DNA polymerase* from *E. coli* is enzymatically cleaved by the _______________.
56. Klenow fragment was first reported by _______________.

57. The mode of DNA replication in which multiple copies of circular DNA molecules are synthesised is known as _______________ replication.
58. An intermediate theta structure is formed during the course of replication of a _______________ DNA molecule.
59. *Topoisomerases* were discovered by _______________.
60. Linking number of DNA is changed by the enzyme _______________.
61. In bacteria, binding of *primase* with *DNA helicase* results in the formation of a complex called the _______________.
62. Each Okazaki fragment is initiated near the _______________ at the RNA primer.
63. All RNA viruses replicate in cytoplasm of the host cell except _______________ and _______________.
64. The DNA strand which is continuously replicated having one primer and is synthesised in 5’⟶3’; direction is known as _______________.
65. Genetically engineered adenovirus called ONYX-015, can replicate only in human cells lacking _______________.
66. *DNA polymerase I* contains _______________ active sites.
67. Multiple DNA replication forks on a strand result in the formation of _______________.
68. The strand of DNA in which synthesis occurs discontinuously having many RNA primers is called _______________.

Answers to Fill in the Blanks

1. Replication	2. 5′⟶3′	3. RNA primer
4. Leading	5. Semiconservative	6. 46
7. Unidirectional, bidirectional	8. 5′	9. Okazaki fragments
10. 3’, 5’	11. *Helicase*	12. 50
13. Autonomous Replication Sequence (ARS)		14. *Telomerase*
15. Replication	16. T-DNA	17. Circular
18. 3′⟶5′, *exonuclease*	19. 3′⟶5′	20. *Reverse transcriptase*
21. 46	22. *Topoisomerase*	23. Shippen (1990)
24. 3′– OH	25. *RNA primase*	26. 3′
27. TTAGGG	28. 5′⟶3′, 3′⟶5′	29. Kornberg enzyme
30. Leading, lagging	31. RNA primers	32. *Helicases*
33. *DNA polymerase I*	34. Repair	35. Okazaki et al.
36. Semiconservative	37. RNA primer	38. Replicon
39. Replication fork	40. Mitochondria, chloroplast	41. 3′⟶5′ *exonuclease*
42. 3′⟶5′ *exonuclease*	43. Single-stranded binding proteins	44. *DNA ligase*
45. DNA repair	46. RNA primar	47. Replicating DNA
48. Autoradiography	49. 5-Bromodeoxyuridine	50. Etoposide, adriamycin
51. 5′⟶3′	52. Replication bubble	53. Replicon
54. *DNA polymerase III*	55. *Protease subtilisin*	
56. Klenow and Henningsen in 1970.		57. Rolling circle
58. Circular	59. J C Wang	60. *Topoisomerases*
61. Primosome	62. Replication fork	63. Orthomyxoviruses, retrovirsuses
64. Leading strand	65. p^{53}	66. Five
67. Replication bubbles	68. Lagging strand	

True or False

1. DNA replication occurs during interphase.
2. Replication of DNA is semiconservative.
3. On the lagging strand synthesis of DNA occurs in 3′⟶5′ direction.
4. *Topoisomerase I* can cut both strands of DNA.
5. Replication of DNA takes place simultaneously on both strands.
6. The nicking of DNA requires energy while resealing does not require energy.
7. The discontinuous DNA replication produces replication forks.
8. The T antigen has *helicase* activity.
9. The growth in lagging strand of DNA is fast.
10. *DNA polymerases* can synthesise DNA even in the absence of RNA primer.
11. DNA replication is bidirectional in bacteria, yeast and mitochondria.
12. *DNA polymerase II* is a very fast enzyme.
13. Enzyme *primase* is required for base pairing.
14. Inhibitors of eukaryotic *topoisomerases* are used as anti-tumour agents.
15. In eggs of *Drosophila*, DNA replication is completed in about 3 minutes.
16. In ϕX174 virus, DNA replication is bidirectional.
17. *Helicase* removes topological links between DNA strands.
18. *DNA polymerase III* forms branching DNA.
19. DNA replication occurs at a specific step in the cell cycle.
20. The replication fork moves in one direction.
21. *DNA polymerase* possesses the ability to form the final product.
22. *DNA polymerase* is a minor enzyme involved in DNA repair.
23. *Helicase* breaks hydrogen bonds between the two DNA strands.
24. *Primase* is activated by *DNA helicase*.
25. Error correction is a property of all *DNA polymerases*.
26. Presence of clamp increases the rate of DNA synthesis.
27. Eukaryotic DNA replication is slow in comparison to prokaryotic DNA replication.
28. In bacteriophage ϕX174, *DNA ligase* converts linear DNA into circular form.
29. In prokaryotes, Okazaki fragments are 1,000 to 2,000 nucleotides long.
30. *DNA topoisomerase* is a reversible enzyme.
31. *DNA polymerases* can add nucleotides on both 3′ and 5′ ends of growing polypeptide chain.
32. According to conservative mode of replication, each strand of DNA contains segments of old and new strands.
33. DNA replication is very fast in nerve cells.
34. Replication of DNA begins at one point and proceeds in both directions.
35. The three main steps in order of DNA synthesis are initiation, elongation and termination.

36. *DNA polymerase* fixes its mistakes by excision repair.
37. *DNA polymerase III* requires a 3′ OH end to add next nucleotide.
38. *DNA ligase* joins 3′– OH to a 5′ monophosphate group.
39. *DNA polymerase* cannot distinguish dTTP from dUTP.
40. *DNA polymerase* alpha is faster than bacterial *DNA polymerases.*
41. *Telomerase* is a ribonucleoprotein.
42. *Helicases* are essential to remove topological links between the parental DNA strands.
43. *DNA polymerase* gamma is responsible for replication, recombination and repair of mitochondrial DNA.
44. *Topoisomerases* have both nuclease and *ligase* activity.
45. RU7 does not affect DNA replication in yeast.
46. Scientists have discovered a small molecule that stops the sliding clamp in its track.
47. *Topoisomerases* are required both for DNA replication and transcription.
48. *Topoisomerase II* does not require energy for its action.
49. The opening of DNA produces two replication forks.
50. *DNA polymerase epilson* binds to the 3′⟶5′ strand in order to create discontinuous segments starting from different primers.

Answers to True or False

1. True	2. True	3. False	4. False	5. True	6. True	7. True	8. True
9. False	10. False	11. False	12. False	13. False	14. True	15. True	16. True
17. False	18. False	19. True	20. True	21. False	22. True	23. True	24. True
25. False	26. True	27. True	28. True	29. False	30. True	31. False	32. False
33. False	34. True	35. True	36. True	37. True	38. True	39. True	40. False
41. True	42. False	43. True	44. True	45. True	46. True	47. True	48. False
49. True	50. True						

Give Reasons

1. Before a cell divides into two daughter cells, the cell must make a copy of cellular DNA.
   – Because after division, each cell must contain a complete set of genetic material.
2. DNA replication is semiconservative.
   – Because each daughter DNA has one old and one new strand.
3. The process of replication of each DNA strand is considerably different.
   – Because the strands of DNA have opposite orientation.
4. The origin of DNA replication tends to be A–T rich.
   – Because A–T base pairs are of two hydrogen bonds, while in G–C pairs there are three hydrogen bonds. Therefore, in comparison to G–C-rich strands, it is easier to separate A–T-rich strands.
5. DNA synthesis is continuous in one strand and discontinuous in the other strand.

- Because *DNA polymerase* is capable of adding new nucleotides only on the 3'end of a DNA strand and both strands of DNA are antiparallel to each other.

6. DNA replication is said to be bidirectional.
   - Because the unzipping takes place in both directions from the origin of replication, creating a replication bubble.
7. DNA of nerve and muscle cells does not replicate.
   - Because these cells do not divide.
8. In *E. coli*, DNA synthesis results in hemi-methylated sequences.
   - Because *E. coli* methylates GATC DNA sequences.
9. *Primase* is of much significance in DNA replication.
   - Because there is no known *DNA polymerase*, which can initiate the synthesis of a DNA strand without an initiator RNA primer, synthesised by the enzyme *primase*.
10. *DNA polymerase* cannot determine sequence specificity.
    - Because *DNA polymerase* has the ability to recognise only the sugar and phosphate part of nucleotides.
11. The *DNA polymerase* can only work on a template that runs in the $3' \longrightarrow 5'$ direction.
    - Because the *DNA polymerase* requires the OH on the 3′ end as an active site. It uses the OH on the 3′ end of a nucleotide to attach phosphate from the 5′ end of the next nucleotide.
12. *DNA polymerase* is considered to be holoenzyme.
    - Because it requires Mg^{++} as co-factor.
12. Replication of eukaryotic chromosomes requires soluble cytoplasmic factors.
    - Because these cytoplasmic factors dissociate histones and other proteins for initiation of replication.
14. If a nucleoside triphosphate is present in the reaction of synthesis of DNA which lacks OH group on the 3′ carbon of the sugar, it becomes added to the growing DNA chain.
    - Because it has the ability to form a phosphodiester bond with the nucleotide at the end of the chain.
15. In the above question, if the nucleoside triphosphate becomes incorporated into DNA, the biosynthesis of DNA is stopped.
    - Because when the nucleoside triphosphate becomes incorporated into DNA, no 3′OH group is available for the formation of a phosphodiester bond with the next nucleotide. Thus, no synthesis of DNA takes place, as no further no chain growth occurs.
16. Unwinding of DNA helix must be rapid at the replication point.
    - Because unwinding of DNA helix creates a torque in the DNA.
17. *Telomerase* is an unusual enzyme.
    - Because it has its own RNA template as a part of its structure.
18. Nicks occur in developing molecules of DNA.
    - Due to removal of RNA primer and synthesis occurs discontinuously in the lagging strand.
19. DNA synthesis requires RNA primer.
    - Because the *DNA polymerases*, which catalyse DNA synthesis, are unable to initiate DNA synthesis and they require 3′ end to initiate reaction.
20. When a partially denatured molecule of DNA having unwind section rich in A–T base pairs is treated with formaldehyde, the single-stranded region becomes stabilised.
    - Because formaldehyde combines with the free amino groups of adenine and prevents the formation of A–T base pairs.

21. A series of proteins are required for unwinding and separation of double-stranded DNA molecule during the course of DNA replication.
    – Because DNA must be single stranded before the process of replication proceeds.
22. Unwinding and separation of DNA strands are essential for DNA replication.
    – Because DNA must be single stranded, before DNA replication can proceed.
23. *DNA ligase* joins DNA fragments.
    – Because it catalyses the formation of a phosphodiester bond between the 3′ end of one nucleotide chain with the 5′ OH group of the newly made Okazaki fragments.

Questions based on Diagram

1. The labelled part P is known as:
   (a) Okazaki fragment
   (b) RNA primer
   (c) *DNA polymerase*
   (d) Point of joining

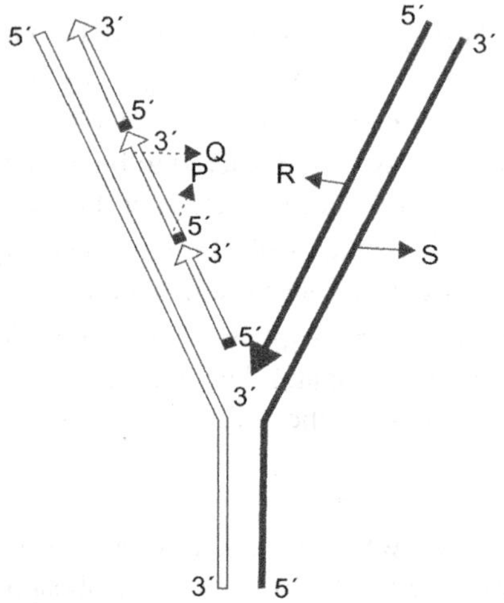

2. In the above diagram, Okazaki fragment is represented by the letter:
   (a) P (b) Q (c) R (d) S
3. The polarity of two strands of DNA (I and II) showing replicating forks is:
   (a) 5′ 3′ and 5′ 3′
   (b) 3′ 5′ and 3′ 5′
   (c) 5′ 3′ and 3′ 5′
   (d) 3’ 5’ and 5’ 3’

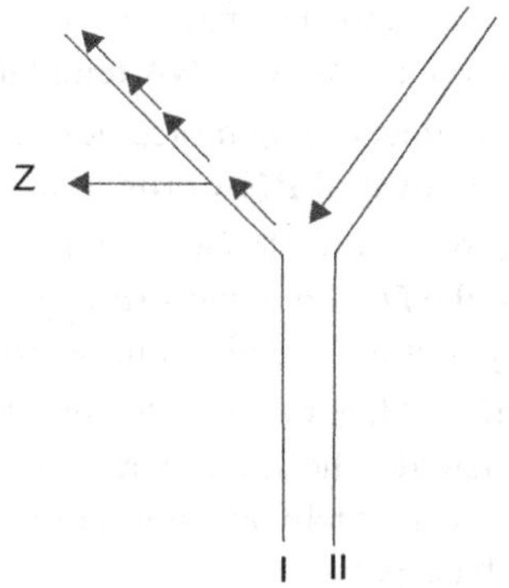

4. In the above diagram, the labelled part Z is known as:
   (a) Lagging strand (b) Leading strand (c) Master strand (d) DNA primer
5. In the diagram showing DNA replication below, the correct mode of replication is:
   (a) I (b) II (c) III (d) IV

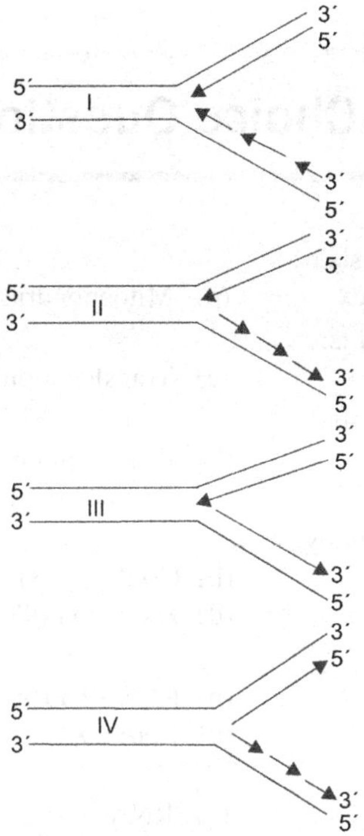

6. The diagram below shows DNA replication using root tip cells of *Vicia fava* with ^{3}H thymidine. Scientists associated with this experiment are:
   (a) Messelson and Stahl
   (b) Cairns
   (c) Taylor, Woods and Hughes
   (d) Nirenberg and Lederberg

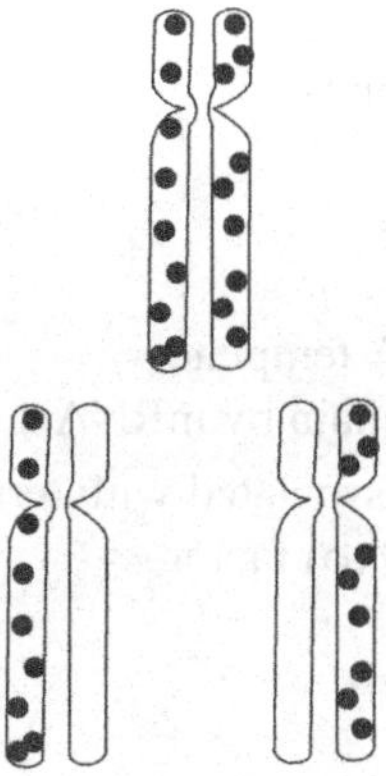

Answers to Questions based on Diagrams

1. (b) 2. (b) 3. (c) 4. (a) 5. (b) 6. (c)

PROTEIN SYNTHESIS

Multiple-Choice Questions

1. The organelle involved in protein synthesis is:
   (a) Ribosome (b) Golgi complex (c) Mitochondria (d) Lysosome
2. The process of protein synthesis in a cell is:
   (a) Transcription (b) Translation (c) Translocation (d) Transduction
3. In molecular biology, 'central dogma' is the:
   (a) Transformation and transduction (b) Transcription and translocation
   (c) Transcription and translation (d) Translocation and transformation
4. The concept of 'central dogma' was given by:
   (a) Watson and Crick (1953) (b) Crick (1958)
   (c) Temin and Baltimore (1970) (d) Khorana (1970)
5. The central dogma of modern biology is:
   (a) DNA → Protein → RNA (b) RNA → Protein → DNA
   (c) DNA → RNA → Protein (d) RNA → DNA → Protein
6. Translation is applicable to synthesis of:
   (a) Protein (b) DNA (c) RNA (d) None
7. Temin and Baltimore are associated with:
   (a) Transcription (b) Translation
   (c) Reverse transcription (d) Biogenesis of ribosomes
8. Enzyme involved in transcription is:
   (a) *DNA polymerase* (b) *RNA polymerase*
   (c) *Aminoacyl-tRNA synthetase* (d) *Peptidyl transferase*
9. Teminism is the:
   (a) Formation of DNA by RNA template (b) Formation of RNA by DNA template
   (c) Formation of polypeptide chain by mRNA (d) Formation of polyribosome
10. Which one of the following is associated with termination of mRNA from DNA template?
    (a) Sigma (σ) factor (b) Rho (ρ) factor (c) Releasing factor (d) *ATPase* factor
11. Transcription does not involve:
    (a) Initiation (b) Elongation
    (c) Termination (d) Activation of amino acids
12. For the first time, transcription was observed in 1960 in:
    (a) T_7 virus (b) SP_3 virus (c) T_7 and SP_3 virus (d) SP_3 and SV_{40} virus
13. Reverse transcription occurs in:
    (a) HIV (b) Higher eukaryotes (c) Retrotransposons (d) All

14. Central dogma is not applicable to:
    (a) Some plants (b) Animal viruses (c) Prokaryotes (d) Retroviruses
15. *RNA polymerase* does not work in the absence of:
    (a) *Guanyl transferase* (b) Catabolite Activator Protein
    (c) Cyclic AMP and Catabolite Activator Protein (d) ATP and *Guanyl transferase*
16. Consider the following statements:
    (A) Many proteins undergo post-translation modification
    (B) The events following biosynthesis of proteins include post-translation modification and protein folding
    (C) Post-translation modification of protein include the formation of disulphide bridge
    (D) Termination of the polypeptide chain occurs, when the P site of the ribosome faces a stop codon
    The correct statements are:
    (a) A and B (b) A, B and C (c) B, C and D (d) All
17. The Kozak consensus sequence occurs in:
    (a) Eukaryotic mRNA (b) Prokaryotic mRNA
    (c) Viral mRNA (d) Mitochondrial mRNA
18. The Kozak consensus has the consensus:
    (a) (gcc) gccRUGGA (b) (gcc) gccRccAUGG
    (c) (gcc) gccRGUGA (d) (gcc) gccRUACA
19. After protein synthesis, carbohydrates or lipids are added to the peptide, which is activated by:
    (a) Acetylation (b) Methylation (c) Phosphoylation (d) All
20. The Kozak consensus sequence plays a major role in the:
    (a) Initiation of transcription process (b) Initiation of translation process
    (c) Termination of polypeptide chain (d) Modification of polypeptide chain
21. Which one of the following statements is incorrect?
    (a) Protein synthesis is an energy consuming process.
    (b) The folding of newly synthesised peptide starts even before it is ready.
    (c) Translation does not work according to an assembly line principle like transcription.
    (d) The synthesis of polypeptide chain occurs catalytically.
22. Lmx_{16} is an example of gene with:
    (a) A weak Kozak consensus sequence (b) A strong Kozak consensus sequence
    (c) No Kozak consensus sequence (d) No role in protein synthesis
23. After synthesis, mRNA migrates from nucleus to the cytoplasm. During this step mRNA goes through different types of maturation including the one called:
    (a) Splitting (b) Splicing (c) Amplification (d) Esterification
24. Which one of the following is a product of transcription?
    (a) mRNA (b) rRNA (c) tRNA (d) All
25. Which one of the following antibiotics inhibits bacterial ribosomes but not eukaryotic ribosomes:
    (a) Tetracycline (b) Streptomycin (c) Chloramphenicol (d) All
26. Which one of the following blocks initiation and causes misreading of mRNA?
    (a) Aminoglycosides (b) Oxazolidinones (c) Lincosamides (d) Macrolides
27. Which one of the following prevents the formation of a peptide bond?
    (a) Macrolides (b) Tetracyclines (c) Chloramphenicol (d) Streptogramines

28. Sometimes *RNA polymerase* selects and inserts incorrect and mismatched ribonucleotides in bacteria and probably in all organisms. When this occurs:
    (a) *RNA polymerase* backs up, removes the incorrect nucleotides
    (b) *RNA polymerase* stops working
    (c) Chain elongation does not take place
    (d) *Ribonuclease* removes the incorrect ribonucleotides
29. Which one of the following is an incorrect match?
    (a) RF-3 stimulates binding of RF-1 and RF-2
    (b) RF-2 recognises the UAA and UGA terminating codons
    (c) RF-1 recognises UAA and UAG terminating codons
    (d) RF-3 recognises UAA, UGA and UAG terminating codons
30. Polyribosomes occur as:
    (a) Free (b) Cytoskeletal bound (c) Membrane bound (d) All
31. The *aminoacyle-tRNA* has ___________ binding sites:
    (a) One (b) Two (c) Three (d) Four
32. The initiation factors IF-1, IF-2 and IF-3 are permanent components of:
    (a) 30S sub-unit of ribosome (b) 50S sub-unit of ribosome
    (c) 70S ribosome (d) None
33. The covalent modification, which occurs after translation, includes:
    (a) Hydroxylation as well as reaction involving attachment of co-enzymes or prosthetic groups (b) Phosphorylation
    (c) Methylation (d) all
34. Translocation is indirectly inhibited by:
    (a) Fusidic acid (b) Tetracycline (c) Cyclohexamide (d) Streptomycin
35. Which one of the following is not a component of *RNA polymerase?*
    (a) α sub-unit (b) β sub-unit (c) ρ factor (d) σ factor
36. Which one of the following is a regulatory sub-unit of the enzyme *RNA polymerase?*
    (a) α sub-unit (b) β sub-unit (c) β' sub-unit (d) σ factor
37. The enzyme responsible for transcribing viral RNA from DNA is:
    (a) *RNA polymerase* (b) *Reverse transcriptase*
    (c) *DNA polymerase* (d) *Topoisomerase*
38. tRNA recognises the *aminoacyl synthetase* enzyme by:
    (a) AA site (b) Anticodon (c) DHU loop (d) TψC loop
39. The number of polypeptide chain in *E. coli RNA polymerase* is:
    (a) One (b) Two (c) Three (d) Five
40. Synthesis of protein does not involve:
    (a) Termination (b) Elongation (c) Transformation (d) Initiation
41. Ribosome is recognised by tRNA with the help of:
    (a) TψC loop (b) DHU loop (c) AA site (d) Anticodon
42. Match column I with column II and select the correct answer using answer codes:

Column I	Column II
(A) CAAT box	1. Initiation of translation
(B) Teminism	2. Gene regulate growth and differentiation

(C) Transcription and translation	3. *Reverse transcriptase*
(D) Kozak consensus sequence	4. Important for Transcription

Answer codes:

	A	B	C	D
(a)	4	3	2	1
(b)	4	2	3	1
(c)	3	4	2	1
(d)	2	1	4	3

43. *RNA polymerase III* catalyses synthesis of:
(a) tRNA (b) mRNA (c) 28S RNA (d) Hn mRNA

44. Synthesis of mRNA proceeds in _______________ direction:
(a) 5′⟶3′ (b) 3′⟶5′ (c) Both (a) and (b) (d) 5′⟶1′

45. RNA polymerase transcribes the DNA strand in which direction?
(a) 5′⟶3′ (b) 3′⟶5′ (c) Both (a) and (b) (d) None

46. Expression of mRNA molecule is likely to be regulated by:
(a) Small nuclear RNA (snRNA) (b) Micro RNA (miRNA)
(c) XISt-RNA (d) All

47. Match column I with column II and select the correct answer using answer codes:

Column I	Column II
(A) Sparsomycin	1. Misreading of the genetic code
(B) Tetracycline	2. Inhibits AA-tRNA binding
(C) Cyclohexamide	3. Inhibits *Peptidyl transferase*
(D) Streptomycin	4. Affects eukaryotic cells

Answer codes:

	A	B	C	D
(a)	3	2	4	1
(b)	3	2	1	4
(c)	2	4	1	3
(d)	4	3	2	1

48. Which one of the following affects 30S sub-unit of ribosome?
(a) Lincosamides (b) Linezolid
(c) Aminoglycosides tetracyclines (d) Macrolides

49. In protein synthesis, premature chain termination is caused by:
(a) Puromycin (b) Chloramphenicol (c) Cyclohexamide (d) None

50. There are no tRNA that recognise the stop codons except:
(a) tRNAs for solenocysteine (b) tRNA for pyrolysine
(c) Supressor tRNAs (d) All

51. Which one of the following statements is incorrect?
(a) EF-Tu is the most abundant protein in *E. coli* cells
(b) EF-Tu cannot bind with $tRNA^{met}$
(c) EF-Tu has a higher affinity for GDP
(d) EF-Tu is active and can bind *aminoacylated-tRNA*

52. The initiation of protein synthesis does not require:
(a) Two ribosomal sub-units – 30S and 50S (b) *Peptidyl transferase*

(c) Three initiation factors – IF-1, IF-2 and IF-3 (d) The initiator F-Met-$tRNA_f^{met}$

53. Match column I with column II and select the correct answer using answer codes:

Column I	Column II
(A) Fusidic acid	1. Blocks dissociation of GDP from EF-Tu after hydrolysis
(B) Kanamycin	2. Inhibits the activity of EF-G by ADP ribosylation
(C) Kirromycin	3. Blocks the dissociation of EF-2 during protein synthesis in Eukaryotes
(D) Diptheriatoxin	4. Causes misreading of the code by interfering with Wobble base pairing

Answer codes:

	A	B	C	D
(a)	3	2	4	1
(b)	2	4	1	3
(c)	3	4	1	2
(d)	4	3	2	1

54. The chain termination leads to:
(a) Release of polypeptide chain (b) Release of tRNA
(c) Dissociation of 70S into 30S and 50S sub-units (d) All

55. In prokaryotes, the initiation site for translation into amino acids is the:
(a) Shine–Dalgarno sequence (b) Kozak consensus sequence
(c) Internal ribosomal entry site (d) 5' end of the mRNA

56. Which one of the following initiation factors is not universally found in all bacterial species?
(a) IF-1 (b) IF-2 (c) IF-3 (d) IF-2 and IF-3

57. Patemine A inhibits ________ *in vivo:*
(a) Eukaryotic transcription (b) Prokaryotic transcription
(c) Eukaryotic translation (d) Prokaryotic translation

58. Polyribosomes were first discovered by:
(a) Jonathan Warner, Paul Knoop and Alex Rich (1963) (b) Skehel and Joklin (1969)
(c) Price and Penman (1972) (d) Rich, Thoma and Klug (1979)

59. Which one of the following ribosomal proteins plays a central role in the energy utilising steps of protein synthesis?
(a) L_5 / L_7 (b) L_8 / L_{12} (c) L_7 / L_{12} (d) L_{12} / L_{34}

60. The starting amino acid in the synthesis of all bacterial polypeptide is the:
(a) Methionine (b) N-formylated methionine
(c) Formylated Lysine (d) Lysine

61. The proteins needed for *peptidyl transferase* activity are:
(a) $L_{2,} L_{4,} L_6$ and L_{10} (b) $L_{4,} L_{6,} L_8$ and L_{12}
(c) $L_{2,} L_{3,} L_{4,} L_5$ and L_{16} (d) $L_{1,} L_{2,} L_3$ and L_5 and L_{16}

62. Initiation factors are proteins that bind to ________ during initiation of the translation:
(a) Smaller sub-unit of ribosome (b) Larger sub-unit of ribosome
(c) mRNA (d) AA-tRNA

63. The F-Met-$tRNA_f^{met}$ enters in ribosome at:
(a) A site (b) P site (c) E site (d) R site

64. Ribosomal association at the end of initiation step is prevented by:
(a) Tobramycin (b) Tobramycin and Kanamycin
(c) Kanamycin and Puromycin (d) Macrolides and Lincosamides

65. The first base, which is recognised for transcription, is that to which sigma (σ) factor attaches and it is:
(a) Generally T (b) Generally C
(c) Generally T and sometimes C (d) Always G or T

66. Protein synthesis does not require:
(a) Ribosomes (b) Primer proteins (c) mRNA (d) rRNA

67. The ion involved in protein synthesis is:
(a) Ca^{++} (b) Mg^{++} (c) Mn^{++} (d) Fe^{++}

68. The process in which the ribosome is engaged is the:
(a) Transduction (b) Transcription (c) Translation (d) Transformation

69. *RNA polymerase* was independently discovered by:
(a) Ochoa and Sam Weiss (1960) (b) Sam Weiss and Jerad Hurwitz (1960)
(c) Kornberg and Jerad Hurwitz (1960) (d) Nirenberg and Mathei (1961)

70. Association of 30S sub-unit of ribosome with 50S sub-unit is prevented by:
(a) IF-1 (b) IF-2 (c) IF-3 (d) IF-3 and Mg^{++}

71. IF-3 is released when:
(a) mRNA binds with 30S sub-unit of ribosome (b) 30S initiation complex is formed
(c) 70S complex is formed (d) Formation of polypeptide chain is completed

72. *RNA polymerase II* catalyses synthesis of:
(a) mRNA (b) Most snRNA (c) Micro RNA (miRNA) (d) None

73. The 5S ribosomal RNA is transcribed by:
(a) *RNA polymerase I* (b) *RNA polymerase II*
(c) *RNA polymerase III* (d) None

74. Sigma (σ) factor dissociates from the holoenzyme, when about ________________ ribonucleotides have been added:
(a) 10 (b) 30 (c) 50 (d) 0

75. α–amanitin blocks *RNA polymerase* of:
(a) Mitochondria (b) Chloroplast (c) Bacteria (d) None

76. Streptomycin:
(a) Binds to a specific protein of 30S sub-unit of ribosome (b) Causes misreading
(c) Causes inhibition (d) All

77. Both *RNA polymerase* and *DNA polymerase* are inhibited by
(a) Actinomycin-D (b) Proflavin (c) Both (a) and (b) (d) Rifampicin

78. The ion needed by both *DNA polymerase* and *RNA polymerase* is:
(a) Mg^{++} (b) Mn^{++} (c) Cl^- (d) Zn^{++}

79. Which one of the following is not associated with translation?
(a) DNA (b) mRNA (c) rRNA (d) tRNA

80. The protein 'Z' is 50 amino acids long. The number of nucleotides present in the mRNA used to make this protein:
(a) 50 (b) 100 (c) 150 (d) 300

81. The major portion involved in central dogma is:
(a) Replication (b) Transcription (c) Translation (d) All

82. The first step in protein synthesis involves:
(a) Binding of mRNA 30S sub-unit of ribosome (b) Joining of two sub-units of ribosomes
(c) Activation of amino acid reaction with ATP (d) Binding of amino acid with tRNA

83. Abrin and Ricin inhibits protein synthesis in eukaryotic ribosome by interacting with:
(a) 60S sub-unit of ribosome (b) 40S sub-unit of ribosome
(c) *Peptidyl transerase* (d) *Peptidyl-tRNA*

84. Puromycin inhibits peptide chain synthesis by interacting with:
(a) mRNA on ribosome (b) *Peptidyl-tRNA* on ribosome
(c) Inactivating *aminoacyl-tRNA synthetase* (d) Inactivating smaller sub-unit of ribosome

85. The ribosomal proteins are insoluble in water at pH ___ under normal condition.
(a) 4·5 (b) 6 (c) 7 (d) Above 7

86. Which one of the following antibiotics acts on 30S sub-unit of ribosome?
(a) Neomycin (b) Amikacin (c) Gentamycin (d) All

87. Premature terminations may be generated by:
(a) Nonsense mutations (b) Frameshift mutations
(c) RNA processing errors (d) All

88. In mammalian cells, *RNA polymerase* binds to the _____ sequence of promoters:
(a) G–C-rich region (b) TATA box
(c) CAAT box (d) A–T-rich region

89. Match the columns I, II and III and select the correct answer using answer codes:

Column I 3′↓ (DNA)	Column II 5′↓ (mRNA)	Column III (Peptide)
(A) TAC	1. UGC	i. Lysine
(B) AGC	2. AUG	ii. Methionine
(C) TTT	3. GGC	iii. Serine
(D) CCG	4. AAA	iv. Glycine

Answer codes:
(a) A, 2, ii (b) B, 1, iii (c) C, 4, ii (d) D, 3, iv

90. 7S L RNA is synthesised by:
(a) *RNA polymerase I* (b) *RNA polymerase II*
(c) *RNA polymerase III* (d) None

91. Consider the following statements:
(A) Actinomycin D has no effect on binding of *RNA polymerase* to DNA
(B) In eukaryotes, elongation proceeds at a rate of about two amino acids per second
(C) Translocation is indirectly inhibited by Fusidic acid
(D) In prokaryotes, elongation proceeds at a rate of 15–20 amino acids per second

The incorrect statements are:
(a) All (b) B and C (c) B and D (d) None

92. The presence of a capped 5' end on the mRNA is required for initiation of protein synthesis, except for a few
(a) Bacterial mRNAs (b) Fungal mRNAs
(c) Viral mRNA (d) All

93. Which one of the following statements is correct about *RNA polymerase III*?
   (a) They are quite abundant in cells (10^4–10^7)
   (b) They are generally short
   (c) They synthesise all tRNAs and 5S rRNAs
   (d) All
94. Which one of the following is essential for *peptidyl transferase* activity?
   (a) 23S rRNA
   (b) 16S rRNA
   (c) 5S rRNA
   (d) mRNA and 16S rRNA
95. Which one of the following is a powerful inhibitor of chain initiation in the process of protein synthesis?
   (a) Sparsomycin (b) Streptomycin (c) Chloramphenicol (d) Puromycin
96. The first nucleotide in the synthesis of RNA chain is:
   (a) Generally pppA or ppG
   (b) Sometimes ppC
   (c) Rarely ppU
   (d) All
97. NuSA protein affects the behaviour of *RNA polymerase* during:
   (a) Elongation of RNA synthesis
   (b) Termination of RNA synthesis
   (c) Both (a) and (b)
   (d) No affect on the behaviour of *RNA polymerase*
98. The polysomes that make haemoglobin contain:
   (a) 1 ribosome
   (b) 4–6 ribosomes
   (c) 50 ribosomes
   (d) More than 100 ribosomes
99. Docking protein plays a role in:
   (a) Transcription (b) Translation (c) Translocation (d) Transduction
100. Which one of the following statements is incorrect?
   (a) Eukaryotic polypeptide chain termination involves GTP hydrolysis.
   (b) Eukaryotic factor EF-2 hydrolyses GTP and catalyses translocation of *amino acyl-tRNA* from the A site to the P site on the ribosome.
   (c) The eukaryotic factor EF-2 corresponds to EF-Tu of prokaryotes.
   (d) Protein synthesis can proceed at a very slow rate in the absence of GTP hydrolysis and elongation factors.
101. The existence of mRNA was first established in experiments with:
   (a) T_2 infeced *E. coli* cells
   (b) T_7 infeced *E. coli* cells
   (c) T_2 infeced with *Bacillus subtilis*
   (d) SV_{40} virus
102. Which one of the following statements is incorrect about Dipthramide?
   (a) Dipthramide is a post-translation modification of translation elongation factor-2.
   (b) It is conserved in all eukaryotes and archaebacteria.
   (c) It is the target of diptheria toxin and is formed in the yeast by the action of five proteins.
   (d) None
103. Which one of the following statements is true about rRNA?
   (a) It does not play a role in protein synthesis.
   (b) It is a part of ribosomes that calls the tRNA molecules.
   (c) It forms the mRNA.
   (d) None
104. Which one of the following statements is correct about tRNA with reference to protein synthesis?
   (a) It forms rRNA.
   (b) Its codons match with the mRNA and call the amino acids to form the protein.
   (c) It forms mRNA.
   (d) It has no role in protein synthesis.
105. Docking protein is responsible for:
   (a) Protein folding
   (b) Polypeptide termination

(c) Association and dissociation of two sub-units of ribosomes
(d) Attaching a ribosome to a membrane by interacting with a signal particle

106. The RNA produced from DNA strand ATTCGG would be:
(a) TAAGCC (b) UAAGCC (c) CAAGCC (d) CCGAAT

107. Consider the following statements:
(A) Transcription occurs in the cell's nucleus
(B) Transcription requires DNA, *RNA polymerase* and ribonucleotides
(C) Transcription does not require ATP
(D) Presence of Mg^{++} is essential for transcription
The correct statements are:
(a) All (b) A, B and C (c) A and B (d) C and D

108. Which one of the following statements is correct about polypeptide?
(a) It is a strand of *RNA polymerase.*
(b) It is a strand of DNA.
(c) It is an incomplete strand of protein (amino acids and enzymes).
(d) None

109. *Corynebacterium diptheriae* alters protein functions in the host by inactivating:
(a) Elongation factor-1(EF-1)
(b) Elongation factor-2(EF-2)
(c) *Peptidyl-transferase*
(d) *Aminoacyl-tRNA synthetase*

110. RL-I is a highly conserved protein which is required for:
(a) Biogenesis of ribosomes
(b) Eukaryotic translation initiation
(c) Prokaryotic translation initiation and biogenesis of ribosomes
(d) Both (a) and (b)

111. Anticodon is present in:
(a) DNA (b) mRNA (c) tRNA (d) rRNA

112. Transcription does not involve:
(a) Initiation
(b) Elongation
(c) Termination
(d) Peptide bond formation

113. Three types of *RNA polymerase* in eukaryotes were discovered by:
(a) Ochowa (b) Kornberg (c) Anderson (d) Nirenberg

114. Initiating amino acid in protein synthesis is:
(a) Leucine (b) Methionine (c) Serine (d) Glycine

115. RNA processing was first observed in:
(a) mRNA (b) rRNA (c) tRNA (d) Not known

116. In which one of the following does transcription take place on both strands of DNA?
(a) *E. coli* (b) TMV (c) Adenovirus (d) Simian virus

117. In Simian virus (SV_{40}) in place of TATA box, _____ rich region is present:
(a) GC (b) AT (c) CT (d) GT

118. Reverse transcription was discovered by Temin and Baltimore in:
(a) Adenovirus (b) Raus sarcoma virus (c) ϕX174 (d) TMV

119. RNA processing was discovered by:
(a) Watson and Crick (1953)
(b) Crick (1958)
(c) Phillip Sharp (1977)
(d) Temin and Baltimore (1970)

120. *RNA polymerase II* transcription initiation requires a ________ conserved sequence:
(a) $TATAA_TAT_A$ (b) GTGATAGC (c) CAAT (d) AGAAGA

121. During mitosis *RNA polymerase* remains bound to:
(a) Nucleolar organiser (b) Centromere
(c) Telomere (d) DNA

122. Match column I to column II and select the correct match using answer codes:

Column I	Column I
(A) Pribnow box	1. Exons
(B) EF–G site	2. Prokaryotes
(C) Coding regions	3. 50S sub-unit of ribosome
(D) Hogness box	4. Eukaryotes

Answer codes:

	A	B	C	D
(a)	4	2	1	3
(b)	2	3	1	4
(c)	2	4	3	1
(d)	4	3	2	1

123. Prokaryotic *RNA polymerase* is composed of α, α', β, β' and σ sub-units in the ratio of:
(a) 2 : 1 : 1 : 1 (b) 1 : 1 : 1: 1 : 1 (c) 1 : 2 : 1 : 1 (d) 2 : 2 : 1

124. Which one of the following acts as barrier between transcription and translation?
(a) Hogness box (b) Pribnow box (c) Nuclear membrane (d) Both (b) and (c)

125. Transcription is:
(a) Asymmetrical (b) Symmetrical (c) Circular (d) Zigzag

126. Translation is:
(a) Circular (b) Directionless (c) Unidirectional (d) Bidirectional

127. Which one of the following is applicable to TATA box?
(a) Anti-terminator (b) Pribnow box (c) Hogness box (d) None

128. In which one of the following is initiation codon Methionine formulated?
(a) Prokaryotes (b) Mitochondria (c) Chloroplast (d) All

129. EF-Tu can bind every AA-tRNA except:
(a) Met-tRNA ^{met} (b) F-Met-tRNA$_f^{met}$ (c) Lys-tRNA (d) Both (a) and (b)

130. Only about _____ percent of heterogenous nuclear RNA (Hn RNA) is converted into mRNA:
(a) 25 (b) 50 (c) 75 (d) 95

131. The 5′ end of the eukaryotic mRNA is marked by a methylated:
(a) Cytosine (b) Guanine (c) Adenosine (d) Uracil

132. The enzyme involved in the formylation of Methionine is:
(a) *Translocase* (b) *Formylase* (c) *Deformylas·e* (d) *Aminopeptidase*

133. The only known mRNA lacking Shine–Dalgarno sequence is found in:
(a) SV_{40} (b) Lambda Pm gene (c) SP_3 (d) TMV

134. Evidence of shortening of heterogenous messenger RNA (Hn mRNA) first came from the study of:
(a) HIV (b) SV_{40} Virus (c) SP_3 Virus (d) Adenovirus

135. The enzyme involved in the elongation of polypeptide chain is:
(a) *Helicase* (b) *Translocase* (c) *Peptidyl-transferase* (d) *Primase*

136. Movement of growing polypeptide chain from A site to P site requires:
(a) Elongation factor-G (b) GTP

(c) Elongation factor-G and GTP (d) GTP and *Peptidyl-transferase*

137. Mutations causing change in purine-rich region of a ribosome-binding site:
   (a) Increases the efficiency of mRNA translation (b) Decreases the efficiency of mRNA translation
   (c) Inactivates mRNA (d) None
138. Within rDNA, the order of transcription is:
   (a) 5′ end → 28S → 18S → 58S → 3′ end (b) 3′ end → 18S → 28S → 5·8S → 3′ end
   (c) 5′ end → 18S → 5·8S → 28S → 3′ end (d) 5′ end → 5·8S → 18S → 28S → 3′ end
139. Which one of the following is a soluble elongation factor?
   (a) EF-Tu (b) EF-Ts (c) EF-G (d) All
140. Mitochondrial *RNA polymerase* contains:
   (a) Single peptide (b) Two peptides
   (c) Three peptides (d) More than three peptides
141. Which one of the following sub-units of bacterial *RNA polymerase* is loosely bound to the rest of the enzyme?
   (a) α (b) β (c) β' (d) σ
142. The three types of eukaryotic *RNA polymerase* can be distinguished by their inhibition sensitivity to:
   (a) Puromycin (b) α-*amanitin* (c) Macrolides (d) Linezolid
143. Cap is lacking in the mRNA of:
   (a) Polio virus (b) HIV (c) ϕX174 (d) *Xenopus* oocytes
144. Which one of the following inhibits bacterial *RNA polymerase?*
   (a) Ampicillin (b) Rifampicin (c) Sparzomycin (d) Actinomycin-D
145. The anticodon of tRNA can establish _______ bonds with the codon in mRNA:
   (a) Disulphide (b) Phosphodiester (c) Hydrogen (d) Hydrophobic
146. In majority of tRNA, _______ is added post-transcriptionally:
   (a) Anticodon loop (b) D loop (c) T loop (d) CAA end
147. The terminating codon is recognised by:
   (a) Ribosomes (b) Proteins called release factors
   (c) tRNA (d) All
148. The process of transcription is similar to replication except:
   (a) Only one DNA strand is transcribed (b) The sugar used is ribose
   (c) Adenine is substituted by Uracil (d) All
149. Which one of the following tRNA is encoded in chloroplast?
   (a) Glutamyl-tRNA (b) Tyrosyl-tRNA (c) Alanine- tRNA (d) Tyrosine-tRNA
150. A virus-infected cell:
   (a) Does not synthesise proteins (b) Synthesises different proteins
   (c) Synthesises proteins at the same rate as an uninfected cell (d) Both (b) and (c)
151. During the process of translation, mRNA binds to:
   (a) P site of ribosome (b) Smaller sub-unit of ribosome
   (c) Larger sub-unit of ribosome (d) A site of ribosome
152. Consider the following statements:
   (A) Antisense RNA codes for protein
   (B) When transcription takes place on both strands of DNA, it is known as complementary DNA transcription

(C) Complementary DNA transcription is found in some mammalian viruses
(D) The snRNA is metabolically unstable and is synthesised by *RNA polymerase III*

The correct statements are:
(a) All (b) B, C and D (c) B and C (d) A and D

153. Antisense RNA affects:
(a) RNA processing (b) Transcription (c) Translation (d) All

154. The snRNA is rich in:
(a) Uridine (b) Cytodine (c) Adenodine (d) Guanodine

155. Christmas tree is applicable to:
(a) *RNA polymerase* (b) *DNA polymerase*
(c) Catabolite activator protein (d) rRNA

156. Which one of the following enzymes is resistant to α-*amanatin?*
(a) *RNA polymerase I* (b) *RNA polymerase II*
(c) *RNA polymerase III* (d) None

157. After the completion of protein synthesis, the ribosomes are separated into sub-units by a dissociation factor that binds to:
(a) 30S sub-unit (b) 50S sub-unit
(c) 70S ribosome (d) Either A or P site of the ribosome

158. Chain elongation factor EF-Tu carries all *aminoacyl-tRNA* to the site of chain elongation except:
(a) Met-tRNA (b) F-Met-$tRNA_f^{met}$ (c) Tyrosine-tRNA (d) Valine-tRNA

159. F-Met-$tRNA_f^{met}$ is carried to the ribosome by:
(a) IF-1 (b) IF-2 (c) IF-3 (d) EF-Ts

160. *Corynebacterium diptheriae* alters protein function in the host by inactivating elongation factor-2 (EF-2) which causes in the host:
(a) Pharyngitis and Wilson's disease (b) Cystic fibrosis
(c) Pharyngitis and pseudomembrane (d) Leigh's syndrome

161. Which one of the following antibiotics binds with the smaller sub-unit of the ribosome, both in prokaryotes and eukaryotes, affecting protein synthesis?
(a) Edeine (b) Kanamycin (c) Diptheria toxin (d) Sparsomycin

162. The initiation factor responsible for the binding of mRNA to the 30S sub-unit of ribosome is:
(a) IF-1 (b) IF-2 (c) IF-3 (d) IF-1 and IF-2

163. The antibiotic which is an analogue of the *aminocyl-tRNA* is:
(a) Puromycin (b) Streptogramin (c) Edeine (d) Sparsomycin

164. Which one of the following is not needed for polypeptide chain elongation?
(a) Functional 70S ribosome (b) *Aminocyl-tRNAs* specified by codons
(c) Mg^{++} and GTP (d) cAMP-dependant protein *kinase*

165. Enhancer sequences were first discovered in:
(a) DNA tumour viruses ($SV_{40,}$ Polyoma) (b) DNA tumour virus
(c) *E. coli* (d) *Neurospora*

166. Which one of the following is associated with translation?
(a) Sigma (σ) factor (b) Rho (ρ) factor (c) Release factor (RF) (d) Omega (ω) factor

167. mRNA cap is essential for:
(a) Amino acid recognition (b) Ribosomal recognition

(c) Binding with *aminocyl-tRNA synthetase* (d) Pairing

168. Match column I with column II and select the correct answer using answer codes:

Column I		Column II
(A) Cryptogram	1.	Component of promoter
(B) CAAT box	2.	The site of DNA on which *RNA polymerase* binds during transcription
(C) Pribnow box	3.	mRNA coded language
(D) Promoter	4.	A conserved region in promoter part of DNA

Answer codes:

	A	B	C	D
(a)	4	3	1	2
(b)	3	1	4	2
(c)	4	2	1	3
(d)	3	1	2	4

169. The enzyme not involved in the process of translation is:
(a) *Peptidyl transferase* (b) *Translocase*
(c) *Transferase* (d) *Aminocyl-tRNA synthetase*

170. Which one of the following is synthesised outside the nucleolus?
(a) 28S rRNA (b) 18S rRNA (c) 5·8S rRNA (d) 5S rRNA

171. Who first showed that nucleic acids are connected with protein synthesis?
(a) Capersson and Brachet (1941) (b) Hoagland and coworkers (1956)
(c) Bernner and Meselson (1961) (d) Nirenberg and Matthei (1961)

172. In prokaryotes, binding of *aminocyl-tRNA* (AA-tRNA) to the ribosome requires:
(a) EF-Tu (b) EF-Ts (c) GTP (d) All

173. The role of mRNA in protein synthesis was first demonstrated by:
(a) Grunberg and Ochoa (1953) (b) Bernner, Jacob and Meselson (1961)
(c) Zamenik (1969) (d) Capecchi and Gussin (1965)

174. Signal hypothesis was proposed by:
(a) Hoagland et al. (b) Hopefield (1974) (c) Blobel et al. (1975) (d) Nierhaus (1982)

175. Eukaryotic ribosomes are able to translate bacterial mRNAs efficiently, provided that a ________ is added enzymatically:
(a) Poly-A tail (b) Cap (c) *Guanylyl transferase* (d) EF-2

176. The process of removal of intron is called:
(a) RNA transcript (b) RNA processing (c) Gene amplification (d) Cleavage

177. RNA processing does not include:
(a) Capping (b) Addition of poly-A tail
(c) Removal of introns (d) Removal of exons

178. The RNA is positioned correctly at the initiation codon (AUG) for the initiation of translation by:
(a) 23S rRNA (b) 16S rRNA
(c) 5S rRNA (d) 23S rRNA and 5S rRNA

179. RNA processing was first observed in:
(a) mRNA (b) rRNA (c) tRNA (d) Not known

180. Consider the following statements about EF-G site:
(a) Located in the 50S sub-unit of ribosome (b) Acts as translocation factor
(c) Transfers tRNA from A site to P site (d) It binds with GTP and carries it to the ribosome

The correct statements are:
(a) All (b) A, B and C (c) B and D (d) B, C and D

181. Which one of the following represents initiation complex in prokaryotes?
(a) AA-tRNA + AMP + Enzyme
(b) 30S mRNA-F-Met-$tRNA_f^{met}$
(c) 70S mRNA-F-Met-$tRNA_f^{met}$
(d) 50S mRNA-F-Met-$tRNA_f^{met}$

182. Peptide bond formation occurs in:
(a) Assembly of ribosomes, mRNA and tRNA
(b) mRNA
(c) tRNA
(d) Ribosomes

183. Rho (ρ) factor is associated with:
(a) Chain elongation
(b) Chain termination
(c) Binding of *RNA polymerase* in DNA
(d) Binding of *aminoacyl-tRNA* to ribosome

184. A research scholar while doing some research work on protein synthesis took a polyribosome containing 10 ribosomes held by an mRNA; each being 27 nucleotides apart from the other. He put this polyribosome in a medium containing all 20 types of amino acids, tRNA and necessary enzymes for the synthesis of proteins. How many amino acids are specified by mRNA in this experiment?
(a) 9 (b) 10 (c) 20 (d) 90

185. The enzyme involved in the first step of biosynthesis of proteins is:
(a) *Translocase*
(b) *Peptidyl transferase*
(c) *Aminoacyl-tRNA synthetase*
(d) *Guanylyl transferase*

186. If a tRNA is mismatched, it will:
(a) Die soon
(b) Be unable to bind to the ribosome
(c) Cause incorporation of an incorrect amino acid into the protein chain
(d) Not incorporate an amino acid into the protein chain

187. Which one of the following statements is correct about eukaryotic factor EF-2?
(a) It corresponds to prokaryotic EF-G
(b) It hydrolyses GTP
(c) It catalyses translocation of the *aminoacyl-tRNA* from the A site to the Psite of the ribosome
(d) All

188. The binding of *RNA polymerase* to DNA involves the:
(a) α sub-unit (b) β sub-unit (c) β' sub-unit (d) σ factor

189. The elongation factor Ts (EF–Ts):
(a) Transports F-Met-$tRNA_f^{met}$ to the ribosome
(b) Forms a complex with *aminoacyl-tRNA* (AA-tRNA) and GTP and carries them to the ribosomes
(c) Catalyses the formation of complex between EF-Tu, AA-tRNA and GTP
(d) Involved in the translocation process

190. The 3′ end of tRNA hangs free which is complementary to the:
(a) The region upstream of AUG of mRNA
(b) The region upstream of UAG of mRNA
(c) The region upstream of GGC of mRNA
(d) The region upstream of AGA of mRNA

191. The ______ binds to a region of the mRNA that precedes the first translated codon:
(a) 28S rRNA (b) 18S rRNA (c) 5·8S rRNA (d) 5S rRNA

192. eIF-2 is active only when it is bound to:
(a) GTP
(b) GDP
(c) ATP
(d) 40S sub-unit of ribosome

193. Which one of the following is not translated into polypeptide?
(a) tRNA (b) 28S RNA (c) 5·8S rRNA (d) All

194. Which one of the following statements is incorrect?
(a) TATA box is surrounded by G–C-rich region
(b) It is comparable to Pribnow box of prokaryotes
(c) CAAT and GC boxes determine the efficiency of transcription
(d) The presence of TATA box is not essential for transcription

195. Protein synthesis accounts for ______ of chemical energy used in the cell:
(a) 30 per cent (b) 50 per cent (c) 90 per cent (d) 95 per cent

196. Protein synthesis is not initiated with formyl-methionine in:
(a) Eubacteria (b) Mitochondria
(c) Archaebacteria (d) Chloroplast and eubacteria

197. mRNA can be simultaneously translated by several ribosomes in:
(a) Prokaryotes (b) Eukaryotes
(c) Both prokaryotes and eukaryotes (d) None

198. *RNA polymerase II* promoter lacks lacks:
(a) Enhancer (b) TATA box (c) CAAT box (d) None

199. Which one of the following enzymes catalyses the first step of biosynthesis of a polypeptide chain?
(a) *Translocase* (b) *Peptidyl transferase*
(c) *Aminoacyl-tRNA synthetase* (d) *RNA polymerase*

200. In which one of the following viruses, transcription occurs at a time, on both strands of DNA?
(a) Simian virus (b) TMV (c) T2 bacteriophage (d) HIV

201. Antisense RNA affects:
(a) Transcription (b) Translation (c) RNA processing (d) All

202. Consider the following statements:
(A) The EF-Tu GTP complex can not recognise the initiator tRNA(F-Met-$tRNA_f^{met}$)
(B) The enzymatic function of EF-G (*Translocase*) is dependant on hydrolysis of GTP
(C) The eukaryotic eIF-2 and eIF-3 are multiple polypeptide chains
(D) In bacteria, it is the 16S rRNA that helps ribosomes to accommodate mRNA

The correct statements are:
(a) All (b) B, C and D (c) A, B and D (d) A and C

203. Who first observed that separate enzymes catalyse the activation of different amino acids?
(a) Kornberg (1969) (b) Hoagland and co-workers (1956)
(c) Zamecnik (1960) (d) Stein et al. (1958)

204. During the formation of *aminoacyl-tRNA*, the overall reaction proceeds with:
(a) A relatively small increase in standard free energy
(b) A relatively small decrease in standard free energy
(c) Neither is there an increase or decrease in standard free energy
(d) An enormous decrease in standard free energy

205. The continuous association and dissociation of bacterial ribosomes during protein synthesis was first observed by:
(a) Kaempfer and co-workers (b) Nomura and co-workers
(c) Crick and Hoagland (d) Zamecnik

206. The polypeptide chain of haemoglobin contains 150 amino acid residues, it is coded by mRNA molecule having_____ nucleotides:
(a) 50 (b) 150 (c) 450 (d) 500

207. In prokaryotes transcription terminates on a region of four to eight consecutive _____ residues:
(a) G (b) C (c) U (d) A

208. In the absence of one of the following, mRNA is unable to form initiation complex with 30S ribosomal sub-unit:
(a) Cap (b) Poly A (c) AAUAAA (d) Both (a) and (b)

209. Which one of the following statements is incorrect?
(a) The rate of transcription of a particular gene is fixed.
(b) Rifampicin prevents initiation of RNA chain but not the elongation of the chain.
(c) Sigma factor carries the specificity for promoter recognition.
(d) Crick and Hoagland gave the first experimental proof that tRNA is a molecular adaptor.

210. Which group of protein is present in four copies in 50S sub-unit of ribosome?
(a) L2–L10 (b) L7–L12 (c) L10–L16 (d) L30–L 34

211. Prior to RNA assembly, 5S rRNA binds to ribosomal protein ______ to form a stable ribonucleoprotein:
(a) L2 (b) L4 (c) L5 (d) L12

212. Which one of the following is associated with conformational changes of ribosome?
(a) Binding of mRNA and tRNA to smaller ribosomal sub-units
(b) Binding of initiation factor to smaller ribosomal sub-units
(c) Association of ribosomal sub-units
(d) All

213. In prokaryotes, which initiation factor is required to locate AUG codon?
(a) IF-1 (b) IF-2 (c) IF-3 (d) IF-1 and IF-2

214. GTP levels have been proposed to be directly coupled to the activity of:
(a) IF-1 (b) IF-2 (c) IF-3 (d) All

215. The initiator tRNA determinants are located in the:
(a) Anticodon stem (b) Acceptor stem (c) Dihydrouridine arm (d) All

216. Consider the following statements:
(A) EF–Tu prevents *aminoacyl* end of the charged tRNA from entering into A site of the ribosome
(B) EF–Tu cannot bind with the $tRNA_f^{met}$
(C) Nearly all the *aminoacyl-tRNA* in the cells are bound by EF-Tu
(D) EF–Tu has higher affinity for GTP
The correct statements are:
(a) All (b) A and B (c) A, B and D (d) C and D

217. Formylation is not necessary for translation initiation in:
(a) *Bacillus subtilis* (b) *Pseudomonas aeruginosa*
(c) *Mycobacterium tuberculosis* (d) *Spirillum minus*

218. Experimentally, more than one isoform of IF-2 has been reported in:
(a) *Bacillus coagulens* (b) *Agrobacterium tumefaciens*
(c) *Bacillus subtilis* (d) *Diplococcus pneumoniae*

219. Rho (ρ) factor binds to a specific region on the RNA chain called:
(a) Hfr (b) Rut (c) Gyr (d) Lex A

220. Self-splicing of RNAs is found in introns of:
(a) Mitochondria (b) Fungi (c) *Tetrahymena* (d) All

221. Whenever there is insufficient free rRNA:
(a) L4 does not bind its own mRNA (b) L4 bind to its own mRNA
(c) L4 bind to tRNA (d) The two ribosomal sub-units do not associate

222. Methylation is common in:
(a) mRNA (b) mRNA and rRNA (c) rRNA and tRNA (d) All

223. Rho (ρ) factor is a:
(a) Protein with *ATPase* activity (b) Protein with *GTPase* activity
(c) Nonprotein (d) Protein that is loosely bound to *RNA polymerase*

224. Bacterial *RNA polymerase* contains two molecules of:
(a) Zinc (b) Calcium (c) Manganese (d) Cobalt

225. A transcription bubble contains:
(a) DNA (b) Nascent DNA (c) *RNA polymerase* (d) All

226. Consider the following statements:
(A) Small nuclear RNAs (snRNAs) complexes with specific proteins to form small nuclear ribonuclear particles (snRNPs) and production of auto-antibodies against snRNPs. This results in a fatal auto-immune disease called systemic lupus erythematosis (SLE)
(B) *Peptidyl transferase* is a ribozyme
(C) *RNAase*-P generates the ends of tRNAs
(D) Roberts and Sharp were awarded the Nobel Prize in 1993 for elucidation of spliceosome activity

The correct statements are:
(a) A and B (b) B and C (c) C and D (d) None

227. The methionine-tRNA was first isolated by:
(a) Hoagland (1957) (b) Robert W Holley (1968)
(c) Baltimore (1975) (d) Paul Berg (1956)

228. Abnormal folding of protein may lead to:
(a) Leigh's syndrome (b) Kearns Sayre syndrome
(c) Prion disease (d) All

229. The initiator tRNA determinants are located in the:
(a) Anticodon stem (b) Acceptor stem (c) Dihydrouridine stem (d) All

230. Which one of the following statements is incorrect about chaperones?
(a) They have ATPase activity. (b) They prevent wrong folding of proteins.
(c) They are the part of mature protein. (d) They help in attainment of tertiary and quaternary structures of proteins.

231. There are some genes which are transcribed into RNA but are not translated into protein products. These products are folded into:
(a) mRNA (b) rRNA (c) rRNA and tRNA (d) None

232. Which one of the following statements is incorrect about puromycin?
(a) It is structurally similar to tyrosine tRNA. (b) It causes release of incomplete peptide.
(c) It acts both in bacterial and mammalian cell. (d) It gets attached to P site of the ribosome.

233. Which one of the following statements is incorrect?
(a) The rate of transcription is increased by enhancers while silencers decrease the rate of transcription.

(b) Enhancers and silencers are located on different chromosomes.
(c) The TATA box is located on the coding strand.
(d) In human beings about 10^5 transcription sites are present on the entire DNA.

234. Match column I, II and III and select the correct answer using answer codes:

Column I (Element)	Column II (DNA sequence)	Column III (Protein-binding factor)
(a) GC box	1. TATAAT	w. Sp1
(b) CAAT box	2. GGGCGG	x. Nuclear factor 1
(c) TATA box	3. CCAAT	y. CAT binding protein
(d) TGG box	4. TGGGCCAA	z. Tf-TI-D

Answer codes:

	A	B	C	D
(a)	2, y	1, z	4, x	3, w
(b)	2, w	3, y,	1, z	4, x
(c)	4, w	2, x	1, y	3, z
(d)	3, y	4, x	2, z	1, w

235. Which one of the following is a ribozyme?
(a) *RNAase* P (b) Spliceosomes (c) *Peptidyl transferase* (d) All

236. Ricin is a highly toxic protein (from castor bean) which inactivates:
(a) 28S rRNA (b) 23S rRNA (c) 18S rRNA (d) 5·8S rRNA

237. Match column I with column II and select the correct answer using answer codes:

Column I	Column II
(A) Introns	1. 50-S sub-unit of ribosome
(B) Hairpin ribozyme	2. Mammalian cells
(C) EF-G site	3. Noncoding region
(D) Kozak consensus	4. Cleave 5' leader sequences

Answer codes:

	A	B	C	D
(a)	3	4	2	1
(b)	2	4	1	3
(c)	3	4	1	2
(d)	4	2	1	3

Answers to Multiple-Choice Questions

1. (a)	2. (b)	3. (c)	4. (b)	5. (c)	6. (a)	7. (c)	8. (b)
9. (a)	10. (b)	11. (d)	12. (c)	13. (d)	14. (d)	15. (c)	16. (b)
17. (a)	18. (b)	19. (d)	20. (b)	21. (c)	22. (a)	23. (b)	24. (d)
25. (d)	26. (a)	27. (c)	28. (a)	29. (d)	30. (d)	31. (b)	32. (d)
33. (d)	34. (a)	35. (c)	36. (d)	37. (b)	38. (c)	39. (d)	40. (c)
41. (a)	42. (a)	43. (a)	44. (a)	45. (b)	46. (b)	47. (a)	48. (c)
49. (a)	50. (d)	51. (d)	52. (b)	53. (c)	54. (d)	55. (a)	56. (c)
57. (c)	58. (a)	59. (c)	60. (b)	61. (c)	62. (a)	63. (b)	64. (b)
65. (c)	66. (b)	67. (b)	68. (c)	69. (b)	70. (c)	71. (b)	72. (d)
73. (c)	74. (a)	75. (d)	76. (d)	77. (c)	78. (a)	79. (a)	80. (c)

81. (d)	82. (c)	83. (a)	84. (b)	85. (c)	86. (d)	87. (d)	88. (b)
89. (a)	90. (c)	91. (d)	92. (c)	93. (d)	94. (a)	95. (b)	96. (d)
97. (c)	98. (b)	99. (c)	100. (c)	101. (a)	102. (d)	103. (b)	104. (b)
105. (d)	106. (b)	107. (c)	108. (c)	109. (b)	110. (d)	111. (c)	112. (d)
113. (a)	114. (b)	115. (b)	116. (d)	117. (a)	118. (b)	119. (c)	120. (a)
121. (a)	122. (b)	123. (a)	124. (c)	125. (a)	126. (c)	127. (c)	128. (d)
129. (b)	130. (a)	131. (b)	132. (b)	133. (b)	134. (d)	135. (b)	136. (c)
137. (b)	138. (c)	139. (d)	140. (a)	141. (d)	142. (b)	143. (a)	144, (b)
145. (c)	146. (d)	147. (b)	148. (d)	149. (a)	150. (d)	151. (b)	152. (c)
153. (d)	154. (a)	155. (d)	156. (a)	157. (a)	158. (b)	159. (b)	160. (c)
161. (a)	162. (c)	163. (a)	164. (d)	165. (a)	166. (c)	167. (b)	168. (b)
169. (c)	170 (d)	171. (a)	172. (d)	173. (b)	174. (c)	175. (b)	176. (b)
177. (d)	178. (b)	179. (b)	180. (a)	181. (c)	182. (d)	183. (b)	184. (d)
185. (c)	186. (c)	187. (d)	188. (b)	189. (c)	190. (a)	191. (b)	192. (a)
193. (d)	194. (d)	195. (c)	196. (c)	197. (c)	198. (d)	199. (c)	200. (a)
201. (d)	202. (a)	203. (b)	204. (b)	205. (a)	206. (c)	207. (c)	208. (a)
209. (a)	210. (b)	211. (c)	212. (d)	213. (b)	214. (b)	215. (d)	216. (a)
217. (b)	218. (c)	219. (b)	220. (d)	221. (b)	222. (c)	223. (a)	224. (a)
225. (d)	226. (d)	227. (d)	228. (c)	229. (d)	230. (c)	231. (c)	232. (d)
233. (b)	234. (b)	235. (d)	236. (a)	237. (c)			

Fill in the Blanks

1. Protein synthesis is the ______________ and ______________ of specific parts of DNA to form protein.
2. Transcription is the synthesis of RNA under the direction of the ______________.
3. Transcription starts with an enzyme called ______________________.
4. The synthesis of protein is known as ________________________.
5. All types of RNA are made by the process known as ______________.
6. Amino acids are monomers, which are polymerised to produce________________.
7. The *aminoacyl-tRNA synthetase* is an enzyme that catalyses the ____________of a specific amino acid.
8. *RNA polymerase* binds to the ______________ end of a gene (promoter) on DNA template strand.
9. Prokaryotic transcription occurs in the ____________________.
10. The *aminoacyl-tRNA synthetase* first binds to ______________ and the corresponding amino acid to form_______________ .
11. During and after synthesis, polypeptide chain often folds to assume so called native __________ and __________ structures.
12. Transcription in archaea is similar to transcription in _________________.
13. In bacteria, RNA transcription stops when the newly synthesised molecule form a _____________ rich hairpin loop followed by a run of _____________.
14. Roger D Kornberg won the Nobel Prize in chemistry, in 2006, for the study of molecular basis of ____________________________ .
15. HIV has an RNA genome that is duplicated into ______________.

16. Some eukaryotic cells contain an enzyme with reverse transcription called_____________.
17. Initiation of protein synthesis in *E. coli* requires initiation factors _____________, _____________ and_____________.
18. The main enzyme involved in the synthesis of DNA from an RNA template is _____________.
19. _____________ is the process of decoding RNA to form polypeptides.
20. At the end of protein synthesis, ribosomes are separated into two sub-units by a dissociation factor, which binds to _____________ sub-unit of ribosome.
21. The EF–G factor is also called _____________.
22. A_____________ causes a GTP binding protein to hydrolyze its GTP into GDP+iP.
23. Lep A exhibits uncoupled _____________ activity.
24. Mutations in Shine–Dalgarno sequence can reduce _____________.
25. Ribosomal protein S1 is only present in _____________ bacteria.
26. Sigma factor is a protein component of _____________.
27. The specific site on DNA where transcription begins is determined by _____________.
28. The _____________ factor inhibits the function of sigma factor.
29. Transcription is initiated at the regions of DNA called_____________.
30. _____________ is the most important step in gene expression.
31. *Peptidyl-tRNA hydrolase* release tRNA from *peptidyl-tRNA* by cleaving the _____________ bond between the peptide and the tRNA.
32. *Aminoacyl-tRNA synthetase* catalyses the bonding between _____________ and _____________.
33. Post-translational modification is the _____________ modification of protein after its _____________.
34. Transcription was first observed in T_7 and SP3 viruses which live in the bacterium _____________ _____________.
35. All ribosomal proteins are associated with the _____________ directly.
36. The Shine–Dalgarno sequence is located at the _____________ terminal region of the mRNA.
37. The process of initiation of RNA synthesis on a DNA template is known as _____________.
38. Temin and Baltimore were awarded the Nobel Prize in 1975 for the discovery of the enzyme _____________.
39. The _____________ is the only tRNA that can bind directly to P site of a ribosome.
40. *Peptidyl transferase* is present in _____________ sub-unit of ribosomes.
41. Puromycin interrupts peptide chain elongation by virtue of its ability to bind at _____________ site of a ribosome.
42. _____________ and _____________ are plant proteins that inhibit protein synthesis in eukaryotes.
43. In a eukaryotic cell, mRNA is synthesised in the _____________ and translated in _____________ in the cytoplasm.
44. The correct folding of polypeptide chain may require another protein called _____________ protein.
45. The collection of ribosome on mRNA is known as _____________ or _____________.
46. The eukaryotic equivalent, the Shine–Dalgarno sequence is called _____________.
47. In eukaryotes about _____________ per cent bases of the rRNA molecules are methylated.
48. Protein-releasing factors recognise the stop codon when they arrive at _____________ site of a ribosome.
49. Protein synthesis begins at the _____________ end of the _____________ structure of mRNA.

50. The enzyme called ______________ adds the correct amino acid to tRNA.
51. __________________ inhibits elongation of an RNA chain.
52. In bacteria there are ______________ initiation factors.
53. The process of shifting of amino acids from A site of ribosome to P site of ribosome is known as ______________.
54. The number of ribosome in a polysome depends on the length of the ______________.
55. The movement of mRNA relative to ribosome is called______________.
56. Spliceosomes are composed of _______________ molecules and a large number of protein molecules.
57. The splicing reactions are catalysed by the ______________.
58. In prokaryotes, the binding small sub-unit of ribosome to the correct place on the mRNA is assigned by a base pairing to a series of bases known as ______________ sequence.
59. The synthesis of each peptide bond utilises__________ high-energy phosphate groups in a completed protein.
60. In prokaryotes , *RNA polymerase* recognises characteristic nucleotide sequences in the promoter region known as ____________ and _________ nucleotide base.
61. The three types of eukaryotic *RNA polymerases* can be distinguished by their sensitivity to _______________.
62. Each anticodon consists of three base and this pairing follows the ______________ and ______________ combination.
63. Most function of the A site is identified on the ______________ sub-unit of the ribosome.
64. In animals, the targets of DNA methylation are almost ______________ which are immediately followed by guanine.
65. According to the Chambon rule, splicing always involves cleavage preceding the ______________ at 5' region, following the ______________.
66. ______________ RNA is the predominant product of transcription both in prokaryotes and eukaryotes.
67. The joining of two sub-units of ribosomes leads to the formation of a groove that accommodates ______________ which is being translated.
68. Most eukaryotic protein encoding genes contain noncoding segments called ______________.
69. Releasing factor (RF-3) catalyses the release of ______________ and ______________ at the end of the termination process.
70. The introns in an enzyme, consisting mainly of RNA, are also called ______________.

Answers to Fill in the Blanks

1. Transcription, translation
2. DNA
3. *RNA polymerase*
4. Translation
5. Transcription
6. Proteins
7. Esterification
8. 3′
9. Cytoplasm
10. ATP, *aminoacyl adenylate*
11. Secondary, tertiary
12. Eukaryotes
13. GC, U's
14. Eukaryotic transcription
15. DNA
16. *Telomerase*
17. IF-1, IF-2, IF-3
18. *Reverse transcriptase*
19. Translation
20. 30S
21. *Translocase*
22. *GTPase* Activating Protein (GAP)
23. *GTPase*
24. Translation

25.	Gram negative	26.	*RNA polymerase*	27.	Sigma (σ) factor
28.	Anti-sigma	29.	Promoter	30.	Initiation
31.	Ester	32.	tRNA, amino acid	33.	Chemical, translation
34.	*Bacillus subtilis*	35.	rRNAs	36.	5′
37.	Repression	38.	*Reverse transcriptase*	39.	Initiator tRNA
40.	50-S	41.	'A'	42.	Abrin, ricin
43.	Nucleus, ribosome	44.	Chaperone	45.	Polyribosome, polysome
46.	Kozak sequence	47.	Two per cent	48.	A
49.	5′, capped	50.	*Aminoacyl-tRNA synthetase*	51.	Streptolydigin
52.	Three	53.	Translocation	54.	mRNA
55.	Translocation	56.	Small nuclear RNA (snRNA)	57.	Small nuclear RNA (snRNA)
58.	Shine–Dalgarno sequence	59.	Four	60.	Pribnow box, TGTTG
61.	α amanitin	62.	A–U, G–C	63.	50S
64.	Cytosine	65.	GU, AG	66.	Ribosomal
67.	mRNA	68.	Introns	69.	RF-1, RF-2
70.	Ribozyme				

True or False

1. Protein synthesis takes place freely in solution.
2. NUSA protein molecule may act on several *RNA polymerase* during the process of transcription.
3. *RNA polymerase* functions only in the presence of DNA.
4. Poly A provides stability to mRNA.
5. Bacterial mRNA is not capped.
6. *RNA polymerase I* is required for the synthesis of HnRNA.
7. The antisense RNA does not code for any protein.
8. DNA is involved in translation.
9. Teminism is similar to transduction.
10. *Glutamyl-tRNA* is encoded in the chloroplast.
11. Antisense RNA affects RNA processing, transcription and translation.
12. The relationship between DNA, RNA and protein is known as Wobble hypothesis.
13. TATA box is also known as Pribnow box.
14. Puromycin inhibits translation in eukaryotes, prokaryotes and mitochondria.
15. Licocin acts as stimulator of transcription in prokaryotes.
16. In transcription, only a single or group of genes is transcribed.
17. In prokaryotes, there is only one type of *RNA polymerase* that catalyses synthesis of all three types of RNA.
18. Initiation factors, viz., IF-1, IF-2 and IF-3 are tightly bound by the ribosomes.
19. The transcription and translation processes provide the correct primary structure of the protein.

20. The rate of error in the synthesis of a polypeptide chain is roughly 1 in every 10,000 amino acids.
21. Termination of the polypeptide chain occurs when P site of the ribosome faces termination codon (UAA, UAG or UGA).
22. Initiation factor IF-1 is the smallest of the initiation factor and is a basic protein.
23. Higher cellular level of IF-1 affects transcription and translation.
24. IF-2 posses a ribosome dependant *GTPase* activity.
25. In bacterial cell, IF-2 is present in two forms, viz., IF-2α and IF-2β.
26. The interaction between mRNA and ribosome is very weak.
27. IF-2 is an acidic protein and is the largest of the initiation factors.
28. Among *RNA polymerases, RNA polymerase I* is the busiest *polymerase.*
29. Some of the tRNAs are synthesised by the *RNA polymerase I.*
30. The genes for rRNA and tRNA are split by introns.
31. The removal of introns and splicing exons is done by spliceosomes.
32. Recently it has been shown that in mammalian cells, 10–15 per cent translation occurs in the nucleus and at least some of this translation occurs when mRNA is still being synthesised by *RNA polymerase.*
33. Ribosomes have no role in the elongation of polypeptide chain.
34. The 30S sub-unit of ribosome attaches with the mRNA which is assisted by IF-1 and IF-2.
35. In archaea, the starting amino acid is the methionine.
36. The 28S rRNA of animal ribosome varies somewhat in size in different species.
37. In majority of organisms, synthesis of protein consumes more energy than any other biosynthetic process.
38. Mg^{++} is needed for the termination of the polypeptide chain.
39. GTP is required for association of 50S sub-unit o ribosome with the initiation complex.
40. Exon part of mRNA has code for polypeptide.
41. Neomycin inhibits interaction between tRNA and mRNA.
42. The introns in the majority of pre-mRNAs begin with GU and with an AG.
43. The central dogma is not applicable in case of all animal viruses.
44. Two high-energy phosphate bonds are needed for the activation of amino acid for the process of protein synthesis.
45. *RNA polymerase I* is located in the nucleoplasm.
46. The hydrolytic step causing release of a polypeptide chain is catalysed by the enzyme *peptidyl-transferase.*
47. A polypeptide chain being translated by an mRNA on ribosome grows from N-terminus to C-terminus.
48. An mRNA molecule is translated by a single ribosome at a time.
49. Toadstool mushrooms are poisonous as they inhibit transcription.
50. Eukaryotic mRNA is metabolically stable in comparison to prokaryotic mRNA.
51. rDNA of amphibian oocytes present a very good example of gene amplification.
52. All mRNAs are polyadenylated.
53. Poly A provides stability to mRNA.
54. All CAAT box lack transcription factor protein.
55. Polycistronic mRNAs are expressed as a sequence of individual proteins only in prokaryotes.
56. Hoagland et al. (1956) showed that separate enzymes catalyse the activation of different amino acids.

57. Transcription initiation is the key point of regulation of gene expression.
58. The TATA box is bounded by a specific protein.
59. The error frequency in synthesis of protein increases with the age of the animal.
60. Prokaryotic ribosomal bases are methylated.
61. Viruses utilise the ribosome of the host cell for their own protein synthesis.
62. In the absence of EF-Tu, no conformational changes occur in the ribosome.
63. The initiation codon lies within the reach of the Shine–Dalgarno sequence.
64. F-Met- $tRNA_f^{met}$ and MetRNA have different anticodon sequences.
65. The GTP forms covalent bonds when it combines with the protein factors in ribosomes.
66. The 70S ribosome undergoes slightly morphological change for the formation of peptide bond.
67. Complimentary DNA transcription is shown by ϕX174.
68. RNA processing is common in both eukaryotes and prokaryotes.

Answers to True or False

1. False	2. True	3. True	4. True	5. True	6. False	7. True	8. False
9. False	10. True	11. True	12. False	13. False	14. True	15. False	16. True
17. True	18. False	19. True	20. True	21. False	22. True	23. False	24. True
25. True	26. True	27. True	28. True	29. False	30. True	31. True	32. True
33. False	34. True	35. False	36. True	37. True	38. False	39. False	40. True
41. True	42. True	43. False	44. True	45. False	46. False	47. True	48. False
49. True	50. True	51. True	52. False	53. True	54. False	55. True	56. True
57. True	58. True	59. True	60. False	61. True	62. True	63. False	64. False
65. False	66. True	67. False	68. False				

Give Reasons

1. Transcription is especially important in terms of gene expression.
   – Because transcription is the point where most of the regulation of gene expression occurs.
2. Dissociation of sub-unit of ribosome is essential for initiation.
   – Because the initial binding (F-Met-$tRNA_f^{met}$) in prokaryotes and Met-tRNA in eukaryotes occurs on small sub-unit of ribosomes.
3. In prokaryotes, protein synthesis starts even before mRNA synthesis is completed.
   – Because prokaryotes lack a nuclear membrane. Thus, there is no separation between genome and ribosome.
4. Sequences of mRNA vary.
   – Because amino acid coding sequences differ. Leader and trailer sequences also differ.
5. The existence of 3′ end of RNA bearing CCA is essential.
   – Because at this end of tRNA, amino acid is attached.

6. The cell needs some way of transferring genetic information from the nucleus to the cytoplasm.
   - Because the synthesis of proteins take place in the cytoplasm (outside the nucleus).
7. The core enzyme *RNA polymerase* by itself is not able to transcribe RNA.
   - Because it transcribes on both strands.
8. The P site of ribosome is so named.
   - Because it binds only to initiator peptidyl-tRNA molecule (except the initiator tRNA), i.e., a tRNA with growing peptide attached.
9. Transcription is asymmetrical.
   - Because, out of two strands of DNA, only one strand is used as a template for transcription.
10. Cap plays an important role in the translation of mRNA.
    - Because in the absence of cap, mRNA is unable to bind with 30S sub-unit of ribosome.
11. Noncoding regions are functionally important.
    - Because their sequence has tended to be conserved through evolution.
12. The TATA box is important for transcription.
    - Because it has been reported that probably its function mainly is to precisely position the start of transcription.
13. N-formyly methionine is unable to form a peptide bond.
    - Because, its end bearing the free amino acid is blocked by the formylation reaction.
14. Several eukaryotic mRNA can be purified.
    - Because they are stable.
15. Each ribosome functionally accommodates two codons at a time.
    - Because each ribosome has two binding sites called A (*aminoacyl*) and P (*peptidyl*) sites.
16. Transfer RNA (tRNA) is so called.
    - Because it carries activated amino acids from the cytoplasm to the site of protein synthesis (ribosomes).
18. Inhibition of translation results in less fertilisation.
    - Because when protein translation is inhibited, capacitation becomes impaired and fertilisation is much less likely.
19. Hydrogen bonds are important for protein synthesis.
    - Because hydrogen bonds are needed for the formation of hydride bridges resulting in twisting of protein molecules in their unusual shapes. Many proteins are used in cells as lock and key and thus without a proper shapes, the keys will not fit into the locks. Ultimately, the proteins will be useless for continuing the process.
20. Prokaryotic mRNA molecules are free to attach ribosomes as they are being synthesised.
    - Because prokaryotes lack nuclear membrane.
21. Appearance of UAA, UAG or UGA codon on mRNA prevents elongation of the polypeptide chain.
    - Because no tRNA molecules bear anticodon to pairs with nonsense codons (UAA, UAG and UGA), so appearance of such codons on mRNA prevent elongation of polypeptide chain.
22. Diptheria toxin inhibits translocation in eukaryotic cells during protein synthesis.
    - Because diptheria toxin enzyme, secreted by diptheria organism, catalyses a reaction between NAD+ and elongation factor-2 (EF-2) in eukaryotic, resulting in the formation of an inactive ADP-ribose-EF-2 complex that inhibits translocation.
23. In the absence of EF-Tu formation of peptide bond is stopped.

– Because in the absence of EF-Tu, no conformational change occurs in the ribosome.

24. The cap of mRNA is protected by the action of *phosphatases* and other nucleases.
    – Because the $m^7G^5ppp^5$ Xp present in cap has no free phosphates.
25. Formation of peptide bonds does not require energy.
    – Because amino acid brought by tRNA is already activated.

Questions based on Diagrams

1. In the diagram showing the flow of precursors into various forms of RNA, name the enzyme marked as I:
   (a) *DNA polymerase*
   (b) *DNA ligase*
   (c) *RNA polymerase*
   (d) *Topoisomerase*

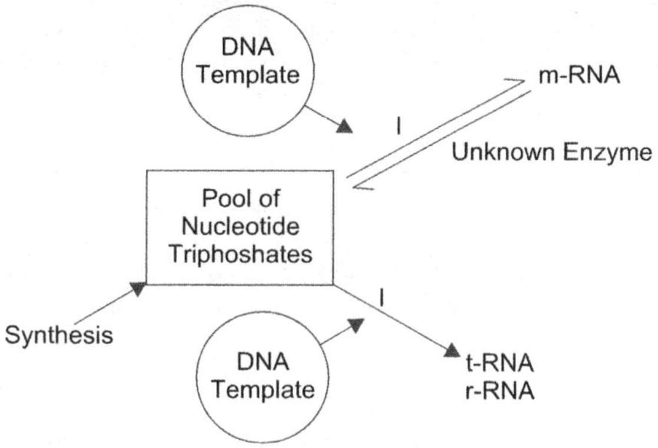

2. The diagram showing transcription of RNA upon a strand of DNA template, the name of site labelled as Q:
   (a) Operator (b) Promoter (c) Regulator (d) Homeobox

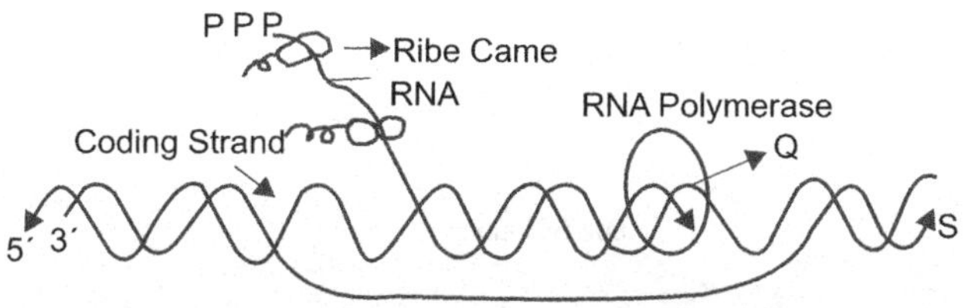

3. In the model showing structure of prokaryotic *polymerase*, which one is not included in the core enzyme and helps in the recognition of promoter site on the DNA?
   (a) α (b) β (c) β′ (d) σ

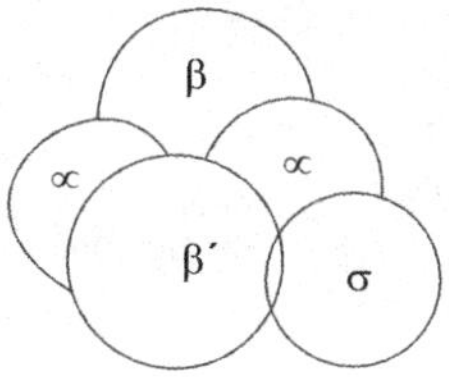

4. In the above diagram, the catalytic centre of *RNA polymerase* is formed by:
(a) α (b) β (c) β and β' (d) α, β', β and σ

5. In the diagram, the labelled sites represent:
(a) I = *peptidyl transferase* site, II = EF-Tu site, III = $L_7 - L_{12}$ site IV= EF-G site
(b) I = $L_7 - L_{12}$ site, II = EF-Tu site, III = *peptidyl transferase* site, IV = EF-G site
(c) I = EF-G site, II = EF-Tu site, III = *peptidyl transferase* site, IV = L_7–L_{12} site
(d) I = Shine–Dalgarno site, II = Conformational site, III = *peptidyl-transferase* site, IV =165r RNA/ site

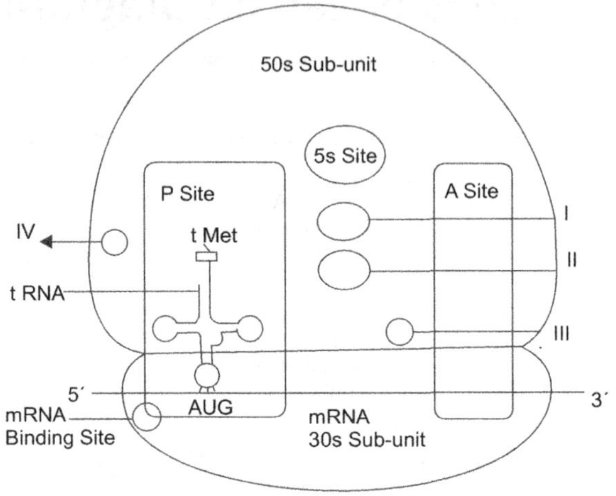

6. In the diagram showing structure of a prokaryotic ribosome, which one is an incorrect match?
(a) EF-G site – Involved in the movement of mRNA through the 30S sub-unit
(b) *Peptidyl-transferase* site – Peptide bond formation
(c) EF–Tu site – Stable site
(d) L_7–L_{12} site – Necessary for *GTPase* activity

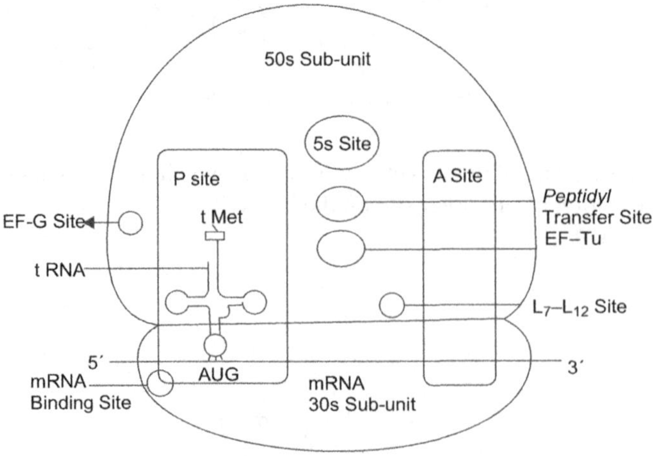

7. In the diagram shown above, the sites concerned with proper binding of *aminoacyl-tRNA* to the ribosome:
   (a) EF-G site (b) *Peptidy-transferase* site
   (c) L_7–L_{12} site (d) EF–Tu site
8. In the diagram showing a DNA segment in a eukaryote, the efficiency of transcription is determined by:
   (a) A (b) B and C (c) C and D (d) B and D

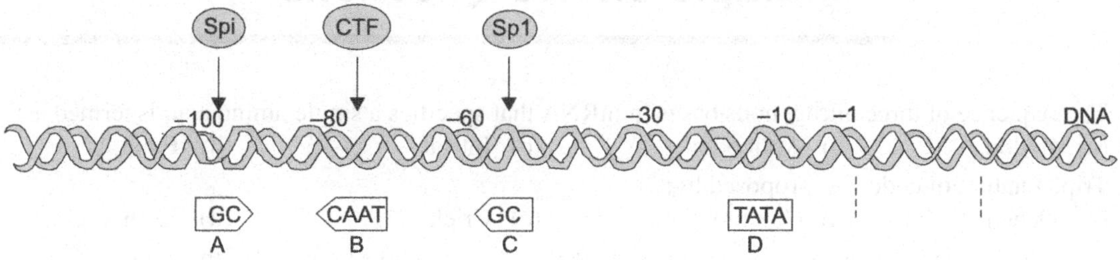

9. In the above diagram__________ aligns *RNA polymerase* at proper site:
   (a) A (b) B (c) C (d) D
10. Identify the site labelled as Z having the following features:
   (A) Located in 50S sub-unit of ribosome (B) Transfers tRNA from A site to P site
   (C) Involved in the movement of mRNA through the 30S sub-unit.
   Answer codes:
   (a) L_7–L_{12} site (b) 5S RNA (c) *Peptidyl-transferase* (d) EF-G site

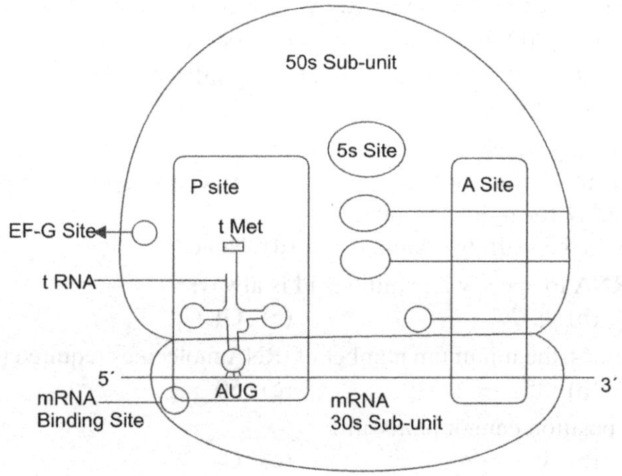

Answers to Questions based on Diagrams

1. (c) 2. (b) 3. (d) 4. (c) 5. (a) 6. (c) 7. (a) 8. (b)
9. (d) 10 (d)

GENETIC CODE

Multiple-Choice Questions

1. The sequence of three nitrogenous bases in mRNA that specifies a single amino acid is termed as:
   (a) Codon (b) Anticodon (c) Intron (d) Exon
2. Triplet nature of code was proposed by:
   (a) Ochoa (b) Gammow (c) Crick (d) Kornberg
3. The effort of deciphering genetic code was made by ______ who was awarded the Nobel Prize:
   (a) Crick (b) Nirenberg (c) Leder (d) Kornberg
4. The wobble concept was proposed by:
   (a) Nirenberg and Lederberg (1964) (b) Crick (1966)
   (c) Weiner and Weber (1973) (d) Ycas (1969)
5. The code letter lacking in mRNA is:
   (a) T (b) C (c) U (d) G
6. During the process of translation tRNA binds by the:
   (a) Ribosome (b) Anticodon (c) Codon (d) All
7. The sequence of three nitrogenous bases in mRNA is called codon, which is related with the formation of:
   (a) Carbohydrates (b) Proteins (c) Fat (d) RNA
8. The first synthetic polyribonucleotide discovered having mRNA activity:
   (a) Poly G (b) Poly C (c) Poly U (d) Poly A
9. Which one of the following statements is incorrect?
   (a) Generally, AUG is the codon for the starting amino acid N-formylmethionine. (b) AUG also specifies internal methionine.
   (c) Rarely GUG acts as the starting codon. (d) None
10. The first codon in mRNA to specify an amino acid is always:
    (a) AUG (b) GUC (c) GUG (d) AGA
11. According to wobble rules, the minimum number of tRNA molecules required to decode universal code is:
    (a) 64 (b) 35 (c) 50 (d) None
12. I (Inosine) at wobble position cannot pair with:
    (a) G (b) U (c) C (d) A
13. Which one of the following is an incorrect match?
    (a) AUG – Methionine (b) GUG –Valine
    (c) GUU – Glycine (d) UUG – Leucine
14. Suppressor genes are found for:
    (a) Frameshift mutations (b) Nonsense codons
    (c) Missense codons and nonsense codons (d) All

15. Which one of the following amino acids is coded by more than six codons?
    (a) Arginine, leucine and valine
    (b) Leucine, arginine and serine
    (c) Isoleucine, serine and tryptophan
    (d) Threonine, valine and serine
16. AUU, AUA and AUC encode amino acid:
    (a) Isoleucine (b) Serine (c) Valine (d) Tryptophan
17. The amino acids specified by single codons are:
    (a) Methionine and phenylalanine
    (b) Tryptophan and serine
    (c) Leucine and serine
    (d) Methionine and tryptophan
18. Which one of the following is an incorrect match?
    (a) AUG – Initiation codon
    (b) UAG – Amber
    (c) UGA – Opal
    (d) UAA – Cysteine
19. The first elucidation of codon was done by:
    (a) Crick et al. (1961)
    (b) Khorana, Holley and Nirenberg (1968)
    (c) Nirenberg and Matthaei (1961)
    (d) Ohno (1967)
20. Which one of the following determines the length of the coding portion of a gene?
    (a) Type of amino acids
    (b) Number of amino acids
    (c) Arrangement of amino acids
    (d) All
21. tRNA is unable to recognise codon:
    (a) UAA (b) UAG (c) UGA (d) All
22. GGG is the code for amino acid:
    (a) Lysine (b) Leucine (c) Glutamic acid (d) Glycine
23. What is correct about silent mutation?
    (a) Changes in nucleotide and amino acid
    (b) Changes in nucleotide but not in amino acid
    (c) Formation of short polypeptide chain
    (d) Changes in one amino acid to another
24. Insertion or deletion of individual nucleotide is applicable to:
    (a) Silent mutation
    (b) Missense mutation
    (c) Nonsense mutation
    (d) Frameshift mutation
25. Which one is never found in wobble position?
    (a) Adenine
    (b) Guanine
    (c) Cytosine
    (d) Uracil
26. Which one of the following statements is incorrect with reference to genetic code?
    (a) Genetic code is degenerate.
    (b) Genetic code is universal.
    (c) Genetic code is overlapping.
    (d) None
27. Which one of the following is an incorrect match with reference the wobble concept?

Bases in anticodon	Base in codon
(a) U	1. A or G
(b) A	2. U
(c) G	3. U or C
(d) I	4. G, U or C

28. Which one of the following works as a stop signal in mitochondria?
    (a) UAG (b) UAA (c) GUC (d) AGG/AGA
29. If there are 60 nucleotides in the gene of transcription and translation, the protein formed would contain:
    (a) 180 (b) 120 (c) 60 (d) 20

30. Consider the following statements:
    (A) Generally, GUG is a codon for valine and is recognised by $tRNA_f^{met}$
    (B) Nonsense mutation changes a codon into a termination codon
    (C) Suppressor genes act by changing the nucleotide sequence of a mutant gene
    (D) During elongation of polypeptide chain, Met-$tRNA_f^{met}$ binds to elongation factor EF–TU
    The correct statements are:
    (a) All (b) B and C (c) A, B and D (d) A, B and D
31. The hypothesis related with the economy of genetic code:
    (a) Wobble hypothesis (b) Null hypothesis
    (c) Gammow hypothesis (d) Wolf hypothesis
32. Diamond code is related with:
    (a) Kornberg (b) Crick (c) Nirenberg (d) Gammow
33. The number of termination codons in mitochondria is:
    (a) 3 (b) 4 (c) 5 (d) 6
34. Ambiguous codon is applicable to:
    (a) GGC (b) GUC (c) GUG (d) AUG
35. How many nitrogenous bases are needed in mRNA to synthesise 30 amino acids?
    (a) 30 (b) 60 (c) 90 (d) 120
36. UGA is the solenocysteine codon in:
    (a) Mammals (b) Fungi and *E. coli* (c) Higher plants (d) All
37. Which of the following is a wobble base pair?
    (a) I–U (b) I–A and I–C (c) G–U (d) All
38. What is correct about G–U wobble base pair?
    (a) Unique chemical properties (b) Unique structural property
    (c) Unique dynamic and ligand-binding properties (d) All
39. UAA and UAG specify glutamine in:
    (a) *E. coli* (b) *Neurospora*
    (c) *Mycoplasma capricolum* (d) *Aspergillus*
40. Wobble hypothesis states that nonstandard base pairing might occur between the nucleotide base in the:
    (a) 5′ position of the anticodon and 3′ position of the codon.
    (b) 3′ position of the anticodon and 5′ position of the codon.
    (c) Both (a) and (b)
    (d) Only 5' position of the anticodon.
41. The genetic code is non-ambiguous, it means:
    (a) All organisms have the same genetic code (b) The same codon may have different meanings
    (c) A particular codon always codes for the same amino acid (d) A nitrogenous base in mRNA is not used for two codons
42. Wobble hypothesis is applicable to:
    (a) Fine structure of gene (b) Codon anticodon matching
    (c) Codon-amino acid matching (d) Evolution of genetic code
43. Synonym codon generally differs in the:
    (a) First base only (b) Second base only
    (c) Third base only (d) Either first, second or third base

44. Which one of the following is not associated with the triplet nature of genetic code?
(a) Anticodons (b) Frameshift mutations
(c) Universal nature of the genetic code (d) RNA homopolymers

45. Multiple codons specifying an amino acid differ by:
(a) Only one base at the third base of codon (b) Only one base at the second base of codon
(c) Only one base at the first base of codon (d) All

46. Which one of the following statements is incorrect?
(a) T H Jukes proposed concept of archetypal code.
(b) The codon UGA specifies amino acid cysteine in *Euplotes octocarinatus.*
(c) Crick based his wobble hypothesis on the chemical structure of tRNA discovered by Robert Holley.
(d) Isoleucine is coded by four codons.

47. The codon recognised by anticodon IGC:
(a) GCA (b) GCU (c) GCC (d) All

48. In which one of the following does codon AUA not code for methionine in the mitochondria?
(a) *Xenopus* (b) *Neurospora* (c) *Drosophila* (d) Mammals

49. The first triplet codons deciphered by Khorana:
(a) Methionine and valine (b) Cysteine and valine
(c) Proline and valine (d) Leucine and valine

50. Which one of the following codons has a similar meaning to that of UGA codon of mitochondria?
(a) UUU (b) UAC (c) UGG (d) UAA

51. What is true about nuclear genetic code and mitochondrial genetic code?
(a) Primitive in nature (b) Similar
(c) Different (d) Evolved in different direction

52. The coding segment stretch of a eukaryotic DNA is known as:
(a) Codons (b) Anticodons (c) Exons (d) Introns

53. An eukaryotic protein contains 105 amino acids, what is the number of nucleotides in the DNA that codes for these amino acids in the protein?
(a) 95 (b) 105 (c) 315 (d) 420

54. The genetic code is transferred from DNA to:
(a) Amino acids (b) mRNA (c) tRNA (d) rRNA

55. The first amino acid in prokaryotic protein is:
(a) Leucine (b) Methionine (c) F-methionine (d) F-serine

56. Most synonym codons differ in position at:
(a) First position (b) Second position (c) Third position (d) Both (b) and (c)

57. Which one of the following statements is incorrect?
(a) Codons rich in UC encode hydrophobic amino acids.
(b) Codon rich in AG encode hydrophilic amino acids.
(c) The codons are assigned randomly.
(d) One of the codons may act as a start signal as well as perform coding of amino acid.

58. There are 1,600 nucleotides in a transcriptional unit, which uses 1,500 nucleotides to make a protein containing 500 amino acids. The best possible explanation of this is:
(a) Due to degeneracy of the code (b) Due to nonsense suppression
(c) It contains many noncoding nucleotides (d) Due to misreading of the frame

59. All 64 possible genetic codons were identified by the end of:
(a) 1958 (b) 1961 (c) 1966 (d) 1975

60. If the number of bases is three, then the number of triplet codons will be:
(a) 5 (b) 15 (c) 64 (d) None

61. 900 nucleotides constitute_______ codons, which code for _________ amino acids.
(a) 90 each (b) 900 each (c) 300 each (d) 300 and 20

62. In *Oenothera*, UGA is a chain termination codon while CGG encodes:
(a) Valine (b) Tryptophan (c) Serine (d) Cysteine

63. If the coding sequence in DNA is CAG, the corresponding codon in mRNA transcribed is:
(a) UCA (b) AAU (c) UUA (d) CGT

64. Which one of the following statements is correct about codons?
(a) A codon is the basic unit of genetic code. (b) A codon consists of two letters.
(c) A codon may code for the same amino acid like other codons. (d) None

65. A hypothetical protein contains only 10 types of amino acids. The smallest possible size of codon having four different bases is:
(a) 20 (b) 30 (c) 2 (d) 1

66. Degeneracy in the genetic code was reported by:
(a) Nirenberg and Ochoa (b) Bernfield and Nirenberg
(c) Nirenberg and Matthaei (d) Crick

67. Which one of the following is an ambiguous codon?
(a) UUU (b) GUG (c) GUC (d) AUG

68. How many nitrogenous bases will be needed to synthesise mRNA codons for 303 amino acids?
(a) 101 (b) 303 (c) 909 (d) 412

69. Pairing between codons and anticodons involves:
(a) Hydrogen bonds (b) Hydrophobic bonds
(c) Ionic bonds (d) Phosphodiester bonds

70. Amino acids proline, glycine, valine and threonine are coded by four codons (each). It shows that genetic code is:
(a) Overlapping (b) Degenerate (c) Ambiguous (d) Universal

71. A gene contains 1,200 nucleotides. What will be the number of amino acids synthesised by this gene and how many tRNA molecules will this gene require for the synthesis of a polypeptide chain?
(a) 20 tRNA and 1,200 amino acids (b) 400 tRNA and 600 amino acids
(c) 20 tRNA and 400 amino acids (d) 400 each

72. An organism was discovered having genetic code similar to organisms on the earth. However, it contains six types of nitrogenous bases instead of four in its DNA and the base sequences are translated as doublets. This type genetic code may code for maximum amino acids:
(a) 64 (b) 12 (c) 36 (d) 96

73. The first experimental proof regarding triplet nature of codon was given by:
(a) Gammow (1954) (b) Crick ad Berner (1961)
(c) Nirenberg and Leder (1964) (d) Crick (1966)

74. By genetic study, it is not known that genetic code is :
(a) Triplet (b) Commaless (c) Overlapping (d) Universal

75. Who first pointed out on the basis of his experiment that an overlapping codon is out of question?
(a) S Brenner (1957) (b) Crick and Berner (1961)
(c) Crick (1966) (d) Yanofsky (1966)

76. Which one of the following statements is incorrect?
(a) The genetic code is the raw material needed by scientists trying to develop a diagnostic test.
(b) Scientists are unable to crack the genetic code of SARS virus.
(c) Yanofsky et al. (1966) pointed out co-linearity between a gene and its polypeptide.
(d) U at 5′ base of anticodon pairs with either A or G of at the 3′ base of codon.

77. Codon AGA codes for arginine in animals and in plants it codes for:
(a) Leucine (b) Serine (c) Arginine (d) Glutamie

78. Triplet code concept was provided by Crick et al. (1961) on their experiments based on:
(a) ϕX174 (b) T_4 bacteriophage (c) SV_{40} (d) *E. coli*

79. Co-linearity of the code has been demonstrated by the studies on:
(a) T_4 mutants (b) HIV (c) TMV (d) *E. coli*

80. Which of the following affect ambiguity of the code?
(a) Change in pH of the system (b) High magnesium ion concentration
(c) Both (a) and (b) (d) Neither (a) nor (b)

81. Missense mutation results in:
(a) Chain termination
(b) Replacement of one amino acid by another amino acid
(c) Change in nitrogenous base but not in the amino acid
(d) Formation of short polypeptide chain

82. Which one of the following is applicable to a biological reading frame?
(a) Non-overlapping (b) Contiguous
(c) Set of three nucleotide codon in DNA or RNA (d) All

83. Nirenberg and Matthaei used __________as their experimental template:
(a) TMV RNA (b) ϕX174 DNA (c) SV40 DNA (d) *E. coli* DNA

84. In general, no codon specifies more than one amino acid with the exception of:
(a) AUG (b) UGA (c) UAG (d) All

85. Which one of the following is responsible for the development of silent mutation?
(a) Degeneracy of the genetic code (b) Non-ambiguous nature of the genetic code
(c) Repair defect (d) Evolution

86. The codon consists of ______________ bases and is a part of:
(a) One, mRNA (b) Three, rRNA (c) Three, tRNA (d) Three, mRNA

87. The tripeptide sequence of arginine, leucine and valine are CGU, UUA and GUC. The sequence of bases in DNA is:
(a) GCGGGTCAG (b) GCAAATCAG (c) GCATCGAGA (d) GCTTTTCAG

88. UAG and UAA code for glutamine in :
(a) *E. coli* (b) *Pramecium*
(c) *Neurospora* (d) *Drosophila* mitochondria

89. The event, which occurs during codon–anticodon pairing:
(a) tRNA is degraded (b) Amino acids form protein
(c) rRNA becomes active (d) Ribosomal sub-units unite

90. The frameshift and nonsense mutations result in the formation of:
(a) Abnormal protein (b) Generally inactive proteins
(c) Shorter proteins (d) All

91. Sickle cell anaemia is an example of:
(a) Missense mutation (b) Nonsense mutation
(c) Silent mutation (d) Frameshift mutation

92. Consider the following statements:
(A) Vernon Ingram (1957) gave the first proof that proteins are coded by genes
(B) According to frozen accident theory, all living beings have descended from a common ancestor
(C) In bacteria, conversion of nonsense codon into a sense codon by tRNA suppressors has been reported
(D) There is an extra base in frameshift suppressor tRNAs
The correct statements are:
(a) All (b) A, B and C (c) A and D (d) C and D

93. Several overlapping genes in different reading frame have been reported in:
(a) Hepatitis B virus (b) Barley yellow dwarf virus
(c) Both (a) and (b) (d) *Neurospora* and SV_{40}

94. Change of codon from GAA to GUA is a characteristic of:
(a) Fanconi anaemia (b) Sickle cell anaemia
(c) Hartnup disease (d) Neiman Pack's disease

95. Nirenberg and Matthaei (1961) were unable to get any result when they were used __________ as synthetic mRNA:
(a) Poly A (b) Poly C (c) Poly U (d) Poly G

96. Replacement of glutamic acid by valine (change of GAA to GUA in haemoglobin, which results in sickle cell anaemia, was first of all reported by:
(a) Tasaki (1953) (b) Veron Ingram (1957) (c) Loewenstein (1960) (d) Eccles (1964)

97. For which one of the following mutations is there no suppressor gene?
(a) Nonsense mutation (b) Frameshift mutation
(c) Missense mutation (d) None of these

98. In mammalian mitochondrial genetic code, the initiating methionine is coded by:
(a) AUG (b) AUU and AUC (c) AUG and AUA (d) All

99. Which one of the following is not applicable to genetic code?
(a) Degeneracy (b) Overlapping (c) Three bases in codon (d) Ambiguity

100. The nucleotide sequence on the DNA template strand that yields the CUA codon for the leucine:
(a) GAT (b) GTT (c) TAG (d) UTG

101. The anticodon UCG of serine tRNA recognises codon:
(a) AGC (b) AGU (c) Both (a) and (b) (d) None

102. In UCGAGU pairing, hydrogen bonding occurs between (according to wobble):
(a) U and C (b) A and G (c) C and G (d) G and U

103. mRNA has following base sequence 5′ AUG GCA AUU AAA UUC UGA 3′. The polypeptide coded by this mRNA is:
(a) Met–Ala–Ile–Lys–Phe (b) Met–Ala–Lys–Ile–Phe
(c) Met–Ileu–Phe–Ala–Lys (d) Met–Phe–Ala–Lys–Ile–val

104. In all nonsense codons, the first base is:
(a) Guanine (b) Uracil (c) Adenine (d) Cytosine

105. What is correct about codon UAA, UAG and UGA?
(a) They specify unspecific amino acids.
(b) Assist in releasing mRNA from ribosome.
(c) Do not code any amino acid.
(d) Recognised by releasing factors and do not specify any amino acid.

106. In some instances, GUG acts as an initiation codon specifying methionine but generally, it codes for:
(a) Leucine (b) Valine (c) Isoleucine (d) Threonine

107. Khorana is associated with deciphering of amino acids:
(a) Cysteine and valine
(b) Cysteine and alanine
(c) Valine and serine
(d) Leucine and isoleucine

108. If a poly U is used as template with sensitive ribosome, the error, which occurs most frequently, is the replacement of amino acid phenylalanine (UUU) by:
(a) Serine (UCU) (b) Leucine (CUG) (c) Isoleucine (AUU) (d) Valine (GUU)

109. Which one of the following is an ambiguous codon?
(a) GUA (b) UUU (c) GUC (d) None

110. A codon, codes for:
(a) RNA (b) Amino acids (c) Fatty acids (d) All

111. Consider the following statements:
(A) UGA is cysteine codon in *Euplotes octocarinatus*
(B) It is an inefficiently read tryptophan codon in *Bacillus subtilis*
(C) In mitochondria, *Mycoplasma* and *Sprioplasma*, UGA is the codon for tryptohan
(D) In evolution, it has served more function than the other code

The correct statements are:
(a) All (b) B and C (c) B, C and D (d) B and D

112. The genetic fault in Huntington's disease is due to:
(a) An extra copy of chromosome 5
(b) A gene having triplet repeats of the bases CAG on chromosome 4
(c) A genetic triplet repeats of the bases CAG on chromosome 7
(d) A genetic triplet repeats of the bases CAG on chromosome 9

113. Which one of the following is an incorrect match?

Amino acids	Number of codes
(a) Methionine	1
(b) Isoleucine	3
(c) Glycine	6
(d) Valine	4

Answers to Multiple-Choice Questions

1. (a)	2. (b)	3. (b)	4. (b)	5. (a)	6. (c)	7. (b)	8. (c)
9. (d)	10. (a)	11. (d)	12. (a)	13. (c)	14. (d)	15. (b)	16. (a)
17. (d)	18. (d)	19. (c)	20. (b)	21. (d)	22. (d)	23. (b)	24. (d)
25. (a)	26. (c)	27. (d)	28. (d)	29. (d)	30. (d)	31. (a)	32. (d)
33. (b)	34. (c)	35. (c)	36. (d)	37. (d)	38. (d)	39. (c)	40. (a)

41. (c)	42. (b)	43. (c)	44. (c)	45. (a)	46. (d)	47. (d)	48. (b)
49. (b)	50. (c)	51. (c)	52. (c)	53. (c)	54. (b)	55. (c)	56. (c)
57. (c)	58. (c)	59. (c)	60. (d)	61. (c)	62. (b)	63. (c)	64. (c)
65. (c)	66. (b)	67. (b)	68. (c)	69. (a)	70. (b)	71. (d)	72. (c)
73. (b)	74. (c)	75. (a)	76. (b)	77. (c)	78. (b)	79. (a)	80. (c)
81 (b)	82. (d)	83. (a)	84. (d)	85. (a)	86. (d)	87. (b)	88. (b)
89. (b)	90. (d)	91. (a)	92. (a)	93. (c)	94. (b)	95. (d)	96. (b)
97. (d)	98. (d)	99. (b)	100. (c)	101. (c)	102. (d)	103. (a)	104. (b)
105. (d)	106. (b)	107. (a)	108. (c)	109. (d)	110. (b)	111. (a)	112. (b)
113. (c)							

Fill in the Blanks

1. A codon is read in the ____________ direction.
2. ______________ coined the term 'triplet code' and 'genetic code'.
3. ________ and ________ are the initiation codons for protein synthesis.
4. AUG initiation codon occurs on __________ end of mRNA.
5. Triplet code is the sequence of ____________ nitrogenous bases in mRNA.
6. The sequence of three nitrogenous bases in tRNA, which binds to a codon of mRNA, is called ____________.
7. There are ___________ punctuation codons in the universal genetic code.
8. Linney et al. (1972) reported overlapping genes in ________.
9. Inosinic acid (I) is a nucleotide that uses ________________ as a base.
10. There are ________ codons for valine.
11. ________________ is the only amino acid having three codons.
12. The three termination codons of the universal genetic code are_____, ______ and _____.
13. In the universal genetic code, UGA is one of the terminator codons, while in mitochondrial genetic code, it codes for the amino acid _____________.
14. A genetic code having more than one codon for the same amino acid is said to be ____________.
15. Yeast tRNA alanine having anticodon GCI can pair with alanine codons _________, _________ and ___________.
16. The anticodon is complementary to ____________.
17. The first base in the anticodon pairs with the _________ base in the codon sequence.
18. According to wobble hypothesis, the base at ________ end of the anticodon is not spatially confined as the first two.
19. In the universal genetic code, UGG specifies amino acid _______ but in mitochondria it acts as a___________.
20. According to wobble rule, a tRNA can only recognise three codons only when ___________ occupies the third position in the anticodon.

21. All amino acids have more than one codon expect ________ and _________________.
22. A nonsense mutation produces _________________ polypeptide chain.
23. The genetic code consists of 64 codons, of which ____ represents terminating codons and rest codes for ________ amino acids.
24. The relationship between DNA, RNA and sequence of amino acids in a polypeptide chain is known as the_________________ .
25. Codons specifying the same amino acids are called _____________ codons.
26. In human mitochondrial genetic code, AUU codes for ________ during initiation but can code for ___________.
27. In human mitochondrial genetic code, AUA codes for __________ instead of _________.
28. ______________ amino acid (Se(c) is the only addition of new amino acid to the genetic code was deciphered since 1960.
29. Adaptor hypothesis was proposed by _________.
30. According to the adaptor hypothesis, adaptor molecules intervene between nucleic acid and _______________ during transaction.
31. Number of codons is six for amino acids ____________, __________ and ___________.
32. GUG has been found to initiate protein synthesis when AUG codon is lost by________.
33. Codon UUU and UUC codes for same amino acid _________________________.
34. __________ and ___________ are common start codons in bacteria and archaea.
35. Animal mitochondria use _________ for methionine and not for isoleucine.

Answers to Fill in the Blanks

1. 5′⟶ 3′	2. Gammow	3. AUG, GUG
4. 5′	5. Three	6. Anticodon
7. Three	8. ϕX174	9. Hypoxanthine
10. Four	11. Isoleucine	12. UAA, UAG and UGA
13. Tryptophan	14. Degenerate	15. GCU, GCC, GCA
16. Codon	17. Third	18. 5′
19. Tryptophan, terminating codon	20. Inosine	21. Methionine, tryptophan
22. Incomplete	23. 3, 20	24. Genetic code
25. Synonym	26. Isoleucine, methionine	27. Methionine, isoleucine
28. Selenium-containing	29. Crick	30. Amino acids
31. Arginine, leucine, serine	32. Deletion	33. Phelylalanine
34. GUG, UUG	35. AUA	

True or False

1. All genetic information is stored as genetic code.
2. Nirenberg et al. were able to determine the sequence of 54 out of 64 codons.

3. The initial nucleotide from which translation starts is known as a codon.
4. A codon alone is sufficient to start the process of translation.
5. The mutation that changes the third base of the codon, generally goes unnoticed.
6. The third base in the anticodon of tRNA may pair with more than one base.
7. Overlapping codons have been reported in many viruses.
8. In fungi, there are four initiating codons.
9. There are some mutant tRNAs having the ability of recognising termination codons.
10. AUA codes for methionine in *Neurospora.*
11. The first code to be deciphered was UUU.
12. UGA acts both as a termination codon and a solenocysteine codon.
13. The mutation that codes for the same amino acid, is known as nonsense mutation.
14. Marshall Nirenberg is best known for breaking the genetic code.
15. Only the third position of some codons is a four-fold degenerate site.
16. Degeneracy of codons minimises the deleterious effect of DNA mutations.
17. Most synonym codons differ in the base at the first position.
18. A single tRNA is able to recognise two or more codons, differing only in the third base.
19. The genetic code is ambiguous.
20. When GUG is used as initiating codon, it codes for methionine.
21. The codon composition of proline is 3C.
22. A suppressor gene acts by changing the nucleotide sequence of a mutant gene.
23. Mutant tRNAs can cause suppression of nonsense as well as missense mutations.
24. Mutation in 30S ribosomal proteins does not affect the accuracy of the reading frame.
25, Intergenic suppressor mutations result in increased frequency of mistakes in reading the genetic code.
26. The standard code currently allows initiation from UUG and CUG in addition to AUG.
27. Streptomycin promotes mistakes in the translation of the genetic code.
28. The genetic code appears to have undergone several changes during the course of long evolutionary period.
29. In an organism having A–T-rich DNA, the most codons have U or A at the third position.
30. Wobble hypothesis provides economy to tRNA.
31. Genetic code was first deciphered by Khorana.
32. The second genetic code is the language of transfer of instructions on the tRNA that specifies which amino acid will be attracted.
33. It has been observed that initiation by of protein synthesis by GUG is less efficient as it has lower affinity for F-met-tRNA.
34. In eukaryotes, the three termination codons are recognised by three release factors.
35. The codon AUG is written as 5′ AUG 3″ and its corresponding anticodon on tRNA as 5′ UAC3′.
36. Amino acids having similar physical properties tend to have similar codons.
37. Violations of the universal code are frequent in nuclear genes.

Answers to True or False

1. False 2. True 3. True 4. False 5. True 6. True 7. True 8. False

9.	True	10.	False	11.	True	12	True	13.	False	14.	True	15.	True	16.	True
17.	False	18.	True	19.	False	20.	True	21.	True	22.	False	23.	True	24.	False
25.	True	26.	True	27.	True	28.	False	29.	True	30.	True	31.	False	32.	True
33.	True	34.	False	35.	False	36.	True	37.	False						

Give Reasons

1. There are 64 codons for 20 different amino acids.
   – Because each codon consists of three nitrogenous bases and there are four different types of nitrogenous bases in DNA and RNA. Thus, four nitrogenous bases combined in three groups 4 × 4 × 4 producing 64 codons.
2. The genetic code is degenerate.
   – Because there are many codons which specify the same amino acid.
3. An mRNA molecule was made by a random combination of adenine, guanine and uracil and placed in a test tube containing all sorts of requirement for protein synthesis. The polypeptide chain thus formed was not so long.
   – Because random combination of adenine, guanine and uracil leads to the formation of termination codon (UAA, UAG or UG(A) in the mRNA and as such when one of these codons appear on the mRNA, they result into the termination of the polypeptide chain. Thus, the polypeptide chain produced is not so long.
4. Sometimes a single tRNA can recognise more than one codon.
   – Because of wobble base at the 5′ end of the anticodon.
5. Silent mutation is unnoticeable.
   – Because it does not cause any change in the sequence of amino acids in a protein.
6. Nonsense and frameshift mutations are generally noticed by organisms.
   – Because they result in shorter or abnormal proteins which are biologically inactive.
7. Missense mutations have a biological effect.
   – Because they change the meaning of a codon changing one amino acid into another.
8. Mitochondrial genetic code shows divergence from the original genetic code.
   – Probably, it is so because mitochondria synthesise only a very limited number of proteins.
9. GUG is an ambiguous codon.
   – Because if it is present at an initiating position, it codes for methionine, otherwise it always codes for valine.
10. While deciphering genetic code using poly A, poly U, poly C and poly G, Nirenberg and Matthaei, could not find any successful result with poly-G mRNA.
    – Because poly G attains secondary structure in solution and is unable to function as a synthetic mRNA. Thus cannot attach to ribosome.
11. All codons are utilised by the organism.
    – Probably, it reduces the effect of harmful mutations.
12. Substitution of a single base due to point mutation does not destroy the biological activity of the resultant protein.

- Because only one amino acid of the polypeptide chain is altered.

13. Genetic code is almost universal.
    - Because one codon always codes the same type of amino acids (apart from a few exceptions) in all types of organisms.
14. There are three termination codons.
    - The existence of more than one termination codon might be a safety measure, so that if one codon fails to function the next will become functional.
15. ϕX174 bacteriophage synthesises 10 proteins, though it contains only 5,375 nucleotides.
    - Because it has been found that in ϕX174 bacteriophage, sometimes genes can overlap. In fact, 600 nucleotides will be needed to synthesise 10 proteins, if overlapping did not exist.

Questions based on Diagrams

1. In the diagram showing an mRNA molecule, involved in protein synthesis, identify the codons labelled I and II.

(a) CCA and UGA or UUU
(b) AUG and UAG or UGA
(c) AUG and UAA or UAG or UGA
(d) GUG and UAA or UAC or UAG

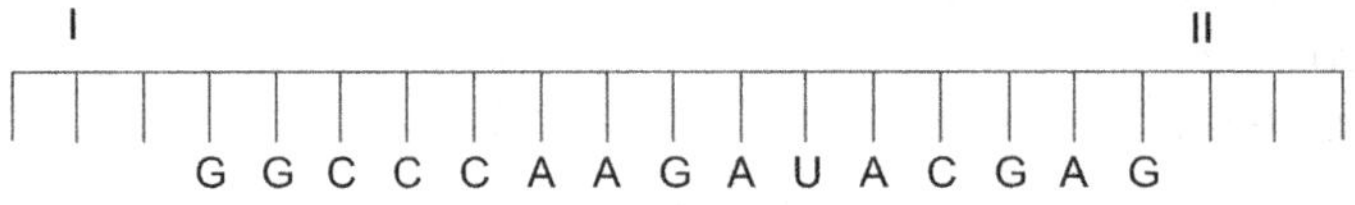

2. According to wobble pairing, which one of the following is an incorrect codon for the anticodon shown in the diagram?

(a) I
(b) II
(c) II
(d) All

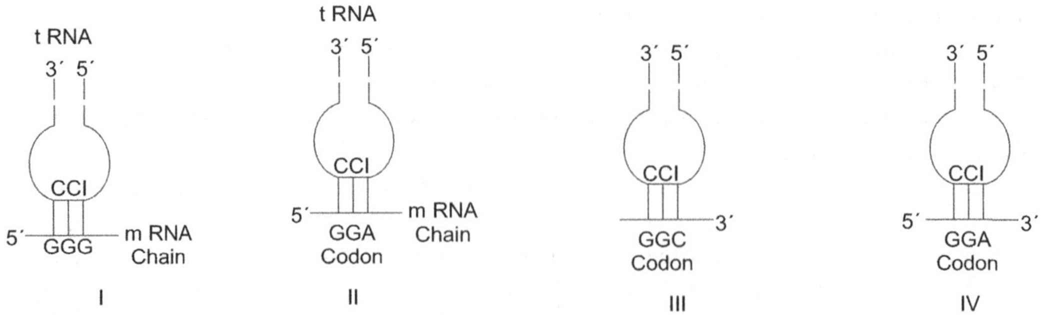

3. In the diagram given below, what will be the correct sequence of amino acids, if the mRNA undergoes the process of protein synthesis?

(a) Arginine, proline, lysine, glycine and proline
(b) Threonine, lysine, histidine, arginine and arginine

(c) Histidine, tyrosine, serine, glycine and arginine
(d) Arginine, arginine, glycine, glycine and glycine

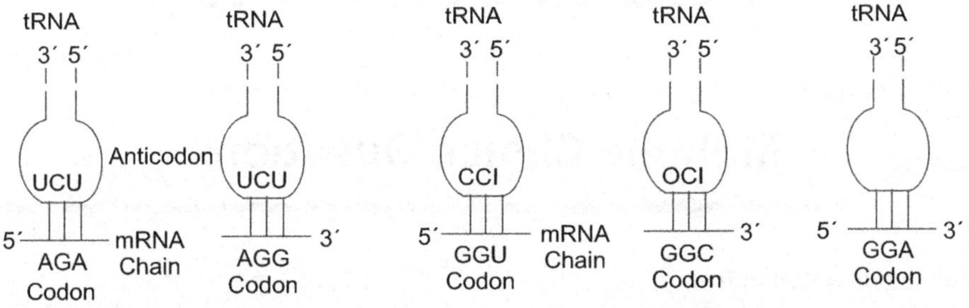

4. In the given diagram showing protein synthesis, identify the anticodon and accordingly the amino acid being added in the polypeptide chain:
(a) UCA and serine
(b) UCA and Valine
(c) UCE and Glycine
(d) UFC and leucine

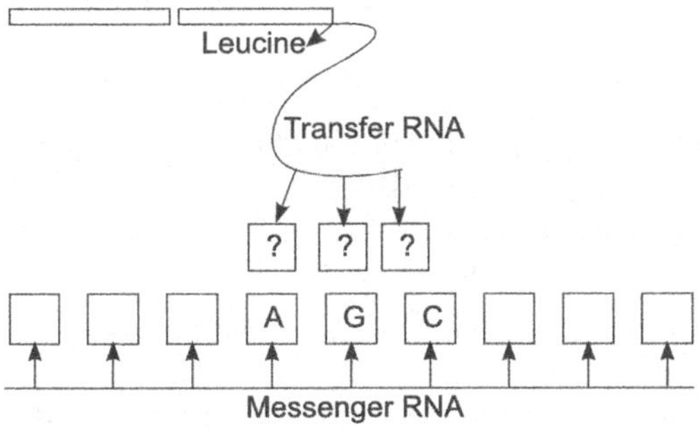

Answers to Question based on Diagrams

1. (a) 2. (a) 3. (d) 4. (a)

GENE REGULATION

Multiple-Choice Questions

1. Gene regulation is essential for:
(a) Prokaryotes (b) Eukaryotes (c) Viruses (d) All
2. Gene regulation drives the process of:
(a) Cellular differentiation (b) Morphogenesis
(c) Both (a) and (b) (d) None
3. Gene regulation is regulated at the stage of:
(a) Transcription and translation (b) RNA transport and mRNA degradation
(c) Chromatin domains (d) All
4. Operons:
(a) Are mainly found in prokaryotes (b) Are found in some eukaryotes
(c) Are found in nematodes (d) All
5. The first operon to be described was the lac operon in:
(a) *Neurospora* (b) *E. coli*
(c) Yeast (d) *Drosophila melanogaster*
6. The first gene regulation to be discovered was:
(a) Lac operon (b) Tryptophan operon (c) L-arabinose operon (d) Histidine operon
7. Which one of the following is a product of the regulator gene?
(a) Inducer (b) Repressor (c) Activator protein (d) None
8. The number of structural genes in lac operon is:
(a) One (b) Two (c) Three (d) Four
9. *Lac Y* codes for:
(a) β-galactosidase (b) β-galactosidase permease
(c) Thiogalactoside transasetylase (d) None
10. Operon model of gene expression in bacteria was proposed by:
(a) Beadle and Tatum (b) Jacob and Monod (c) Nirenberg (d) Watson and Crick
11. Consider the following statements:
(a) An operon is a cluster of genes that is transcribed as a single mRNA
(b) An operon model comprises two classes of genes called structural genes and regulator genes
(c) Expression of structural genes is regulated by regulatory proteins
(d) In the absence of a repressor, RNA polymerase can bind to promoter and initiate transcription
The correct statements are:
(a) All (b) A, B and C (c) B and C (d) C and D
12. Which one of the following appears to be necessary for lactose catabolism?
(a) *Lac Z* (b) *Lac Z and lac Y* (c) *Lac Z and lac A* (d) *Lac Y and lac A*

13. RNA polymerase recognises only if __________ is already bound to it:
(a) CAP (b) cAMP (c) CAP–cAMP complex (d) cAMP level is high
14. The inducible operon that requires positive control mechanism for expression is:
(a) Galactose (b) Maltose and lactose (c) Arabinose (d) All
15. Which one of the following enzyme is not involved in the utilisation of lactose?
(a) Tranacetylase (b) Transformylase (c) β-galactosidase (d) Lacpermease
16. The trytophan operon codes for __________ enzymes:
(a) 7 (b) 5 (c) 3 (d) 2
17 Match column I with column II and select the correct answer using answer codes:

Column A	Column B
A. *Gene lac I*	1. β-galactose transacetylase
B. *Gene lac Z*	2. β-galactose permease
C. *Gene lac Y*	3. Repressor
D. *Gene lac A*	4. β-galactosidase

Answer codes:

	A	B	C	D
(a)	4	1	3	1
(b)	3	4	1	2
(c)	2	3	4	1
(d)	3	4	2	1

18. The inducer binds with:
(a) Repressor and prevents it from binding with the promoter (b) Repressor and prevents it from binding with the operator
(c) Operator (d) Structural genes
19. The operon works in cooperation with:
(a) A regulator gene (b) A promoter gene
(c) A repressor protein and an inducer substance (d) All
20. The stretch of DNA that comprises operator, structural genes and functions as a regulating unit in transcription is called:
(a) Promoter (b) TATA box (c) Operon (d) Regulatory unit
21. Transcription of the *lac* genes is influenced by:
(a) The presence of inducer (b) Absence of inducer
(c) Both (a) and (b) (d) None
22. Consider the following statements:
(A) Catabolite repression is found in those operons, which are involved in the degradation of compounds used as a source of energy
(B) In bacteria, there is an inverse relationship between glucose levels and cyclic AMP
(C) The activities of bacterial enzymes cannot be regulated by epigenetic modifications
(D) Tryptophan suppresses the expression of the *trp* genes
The incorrect statements are:
(a) A and B (b) B and C (c) C and D (d) D
23. Which one of the following acts as a positive regulator?
(a) Sigma factor (b) N-gene products of lambda phage
(c) CAP-cAMP (d) All

24. Repressor is a:
    (a) Polypeptide (b) Disaccharide (c) Phospholipid (d) Phosphoprotein
25. Consider the following statements:
    (A) CAP protein acts both as activator and repressor
    (B) One repressor always controls one promoter
    (C) *Lex A* repressor of *E. coli* controls one operon
    (D) The rate of synthesis of many proteins is not controlled
    The incorrect statements are:
    (a) A and B (b) B and C (c) C and D (d) A and D
26. Which one of the following conditions induces synthesis of cAMP?
    (a) Absence of carbon source (b) Amino acid starvation
    (c) Folate starvation (d) None
27. The most direct way to control expression of a gene is to regulate:
    (a) The rate of transcription (b) The rate of translation
    (c) Both (a) and (b) (d) Activation and inactivation of repressor
28. In eukaryotes, genes are not linked in operons except:
    (a) *D. melanogaster* (b) *C. elegans* (c) Yeast (d) *Trypanosoma cruzi*
29. A repressor binds anywhere in the DNA using:
    (a) Hydrogen bonds (b) Ionic interactions
    (c) Both (a) and (b) (d) Hydrophobic interaction and hydrogen bonds
30. Primary transcription in eukaryotes contains the transcript of a single gene except for:
    (a) Nematodes (b) Certain molluscs (c) Helminthes (d) Insects
31. The first repressible operon to be discovered was:
    (a) *Trp* operon in *E. coli* (b) *Lac* operon in *E. coli*
    (c) His operon in *E. coli* (d) *Ara* operon in *E. coli*
32. The *lac* operon is regulated by:
    (a) Availability of lactose (b) Availability of glucose
    (c) Many factors (d) All
33. Which one of the following is a repressible gene?
    (a) *Lac* operon (b) *Ara* operon (c) *Trp* operon (d) None
34. Which one of the following determines whether or not genes of an operon are transcribed?
    (a) Regulator gene (b) Promoter gene (c) Operator gene (d) Structural gene
35. Technically which one of the following is not a part of an operon?
    (a) *Lac I* gene (b) *Lac Z* gene (c) *Lac Y* gene (d) *Lac A* gene
36. Which one of the following statements is incorrect about *lac I* gene?
    (a) It is a controlled by a separate promoter. (b) It is expressed all the time or constitutively.
    (c) It encodes a protein called repressor. (d) It has no function in operon function.
37. An operator:
    (a) Is the short region of DNA (b) Lies partially within the operator
    (c) Interacts with regulatory protein (d) All
38. Which one of the following statements is correct?
    (a) A mutation in promoter gene totally prevents the attachment of mRNA polymerase to the promoter site.
    (b) A mutation in promoter gene increases the attachment of RNA polymerase to the promoter site.

(c) A mutation in promoter gene decreases the attachment of RNA polymerase to the promoter site.
(d) All

39. In which one of the following is the promoter site located entirely within the operator?
(a) Yeast (b) *Neurospora* (c) Lambda phage (d) *E. coli*

40. Gene regulation mechanism is lacking in:
(a) Viruses (b) *Neurospora* (c) Archaea (d) None

41. In eukaryotes, gene regulation occurs at:
(a) The levels of transcription (b) RNA processing and mRNA longevity
(c) The levels of translation (d) All

42. The activity of operator gene is controlled by:
(a) Structural genes (b) Promoter gene (c) Regulator gene (d) Constitutive gene

43. In which one of the following genes can mutation create the *lac* genotype in which cells cannot utilise lactose:
(a) *Lac Z* (b) *Lac Y* and *lac A* (c) *Lac A* (d) *Lac* Z and *lac Y*

44. A gene, which is continually expressed, is called:
(a) Supplementary gene (b) Constitutive gene
(c) Complementary gene (d) Operon gene

45. Which one of the following statements is incorrect?
(a) The *lacY* mutants are unable to take lactose from the medium (b) The *lac* Z^- mutants abolish enzyme activity
(c) In *lac A* cells, no defect is identifiable (d) None

46. The three structural genes of *lac* operon are organised as a unit:
(a) *LacZ–lacA–lacY* (b) *LacA–lacy–lacZ*
(c) *LacY–lacZ–lacA* (d) *LacZ–lacY–lacA*

47. Which one of the following statements is incorrect?
(a) *Lac I* gene is expressed from its own promoter.
(b) The *lac I* promoter is a strong promoter.
(c) Lactose repressor binds to the operator as well as inducer.
(d) Regulator genes code for repressors, which can act potentially at many other locations in the genome.

48. Which one of the following is not applicable to *trp* operon?
(a) Attenuation (b) Silent allele
(c) Hairpin structure (d) Premature termination

49. Match column I with column II and select the correct answer using answer codes:

Column I	Column II
(A) *Trp* operon	1. Homoeotic mutation
(B) Operator	2. Attenuation
(C) Sensor genes	3. Binding site for repressor
(D) Antennapedia	4. Respond to certain hormone receptor proteins

Answer codes:

	A	B	C	D
(a)	2	4	3	1
(b)	2	3	4	1
(c)	1	4	2	3
(d)	4	1	3	2

50. According to the Britten–Davidson model, the regulation of gene occurs at the level of:
(a) Transcription (b) RNA processing (c) Translation (d) mRNA longevity

51. Which one of the following is applicable to the Britten–Davidson model?
(a) Sensor genes (b) Integrator genes (c) Producer genes (d) All

52. According to the Britten–Davidson model, the regulation of sets of structural genes occurs by means of moderately repetitive:
(a) Operator genes (b) Promoter genes (c) Regulator genes (d) All

53. According to the Britten–Davidson model of gene regulation, which of the following genes are considered analogous to structural genes?
(a) Producer genes (b) Receptor genes (c) Integrator genes (d) Sensor genes

54. Which one of the following is different?
(a) Receptor genes (b) Integrator genes (c) Producer genes (d) Sensor genes

55. Consider the following statements:
(A) In eukaryotes, polycistronic mRNA is present
(B) In eukaryotes, gene regulation is either short term or long term
(C) Short-term gene regulation is the control of enzyme synthesis by steroid hormones
(D) Long-term gene regulation involves regulation during development and differentiation

The correct statements are:
(a) All (b) A, B and C (c) B, C and D (d) C and D

56. Which one of the following does not function in *trp* operon?
(a) Glucose (b) CAP (c) cAMP-CAP (d) All

57. Constitutive genes are applicable to:
(a) Operator genes (b) Overlapping genes (c) Housekeeping genes (d) Selfish genes.

58. Circular DNA with two promoters is found in:
(a) Polyma virus (b) SV_{40}
(c) Both polyma virus and SV_{40} (d) ϕX174

59. Repressor protein is bound to the DNA site mainly by:
(a) Hydrogen bonds (b) Electrostatic forces
(c) Both (a) and (b) (d) Hydrophobic interactions and covalent bonds

60. Enhancer is present within the transcribed portion of the gene in:
(a) Immunoglobulin (b) Thyroxine (c) Collagen (d) All

61. Enhancer was first identified in:
(a) Polyoma virus (b) SV_{40} (c) ϕX174 (d) Herpes virus

62. Antisense RNA does not affect:
(a) Translation (b) Transcription (c) RNA processing (d) Operon activity

63. A change in a segment containing TATA causes:
(a) Increase in transcriptional level (b) Decrease in transcriptional level
(c) Appearance of new mRNA species having starting sites displaced from the normal ones and decrease in transcriptional level
(d) Increase in transcriptional level and appearance of new mRNA species having starting sites similar to those of normal sites

64. Pseudogenes:
(a) Are duplicated genes (b) Have become inactivated by mutations

(c) Are considered dead by products of evolution (d) All

65. Which one of the following is a post-transcriptional process?
(a) The terminal addition of CCA (b) Removal of 5′ and 3′ extensions
(c) Methylation of some bases (d) All

66. In eukaryotes, heat shock proteins are controlled at:
(a) Transcriptional level (b) Translational level
(c) Both transcriptional and translational levels (d) Post translational level

67. Enhancers can be located:
(a) Upstream of the gene they regulate (b) Downstream of the gene they regulate
(c) Within the gene they control (d) All

68. Which one of the following prevents a gene from being influenced by the activation or repression of its neighbours?
(a) Silencers (b) Insulators (c) Enhancers (d) Operator

69. Which one of the following genes is involved in routine metabolic functions?
(a) Housekeeping genes (b) Regulator genes
(c) Operator genes (d) Both (b) and (c)

70. HACNS1 is a gene:
(a) Enhancer (b) Silencer (c) Regulator (d) Insulator

71. A classic example of an attenuated operon is the:
(a) *Lac* operon (b) *Trp* operon (c) *Ara* operon (d) All

72. The most important mode for control of eukaryotic gene expression is:
(a) Epigenetic control (b) Transcriptional initiation
(c) Transcript ability (d) Translational initiation

73. Deficiencies of $MeCP_2$ protein causes:
(a) Rett syndrome (b) Lesch–Nyhan syndrome
(c) Noorie's disease (d) Ehlers–Danlos syndrome

74. Epigenesis play a key role in the:
(a) Regulation of gene expression (b) Control of gene expression
(c) Maintenance of gene regulation (d) All

75. In which one of the following has it been shown that control of one gene was exerted by the small non-coding RNA product of another gene?
(a) *D. melanogaster* (b) *C. elegans* (c) *Neurospora* (d) *E. coli*

76. In the *trp* operon, an attenuator region lies between the:
(a) Operator and promoter (b) Promoter and the first structural gene
(c) Operator and the first structural gene (d) Structural genes

77. The stem-loop structure causes:
(a) Transcription to start (b) Transcription to terminate
(c) Suppresses transcription (d) None

78. Inversion control of gene expression has been reported in:
(a) *Neurospora* and bacteria (b) *Neurospora* and *D. melanogaster*
(c) Bacteria (d) Bacteria and bacterial viruses

79. In which one of the following is the transposed gene activated by the removal of its promoter from repressing the neighbouring sequences:

(a) *Trypanosoma cruzi* (b) *Neurospora*
(c) *E. coli* (d) Yeast mating-type switch

80. Gene activation by promoter addition and enhancer addition occurs in:
(a) Mating-type switch in yeast (b) Trypanosome antigenic variation
(c) Immunoglobulin gene formation in mammals (d) Phase variation in *Salmonella*

81. The number of structural genes in histidine operon of *Salmonella typhimurium* is:
(a) 3 (b) 5 (c) 7 (d) 9

82. Which one of the following is applicable to lac operon?
(a) Overlapping genes (b) Selfish genes
(c) Inducible operon (d) Repressible operon

83. A clear case of development regulation can be seen in human genes for:
(a) Insulin (b) Haemoglobin (c) Sex hormones (d) Growth hormone

84. Which one of the following statements is correct about regulatory proteins?
(a) They are allosteric proteins having two binding sites.
(b) They exert negative as well as positive controls.
(c) They act as switches for genes.
(d) All

85. Changes in the Pribnow box:
(a) Affects transcription (b) Can enhance promoter activity
(c) Can decrease promoter activity (d) All

86. Which one of the following can be directly affected by a mutation in the *I* gene?
(a) Repressor (b) Operator (c) Promoter (d) Inducer

87. RNA polymerase moves to structural genes from the promoter when:
(a) There is no inducer (b) There is repressor on the operator
(c) There is no repressor on the operator (d) Gene I is functional

88. A *trp* operon is:
(a) Inducible (b) Repressible and under positive control
(c) Repressible and under negative control (d) All

89. Repressor binds to:
(a) Promoter gene (b) Operator gene (c) Regulator gene (d) Silencer

90. Which one of the following is applicable to regulator genes in an operon model?
(a) Inhibitors (b) Repressors (c) Regulators (d) All

91. A set of closely located genes involved in the regulation of metabolic pathways in prokaryotes:
(a) Housekeeping genes (b) Overlapping genes
(c) Operon (d) Regulatory genes

92. The promoter regions in most eukaryotic gene bears:
(a) TATA box (b) GC box (c) CAAT box (d) All

93. Which one of the following statements is incorrect?
(a) CAAT box functions in both 5′→3′ or 3′→5′ orientation.
(b) TF II D is the first transcriptional factor to bind close to the promoter.
(c) Abnormal gene regulation causes formation of tumour.
(d) The steroid hormones act both at the transcriptional as well as translational level.

94. What is incorrect about enhancers?
 (a) Are relatively large elements
 (b) Are position-independent
 (c) Function in either orientation
 (d) Need to reside 5′ to a gene
95. Presence of methyl group in the major groove of DNA regulates gene expression by:
 (a) Increasing the affinity of DNA for specific regulatory proteins
 (b) Decreasing the affinity of DNA for specific regulatory proteins
 (c) Has no effect on gene expression
 (d) Both (a) and (b)
96. The first evidence came from the studies of gene regulation in:
 (a) Yeast (b) *Neurospora* (c) *Salmonella* (d) *E. coli*
97. Lactose operater contains:
 (a) 12 base pairs (b) 24 base pairs (c) 30 base pairs (d) 45 base pairs
98. Match column I with column II as per the Britten–Davidson model of gene regulation:

Column I	Column II
(A) Sensor site	1. Comparable to promoter gene of operon
(B) Producer gene	2. Comparable to operator gene of operon
(C) Integrator gene	3. Comparable to structural gene of operon
(D) Receptor site	4. Regulates activity of an integrator gene

Answer codes:

	A	B	C	D
(a)	4	2	3	1
(b)	2	3	4	1
(c)	4	3	1	2
(d)	3	4	2	1

99. Homoeo box is applicable to:
 (a) Antisense RNA
 (b) Repetitive DNA
 (c) Several genes having similar base pair sequence of a short segment
 (d) Several genes having different pair sequence of a long segment
100. Post-transcriptional gene regulation does not involve:
 (a) Methylation
 (b) RNA processing
 (c) RNA lifespan
 (d) Polypeptide modification
101. Housekeeping genes:
 (a) Have common proximal elements
 (b) Are not recognised by activator proteins found in cells
 (c) Are cell-specific
 (d) All
102. Transcription factors bind to:
 (a) Promoter (b) Enhancer (c) Silence DNA (d) All
103. CpG sites are rare in eukaryotic genome except in regions near the:
 (a) Promoter (b) Enhancer (c) Silencer (d) All
104. A regulatory sequence is:
 (a) Promoter
 (b) Enhancer
 (c) A segment of DNA where regulatory proteins bind preferentially
 (d) All
105. A mutation in the promoter:
 (a) Totally prevents attachment of RNA polymerase
 (b) Causes increase in RNA polymerase attachment

(c) Causes decrease in RNA polymerase attachment (d) All

106. In lambda, the Q regulator is a:
   (a) Terminator factor (b) Anti-terminator factor
   (c) Lysogenic repressor (d) Antilysogenic factor
107. Which one of the following is a regulatory protein?
   (a) Cro (b) CAP activator (c) λ repressor (d) All
108. In which one of the following bacterial viruses, the sequential gene expression is controlled by modifying the promoter specificity of RNA polymerase:
   (a) *T4* (b) *T4* and *T7* (c) *SPO*1 (d) All
109. Consider the following statements with reference to lambda (λ) phage:
   (A) The process of transcription is initiated as soon as it infects bacteria with the help of two promoters, PR in the right side and PL on the left side
   (B) The process of transcription occurs on only one strand
   (C) Out of two early genes, gene *cro* being transcribed on the left side while gene *N* transcribed on right side
   (D) The repressor protein coded by regulator gene *CI* regulates its own synthesis
   The correct statements are:
   (a) A, B and C (b) A and D (c) B and C (d) All
110. Phosphorylation of protein does not affect:
   (a) Transcription (b) Translation
   (c) Cell differentiation and cell transformation (d) All
111. Which one of the following is an incorrect match?
   (a) Homeotic genes – Determine the body plan
   (b) Kozak's scanning hypothesis – Reading of mRNA by ribosomes in prokaryotes
   (c) Methylation and demethylation – Suppression and expression of gene activity
   (d) Masked mRNA – mRNA for fibronectin in *Xenopus*
112. In which one of the following gene regulation occurs through gene arrangement?
   (a) Human immune system involving antibody production
   (b) Ribonucleotide reductase in eggs of sea urchins
   (c) Heat shock elements of eukaryotes
   (d) Formation of interferon
113. Krebs and Fischer were awarded the Nobel Prize in 1992 for physiology and medicine for their work on:
   (a) Regulation of cholesterol metabolism
   (b) Reversible protein phosphorylation as a biological regulatory mechanism
   (c) Study of cell surface receptor for low-density lipoprotein
   (d) Role of homeobox in gene regulation
114. Which one of the following statements is correct?
   (a) In bacteriophage T4, late genes are concerned with the synthesis of DNA
   (b) In phage *SPO*1, bacterial *RNA polymerase* is used for early gene expression
   (c) In lambda (λ) phage gene Q acts as terminator factor
   (d) Bacteriophage T4 lacks non-essential genes
115. In bacteriophage T7, the phage enzymes are responsible for the expression of:
   (a) Class I genes (b) Class II genes
   (c) Class II and Class III genes (d) Class I, Class II and Class III genes

116. The expression of structural gene is influenced by the:
(a) Specific regulatory gene
(b) Presence of the inducer
(c) Absence of the inducer
(d) All

117. The DNA sequences of the glucocorticoid response element are:
(a) 3′ ACAGAc nnn ACAGAc
(b) 5′AGAACA nnn TGTTCT3′
(c) 3′ TCTTGT nnn ACAAGA5′
(d) Both (b) and (c)

118. Which one of the following statements is incorrect about the insulin-like growth factor2 (*IGF*2)?
(a) In humans, the *IGF*2 allele, inherited from the father is expressed.
(b) The allele inherited from the mother is not expressed.
(c) If both alleles start expressing in a cell, the cell may develop into a cancer cell.
(d) None

119. Genomic imprinting:
(a) Is an epigenetic form of gene regulation
(b) Results only in the copy inherited from the mother to function
(c) Results only in the copy inherited from the father to function
(d) All

120. Imprinted genes are found in:
(a) Eutherians
(b) Metatherians
(c) Prototherians and plants
(d) All

121. Which one of the following disorder is due to genome imprinting?
(a) Prader–Willi syndrome
(b) Angelman syndrome
(c) Beckwith–Wiedemann syndrome
(d) All

122. The minimum part of the promoter required to initiate transcription correctly:
(a) Distal promoter (b) Proximal promoter (c) Core promoter (d) More than one

123. RNA can regulate gene expression in:
(a) Yeast and viral infection
(b) *Drosophila*
(c) Mammals
(d) All

124. Gene regulation is responsible for:
(a) Versatility and adaptability of an organism
(b) Cellular differentiation
(c) Morphogenesis
(d) All

125. Lac genotype can be created by mutations in:
(a) *Lac Z* gene
(b) *Lac Y* gene
(c) Either *lac Z* or *lac Y* genes
(d) *Lac Z*, *lac Y* and *lac A* genes

126. Intergenic exclusion is applicable to:
(a) DNA replication
(b) DNA rearrangements
(c) Somatic cell hybridisation
(d) DNA repair mechanism

127. In lambda (λ) phage, lysogenic growth is favoured by:
(a) C II protein (b) Cro protein (c) Q protein (d) All

128. Lytic growth in lambda (λ) phage is favoured by:
(a) *O* protein (b) *Q* protein (c) *C II* protein (d) *Cro* protein

129. Several promoters and terminators are involved in the level of gene expression in:
(a) ϕX174 (b) G_4 (c) M_{13} (d) All

130. Which one of the following lactose operon genotypes in *E. coli* will not be able to produce β-galactosidase, even when lactose is present?

(a) *Lac I^- lac P^+ lac O^c lac Z^+*
(b) *Lac I^+ lac P^+ lac O^c lac Z^+*
(c) *Lac I^+ lac P^+ lac O^+ lac Z^+*
(d) *Lac I^- lac P^+ lac O^+ lac Z^+*

131. Consider the following statements:
(A) Promoter mutants have no effect on the synthesis of enzymes responsible for lactose metabolism
(B) The *lac OC* mutants do not synthesise enzymes for lactose metabolism, even when lactose is present
(C) The *lac I* gene is not a totally independent transcriptional entity
(D) The structural genes of lac operon are loosely linked

The incorrect statements are:
(a) All (b) A, B and C (c) B and C (d) A and D

132. Regulans and stimulons are related to:
(a) Promoters (b) Operators (c) Cistrons (d) Mutons

133. Methylation regulates gene activity by adding methyl group to:
(a) Guanine to CG doubles in DNA
(b) Adenine to AT doublets in DNA
(c) Cytocine to GC doublets in DNA
(d) All

134. All are binding domains except:
(a) Zinc finger (b) Leucine Zipper (c) Intron-exon-intron (d) Helix-turn-helix

135. Which one of the following is trans-acting?
(a) Promoter (b) Regulator (c) Terminator (d) Operator

136. Which one of the following is an incorrect match?

	Transcription factor	DNA-binding domain
(a)	Fos	Leucine Zipper
(b)	GR	In finger
(c)	CREB	Leucine Zipper
(d)	SRF	Zn finger

137. Match column I with column II and select the correct answer using answer codes:

Column I (Repressor)	Column II (Sites of action)
(A) T4 *DNA polymerase*	1. Ribosome binding site
(B) T4 p32	2. Shine–Delgarno sequence
(C) R17 coat protein	3. Initiation codon
(D) T4 RegA	4. Single stranded 5′ leader

Answer codes:

	A	B	C	D
(a)	4	2	2	3
(b)	2	4	1	3
(c)	3	2	3	1
(d)	2	4	3	1

138. The partial diploid with genotype $O^cZ^+Y^-/OZ^-Y^+$ will synthesise:
(a) Only β-galactoside permease
(b) Only β galactosidase
(c) Both (a) and (b)
(d) None

139. Which one of the following statements is incorrect?
(a) Operator and promoter are dominant.
(b) In prokaryotes, gene expression does not occur through DNA rearrangement.
(c) A mutation in operator gene does not permit repressor function by acting on operator site.
(d) Eukaryotic genes can be translated in bacteria.

140. Polarity mutations in the Z locus:
   (a) Affect β-galactosidase synthesis
   (b) Reduce the quantity of permease
   (c) Reduce the quantity of acetylase
   (d) All

Answers to Multiple-Choice Questions

1.	(d)	2.	(c)	3.	(d)	4.	(d)	5.	(b)	6.	(a)	7.	(b)	8.	(c)
9.	(c)	10.	(b)	11.	(a)	12.	(b)	13.	(c)	14.	(d)	15.	(b)	16.	(b)
17.	(d)	18.	(b)	19.	(d)	20.	(c)	21.	(c)	22.	(c)	22.	(d)	24.	(a)
25.	(b)	26.	(a)	27.	(a)	28.	(b)	29.	(c)	30.	(a)	31.	(a)	32.	(d)
33.	(c)	34.	(c)	35.	(a)	36.	(d)	37.	(d)	38.	(d)	39.	(c)	40.	(d)
41.	(d)	42.	(c)	43.	(d)	44.	(b)	45.	(d)	46.	(d)	47.	(b)	48.	(b)
49.	(b)	50.	(b)	51.	(d)	52.	(c)	53.	(a)	54.	(c)	55.	(c)	56.	(d)
57.	(c)	58.	(b)	59.	(c)	60.	(a)	61.	(b)	62.	(c)	63.	(c)	64.	(d)
65.	(d)	66.	(c)	67.	(d)	68.	(b)	69.	(a)	70.	(a)	71.	(b)	72.	(b)
73.	(a)	74.	(d)	75.	(b)	76.	(c)	77.	(b)	78.	(d)	79.	(d)	80.	(c)
81.	(d)	82.	(c)	83.	(b)	84.	(d)	85.	(d)	86.	(a)	87.	(c)	88.	(c)
89.	(b)	90.	(b)	91.	(c)	92.	(d)	93.	(d)	94.	(d)	95.	(d)	96.	(d)
97.	(b)	98.	(c)	99.	(c)	100.	(a)	101.	(a)	102.	(d)	103.	(a)	104.	(d)
105.	(d)	106.	(b)	107.	(d)	108.	(d)	109.	(b)	110.	(d)	111.	(b)	112.	(a)
113.	(b)	114.	(b)	115.	(c)	116.	(d)	117.	(d)	118.	(d)	119.	(d)	120.	(d)
121.	(d)	122.	(c)	123.	(d)	124.	(d)	125.	(c)	126.	(b)	127.	(a)	128.	(d)
129.	(d)	130.	(c)	131.	(a)	132.	(b)	133.	(d)	134.	(c)	135.	(b)	136.	(d)
137.	(b)	138.	(b)	139.	(b)	140.	(d)								

Fill in the Blanks

1. Operons are typed as ____________________ or repressible.
2. The enzymes synthesis of which can be induced by adding substrate are called ________________ enzymes.
3. An active repressor can be made inactive by adding ____________________.
4. A natural inducer of the lactose operon is ____________________.
5. The binding of repressor to the operator is regulated by the ____________________.
6. When a repressor binds to a metabolite, it is called a ____________________.
7. Inactive repressor + Co-repressor ____________________.
8. Lactose ____________________ glucose + Galactos.
9. Two proteins which are involved in the regulation of lactose operon are ________________ and ________________.
10. An operon consists of an ____________________ and structural genes.
11. The most important example of gene regulation at transcriptional level is the ____________________.
12. Transcription is controlled by ____________________ regulatory mechanism.
13. Lac operon expression is regulated by the ____________________.
14. Promoter region is the binding site for ____________________ *polymerase.*

15. A functional gene product may be an RNA or a ________________.
16. An operon contains ________________ structural genes.
17. The promoter is recognised by ________________.
18. A repressor or activator can bind to an ________________.
19. ________________ pumps lactose into the cell.
20. Operons are either ________________ or ________________ according to control mechanism.
21. Trytophan operon of *E. coli* is an example of ________________ operon.
22. In the trytophan operon, ________________ acts as a co-repressor.
23. Repressible operons generally function in the ________________.
24. Premature termination of transcription is called ________________.
25. The *lac* γ gene is immediately followed by ________________.
26. The *E. coli* operon for galactose utilisation contains ________________ promoters.
27. The enzymes of arabinose operon are controlled by a protein (arac) which acts either ________________ or ________________.
28. The induction of ara operon depends on the regulatory effects of ________________ and ________________.
29. Operons were discovered by ________________ in bacteria.
30. Operons occur primarily in ________________ and ________________.
31. Attenuation was discovered by ________________ in *E. coli*.
32. A trans-acting factor is a protein that regulates ________________ by binding to the operator or promoter.
33. ________________ is a common method of gene silencing.
34. Regulation of lac operon by repressor is called ________________ control.
35. The *lac I* gene is transcribed ________________ from its own promoter
36. A functional gene product may be ________________ or a ________________.
37. The repressor blocks transcription of a gene when the enzyme ________________ is not needed by *E. coli*.
38. Multiple operons controlled by the same repressor or activator are called a________________.
39. The process of producing biologically functional molecules of either RNA or protein is called ________________.
40. Induction occurs at the level of ________________.
41. Short-term gene regulation operates at ________________ level.
42. The cAMP–CAP complex acts as an ________________ repressor of CAP gene expression by binding to the CAP proter .
43. The process by which the genotype coded in the genes are exhibited by phenotype is called ________________.
44. An operon includes a special segment of genes that are regulators of the protein synthesis but do not code for protein. They are called the ________________ and ________________.
45. The operator is part of DNA to which the ________________ bind.
46. The regulator genes code for a ________________ protein that binds to ________________.
47. Lambda has only ________________ immediate early genes that are transcribed independently by the host *RNA polymerase*.
48. The binding of the purifier repressor protein to DNA was first characterised by a ________________ ________________.

49. Synthesis of in chicken is an example of induction of transcription by a hormone.
50. The synthesis of transposase enzyme in Tn_{10} is regulated through the synthesis of ________________.
51. The three phages of gene expression in T4 are ______________, ____________ and ________________.
52. According to the gene battery model, a set of structural genes are controlled by one sensor site called ________________.
53. Hormones influence target cells by activating ________________.
54. Regulation of p32 protein coded by bacteriophage T4 gene 32 is an example of ________________ regulation.
55. A ________________ continuously expresses the gene for the antibody it synthesises.
56. The glococorticoid receptor is a ________________ transcription factor.
57. The response element is a part of the ________________ of a gene.
58. The genes whose expression is determined by the parents that contributed them are called ________________ genes.
59. ________________ is the reason that parthenogenesis does not occur in mammals.
60. Genomic imprinting is an epigenetic form of ________________.
61. An epigenetic change alters the phenotype without changing the ________________.
62. RNA Immuno Precipitaton in chip (RIP) (RIP-Chip) technique is used to study ________________ gene regulation.
63. ________________ is a disease associated with promoter elements.
64. Gene expression is also known as ________________.
65. The lac operon consists of three________________ and the adjacent ________________.and ________________.
66. ________________ inducers are molecules that induce enzyme synthesis but are not metabolised.
67. Regulator genes are responsible for controlling the expression of ________________ genes.
68. The repressor of the SOS operon is coded by the wild type ________________ gene.
69. Gene enhancer comes into action when ________________ bind to it.
70. In lambda (λ) phage, genes ________________ and ________________ are responsible for determining the direction of development.
71. Phage MU replicated by ________________.
72. Gene regulation in higher eukaryotes is either ________________.or ________________.
73. Polygenic mRNA are common in bacteria and ________________.
74. In the ________________ operon there is overlapping of promoter and operator sites.
75. Trypanosomes variable surface antigene glycoprotein (VSG) production is an example of gene regulation through ________________.
76. ________________ are the regions of eukaryotic DNA that encode a polypepetide product.

Answers to Fill in the Blanks

1. Inducible
2. Inducible
3. Inducer
4. Allolactose
5. Inducer
6. Co-repressor
7. Active repressor
8. β galoctosidase
9. Lac repressor, catabolite gene activator protein
10. Operator gene
11. Lac operon
12. Positive and negative

13. Lac repressor
14. RNA
15. Protein
16. One or more
17. *RNA polymerase*
18. Operator
19. β galactosidepermease
20. Inducible or repressible
21. Repressible
22. Trytophan
23. Anabolic pathway
24. Attenuation
25. *Lac A* gene
26. Two
27. Positively, negatively
28. Ara C protein, CAP
29. Jacob and Monod (1961)
30. Prokaryotes, nematodes
31. Yanofsky
32. Transcription
33. Methylation of DNA
34. Negative
35. Constitutively
36. RNA, protein
37. Beta galactosidase
38. Regulon
39. Gene expression
40. Transcription
41. Transcriptional level
42. Autogenous
43. Gene expression
44. Promoter, operator
45. *RNA polymerase*
46. Repressor, operator
47. Two
48. Filter assay binding
49. Ova albumin
50. Anti-sense RNA
51. Early genes, quasi-late genes, late genes
52. Battery
53. Gene transcription
54. Autogenous
55. Plasma cell
56. Zinc finger
57. Promoter
58. Imprinted
59. Imprinting
60. Gene regulation
61. Genotype
62. Post transcriptional
63. Beta (β)-thalassemia
64. Protein expression
65. Structural genes operator
66. Gratuitous
67. Structural
68. LexA
69. Cell-specific protein(s)
70. *Cro, CII*
71. Transposition
72. Short term, long term
73. Bacteriophages
74. Lac
75. Gene rearrangement
76. Exons

True or False

1. Catobolite activator protein is a dimeric.
2. The repressor is an allosteric protein.
3. The gene for DNA-dependent *RNA polymerase* is not a part of the operon.
4. Operons are a very useful tool for protein regulation in bacteria.
5. Genes of some operons may lack a clear functional relationship.
6. Cyclic AMP directly promotes *lac* mRNA synthesis.
7. The arabinose C protein changes the activity of *RNA polymerase*.
8. CAP has no effect on transcription until cAMP is bound to it.
9. CAP protein also acts as repressor.
10. Synthesis of constitutive proteins is under direct environmental control.
11. Operator constitutive mutations are cis dominant.
12. Lac repressor is homotetramer.
13. Operons are more frequent in higher organisms.
14. The trytophan operon is inhibited by tryptophan.
15. Genes for constitutive enzymes are always on.
16. Adaptive genes may be either inducible or repressible.
17. A cis-acting factor is a DNA sequence that controls adjacent genes.

18. The operator is a cis-acting factor.
19. Eukaryotic genes have only one enhancer.
20. Catabolite expression is an example of positive regulation.
21. Gene expression can be regulated through stability of mRNA.
22. Attenuators common in inducible operon.
23. Steroid hormones generally act at the post-transcriptional level.
24. In *trp* operon, the transcript forms a hairpin structure resulting in premature termination.
25. The *trp* operon is similar to operon.
26. Antisense RNA codes protein.
27. *Tetrahymena thermophilia* has a single intron.
28. Mutation in the TATA box reduces transcription.
29. Lactose is not the real inducer of the *lac* operon.
30. The induction of *ara* operon depends on the regulatory effects of *ara* C protein and CAP.
31. Exons created by Junk DNA are linked with gene regulation.
32. Enhancers act on the promoter region itself.
33. An enhancer may be execised and inserted elsewhere in the chromosome.
34. The reversal of orientation of enhancer affects its function.
35. There are several operons in *E. coli* that contain overlapping sequential elements.
36. Repression of the *trp* operon is comparatively weak.
37. Operator region is a palindrome.
38. Natural antisense RNAs may regulate translation.
39. Nonprotein coding genes are translated into protein.
40. Arginine operon is a repressible operon system.
41. The viral genes are expressed in genetically preprogrammed sequences.
42. Repression does not occur at the level of transcription.
43. The binding of cAMP-CAP to the operon does not affect the binding of a repressor.
44. Arabinose C protein changes the activity of *RNA polymerase*.
45. Tryptophan is a co-repressor.
46. The *lac* and λ repressors are transcriptional repressors.
47. Cyclic AMP directly promotes lac mRNA synthesis.
48. Transcriptional enhancers are tissue specific.
49. Detectable methylation is lacking in insects.
50. Isopropylthiogalactoside is a powerful repressor.
51. In the Jacob and Monad system, the regulation of the operator is reversible.
52. A single repressor may affect synthesis of many enzymes.
53. Twins splitting from an early embryo have more similar methylation patterns than the twins from a later split.
54. Epigenetic patterns in dividing cells are more error prone.
55. Tight histone packing prevents *RNA polymerase* from transcribing DNA.
56. Silencers block translation.
57. Some promoter mutations increase the requirement of cyclic AMP.
58. Almost all phage genes can be expressed using transcription apparatus of the host cell.

59. The infection of phage T_4 depends on a mechanical link between replication and late gene expression.
60. Antirepressor is needed for lytic infection.
61. The PE transcript contains the antisense strand of the *cro* gene.
62. Presence of TATA box is not essential for transcription.
63. Silencer functions respective to its position and orientation relative to gene, whose expression it controls.
64. Histones are not involved in repression of gene activity.
65. Nonhistones play a positive regulatory role in eukaryotes.
66. Ubiquitin plays a key role in the regulation of heat shock genes.
67. CAP binding site is common to *lac*, *gal* and *ara* operons.
68. The antibiotic-resistant gene in bacteria only expressed when an antibiotic is present in the environment.
69. The expression of the structural gene is influenced by the presence or absence of the co-repressor.
70. Imprinted genes violate the usual rule of inheritance, which is that both alleles in a heterozygote are equally expressed.
71. Remodelling of chromatin has no effect on gene expression.
72. Cis-acting mutations in the leader region influence transcription of the operon.
73. Attenuation has an eight-to-tenfold effect on transcription.
74. Gene expression by DNA rearrangements requires induction.
75. The *lac* repressor of *E. coli* controls only one enzyme.
76. Exercise affects gene expression.
77. MecP2 protein acts as a transcriptional activator and repressor depending upon the context.

Answers to True or False

1. True	2. True	3. True	4. True	5. True	6. False	7. False	8. True
9. True	10. False	11. True	12. True	13. False	14. True	15. True	16. True
17. True	18. True	19. False	20. True	21. True	22. False	23. True	24. True
25. False	26. False	27. True	28. True	29. True	30. True	31. True	32. False
33. True	34. False	35. True	36. True	37. True	38. True	39. False	40. True
41. True	42. False	43. True	44. True	45. True	46. True	47. False	48. True
49. True	50. False	51. True	52. True	53. True	54. True	55. True	56. False
57. False	58. False	59. True	60. True	61. True	62. False	63. False	64. False
65. False	66. True	67. True	68. True	69. False	70. True	71. False	72. True
73. True	74. False	75. False	76. True	77. True			

Give Reasons

1. The lac regulatory protein is called a repressor.
   – Because it keeps away RNA polymerase from transcribing the structural genes.
2. When lactose is present, the *lac* genes are expressed.
   – Because allolactose binds to the *lac* repressor protein and keeps it from binding to the operator.

3. The bacterial mRNA synthesised by operon is called polycistronic.
   – Because it carries information for more than one type of protein.
4. Stem loop structure is called the terminator or attenuator.
   – Because it causes transcription to terminate.
5. Both repression and induction are examples of negative control.
   – Because repressor proteins turn off transcription.
6. Activator enhancer complex is unique in eukaryotes.
   – Because they normally have to be activated to start protein synthesis, which involves the use of transcription factors and RNA polymerase.
7. Two complete female genome cannot produce viable young.
   – Because of the imprinted genes.
8. Gene regulation may serve as a raw material for evolutionary change.
   – Because control of timing, location and amount of gene expression may have a profound effect on the function of the genes in the organism.
9. When estrogen is injected into chickens, only oviduct synthesises mRNA.
   – Because only oviduct has hormone receptor for estrogen and all other tissues lack the hormone receptors.
10. The *trp* operon is quite different from the *lac* operon.
    – Because tryptophan acts directly in repression and not as an inducer.
11. About more than 95% of the RNA synthesis by RNA polymerase never reaches the cytoplasm.
    – Mainly because of removal of introns which accounts for 80% of the total bases.
12. Usually the promoter of pseudogene is inactive.
    – Due to lack of selective pressure to maintain a functional gene product.
13. Whenever the intracellular cAMP level is low, the *lac* operon is not functional even in the presence of glucose.
    – Because CAP protein is unable to bind to the promoter region due to low level of intracellular cAMP.
14. Gratuitous inducers are very useful.
    – Because they remain in the cell in their original form.
15. The genes *CI*, *CII*, *CIII* are so named.
    – Because of production of clear plaques in mutant of these genes due to presence of only lysed cells.
16. Gene expression occurs by DNA rearrangements.
    – Because such arrangements:
    (a) Occur mostly at very low rates
    (b) Allow a cell to express only one of a set of genes at a time
    (c) Are uninduced and are sometimes reversible
17. Regulation of gene expression is essential.
    – The living organisms need a variety of proteins for their cellular activities. All proteins are not needed at a time. There are specific proteins which are needed at specific stage of life cycle. So, there must be a controlling mechanism for the expression of a particular set of genes at a particular stage and for this reason regulation of gene expression is essential.
18. The regulatory substance produced by the gene I^+ is called repressor.
    – Because it represses the function of lac structural genes in the absence of inducers.
19. The presence of repressor has no effect if one transcription is initiated.
    – Because the operator binding site is not available to the repressor.

HUMAN GENOME PROJECT

Multiple-Choice Questions

1. The Human Genome Project began in the year:
   (a) 1985 (b) 1990 (c) 1995 (d) 2000
2. The Human Genome Project was headed by:
   (a) Craig Venter (b) James D Watson (c) Francis Collins (d) Water Fiers
3. The haploid human genome contains over_______________ billion base pairs:
   (a) 2 (b) 3 (c) 5 (d) 7
4. Which one of the following animals has the same genes and regulatory gene sequences as humans?
   (a) Golden poison frog (b) Turtle
   (c) Puffer fish (d) Dogfish
5. The first working draft on the Human Genome Project was released in:
   (a) 1995 (b) 2000 (c) 2002 (d) 2005
6. The human genome has __________ per cent of repeat sequences:
   (a) 3 (b) 7 (c) 11 (d) 50
7. Which one of the following human chromosomes contains the highest number of genes?
   (a) 1 (b) 5 (c) 7 (d) X
8. All human cells contain a complete set of genome, except:
   (a) Neutrophils (b) Matured red blood cells
   (c) Sperms (d) Liver cells
9. Which one of the following molecular markers is used in the mapping of human genome?
   (a) RFLP (b) VNTRs (c) STSs and SNPs (d) All
10. Which one of the following is used to make millions of exact copies of DNA?
    (a) RFLP (b) STR analysis
    (c) PCR analysis (d) Y-chromosome analysis
11. Consider the following statements:
    (A) The genome of an organism is complete set of DNA
    (B) The smallest known genome contains about 6,00,000 DNA base pairs (a free-living bacterium)
    (C) The human genome contains about 20,000–25,000 genes
    (D) Genes comprise only 2 per cent of the human genome
    The correct statements are:
    (a) A and B (b) B and D (c) A, B and D (d) All
12. Mitochondrial DNA analysis can be used to examine the DNA from samples that can not be analysed by:
    (a) STR (b) RFLP (c) Both (a) and (b) (d) None
13. The technique used for analysing the variable length of DNA fragments:
    (a) RFLP (b) STR (c) PCR analysis (d) None
14. The Los Almos Centre for Human Genome Studies was established in:

(a) 1980 (b) 1988 (c) 1990 (d) 1995

15. The first vertebrate whose genome has been sequenced completely after human is:
(a) Chimpanzee (b) Puffer fish (c) Krait (d) Dogfish

16. The first plant whose genome has been sequenced is:
(a) *Arabidopsis thaliana* (weed) (b) *Feronia limonia*
(c) *Carum curvi* (d) *Oryza sativa*

17. The goal of International HapMap Project is:
(a) To identity patterns of Single Nucleotide Polymorphisms (SNPs)
(b) To make comparative analysis of genome sequences
(c) To generate biological data from various human races
(d) To prepare metabolic finger printing

18. The genome of an individual is unique, except for:
(a) Identical twins (b) Cloned organisms
(c) Both (a) and (b) (d) Members of the same family

19. Junk DNA comprises 50 per cent of human genome, which may participate to:
(a) Modify the existing genes (b) Reshuffle the existing genes
(c) Create new genes (d) All

20. DNA-based gene test is not available for:
(a) Cystic fibrosis (b) Fanconi anaemia
(c) Marfan syndrome and fragile X syndrome (d) None

21. Though the full human DNA sequence is known still we don't know the:
(a) Gene regulation (b) Gene number, exact locations and functions
(c) Genes involved in complex traits and multigene disease (d) All

22. Genome of which one of the following animals has been completely sequenced:
(a) *Drosophila melanogaster* (b) *Caenorhabditis elegans*
(c) *Saccharomyces cervisiae* (d) *All*

23. In humans only about 1.5 per cent of the genome codes for proteins while the rest comprises:
(a) Junk DNA (b) Regulatory sequences
(c) RNA genes and introns (d) All

24. Which one of the following is used for certain genealogical DNA tests?
(a) Trinucleotide repeats (b) Short tandem repeats
(c) Microsatellite (d) Minisatellite

25. The repeated sequences of nucleotides between 10 and 60 are known as:
(a) Short tandem repeats (b) Minisatellite
(c) Microsatellite (d) Variable number tandem repeats

26. The first two human chromosomes to be sequenced were chromosomes:
(a) 15 and 21 (b) 17 and 22 (c) 21 and 22 (d) 20 and 22

27. The GC content in chromosome 22 is:
(a) 40 per cent (b) 48 per cent (c) 42 per cent (d) 60 per cent

28. The Human Diversity Project was suggested in:
(a) 1995 (b) 1997 (c) 2000 (d) 2003

29. About _________ per cent of human genome remains unsequenced:
(a) 2 (b) 4 (c) 8 (d) 10

30. The ratio of germ line mutation in males versus females is:
(a) 1:1 (b) 2:1 (c) 1:2 (d) 3:1
31. Which one of the following animals has the same gene regulatory sequences as humans?
(a) *Drosophila melanogaster* (b) *Fugu rubripes*
(c) *Caenorhabditis elegans* (d) *Scoliodon*
32. Which one of the following cells were used by scientists working on the Human Genome Project?
(a) WBCs (b) RBCs (c) Sperms (d) None
33. The number of hypervariable regions in human genome is approximately:
(a) 10 (b) 50 (c) 100 (d) 500
34. Which one of the following markers has been used in mapping the human genome?
(a) Variable number tandem repeats (b) Restriction fragment length polymorphisms
(c) Sequences-tagged sites and single nucleotide polymorphisms (d) All
35. The human chromosome having the maximum number of polymorphic markers?
(a) 15 (b) 22 (c) 20 (d) X
36. Which one of the following human chromosome lacks polymorphic markers?
(a) X (b) Y (c) 22 (d) None
37. About ___________ million locations have been identified in the human genome, where single nucleotide base DNA differences occur:
(a) 1 (b) 1.4 (c) 2.4 (d) 3.5
38. Genomic scale technology is being used to study:
(a) Gene families from a large number of species (b) To compare entire genomes
(c) Variation among individuals (d) All
39. The Institute for Genomic Research has been founded by:
(a) Craig Venter (b) Weber James (c) A Peter (d) G Philip
40. A single set of genome is lacking in:
(a) Bacteria and viruses (b) Mitochondria
(c) Humans (d) Archaea
41. The first multicellular eukaryote whose genome was sequenced is:
(a) *Homo sapiens* (b) *Caenorhabditis elegans*
(c) *Drosophila melanogaster* (d) *Saccharomyces cervisiae*
42. Human genome contains __________ genes:
(a) 30,000 (b) 3,00,000 (c) 8,00,000 (d) 10,00,000
43. James Watson was the first director of the Human Genome Project. In 1993, he was replaced by:
(a) J Craig Venter (b) Francis Collins (c) Frederick Sanger (d) Craig C Mellio
44. The papers dealing with sequence and analysis of the human genome were published in:
(a) January 2000 and April 2002 issues of *Nature and Science*
(b) February 2001 and April 2003 issues of *Nature and Science*
(c) February 2000 and April 2003 issues of *Annals of Human Biology*
(d) March 2002 and April 2003 issues of *Human Genetics*
45. The disease causing change may be:
(a) As small as the substitution/deletion (b) Addition of a single base pair
(c) As large as a deletion of thousands of bases (d) All
46. Consider the following statements:

(A) Human genes are evenly distributed across the chromosome
(B) There are gene-rich and gene-poor regions in each human chromosome
(C) Gene finding refers to Computational Biology
(D) Most human genes lack multiple exons, and human introns are frequently much shorter

The incorrect statements are:

(a) A (b) A and C (c) A and D (d) C and D

47. In human beings, the percentage of identical DNA is:
(a) 25 per cent (b) 50 per cent (c) 75 per cent (d) 99.9 per cent

48. During the course of the Human Genome Project, scientists identified genes for:
(a) Neurofibromatosis (b) Cystic fibrosis and inherited form of breast cancer
(c) Huntington disease (d) All

49. Match column I with column II and select the correct answer using answer codes:

Column I	Column II
(A) Y chromosome and mt DNA	1. 1–6 bp
(B) *SRY* gene	2. Uniparental inheritance
(C) Dystrophin	3. Longest gene
(D) Micro satellites	4. Smallest gene

Answer codes:

	A	B	C	D
(a)	4	2	3	1
(b)	2	4	3	1
(c)	3	4	1	2
(d)	4	1	3	2

50. Human Genome Project also decoded the genome of __________ in order to study genetic similarities between species:
(a) *E. coli* (b) Fruit fly (c) A nematode worm (d) All

51. The first sequence of human individual genome published was that of:
(a) James D Watson (b) James D Watson and J Craig Venter
(c) J Craig Venter (d) Watson and Philip

52. The potential use of genome research includes:
(a) Bioarchaeology, anthropology, evolution and human migration
(b) Molecular medicine and risk assessment
(c) Energy sources and environmental applications
(d) All

53. Computerised gene maps make gene hunting:
(a) Faster (b) Cheaper
(c) More practical for almost any scientist (d) All

54. Which one of the following is a correct sequence from the smallest to the largest unit?
(a) Chromosome – Genome – Gene – Nucleotides
(b) Genome – Chromosome – Nucleotide – Gene
(c) Nucleotide – Gene – Chromosome – Genome
(d) Gene – Genome – Nucleotide – Chromosome

55. The largest known genome is that of:
(a) *Amoeba dubia* (b) *E. coli*
(c) *Caenorhabditis elegans* (d) *Homosapiens*

56. The term 'genome' was coined by:
    (a) M H F Wilkins (1956)
    (b) J Lejeune (1964)
    (c) Hans Winkler (1920)
    (d) J M Hill (1970)
57. The Sanger Institute is engaged upon collaborative projects to sequence all or part of chromosome:
    (a) 1, 9, 11, 13, 19, 21 and X
    (b) 9, 12, 14, 17, 21, 22 and Y
    (c) 1, 6, 9, 10, 11, 13, 20, 22 and X
    (d) 5, 9, 10, 11, 21, 22 and X
58. LINE1 repeat elements cover ___________ of the X chromosome:
    (a) Entire (b) One fourth (c) One third (d) Half
59. Chromosome 1 constitutes ___________ per cent of the human genome:
    (a) 3 (b) 5 (c) 8 (d) 10
60. Which one of the following human chromosome harbours the largest transfer RNA?
    (a) 1 (b) 6 (c) 10 (d) 22
61. In the human genome, the most polymorphic gene on chromosome 6 is:
    (a) *HLA-B* (b) *BDNF* (c) *BRAC2* (d) *RHO*
62. The largest interferon gene cluster in the human genome is found in the chromosome:
    (a) 6 (b) 9 (c) 11 (d) 19
63. Genomics includes:
    (a) Sequencing of genome
    (b) Mapping of genome
    (c) Function of genome
    (d) All
64. Which one of the following is not a nucleic acid database?
    (a) PIR (b) CSD (c) EMBL (d) Gene Bank
65. Which one of the following is an incorrect match?
    (a) European Bioinformatics Institute – United Kingdom
    (b) Genome NET – Germany
    (c) National Centre for Biotechnology Information – United States of America
    (d) Swiss Institute of Bioinformatics – Switzerland
66. PRODOM is a:
    (a) Nucleic acid database
    (b) Technique for protein detection
    (c) Protein database
    (d) DNA chip
67. The branch of life sciences that deals with the study of information technology in the field of molecular biology is known as:
    (a) Bioengineering
    (b) Biotechnology
    (c) Bioinformatics
    (d) Biomedical informatics
68. Computational Biology is concerned with mathematically identifying stretches of sequences:
    (a) Protein-coding genes
    (b) Regulatory genes
    (c) RNA genes
    (d) All
69. Which one of the following organism has very little noncoding DNA as compared to other organisms; and also has small compact genome?
    (a) *Drosophila melanogaster*
    (b) *Caenorhabditis elegans*
    (c) *Takifugu rubripes*
    (d) *Mus musculus*
70. Which one of the following is not applicable to genomics?
    (a) Proteome (b) Transcriptome (c) Genotype (d) Phenotype
71. DNA microarrays are used to:
    (a) Detect single nucleotide polymorphisms
    (b) Genotyping

(c) Measure changes in gene expression (d) All

72. Microarray technology evolved from:
(a) Northern blotting (b) Southern blotting (c) cDNA (d) PCR

73. Which one of the following is an intragenome phenomenon?
(a) Epistasis (b) Pleiotropy (c) Heterosis (d) All

74. Genomics is primarity related with:
(a) Structure of genome (b) Function of genome
(c) Evolution of genome (d) All

75. Which one of the following is different?
(a) STS (b) SNPs (c) SAS (d) VNTR

76. The smallest number of genes required for an organism's survival is:
(a) 100–250 (b) 300 (c) 300–500 (d) 700–1000

77. Which one of the following organism has the minimum number of genes?
(a) *Haemophilus influenza* (b) *Mycoplasma genitalium*
(c) *E. coli* (d) *Drosophila melanogaster*

78. DNA microarrays are used to analyse a sample for:
(a) Pattern of gene expression (b) Gene variations
(c) Mutations (d) All

79. Which one of the following is applicable to DNA fingerprinting?
(a) DNA profiling (b) DNA typing (c) DNA testing (d) All

80. The DNA fingerprinting technique was first reported by:
(a) Roderick (1986) (b) Singer (1950) (c) Alec Jeffreys (1985) (d) Francoeur (1988)

81. The most desirable way to collect sample for DNA fingerprinting is:
(a) Blood (b) Buccal swab (c) Bone marrow (d) Semen

82. DNA fingerprinting determines whether two DNA samples are from the:
(a) Same person (b) Related people (c) Nonrelated people (d) All

83. DNA fingerprinting is used to resolving:
(a) Criminal cases (b) Clarifying paternity (c) Immigration arguments (d) All

84. The first criminal caught using DNA fingerprinting:
(a) Gay Dotson (1998) (b) H Cooley (2001)
(c) Westergard (1995) (d) Colin Pitchfork (1987)

85. DNA fingerprinting is a technique of identifying:
(a) Some specific traits (b) A species
(c) A specific individual (d) None

86. DNA fingerprinting uses a specific type of DNA sequence known as:
(a) Palindromic sequence (b) Microsatellite
(c) Chimeric DNA (d) cDNA

87. Which one of the following statements is incorrect?
(a) Microsatellites tend to be highly variable.
(b) Short pieces of DNA that repeat many times in a given person's DNA is known as telli microsatellite.
(c) It is easy to identify a person by comparing the number of microsatellites in a given area.
(d) DNA fingerprinting is not the same in every cell, tissue, and blood of an individual.

88. The probability that a DNA fingerprint obtained will have its other matching fingerprint in the unrelated population is one in:

(a) One thousand (b) Ten thousand (c) One hundred million (d) One hundred billion

89. What is incorrect about Amplified Fragment Length Polymorphism–Polymerase Chain Reaction (AFLP–PCR):
(a) AFLP–PCR was first described by Pieter Vos et al. (1995)
(b) AFLP–PCR can only detect dominant genetic markers
(c) It can report whether an individual is homozygous for a given marker
(d) It cannot permit simultaneous amplification of multiple genomic DNA fragments

90. Hypervariable regions are also known as:
(a) Satellite DNA (b) Minisatellite DNA (c) Macrosatellite (d) All

91. DNA fingerprinting cannot be obtained from:
(a) Vaginal fluid (b) Bones that have been buried for a long time
(c) Roots of hair (d) None

92. Which one of the following statements is incorrect?
(a) The number of repeating sequences of DNA in a given position on a chromosome is different for each person.
(b) Bacteria have a unique DNA pattern.
(c) DNA fingerprinting helps in knowing genography.
(d) Highly repetitive DNA sequence includes mobile elements.

93. In India, first DNA fingerprinting for paternity dispute was carried out in:
(a) 1989 in Chennai (b) 1989 in Mumbai (c) 1992 in New Delhi (d) 1994 in Kolkata

94 Which one of the following can be used for DNA fingerprinting?
(a) Sperms (b) WBCs (c) RBCs (d) Vaginal fluid

95. Which one of the following statements is incorrect?
(a) Microsatellite DNA is composed of tandem repeats of two nucleotide pairs.
(b) Minisatellite DNA is also called Variable Number Tandem Repeats (VNTRs).
(c) Minisatellite does not contribute to the function of gene.
(d) Variable number tandem repeats are not dispersed throughout genome.

96. The most common method of DNA fingerprint today is based on polymerase chain reaction which uses:
(A) Long tandem repeats (b) Short tandem repeats
(c) Both (a) and (b) (d) None

97. Which one of the following is an incorrect statement?
(a) DNA fingerprinting uses a specific type of DNA sequence called microsatellites.
(b) Microsatellites are long pieces of DNA.
(c) In a given area microsatellites tend to be highly variable making them ideal for DNA fingerprinting.
(d) A person can be easily identified by comparing a number of microsatellites in a given area.

98. Consider the following statements:
(A) Variable tandem repeats are results of genetic inheritance
(B) VNTRs are evenly distributed across all human populations
(C) A given VNTR has a stable probability of occurrence
(D) VNTR may vary depending on the individual's genetic background

The correct statements are:
(a) A and B (b) B and C (c) A and D (d) All

99. Tandem repeats do not include:
(a) Minisatellites (b) Microsatellites (c) Satellites (d) Transposons

100. In humans, most satellites are located in the:
   (a) Centromere (b) Telomere (c) Chromatid (d) All
101. The telomere contains tandem repeated sequence of:
   (a) AAATIG (b) GGGTTA (c) GGGCCA (d) CCCGGT
102. The best way of determining paternity is the:
   (a) Counting of chromosome number (b) Testing of blood group
   (c) DNA fingerprinting (d) Pedigree analysis
103. Which one of the following has a variety of STRs?
   (a) Barley (b) Rice (c) Maize (d) Wheat
104. Which one of the following is used for amplification of DNA in the process of DNA fingerprinting?
   (a) Southern blotting (b) Northern blotting
   (c) Polymerase chain reaction (d) All
105. When one of the following is an incorrect match?
   (a) Barley – Only AT type of STR is found
   (b) Sequence tagged sites – DNA length of 100–500 bp that are unique in genome
   (c) England – DNA fingerprinting was first used
   (d) *Alu* repeats – Lacking in human genome
106. In variable number of tandem repeats, variation may occur due to:
   (a) Mutations (b) Small deletions (c) Small insertions (d) All
107. Which one of the following is not associated with DNA fingerprinting?
   (a) Polymerase chain reaction (b) Somatic cell hybridisation
   (c) Southern blotting (d) Restriction fragment length polymorphism
108. DNA fingerprinting was first used in England to identify:
   (a) Rapist (b) Criminal
   (c) Parentage of an immigrant (d) Parental dispute
109. In which one of the following countries is the 13 core STR loci being used to generate the genetic profile of an individual?
   (a) United Kingdom (b) United States of America
   (c) India (d) Germany
110. What is incorrect about microsatellites?
   (a) They are molecular markers. (b) Typically they are neutral.
   (c) They are co-dominant. (d) Consist of repeating units of more than 10 base pairs.
111. Microsatellites are:
   (a) Polymorphic loci (b) Present in nuclear DNA
   (c) Organelle or DNA (d) All
112. What is correct about satellite DNA?
   (a) Has usual base composition (b) Quite uniform within a species
   (c) Never found in heterochromatic region (d) Shows no variability within the closely related species
113. Which one of the following is related to DNA fingerprinting?
   (a) Allelic exclusion (b) Null allele
   (c) Gene amplification (d) Gene knockout
114. Which one of the following is a correct match?

(a) DNA fingerprinting – DNA profile
(b) Expression analysis – Expression profiling
(c) Gene finding – Computational Biology
(d) All

115. The institute associated with DNA fingerprinting:
(a) Indian Institute of Science, Bangalore
(b) Bose Institute, Kolkata
(c) Centre for Cell and Molecular Biology, Hyderabad
(d) All India Institute of Medical Sciences, New Delhi

116. *Bkm*-derived probe for DNA fingerprinting has been developed by:
(a) T Sharma (b) Lalji Singh (c) A Mukherjee (d) S P Roychaudhary

117. *Bkm*-derived probe for DNA fingerprinting is being extensively used for:
(a) Paternity determination
(b) Forensic investigation
(c) Seed stock verification
(d) All

118. Which of the following tools were first used for translating DNA sequences into proteins?
(a) Sequence assembly program
(b) Protein engine and *Transeq*
(c) *Paralign*
(d) *Patslatch*

119. Which one of the following allows for comparison of gene expression profiles between two mRNA samples?
(a) Mass spectrometry
(b) Microarray experiments
(c) Polyacrylamide gel electrophoresis
(d) Massively parallel signature sequencing

120. The objective of sequence alignment is to match the:
(a) Most similar elements of two sequences
(b) Different elements of two sequences
(c) Most similar elements of the same sequence
(d) Protein and DNA of different sequences

121. Genomics is helpful in:
(a) Intergenomic comparison
(b) Identifying sequence patterns of regulatory sites and gene regulations
(c) Identification of open reading frame sequences
(d) All

122. Who established genomics?
(a) Alwine (b) W Gilbert (c) Herbert Buyer (d) Tattersall Smith

123. The first man to determine the sequence of a gene:
(a) Rosenberg and Court (1979)
(b) John and Milkos (1980)
(c) Walter Fiers (1972)
(d) Forbes D J (1992)

124. The first genome to be sequenced was that of:
(a) Bacteriophage (b) Bacteria (c) Yeast (d) *Drosophila*

125. Expression profiling is applicable to:
(a) DNA fingerprinting
(b) Genomics
(c) Transcriptomics
(d) Metabolomics

126. In human body nearly ___________ can be found:
(a) 3,500 food components
(b) 2,500 metabolites
(c) 1,200 drugs
(d) All

127. Genome sequencing of which one of the following is yet to taken:
(a) *E. coli* (b) *Oryza sativa* (c) *Pan troglodytes* (d) None

128. Which one of the following scientists is not associated with the Human Genome Project?
(a) Ari Patrinos (b) Brenner (c) Craig Venter (d) Francis Collins

129. In which one of the following cells can microsatellites be gained or lost at higher rate, during one mitosis division:
(a) Liver cells (b) Brain cells (c) Tumour cells (d) Kidney cells

130. The most common microsatellite in humans is a:
(a) Dinucleotide repeat of CA (b) Trinucleotide repeat of CA
(c) Trinucleotide repeat of GC (d) Dinucleotide repeat of AT

131. Minisatellites consist usually of ___________-rich variant repeats.
(a) AT (b) CT (c) GC (d) GA

132. Consider the following statements:
(A) For forensic purpose, 4–6 different VNTR loci are analysed
(B) One VNTR in humans is a 17 bp sequence of DNA repeated between 70 to 450 times in the genome
(C) The total number of base pairs at this locus could vary from 1140 to 76750
(D) One VNTR is inherited from each parent
The correct statements are:
(a) All (b) B, C and D (c) A and C (d) B, C and D

133. VNTR can be isolated from an individual:
(a) Chromosome (b) DNA (c) DNA and RNA (d) All

134. Genome of viruses is in:
(a) Single chromosome (b) In the DNA
(c) In the DNA or RNA (d) All

135. Which one of the following is related to the Minimal Genome Project?
(a) ϕX174 (b) *Mycoplasma laboratorium*
(c) *E. coli* (d) *C. elegans*

136. Consider the following statements:
(A) *Eta globin* gene is located on chromosome 11 (B) It has no start codon (AUG)
(C) It has several stop codons (D) It is unable to form mRNA and protein
The correct statement is:
(a) All (b) C and D (c) A and B (d) None

137. Pyknons are:
(a) Variable length pattern within DNA sequences (b) Having identically conserved copies
(c) Multiplies above as expected by chance (d) All

138. What is incorrect about pyknons?
(a) Reported by Rigoukos et al. (2006). (b) Pyknons are most common in the human genome.
(c) Pyknons form one sixth of the human intergenic and intronic regions. (d) Pyknons are nonfunctional.

Answers to Multiple-Choice Questions

1. (b)	2. (b)	3. (b)	4. (c)	5. (b)	6. (d)	7. (a)	8. (b)
9. (d)	10. (c)	11. (d)	12. (c)	13. (a)	14. (b)	15. (b)	16. (a)
17. (a)	18. (c)	19. (d)	20. (d)	21. (d)	22. (d)	23. (d)	24. (b)
25. (b)	26. (c)	27. (b)	28. (b)	29. (c)	30. (b)	31. (b)	32. (a)
33. (c)	34. (d)	35. (c)	36. (d)	37. (b)	38. (d)	39. (a)	40. (c)

41. (d)	42. (a)	43. (b)	44. (b)	45. (d)	46. (c)	47. (d)	40. (d)
49. (b)	50. (d)	51. (b)	52. (d)	53. (d)	54. (c)	55. (a)	56. (c)
57. (c)	58. (c)	59. (c)	60. (b)	61. (a)	62. (b)	63. (d)	64. (a)
65. (b)	66. (c)	67. (c)	68. (d)	69. (c)	70. (d)	71. (d)	72. (b)
73. (d)	74. (d)	75. (c)	76. (b)	77. (b)	78. (d)	79. (d)	80. (c)
81. (b)	82. (d)	83. (d)	84. (d)	85. (c)	86. (b)	87. (d)	88. (d)
89. (c)	90. (b)	91. (d)	92. (d)	93. (a)	94. (a)	95. (d)	96. (b)
97. (b)	98. (c)	99. (d)	100. (a)	101. (b)	102. (c)	103. (b)	104. (c)
105. (d)	106. (d)	107. (b)	108. (a)	109. (b)	110. (d)	111. (d)	112. (b)
113. (c)	114. (d)	115. (c)	116. (b)	117. (d)	118. (b)	119. (b)	120. (a)
121. (d)	122. (d)	123. (c)	124. (a)	125. (c)	126. (d)	127. (d)	128. (b)
129. (c)	130. (a)	131. (c)	132. (a)	133. (b)	134. (c)	136. (b)	136. (a)
137. (d)	138. (d)						

Fill in the Blanks

1. Genome includes both genes as well as __________ sequences of the DNA.
2. The hereditary information encoded in the DNA of an organism is called the__________.
3. The first eukaryote whose genome was sequenced is __________.
4. Almost all studies of human genetic variation have focused on __________.
5. Cystic fibroses is caused by mutations in the __________ gene.
6. The Human Genome Project could not analyse __________ areas.
7. The Human Genome Project was $ __________ billion, publicly funded project.
8. The Human Genome Project was led at the __________ by the National Human Genome Research Institute.
9. The first human chromosome to be sequenced was chromosome __________.
10. VNTRs are separated through __________.
11. The first draft on an individual's human genome was published in the year __________.
12. The two general methods developed for mapping human genome were __________ and __________ using __________.
13. A collection of Bacterial Artificial Chromosome (BAC) clones containing the entire human genome is called __________.
14. __________ per cent of our genes are similar to a mouse.
15. Complete human genome is present in every cell, except __________.
16. The governmental department involved in initiating the Human Genome Project was __________.
17. Genetic polymorphisms occur on average every __________ nucleotides.
18. The first organism to have its genome sequenced was __________.
19. Funding of the Human Genome Project is met by __________ and __________.
20. The first draft on the Human Metabolome Project was completed in January 2007 by __________.
21. The Human Genome Project was established to map and to __________ the human genome.
22. The largest gene in human beings is dystrophin containing __________ Kbp.

23. The smallest gene in human beings consists of only ____________________ bp.
24. The term 'genomics' was coined by ____________________.
25. Genomics is divided into ____________________ genomics and ____________________.
26. The entire protein component of a given organism is called ____________________.
27. The term 'proteome' was coined by ____________________.
28. ____________________ are used to search the genome of organisms.
29. The technique by which an individual can be identified at molecular level is known as ____________________.
30. The sequence of ____________________ in DNA is similar to a DNA fingerprint.
31. Agrose gel is made from ____________________.
32. Short Tandem Repeats (STRs) are analysed using ____________________.
33. ____________________ is used for DNA fingerprint of highly degraded materials.
34. In order to analyse a southern blot, a radioactive genetic probe is used in ____________________ with the DNA in question.
35. To determine, whether a person has a particular VNTR, a ____________________ is performed and then the southern blot is ____________________.
36. The DNA differences in between two different individuals of the genome can be detected by ____________________ and is called ____________________.
37. Minisatellites are located in ____________________ regions.
38. In the human germ cell, the size of a telomere is about ____________________ kb.
39. Alphoid DNA is located at the ____________________ of all chromosomes.
40. Restriction fragment length polymorphisms are helpful in ____________________.
41. For DNA fingerprinting, DNA is cut into fragments of different length using ____________________.
42. Usually ____________________ of DNA is sufficient for preparing DNA fingerprinting.
43. The short tandem repeats in use today for forensic analysis are all ____________________ or ____________________ nucleotide repeats.
44. Microsatellites can be identified and amplified by ____________________.
45. A common method for DNA fingerprinting is____________________.
46. Short tandem repeats (a new method of DNA fingerprinting) analyses DNA segments for the number of repeats at ____________________ specific DNA sites.
47. The first human minisatellite was discovered by ____________________.
48. Usually ____________________ technology is used for the preparation of most DNA microarrays.
49. In reverse genetics, the functional study of a gene starts with the ____________________.
50. Variable tandem repeats are divided into two groups, viz., ____________________ and ____________________.
51. Amplified fragment length polymorphism is a modified technique of ____________________.
52. ____________________ is the 'Father of Bioinformatics'.
53. Expression analysis is the measuring gene expression via ____________________.
54. The largest project ever undertaken in life sciences is the ____________________.
55. Study of genomes of organisms is called____________________.
56. Study of genomics in relation to nitrogen utilisation and assimilation in organisms is called ____________________.
57. Proton (1H) NMR can detect any metabolite containing____________________.

58. Qualitative and quantitative analysis of metabolites is called ____________________.
59. The regions where a particular sequence of nucleotides or amino acids is indicative of the function in the protein are called ____________________.
60. One type of minisatellite is called ____________________.

Answers to Fill in the Blanks

1. Noncoding sequences
2. Genome
3. *Saccharomyces cervisiae*
4. Single Nucleotide Polymorphisms (SNPs)
5. CFTR
6. Heterochromatic
7. 3
8. National Institute of Health (NIH)
9. 22
10. Electrophoresis
11. 2007
12. Whole genome shot gun method, Standard method, molecular markers
13. Bacterial artificial chromosome library
14. 90
15. Red blood cells
16. Department of Energy
17. 300–500
18. *Haemophilus influenza*
19. National Institute of Health, Department of Energy (USA)
20. David Wishart
21. Sequences
22. 24,000
23. 24
24. Rodericks (1986)
25. Structural, functional
26. Proteome
27. Mare Wilikins (1994)
28. Computer programs
29. DNA fingerprinting
30. Nucleotides
31. Sea weed
32. Polymerase Chain Reaction
33. mtDNA
34. Hybridisation reaction
35. Southern blot, probe
36. Minisatellites DNA, Number of Tandem Repeats (VNTR)
37. Noncoding
38. 15
39. Centromere
40. DNA fingerprinting, gene mapping
41. Restriction enzymes
42. One microgram
43. Tetra, penta
44. Polymerase chain reaction
45. Restriction Fragment Length Polymorphism (RFLP)
46. 13
47. Wyman and White (1982)
48. Robotic
49. Gene sequence
50. Minisatellite, microsatellites
51. Random amplified polymorphic DNA
52. Hwa Lim
53. cDNA
54. Human Genome Project
55. Genomics
56. Nitrogenomics
57. Hydrogen
58. Metabolomics
59. Domians
60. Variable number of tendom repeats

True or False

1. The Human Genome Project is an international collaborative research programme.
2. The Human Genome Project was started on 1 October 1990 and was completed in 2003.
3. More refined sequence of the Human Genome Project becomes available in 2006 with minor corrections.
4. The human genome contains 3164.7 million basepairs.
5. As per the Human Genome Project, A–T regions are gene-dense regions while G–C regions are gene-poor regions.
6. The Human Genome Project sequence data are available to every one.
7. The Human Genome Project has been compared to the discovery of antibiotics.
8. The human genome reference sequence is person-specific genome.

9. 'Human Genome' as a book contains 46 chapters called chromosomes.
10. In the Human Genome Project, samples were collected mainly from races with American, European, African and Asian ancestry.
11. The Humane Genome Project is limited to studying only the human genome.
12. Duplications play a major role in shaping the genome.
13. Our DNA is 98 per cent identical to chimpanzees.
14. The onion genome is 20 times the size of the human genome.
15. In the human genome, each gene is expressed in every cell.
16. Exonic structural variations may have a direct influence on gene products.
17. The size of the genome affects the complexity of the organism.
18. 12,000 letters of DNA are decoded by the Human Genome Project every second.
19. Chromosome 9 in human beings is highly structurally polymorphic.
20. All human genes are expressed during development.
21. Each chromosome contains gene-rich and gene-poor regions.
22. mtDNA has a less rate of variation in comparison to nuclear DNA.
23. SPSS is a major tool for statistical analysis of data for genetic studies.
24. Most SNPs have no physiological effect.
25. Satellite DNA is used in DNA fingerprinting.
26. Short tandem repeats are very common.
27. Only one tenth of one per cent of DNA differs from one person to another.
28. cDNA sample is used for genotypic analysis.
29. It is not possible to identify an individual by auto-antibody fingerprint.
30. Transcriptome is the set of all mRNA as well as DNA molecules.
31. The genome is relatively fixed for a given cell line.
32. A given VNTR has a stable probability of occurrence.
33. A common procedure for DNA fingerprinting is Restriction Fragment Length Polymorphism (RFLP).
34. The techniques used in DNA fingerprinting have no applications in paleontology and archaeology.
35. DNA fingerprinting is used to diagnose inherited disorders.
36. DNA fingerprinting is a very quick method to compare DNA sequences of two living organisms.
37. The transcriptome is relatively fixed.
38. The effect of gene knockout is permanent.
39. The function of all genes in a gene family can be investigated through reserve genetics.
40. In eukaryotes, proteome is smaller than genome.
41. Hypervariable regions have constant number of Variable Number of Tandem Repeats (VNTRs).
42. In 'Human Genome' as a book, each word is written in letters called bases.
43. 2DPAGE (Polyacrylamide Gel Electrophoresis) is a method of proteome analysis.
44. Y chromosome lacks polymorphic markers.
45. Transcriptome is not the same in all cells.
46. The repeated stretches of DNA sequences are generally the same from person to person but the number of times they are repeated tends to vary.
47. In VNTR, the repeated sequence is shorter.
48. VNTR with very short repeat blocks is quite stable.

49. Repetitive DNA forms over 40 per cent of the human genome.
50. Damaged DNA may result in microsatellite instability.
51. Simple sequence repeats cannot be used as genetic markers.
52. The number of VNTR in humans varies from 7 to 40.
53. Temple Smith (1991) coined the term Bioinformatics.
54. Human pyknons are also present in other genomes.
55. VNTR loci are not similar between closely related individuals.

Answers to True or False

1. True	2. True	3. True	4. True	5. False	6. True	7. True	8. False
9. False	10. True	11. False	12. True	13. True	14. True	15. False	16. True
17. False	18. True	19. True	20. False	21. True	22. False	23. True	24. True
25. False	26. True	27. True	28. False	29. False	30. False	31. True	32. False
33. True	34. False	35. True	36. True	37. False	38. True	39. True	40. False
41. False	42. True	43. True	44. False	45. True	46. True	47. False	48. False
49. True	50. True	51. False	52. True	53. True	54. True	55. False	

Give Reasons

1. The knowledge obtained from the Human Genome Project is applicable to everyone.
   - Because all human beings share the same basic set of genes and genomic regulatory regions that control the development and maintenance of their biological structure and processes.
2. DNA fingerprinting is useful.
   - Because DNA structure cannot be changed.
3. DNA fingerprinting can be used to indicate paternity and maternity.
   - Because DNA patterns are inherited from parents.
4. Short tandem repeats is used to discriminate between unrelated individuals.
   - Because different unrelated individuals have different number of repeats.
5. Microsatellites are also known as Short Tandem Repeats (STRs).
   - Because a repeat unit consists of only 1 to 6 base pairs and the whole repetitive region is less than 150 base pairs.
6. Variable Number Tandem Repeats (VNTRs) are not distributed evenly across all human populations.
   - Because VNTRs are results of genetic inheritance.
7. It is easier to prepare DNA cleanly from sperm than other cell types.
   - Because sperm has higher ratio of DNA to protein and purification can be done due to smaller volume.
8. Sperm is suitable for DNA sequencing.
   - Because it contains equal number of cells with X chromosome (female) and Y chromosome (male).
9. Microsatellites are more suitable for PCR analysis.
   - Because they are smaller in size.

MOLECULAR GENETICS OF CANCER

Multiple-Choice Questions

1. Cancer cells are not characterised by:
   (a) Changes in plasma membrane
   (b) Disappearance of gap junctions
   (c) Higher glycolytic activity
   (d) No changes in the fibronectin
2. Consider the following statements:
   (A) Uncontrolled growth is the main characteristic of cancer cells
   (B) Cancer cells are structurally and biochemically different from normal cells
   (C) In cancerous cells there is reduction in gangliosides
   (D) The cancerous tumours are monoclonal
   The incorrect statement is:
   (a) D (b) C (c) B (d) A
3. Which one of the following genes is not a member of the *C-ras* family?
   (a) M (b) N (c) H (d) K
4. Oncogene *myc* is responsible for:
   (a) Sarcoma (b) Carcinoma (c) Myelocytoma (d) All
5. Reversible phosphorylation is shown by tumour-suppressor gene:
   (a) *RB* (b) *NF-1* (c) *WT-1* (d) P^{53}
6. Which one of the following has been derived from the entire *C-onc* genes having no missing regions?
   (a) *ras* gene (b) *myc* and *sis* genes (c) *mos* gene (d) All
7. In retinoblastoma there is a deletion in the:
   (a) Short arm of chromosome 13
   (b) Long arm of chromosome 13
   (c) Long arm of chromosome 15
   (d) Short arm of chromosome 9
8. Philadelphia chromosome is associated with:
   (a) Bladder cancer
   (b) Retinoblastoma
   (c) Burkitt's lymphoma
   (d) Chronic myeloid leukaemia
9. Cancer cells are characterised by:
   (a) Immortalisation (b) Metastasis (c) Transformation (d) All
10. Consider the following statements:
    (A) Cancerous cells need a solid surface for growth
    (B) Cancerous cells have strong dependence on serum
    (C) Cancerous cells form a thick mass of cells
    (D) Cancerous cells may induce tumours
    The correct statements are:
    (a) A, B and C (b) B, C and D (c) B and C (d) C and D

11. Which one of the following activates protoncogenes?
    (a) Amplification of DNA sequences (b) Insertions
    (c) Translocations (d) All
12. Oncoproteins and their corresponding proto-oncoproteins are located in the:
    (a) Cytoplasm (b) Nucleus (c) Nuclear membrane (d) All
13. Formation of tumour is:
    (a) Spontaneous (b) Induced by viruses
    (c) Induced by carcinogens (d) All
14. Sometimes genes present in the host are not oncogenic, they are known as:
    (a) Antioncogenes (b) Protoncogenes
    (c) Tumour suppressor genes (d) Metastatic genes
15. Chromosome aberrations are particularly important in the diagnosis, progression and treatment of:
    (a) Acute and chronic leukaemia (b) Small round cell tumours
    (c) Central nervous system tumours (d) All
16. Consider the following statements:
    (A) All protooncogenes have *introns*
    (B) Viral oncogenes lack *introns*
    (C) When a proto-oncogene undergoes somatic mutation, control of cell growth fails in the cell and may result in cancer
    (D) *Src* oncogene is present in the Rous Sarcoma virus

    The incorrect statements are:
    (a) A, B, C and D (b) A and B (c) B and D (d) None
17. Which one of the following is not used in the treatment of cancer?
    (a) Docetoaxel (b) Vincristine (c) Colchicine (d) Paclitaxel
18. The transformation of a normal cell to a cancer may start with changes in the:
    (a) Cytoplasm (b) Nucleus (c) Mitochondria (d) Fibronectin
19. Oncogenes were first discovered in certain retroviruses of ____________ origin:
    (a) Avian (b) Rodent (c) Feline/primate (d) All
20. A change in the *ras* oncogene from guanine to cytosine is frequently associated with:
    (a) Leukaemia (b) Bladder cancer (c) Lung cancer (d) All
21. Which one of the following statements is incorrect?
    (a) Tumour-suppressor genes are present in the human genome.
    (b) *Rb* gene product interacts with protein E2F.
    (c) The *Rb* gene product is only active when it is not phosphorylated by a *kinase*.
    (d) The mutant *Rb* gene product is not phosphorylated.
22. Benign tumour:
    (a) Does not invade other tissues (b) Does not form metastasis
    (c) Stops growing by itself (d) All
23. Mutation of P^{53} may cause:
    (a) Brain tumour (b) Soft tissue sarcoma (c) Osteosarcoma (d) All
24. Cancer-inducing viruses carry cancer-inducing genes called:
    (a) Proto-oncogenes (b) Oncogenes
    (c) Anti-oncogenes (d) Tumour-suppressor genes

25. Knudson two hit hypothesis is associated with:
    (a) AIDS (b) TB (c) Leprosy (d) Cancer
26. Approximately ________________ genes in the human genome have been associated with cancer:
    (a) 5,000 (b) 20,000 (c) 35,000 (d) 50,000
27. Basal nervous syndrome is:
    (a) Also called Gorlin syndrome
    (b) Caused by a tumour-suppressor gene, called *PTCH*, located on chromosomes 9
    (c) Mutations in this may increase the risk of ovarian cancer
    (d) All
28. Tumour-suppressor genes:
    (a) Generally control cell growth
    (b) Generally control cell death
    (c) Both copies of this gene must be mutated before a person develops cancer
    (d) All
29. Which one of the following genes is not implicated in cancer?
    (a) Oncogenes
    (b) Tumour-suppressor genes
    (c) Cell death genes
    (d) C-genes
30. What is incorrect about cancer cells?
    (a) Loss of apoptosis
    (b) Loss of senescence
    (c) Loss of ability to repair genetic error
    (d) Loss of capacity of angiogenesis
31. Proto-oncogenes are normally carefully regulated during:
    (a) Development
    (b) Tissue repair
    (c) Cell replacement
    (d) All
32. Which one of the following statements is incorrect?
    (a) Cancer is the most dangerous form of neoplasia.
    (b) Majority of cancers are sporadic.
    (c) Breast cancer is a familial cancer.
    (d) None
33. Which one of the following is the best-known example of failure of programmed cell death in cancer?
    (a) T-cell leukaemia
    (b) B-cell leukaemia
    (c) Platelet leukaemia
    (d) All
34. Malignant tumours generally contain:
    (a) Undifferentiated cells
    (b) Sometimes large nuclei
    (c) Sometimes large nucleoli
    (d) All
35. Oncogene theory has been given by:
    (a) Freeman et al. (1971)
    (b) Holley (1969)
    (c) Steel (1972)
    (d) Hubner and Todaro (1969)
36. Which one of the following types of gangliosides, predominantely are found in tumour cells?
    (a) GM la (b) GM1 (c) GM2 (d) GM3
37. For transformed cells, which one of the following is not true?
    (a) Myosin-like filaments disappear
    (b) Microfilament fibres disappear
    (c) Anodic mobility is generally higher
    (d) Well-developed electrical communication
38. Which one of the following is used to select transformed cells from normal cell population?
    (a) Loss of anchorage
    (b) Lower serum requirement
    (c) Loss of contact inhibition
    (d) Disaggregation of microtubules

39. Hodgkin's disease is an example of:
(a) Leukaemia (b) Sarcoma (c) Carcinoma (d) Lymphoma
40. Which one of the following is an anti-promoter of cancer?
(a) Cabbage (b) Cauliflower (c) Selenium (d) All
41. Which one of the following cancer secretes abundant amount of prostaglandins?
(a) Breast cancer (b) Prostate cancer (c) Blood cancer (d) Bone cancer
42. Anticancer effect is shown by:
(a) Indoles (b) Isothiocynates (c) Flavones (d) All
43. Which one of the following is an inhibitor of cancer?
(a) Bean (b) Tomato (c) Onion (d) Pea
44. Majority of human cancers are:
(a) Carcinoma (b) Sarcoma (c) Lymphoma (d) Leukaemia
45. Which one of the following is principally made up of connective tissue?
(a) Sarcoma (b) Carcinoma (c) Lymphoma (d) Leukaemia
46. Epstein Barr virus has been consistently isolated from:
(a) Neurological degeneration (b) Burkitt's lymphoma
(c) Osteosarcoma (d) Chondrosarcoma
47. *C-myc* gene contains:
(a) One exon (b) Two exons (c) Three exons (d) Five exons
48. Which one of the following tumour-suppressor genes stimulates GTPase activity?
(a) p^{53} (b) *NF-1* (c) *DCC* (d) *WT-1*
49. EIA is an oncoprotein located in the:
(a) Nucleus (b) Cell membrane (c) Cytoplasm (d) No definite location
50. Chromosomal abnormality associated with cancer was discovered by:
(a) Hubner and Todaro (1969) (b) Nowell and Hungerford (1960)
(c) Janet Rowley (1971) (d) Freeman et al. (1971)
51. Chromosomal rearrangement can cause cancer by:
(a) Forming a hybrid gene (b) Disregulation of a gene
(c) Both (a) and (b) (d) None
52. The chromosomal region *12q13-q14* is highly amplified 2n many:
(a) Carcinomas (b) Sarcomas (c) Lymphomas (d) Leukemias
53. Binding protein MCM2 when amplified:
(a) Prevents p^{53} from regulating cell growth (b) Initiates p^{53} to regulate cell growth
(c) Initiates *EWSR1* to stop cell growth (d) Prevents *RET* oncogenes to form hybrid gene
54. Chromosomes 1q and 16q are commonly lost in:
(a) Lymphomas (b) Solid tumour cells (c) Leukaemias (d) All
55. As tumours develop, the:
(a) Genome become unstable and complex (b) Centromeres might be duplicated
(c) Telomeres can erode (d) All
56. Deletions, inversions and translocation are commonly detected in chromosomal region 9q21 in:
(a) Leukaemias (b) Melanomas (c) Gliomas (d) All
57. Which one of the following statements is incorrect?
(a) Gene duplications and increase in gene copy numbers may contribute to cancers.

(b) Haematological cancers cannot be detected early.
(c) Presence of abnormal centromeres in tumours may cause genome instability.
(d) Measuring of *ERBB2* gene copy number may be a diagnostic tool for breast cancer.

58. Which one of the following is not used to treat cancer?
(a) Docetaxel (b) Vindesine (c) Colchicine (d) Vinblastine

59. Presently cytogenetic study is not used to make treatment decisions in:
(a) Lymphoma (b) Carcinoma (c) Sarcoma (d) Leukaemia

60. Protooncogenes produce:
(a) Mitogens (b) Hormones
(c) Cell surface receptors and transcription factors (d) All

61. Which one of the following genes is involved in multiple translocations?
(a) *ETV6* (b) *MLL* (c) *NUP98* (d) All

62. Cancer may develop:
(a) At any age (b) In any tissue (c) In any organ (d) All

63. Which one of the following oncogenes is amplified in neuroblastoma?
(a) *C-myc* (b) *N-myc* (c) *C-erb B-2* (d) None

64. Which one of the following parasites has been linked to bladder cancer?
(a) *Ascaris megalocephala* (b) *Taenia solium*
(c) *Wuchereria bancrofli* (d) *Schistosoma haematobium*

65. The development of cancer involves multiple mutations in:
(a) Proto-oncogenes (b) Tumour-suppressor genes
(c) DNA repair genes (d) All

66. The cancer with lowest incidence is the:
(a) Skin cancer (b) Kidney cancer (c) Thyroid cancer (d) Colorectal cancer

67. Oncoproteins and proto-oncoproteins are not present in the:
(a) Cytoplasm (b) Nucleus (c) Nuclear membrane (d) None

68. γ (gamma) interferon:
(a) Is a lymphokine with anti-tumour activity (b) Stimulates formation of granulocytes
(c) Produced by activated lymphocytes (d) Lacks anti-tumour activity

69. The first biochemical difference between normal cells and cancer cells was discovered by:
(a) Todaro and Green (b) Otto Warburg
(c) Pollack (d) Harris and Watkins

70. *C-src* gene is present in:
(a) Insects (b) Aves (c) Rat and man (d) All

71. Which one of the following oncogene is not located in the plasma membrane?
(a) *src* (b) *yes* (c) *mos* (d) *kit*

72. Cancer cells may contain:
(a) Structurally abnormal mitochondria (b) Structurally abnormal Golgi complex
(c) Poorly developed endoplasmic reticulum (d) All

73. Neoplasms are characterised by:
(a) Continued cell divisions (b) Abnormal biochemical behaviour
(c) Abnormal migration (d) All

74. Warburg effect is the excess consumption of:
(a) Glucose (b) Fat (c) Amino acids (d) None
75. Burkitt's lymphoma is a cancer of:
(a) Lung (b) Kidney (c) B cells (d) T cells
76. Which one of the following tests is widely accepted as an indication of carcinogenic nature of the substance?
(a) CIB test (b) Schick test (c) Ames test (d) Vidal test
77. Benzene induces:
(a) Liver cancer (b) Bladder cancer (c) Leukaemia (d) Stomach cancer
78. Which one of the following oncogenes is activated to make tumour by the insertion of retroviral genome?
(a) *int-1* (b) *int-2* (c) *c-myb* and *c-mos* (d) All
79. The factor which contributes to cancer pathogenesis is:
(a) Cell cycle activation of quiescent cells (b) Failure of apoptosis
(c) Inactivation of tumour-suppressor genes (d) All
80. In the US, the most common cancer is the:
(a) Lung cancer (b) Liver cancer (c) Prostate cancer (d) Breast cancer
81. Sezary's syndrome is a type of:
(a) T-cell lymphoma (b) B-cell lymphoma
(c) Cutaneous lymphoma (d) Bladder cancer
82. Which one of the following mutations can cause cancer?
(a) Random gene mutation (b) Inherited gene mutation
(c) Both (a) and (b) (d) None
83. CA-125 is a protein which is more prevalent in:
(a) Ovarian cancer (b) Bladder cancer (c) Prostate cancer (d) Breast cancer
84. Cancer can be treated by:
(a) Radiation and chemotherapies (b) Surgery
(c) Immunotherapy and monoclonal antibody therapy (d) All
85. Which one of the following body fluid is used to test the presence or absence of cancer?
(a) Pleural and pericardial fluid (b) Spinal fluid
(c) Sputum and urine (d) All
86. Experimentally it has been shown that alpha-tocopherol succinate seems to inhibit:
(a) Brain tumour (b) Leukaemia (c) Prostate cancer (d) All
87. Ewing's sarcoma is the result of a translocation between chromosomes:
(a) 10 and 22 (b) 11 and 22 (c) 8 and 14 (d) 2 and 8
88. Wilm's tumour is a neoplasm of:
(a) Liver (b) Kidney (c) Lung (d) Skin
89. The first conclusive evidence for cancer was published in:
(a) 1970 (b) 1980 (c) 1990 (d) 1997
90. Deletion of part or all of chromosome 9 are found in:
(a) Lung cancer (b) Prostate cancer (c) Bladder cancer (d) All
91. Which one of the following statements is incorrect?
(a) A malignant tumour is cancer. (b) Cysts and polyps are malignant tumours.
(c) Benign tumour is not cancer. (d) Cancer is a genetic disease but rarely is it inherited

92. The common area of occurrence of Ewing's sarcoma is:
    (a) Lung, kidney, pelvis and ribs
    (b) Pelvis, femur, humerus and ribs
    (c) Blood, kidney, pelvis and pharynx
    (d) Ovary, mammary glands, pelvis and ribs
93. Which one of the following is a correct match?

	Virus	Cancer
(a)	Epstein–Barr	Burkitt's lymphoma
(b)	Luck's disease virus	Breast cancer in women
(c)	Rous Sarcoma virus	Sarcoma in chicken
(d)	Herpes Simplex II	Cancer of uterus cervix in women

94. Which one of the following genes was discovered as the transforming gene of the highly oncogenic Abelson murine leukaemia virus?
    (a) a*bl* gene (b) *ras* gene (c) *mys* gene (d) *mos* gene
95. Homogenously Staining Regions (HSRs) and Double Minute Chromosomes (DMs) within chromosomes are related karyotypic abnormalities shown by:
    (a) Normal cells (b) Sex cells (c) Tumour cells (d) Degenerated cells
96. Consider the following statements:
    (A) Epistein–Barr Virus (EBV) causes mononucleosis
    (B) Mononucleosis is communicated through sexual contact
    (C) Mononucleosis is a disease of excessive lymphoid proliferation
    (D) Mononucleosis involves translocation between chromosome 6 and 22

    The incorrect statements are:
    (a) A, B and C (b) B, C and D (c) B and C (d) B and D
97. What is incorrect about retinoblastoma?
    (a) Deletion of chromosome 13 is associated with retinoblastoma.
    (b) Retinoblastoma arises when one copy of the gene *Rb-1* is delected.
    (c) Retinoblastoma arises when both copies of the gene *RB-1* is deleted or mutated.
    (d) Retinoblastoma involves tumour on the retina.
98. In small cell lung carcinoma, there is deletion of chromosome:
    (a) 3p (b) 13q (c) 11p (d) 15q
99. In Aniridia Wilm's tumour, there is deletion of chromosome:
    (a) 9p (b) 5q (c) 11p (d) 15q
100. In Burkitt's, translocation invariably involves chromosome 8 and chromosome:
    (a) 22 (K light chain genes)
    (b) 2 (λ light chain genes)
    (c) 14 (Heavy g chain genes)
    (d) All
101. The proto-oncogene involved in Burkitt's lymphoma is:
    (a) C-*myc* (b) N-*ras* (c) *C-erB* (d) *C-abl*
102. What is incorrect about humans *C-myc*?
    (a) Lies on the long arm of chromosome 8, at band q24
    (b) Spans about 7 kilobase pairs
    (c) Contains three exons
    (d) The first exon specifies a long untranslated region of the mRNA, while the remaining two exons encode the *myc* protein
103. Chronic myelogenous leukaemia contains two abnormal chromosomes which are:
    (a) The short-ended 22 (Ph^1) and an abnormal chromosome 9 called $9q^+$

(b) The short-ended 9(Ph^1) and an abnormal chromosome 22 called $22q^+$
(c) The short-ended chromosome 22 (Ph^1) and an abnormal chromosome 8 called 8 q^+
(d) The short-ended chromosome 22 (Ph^1) and an abnormal chromosome 2 called $2q^+$

104. In chronic myelogenous leukaemia, the translocation break points involved in generating Ph^1 and 9q+ always occur in the:
(a) q^{12} on chromosome 22 and q^{34} on chromosome 9
(b) q^{34} on chromosome 9 and q^{11} on chromosome 22
(c) q^{16} on chromosome 9 and q^{12} on chromosome 22
(d) q^{18} on chromosome 9 and q^{11} on chromosome 22

105. Which one of the following oncogenes produces epidermal growth factor receptor?
(a) *erb-B* (b) *ras* (c) *sis* (d) *abl*

106. Oncogene *sis* is derived from a gene that encodes:
(a) The B-chain of the platelet derived growthr facto
(b) Guanine–nucleotide-binding protein
(c) Tyrosine-specific protein *kinase*
(d) Thyroid hormone receptor

107. What is incorrect about nondefective cancer causing retroviruses?
(a) Lack oncogenes
(b) Lack ability to transform cells in culture
(c) Mainly induces leukaemias
(d) Have short latent period

108. All cancer causing retroviruses are defective except:
(a) Rous sarcoma virus
(b) Avian myeloblastosis virus
(c) Reticulo-endothelial virus
(d) Abelson murine leukaemia virus

109. The Philadelphia chromosome in chronic myelogenous leukemia results from a translocation between chromosomes 9 and 22 which joins the 5′ half a gene called *bcr* to the 5′ side of the:
(a) C-myc oncogene (b) ras oncogene (c) C-erB oncogene (d) abl oncogene

110. The *C-onc* which causes Kaposi's sarcoma:
(a) *neu* (b) *met* (c) *KS* (d) *mos*

111. Cancer cells contain ________ times phosphotyrosine than the normal cells:
(a) 5 (b) 10 (c) 20 (d) 25

112. Match column I with column II and select the correct answer using answer codes:

Column I (Disease)	Column II (Chromosomal rearrangement)
(A) Neuroblastoma	1. t(14;18)(q32.3, q21.3)
(B) Glioma	2. del(1) (p32-36)
(C) Malignant lymphoma	3. t(q 22)(q34;q11)
(D) Chronic myeloidleukaemia	4. del(22)(q11)

Answer codes:

	A	B	C	D
(a)	4	1	2	3
(b)	2	4	1	3
(c)	3	4	2	1
(d)	4	3	1	2

113. Who predicted that normal cells contain enemies in the form of oncogenes with in themselves?
(a) Beerman (1952)
(b) Huebner and Todaro (1968)
(c) Gall and Atherton (1974)
(d) Goldberg (1983)

114. The *src* gene of Rous sarcoma virus codes for a plasma membrane bound protein kinase which specifically phosphorylates amino acid:
 (a) Tyrosine (b) Threonine (c) Serine (d) None
115. Which one of the following statements is incorrect?
 (a) Mycosis fungoides is a rare form of T-cell Hodgkin's syndrome.
 (b) It was first described by Jeans Louis Mare Alebert.
 (c) It generally affects skin.
 (d) It is related to a fungus.
116. Variations of which of the following genes increase the risk of developing breast cancer:
 (a) *NBN* (b) *ATM* and *AR* (c) *CDH1* and *PTEN* (d) All
117. Consider the following statements:
 (A) The mutations in RET proto-oncogene can be inherited and may result in multiple endocrine neoplasia II
 (B) Tumour-suppressor genes are recessive at the cellular level
 (C) A mutation in tumour-suppressor gene can be inherited
 (D) In inherited cancer, one mutation is passed from the parent, but the rest acquired
 The incorrect statements are:
 (a) A and B (b) C and D (c) A and D (d) None
118. Two hit theory is associated with:
 (a) Cancer (b) AIDS (c) Immunoglobulin (d) Leprosy
119. Tumour-suppressor genes:
 (a) Are present in our cells (b) Control the process of cell growth
 (c) Control the process of cell death (d) All
120. Which one of the following concerns may be caused by inherited mutation?
 (a) Pancreatic (b) Prostate (c) Testicular (d) All
121. Li–Fraumeni syndrome is a rare:
 (a) Autosomal dominant disorder (b) Autosomal recessive disorder
 (c) X-linked dominant disorder (d) X-linked recessive disorder
122. Li–Fraumeni syndrome is related with germline mutation of tumour-suppressor genes:
 (a) *RB* (b) P^{53} (c) *NF-1* (d) *erb-A*
123. The neoplastic transformation involves the random activation of oncogenes and or the silencing of tumour-suppressor gene. This is known as:
 (a) Sewall Wright effect (b) Knudson two hit theory
 (c) Gene conversion hypothesis (d) Gene activation and inactivation theory
124. Cancer-related genetic aberrations lead to:
 (a) Altered proliferation (b) Survival
 (c) Differentiation (d) All
125. The oncogene products are similar to:
 (a) Growth factor (b) Hormone receptor
 (c) Analogue of nuclear receptor (d) All
126. Proto-oncogenes are activated by:
 (a) Mutation (b) Amplification of DNA
 (c) Insertion translocation (d) All

Answers to Multiple-Choice Questions

1. (d)	2. (b)	3. (a)	4. (d)	5. (a)	6. (d)	7. (b)	8. (d)
9. (d)	10. (d)	11. (d)	12. (d)	13. (d)	14. (b)	15. (d)	16. (d)
17. (c)	18. (b)	19. (d)	20. (b)	21. (d)	22. (d)	23. (d)	24. (b)
25. (d)	26. (c)	27. (d)	28. (d)	29. (d)	30. (d)	31. (d)	32. (d)
33. (b)	34. (d)	35. (d)	36. (d)	37. (c)	38. (a)	39. (d)	40. (d)
41. (a)	42. (d)	43. (a)	44. (a)	45. (a)	46. (b)	47. (c)	48. (b)
49. (a)	50. (b)	51. (c)	52. (b)	53. (a)	54. (b)	55. (d)	56. (d)
57. (b)	58. (c)	59. (a)	60. (d)	61. (d)	62. (d)	63. (b)	64. (d)
65. (d)	66. (c)	67. (d)	68. (a)	69. (b)	70. (d)	71. (c)	72. (d)
73. (d)	74. (a)	75. (c)	76. (c)	77. (c)	78. (d)	79. (d)	80. (c)
81. (c)	82. (c)	83. (a)	84. (d)	85. (d)	86. (d)	87. (b)	88. (b)
89. (d)	90. (c)	91. (b)	92. (b)	93. (b)	94. (a)	95. (c)	96. (d)
97. (b)	98. (a)	99. (c)	100. (d)	101. (a)	102. (b)	103. (a)	104. (b)
105. (a)	106. (a)	107. (d)	108. (a)	109. (a)	110. (c)	111. (b)	112. (b)
113. (b)	114. (a)	115. (d)	116. (d)	117. (d)	118. (a)	119. (d)	120. (d)
121. (a)	122. (b)	123. (b)	124. (d)	125. (d)	126. (d)		

Fill in the Blanks

1. A class of disease in which cells are characterised by uncontrolled growth is called ________________.
2. Almost all cancer is caused by abnormalities in the ____________________ of the transformed cells.
3. ____________________ is the abnormal proliferation of genetically altered cells.
4. Neoplasm may be ____________________ or ____________________.
5. Cancer is fundamentally a ____________________ disease.
6. Cancer starts and progresses because of changes in the____________________ of certain genes.
7. ____________________ and metastasis are necessary for a tumour to be a cancer.
8. Tumours are generally of two types, viz., ____________________ and ____________________.
9. Nearly almost all human cancers are initiated by ____________________.
10. ____________________ are genes that cause cancer.
11. Mixed malignant tumours have both ____________________ and ____________________ tissues.
12. Janet Rowley (1973) discovered that the Philadelphia chromosome is actually formed from a specific translocation of DNA between chromosome 9 and ____________________.
13. Sometimes, the same ____________________ gene is present in multiple types of cancers.
14. Chromosomal changes are highly ____________________ in different cancers.
15. Karyotype analysis based on ____________________ or ____________________ banding techniques have been widely applied to characterisation of cytogenetic abnormalities in tumour cells.
16. Only ____________________ tumours are capable of invading other tissues.
17. Carcinomas originates in ____________________ cells.
18. Leukaemia and lymphoma arise from cells having ____________________ origin.

19. Retinoblastoma in young children is an ____________________ cancer.
20. ____________________ tumours do not spread to other parts of the body.
21. Teratoma beings within ____________________ cells.
22. Inheriting certain mutation in the __________________ gene makes a woman more susceptible to breast cancer.
23. Telomerase mutations make a cell to divide____________________.
24. Cancer that begins in the basal cell of the skin is called ____________________.
25. Cancer that begins in the cells of the immune system is called ____________________.
26. All tumourigenic RNA viruses belong to the ____________________ group.
27. The first known tumour virus, ____________________, was discovered in 1911.
28. All cancer-causing agents are called ____________________.
29. There are ____________________ types cancer-causing retroviruses.
30. The viral oncogene present in Rous sarcoma virus is the____________________ gene.
31. For each type of cancer, there is ____________________ dominantly inherited form of mutation.
32. ____________________ is the invading of surrounding tissues by the transformed cells.
33. Carcinogenesis occurs in two stages called ____________________ and ____________________.
34. In the initiation phase normal cells exposed to a carcinogenic agent are irreversibly altered to a state called ____________________ state.
35. ______________ cells when stimulated by a promoting agent undergoes cancerous type of growth.
36. The initiation effect of carcinogen arises from its ability to chemically modify ____________________.
37. The phase of cancer induction during which carcinogens act on DNA is called the _______________.
38. The phase of carcinogenesis during which uncontrolled cell division is stimulated is called ____________________.
39. Initiation of tumour in an organism is called____________________.
40. ____________________ tumour does not invade surrounding tissues.
41. ____________________ is the cancer of glands.
42. Leukaemia is due to excess of ____________________.
43. ____________________ was the first specific chromosome abnormality to be associated with cancer.
44. ____________________ were one of the first tumours shown to harbour Homogenously Staining Regions (HSRs) and Double Minutes (DMs).
45. Burkitt's kymphoma is a cancer of B cells which is common in ____________________.
46. Gliomas are cancers of ____________________ cells.
47. Burkitt's lymphosia is characterised by specific translocations involving $8q^{24}$ and _____________, ____________________ or ____________________.
48. In Burkitt's lymphoma, an oncogene ____________________ which is normally located at $8q^{24}$ is transferred to 14q32 in a majority of cases.
49. The basic defect in cancer is the ____________________molecule.
50. The oncogenes carried by viruses are called ____________________ and loci with homologous sequences found in the host genes are called ____________________ genes.
51. The C-one genes are usually interrupted by ____________________.
52. The chromosomal rearrangement in Ewing sarcoma is____________________.

53. The B-cell malignancy characterised by specific chromosomal translocations involving $8q^{24}$ and either ____________________, ____________________ or ____________________.
54. The increased version of oncogenes causes ____________________.
55. The normal cellular version of the gene is called a ____________________.
56. The ____________________ protein was the first tumour-suppressor protein to be discovered.
57. Von Hippel-Lindau syndrome (VHL) is a ____________________ inherited hereditary cancer.
58. Von Hippel syndrome is caused due to the mutation of ____________________ gene on the short arm of the chromosome 3 ($3p^{26}$-p^{25}).
59. In cancer cells, two major types of chromosomal abnormalities are found called ____________________ rearrangement and ____________________ rearrangement.
60. The chromosomal rearrangement in acute lymphatic leukaemia is ____________________.
61. T-cell lymphoma may result by the rearrangement of the T-cell receptor alpha chain gene from ____________ to the locus of the immunoglobulin heavy chain genes on ____________ by paracentric inversion of the region $14q^{11}$, $14q^{32}$.
62. ____________________ test is a widely used test to detect possible chemical carcinogens.
63. Chicken cellular *src* pro to-oncogene contains ______________ introns separating ______________ coding sequences.
64. The viral oncogenes and cellular protocogenes are distinguished by using prefixes ______________ and ____________________.
65. The chromosomal arrangement in ovarian carcinoma is ____________________.

Answers to Fill in the Blanks

1. Cancer	2. Genetic material	3. Neoplasm
4. Benign, malignant	5. Genetic	6. DNA sequence
7. Malignant behaviour	8. Benign, malignant	9. Somatic mutations
10. Oncogenes	11. Ectodermal, endodermal	12. 22
13. Hybrid	14. Variable	15. G, R
16. Cancerous	17. Epithelial	18. Heamatopoeitic
19. Inherited	20. Benigin	21. Germ
22. *BRCA1*	23. Indefinitely	24. Basal cell carcinoma
25. Lymphoma	26. Retrovirus	27. Rous sarcoma virus
28. Carcinogens	29. Two	30. *src*
31. One or more	32. Metastasis	33. Initiation, promotion
34. Proneoplastic	35. Proneoplastic	36. DNA
37. Initiation	38. Promotion	39. Oncogenesis
40. Benign	41. Adenoma	42. White blood corpuscles
43. Philadelphia chromosome	44. Human neuroblastomas	45. Africa
46.. Glial	47. $14q^{32}$, $2p^{11}$, $22q^{11}$	48. *myc*
49. DNA	50. v-*onc*, c-*onc*	51. Introns
52. t(11;22)(q^{24}. q^{12})	53. $14q^{32}$, $2p^{11}$, $22q^{11}$	54. Cancer
55. Protooncogene	56. Retinoblastoma	57. Dominantly
58. VHL	59. Balanced chromosomal, imbalanced chromosomal	
60. t(1;3)(p^{36}. q^{21})	61. $14q^{11}$, $14q^{32}$	62. Ames
63. 11, 12	64. *v*, *c*	65. t(6;14)(q^{21}., q^{24})

True or False

1. Cancer comprises a group of diseases.
2. Cancer cells divide to form other cancer cells.
3. Many cancer cells have high level of cyclic AMP.
4. Generally malignant cells have higher glycolytic activity.
5. Cancer cells lack gap junctions.
6. Surface proteins of the cancer cells limit cell growth.
7. Oncogenic adenoviruses and papovaviruses have much smaller number of genes.
8. Retinoblastoma patients lack middle piece of chromosome13.
9. Mustard gas is a carcinogen.
10. It is possible to isolate oncogenes from tumours cell DNA.
11. SV_{40} causes cancer in man but not in laboratory animals.
12. Cancer was first recorded in ancient Egypt.
13. Elastography is used to classify tumours.
14. I^{131} is used in the treatment of thyroid cancer.
15. Ewing's sarcoma is a rare disease in which cancer cells are found in the bone or in soft tissues.
16. Ewing's sarcoma is the primitive name of neuroectodermal tumour.
17. Mutation in p^{53} gene occurs in approximately 50 per cent of all cancers.
18. The mutant p^{53} protein has the ability to control uncontrolled cell division.
19. Barley plants have anti-tumour property.
20. Talcum powder kills lung cancer cells.
21. Generally oncogenes are dominant.
22. Fine Needle Aspiration (FN(A) biopsy is related with cancer.
23. p^{53} is the most important tumour-suppressor gene.
24. Oncogenes are present in all cells.
25. In yeast, there are three *ras* genes, viz., *RAS* 1, *BAS* 2 and *RAS* 3.
26. There are enormous amplifications of specific DNA sequences in some cancer cells.
27. Chromosome of blood cells can be analysed easily.
28. Specific chromosome abnormalities may cause specific oncogene activation.
29. Different papilloma viruses have similar genome organisation.
30. Neuroblastomas lack the capacity to undergo progression with time from less malignant to more malignant form.
31. Chronic myelogenous leukaemia contains two abnormal chromosomes.
32. Feline leukaemia viruses (FeLv) are spread horizontally from cat to cat.
33. In many families there is a hereditary tendency of cancer.
34. Proto-oncogenes lack intervening sequences.

35. The *v-src* oncogene of Rous sarcoma virus is 1,700 nucleotides long.
36. Transformed cells are unable to grow in lack of attachment to a solid substratum.
37. Among RNA viruses only retroviruses are known to cause cancer.
38. People with Down syndrome have a predisposition to leukaemia.
39. Leukaemia has been reported in individuals with sex chromosomal mosaicism.
40. Mutations in antioncogenes can be transmitted dominantly.
41. Beans are cancer inhibitors.
42. Selenium induces breast cancer.
43. All oncogenes code for *Tyr protein kinase*.
44. The *erb-B* oncogene codes for the Platelet Derived Growth Factor (PDGF).
45. Carbon dioxide laser is used for removal of brain tumour.
46. Inherited mutations in *BRCA*1 and *BRCA*2 genes lead to early onset of breast cancer.
47. Deficiency of folic acid and B-carotene contributes to cervical cancer.
48. Necrosis is not a sign of excessive tumour growth.
49. Mutations in *BRCA1* and *BRCA2* genes are inherited in an autosomal dominant pattern.
50. Acquired mutations are always involved in causing cancer.
51. Mutations in proto-oncogenes are usually acquired.
52. All the people who inherit a mutation in a tumour-suppressor gene, proto-oncogene or DNA repair gene will develop cancer.
53. Most of the genea associated with hereditary cancer are tumour-suppressor genes.
54. Most of mutations in tumour-suppressor genes are inherited.
55. Multiple Endocrine Neoplasia I (MENI) tumours may be noncancerous (benign) or cancerous (malignant).
56. In cancer cells DNA is not damaged.
57. Bishop and Varmus (1976) pointed out that oncogenes are defective proto-oncogenes and they were awarded the Nobel Prize in 1989 for this discovery.
58. Malignant cells have generally shorter generation time.
59. *Ras* protein regulates cell division.
60. Activation of telomerase in tumours permit continuous proliferation of tumours.
61. Hair dyes cause bladder cancer.
62. Viral infection is necessary for tumour formation.
63. Viral oncogenes and cellular oncogenes are almost the same.
64. Viral oncoggenes are single exons.
65. Oncogenes of DNA tumour viruses have cellular counterparts.
66. *ras* proteins play an essential role in controlling the mammalian cell cycle.

Answers to True or False

1. True	2. True	3. False	4. True	5. True	6. False	7. True	8. True
9. True	10. True	11. True	12. True	13. True	14. True	15. True	16. True
17. True	18. False	19. True	20. True	21. True	22. True	23. True	24. True
25. False	26. True	27. True	28. True	29. True	30. False	31. True	32. True

33. True	34. False	35. True	36. False	37. True	38. True	39. True	40. True
41. True	42. False	43. False	44. False	45. True	46. True	47. True	48. False
49. True	50. True	51. True	52. False	53. True	54. False	55. True	56. True
57. True	58. True	59. True	60. True	61. True	62. False	63. True	64. True
65. False	66. True						

Give Reasons

1. Some cancers can be passed on to the next generation.
   – Because they occur in the germ cells.
2. Some cancers may occur more commonly in a particular family than in others.
   – Because such cancers occur in the germ cells and thus that family has inherited a predisposition to develop a particular type of cancer.
3. Some individuals who have inherited germ line tumour-suppressor gene mutation may never develop cancer.
   – Because there is no second mutation necessary to activate the function of the gene and to start the process of tumour formation.
4. In basal cell nervous syndrome, mutations can be inherited from the father or the mother's side of the family.
   – Because the gene responsible for basal cell nervous syndrome is not located on the sex chromosomes.
5. Genetic damage that occurs in cancer cells sets up a vicious cycle.
   – Because it reduces the cell's ability to repair future genetic damage.
6. Some tumour-suppressor genes are popularly called gatekeepers.
   – Because they normally regulate and control cell division and thus block the growth of the cancers.
7. Tumour cells have easier agglutination by lectins.
   – Because of increased mobility of the surface proteins in transformed cells.
8. Early detection of cancer is very important.
   – Because the earlier the cancer is detected, the chances for it being cured are high.
9. Cancer more frequently occurs in older persons.
   – Because development of cancer involves multiple mutations in several genes and accumulation of mutations in such genes takes many years.
10. Oncoproteins are so named.
    – Because they have the ability to bind guanine residues.
11. Tumour cells secrete more lactic acid.
    – Because they consume more glucose.
12. Occasionally cancer may not develop until many years after exposure to a carcinogenic agent.
    – Possibly because of the long time taken by the promotion phase.
13. Cytogenetic testing of cancer cells usually takes about three weeks.
    – Because cancer cells must grow in laboratory dishes for about two weeks before their chromosomes are ready to be viewed under a microscope.

14. Barley plants have anti-tumour property.
    - Because barley plant contains alpha tocopherol succinate, which has anti-tumour action.
15. Cancer cells are easily damaged by radiation in comparison to normal cells.
    - Because cancer cells have a very high rate of division.
16. Early chromosome preparations were not suitable for identifying all the human chromosomes.
    - Because in the early preparations of chromosome, the only distinguished characteristics available were length and centromere location.
17. Cancer causes death.
    - Because cancer cells multiply indefinitely. As a result, their number increases day by day demanding essentially more nutrients available in the body. And hence, normal cells gradually suffer from nutritive death.
18. Oncogenes are said to be dominant at cellular level.
    - Because a mutation in just one of the two copies of a particular proto-oncogene is sufficient to cause a change in cell growth, leading to the formation of a tumour.
19. In transformed cells, oncogenic proteins are made in large quantities.
    - Due to the function of oncogenes like active viral genes.
20. Tumour-suppressor genes are recessive at cellular level.
    - Because both copies of a specific tumour-suppressor gene need to be mutated in order to cause change in cell growth and formation of tumour.

IMMUNOGENTICS

Multiple-Choice Questions

1. Consider the following statements:
(A) An antigen is a large complex molecule
(B) All foreign proteins are antigens
(C) A complex antigen may have many antigenic determinant sites
(D) Proteins are more antigenic
The correct statements are:
(a) All (b) A, B and C (c) A and C (d) A and D
2. The T and B stem cells originate from:
(a) Bone marrow (b) Yolk sac (c) Foetal liver (d) All
3. An average human body contains approximately ______ antibodies and ______lymphocytes:
(a) 10^{10}, 10^{15} (b) 10^{20}, 10^{12} (c) 10^{12}, 10^{10} (d) 10^{25}, 10^{50}
4. Which one of the following is a lymphoepithelial organ?
(a) Thyroid (b) Parathyroid (c) Thymus (d) Adrenal
5. Lymphopoiesis in thymic grafts has been found to be proportional to the extent of ________________ in graft:
(a) Epithelial tissue (b) Connective tissue (c) Muscular tissue (d) Both (b) and (c)
6 Which one of the following is applicable to lymphoid tissues?
(a) Production of lymphocytes (b) Processing of lymphocytes
(c) Storage of lymphocytes (d) All
7. The site of maturational processing of T lymphocytes:
(a) Bone marrow (b) Thymus (c) Spleen (d) Lymph nodes
8. The site of maturational processing of B lymphocytes:
(a) Bone marrow (b) Adenoids (c) Spleen (d) Tonsils
9. Match column I with column II and select the correct answer using answer codes:

Column I	Column II
(A) Metchnikov	1. Radioimmunoassay
(B) von Behring	2. Rabies vaccine
(C) Pasture	3. Phagocytosis
(D) Rosalyn Yalow	4. Introduced therapy with antibodies

Answer codes:

	A	B	C	D
(a)	3	4	2	1
(b)	4	2	1	3
(c)	2	1	4	3
(d)	3	2	4	1

10. Which one of the following is a correct match?
    (a) Neutrophils – Highly mobile
    (b) Basophils – Destroy parasitic worms
    (c) Eosinophils – Release heparin and histamine
    (d) T lymphocytes – Transformed into plasma cells
11. Lactoferrin is secreted by:
    (a) Eosinophils (b) Basophils (c) Neutrophils (d) Lymphocytes
12. Lactoferrin is a protein, which binds tightly with __________, making it unavailable for use of bacteria:
    (a) Zinc (b) Iron (c) Copper (d) Calcium
13. Antibodies are termed as:
    (a) α globulins (b) β globulins (c) γ globulins (d) Albumins
14. Which one of the following is the slowest moving protein?
    (a) Albumins (b) α globulins (c) β globulins (d) γ globulins
15. Which one of the following is applicable to the functioning of the immune system?
    (a) Retention of the memory
    (b) Learning
    (c) Recognising
    (d) All
16. B lymphocytes were first discovered in:
    (a) Thymus gland
    (b) Lymph nodes
    (c) Spleen
    (d) Bursa Fabricius of birds
17. Which one of the following is not applicable to T lymphocytes?
    (a) Synthesis of antibodies
    (b) Cell-mediated immunity
    (c) Regulation of immune response
    (d) Proliferation of killer T cells
18. Which one of the following is incorrect about gamma (γ) interferon?
    (a) Heat labile and nondialysable
    (b) Nonglobulin protein
    (c) Resistant to DNAase and RNAase
    (d) Destroyed by trypsin
19. Antibodies are formed against:
    (a) Nucleic acids (b) Polysaccharides (c) Galactolipids (d) All
20. Who is the founder of immunology?
    (a) Edward Jenner (b) Henry H Dale (c) Louis Pasture (d) Paul Ehrlich
21. Paul Ehrlich was awarded the Nobel Prize for:
    (a) Discovery of phagocytosis
    (b) Explaining antibody production
    (c) Introduced therapy with antibodies
    (d) Immunisation procedures
22. The effect or cells are:
    (a) T cells (b) B cells (c) Both T and B cells (d) K cells
23. T cells mediate effective immunity against:
    (a) Almost all parasites
    (b) Myobacteria
    (c) Many viruses
    (d) All
24. On stimulation by antigens, T cells secrete:
    (a) Lymphokines (b) Pertussis toxin (c) Cadaverine (d) Calbindin
25. Which one of the following statements is incorrect about IgG?
    (a) It is a four-peptide unit.
    (b) It contains two light chains and two heavy chains.
    (c) For a particular amino acid the light chains and heavy chains are not identical.
    (d) The light and heavy chains are joined together by S–S bonds.
26. Purified IgG is:

(a) S shaped (b) Q shaped (c) Y shaped (d) C shaped

27. The structure of immunoglobulin molecules has been proposed by:
(a) Edelmand Porter (1962) (b) Rosalyn Yalow (1977)
(c) Ogston (1965) (d) Enders and Weller (1954)

28. An antigen-binding site on IgG molecule is constituted by:
(a) The variable domains of light chains (b) The variable domains of heavy chains
(c) The variable domains of both light and heavy chains (d) The constant domains of both light and heavy chains

29. There are _________ antigen-binding sites for each antibody in a IgG molecule:
(a) One (b) Two (c) Three (d) More than three

30. In an IgG molecule, the heavy chain is composed of a specific polypeptide chain, the:
(a) α chain (alpha (α) (b) δ chain (delta (δ)
(c) ε (epsilon) chain (d) γ (gamma (γ) chain

31. Helper T cells are:
(a) Phagocytic (b) Activate B cells in contact with antigen
(c) Activate killer T cells (d) All

32. In an IgG molecule, the valency for antigen-binding site is:
(a) 2 (b) 3 (c) 4 (d) 1

33. Five S–S loops are not found in:
(a) IgM (b) IgD (c) IgE (d) IgG

34. Levels of which one of the following immunoglobulins is raised in several tropical parasitic diseases?
(a) IgA (b) IgM (c) IgD (d) IgE

35. The most abundant immunoglobulin blood is:
(a) IgM (b) IgG (c) IgE (d) IgD

36. Which one of the following immunoglobulins is present in milk and tears?
(a) IgA (b) IgD (c) IgE (d) IgG

37. The major immunoglobulin involved in defence in the initial stages of infections is:
(a) IgA (b) IgD (c) IgE (d) IgG

38. Which one of the following immunoglobulins was discovered from a myeloma protein?
(a) IgG (b) IgM (c) IgD (d) IgA

39. Which one of the following immunoglobulins is present as 7S monomere in serum?
(a) IgM (b) IgA (c) IgD (d) IgE

40. The major protective antibody in newborns is:
(a) IgG (b) IgM (c) IgD (d) IgA

41. In 1984, the Nobel Prize was awarded to Niels K Jeme for his work on:
(a) Preparation of antihistaminic drug (b) Idiotypes as well as theories on antibody production
(c) Genetics of cancer (d) Transplantation research

42. Which one of the following is not applicable to IgE?
(a) Cytophilic antibodies (b) Mediate allergy
(c) Hypersensitivity and anaphylaxis (d) Lack ability to fix on mast cells and basophils

43. The immunoglobulin related with Rh isoimmunisation is:
(a) IgG (b) IgM (c) IgE (d) IgD

44. Natural antibodies are applicable to:
(a) IgD (b) IgE (c) IgM (d) IgG

45. Aagenase syndrome is:
(a) Autosomal dominant (b) Autosomal recessive
(c) Sex-linked dominant (d) Sex-linked recessive

46. J chain is a protein component of the antibodies:
(a) IgA (b) IgG and IgM (c) IgM and IgA (d) IgM and IgD

47. Antibody opsonisation is the:
(a) Process of formation of antibody (b) Process by which a pathogen is marked for ingestion and destruction by phagocyte
(c) Rearrangement of DNA for the formation of antibody (d) All

48. Animal immunogenetics is based on the use of:
(a) Recombinant strains (b) Inbred strains
(c) Congenic strains (d) All

49. The antigenic determinants present on immunoglobulins are:
(a) Isotypes (b) Allotypes (c) Idiotypes (d) All

50. In the mouse germ line, there are about _________ variable chain genes:
(a) 50 (b) 100 (c) 200 (d) 500

51. Immunoglobulins are heterogenous with respect to:
(a) Antigencity (b) Function (c) Size and charge (d) All

52. Consider the following statements:
(A) Each lymphocyte produces more than one type of lymphocyte
(B) Somatic mutation may occur in both variable regions and constant regions of an antibody molecule
(C) In the expression of immunoglobulin genes, RNA processing does not play any role
(D) The immune system does not work by clonal selection
The correct statements are:
(a) All (b) A, B and C (c) B, C and D (d) None

53. Which one of the following is not an immunoglobulin (Ig) gene family?
(a) Lambda (b) Kappa (c) Sigma (d) Rho

54. The immunoglobulin heavy chain variable region gene contains:
(a) Variable segment and diversity segment (b) Diversity segment
(c) Joining segment (d) Variable segment, joining segment and diversity segment

55. Which one of the following is not applicable to IgG isotypes?
(a) Opsonisation (b) Fixation to macrophages
(c) Complement fixation (d) External secretion

56. 72 F fusion protein is:
(a) Used in cell fusion (b) A tuberculosis vaccine
(c) A malarial vaccine (d) Oncoproteins

57. Each immunoglobulin family contains:
(a) Only v gene (b) Only c gene
(c) Both v and c genes, located on a single chromosome (d) Both v and c genes, located on different chromosomes

58. Fusion of immune cells with mouse myeloma is done through:
(a) Aminopterine (b) Polyethylene glycol (c) 8-azaguanine (d) None
59. In an IgG molecule, which one of the following bears a binding site for macrophages?
(a) C_1 (b) C_2 (c) C_3 (d) V_H
60. Multiple myeloma is characterised by:
(a) Proteinuria (b) Paraproteinemia (c) Lytic bone lesions (d) All
61. The idiotypic determinants are located in the:
(a) Variable part of the antibody (b) Constant part of the antibody
(c) Both (a) and (b) (d) Hinge region
62. Complement system was detected by:
(a) Jules Bordet (1919) (b) Charles R Richet (1913)
(c) Albert Szent–Gyrogyi (1937) (d) Gerhard Domagk (1939)
63. Small molecules when become covalently conjugated proteins are called:
(a) Epitopes (b) Haptens (c) Teratogens (d) Effectors
64. IgG isotypes are not associated with:
(a) Opsonisation (b) Fixation to macrophages and membrane transport
(c) Complement fixation (d) Hypersensitivity
65. Which one of the following is responsible for antigenic variation?
(a) Gene mutation (b) Switching (c) Recombination (d) All
66. Which one of the following is an incorrect match?
(a) External secretions – IgA (b) Placental transfer – IgG
(c) Agglutination – IgD (d) Fixation to mast cells and basophils – IgE
67. Which one of the following complement is mainly synthesised by hepatocytes?
(a) C_1 (b) C_2 (c) C_3 (d) C_9
68. The secretory piece of IgA is produced in:
(a) Bone marrow (b) Liver (c) Spleen (d) Pancreas
69. The cells involved in immune responses are:
(a) Lymphocytes and macrophages (b) Lymphocytes and Langerhans cells
(c) Monocytes and macrophages (d) All
70. The inability of the immune system to respond to an infection is known as:
(a) Antigenic determinant (b) Epitope
(c) Antigenic escape (d) Two hit hypothesis
71. The cellular response is important in the reaction:
(a) Viruses and fungal infections (b) Malignant cells
(c) Transplanted tissues (d) All
72. The phagocytic activity does not interfere with:
(a) Deactivation of toxins (b) Bacterial cell divisions
(c) Trapping viral particles (d) All
73. The antigen receptors of helper T cells are similar to those of:
(a) Cytotoxic T cells (b) Suppressor T cells (c) Both (a) and (b) (d) None
74. Antibody idiotype is not determined by:
(a) Junctional diversity (b) Somatic hypermutations

(c) Gene arrangement (d) Somatic cell hybridisation

75. In humans, there are ________ isotypes of immunoglobulin:
(a) 3 (b) 7 (c) 9 (d) 11

76. Consider the following statements:
(A) The T cell receptor is similar in its structure to the immunoglobulins
(B) The T cell receptor consists of two chains: α and ß
(C) The T cell receptor α chain gene cluster is located on chromosome 14 while the β cluster gene is on chromosome 7
(D) Both α and β chains of T receptor undergo somatic recombination to create different T-cell specificities
The correct statements are:
(a) All (b) None (c) A, B and C (d) C and D

77. The first antibodies produced by a clone of differentiated lymphocytes are a class of:
(a) IgM (b) IgG (c) IgE (d) IgD

78. Kveim test is a skin test, which is used to detect:
(a) Cancer (b) Sarcoidosis (c) AIDS (d) Leprosy

79. The kappa and lambda chain gene cluster lack:
(a) Junctional genes (b) Diversity genes (c) Variable genes (d) Constant gene

80. Which one of the following immunoglobulins specially acts as opsonins?
(a) IgG (b) IgD (c) IgE (d) IgM

81. The different behaviour of immunoglobulin molecules is due to their:
(a) J chain (b) Heavy chain constant regions
(c) Heavy chain variable regions (d) Both (a) and (c)

82. Natural killer cells:
(a) Are usually null cells (b) Have CD 16 as surface markers
(c) Bind to Fc portion of the IgG molecule (d) All

83. The antibody isotype of a B-cell changes during cell:
(a) Development (b) Development and activation
(c) Secretion and activation (d) Maturation and secretion

84. Consider the following statements with reference to the complement system:
(a) They are a group of serum proteins (b) They have the structure of immunoglobulins
(c) Their numbers increase after immunisation (d) In mammals, a group of nine proteins forms the complement system
The correct statements are:
(a) All (b) A, C and D (c) B and C (d) A and D

85. Which one of the following protein components of the complement system is a convertase that always remains bound to the original antigen-antibody complex?
(a) C_3 (b) C_5 (c) C_7 (d) C_9

86. Which one of the following antibodies activates the same complement system?
(a) IgM and IgD (b) IgD and IgA (c) IgM and IgE (d) IgG and IgM

87. The antibodies involved in viral neutralisation are:
(a) IgM and IgG (b) IgG and IgD (c) IgG and IgE (d) IgD and IgA

88. An antigen directly determines the shape of an antibody. This is known as:
(a) Somatic theory (b) Germ line theory (c) Instructive theory (d) Switching theory

89. Which one of the following T cells of humans contain T_4 antigen?
(a) Helper T cells (b) Cytotoxic T cells (c) Suppressor T cells (d) All

90. During the process of opsonisation:
(a) Antigens are bound by antibody (b) Antigens are bound by complement molecules
(c) Antigens are activated (d) Both (a) and (c)

91. Fc region of immunoglobulin is not involved in:
(a) Cell lysis (b) Degranulation of mast cells
(c) Opsonisation (d) Secretion

92. Variable domian is present in:
(a) Each heavy chain of the antibody (b) Each light chain of the antibody
(c) Each light and heavy chain of each antibody (d) None

93. Which one of the following provides the explanation for the genetic basis of immunological diversity and specificities?
(a) Comparison of amino-acid sequences (b) Gel electrophoresis
(c) Mancini technique (d) Gene cloning

94. Consider the following statements:
(A) The lambda gene family comprises 4C regions
(B) There are approximately 30V regions in lambda gene family
(C) The kappa light gene family lacks C region
(D) In kappa light gene family, there is only one exon for J region

The correct statements are:
(a) All (b) A, B and C (c) A and B (d) C and D

95. Which one of the following antigenic determinants, present in immunoglobulins, are found in all individuals?
(a) Isotypes (b) Allotypes
(c) Idiotypes (d) Allotypes and isotypes

96. J chain is a protein component of:
(a) IgM (b) IgM and IgA (c) IgD (d) IgA and IgD

97. Match column I with column II and select the correct answer using answer codes:

Column I (Complement Components)	Column II (Clinical association)
(A) C1	1. Raynaud's phenomenon
(B) C4	2. Hereditary angioedema
(C) C5	3. SLE like syndrome
(D) C7	4. Gonococcal and recurrent infections

Answer codes:

	A	B	C	D
(a)	2	3	4	1
(b)	3	2	3	1
(c)	4	3	2	1
(d)	3	4	1	2

98. Neutralisation of herpes-type viruses by antibody is enhanced by component of complement system:
(a) C_1, C_4 and C_2 (b) C_3 and C_5 (c) C_4 and C_5 (d) C_1, C_4 and C_5

99. It is an autoimmune disease in which antibodies become deposited on the membranes of both lungs and kidneys:

(a) CREST syndrome
(b) Goodpasture syndrome
(c) Nezelof syndrome
(d) Alper syndrome

100. The antibodies having the ability to polymerise:
(a) IgG
(b) IgG and IgM
(c) IgA and IgM
(d) IgM and IgA

101. Which one of the following is incorrect about severe combined immunodeficiency (SCI(D)?
(a) A marked reduction in immunoglobulin levels.
(b) ABO isoagglutinins are present.
(c) Both cellular and humoral responses are absent.
(d) Lymphocyte count is reduced.

102. Which one of the following syndrome is autosomal recessive in which precursors of T cells fail to migrate from bone marrow to thymus?
(a) Nezelof syndrome
(b) DiGeorge syndrome
(c) Sturge–Weber syndrome
(d) Klippel–Feil syndrome

103. Which one of the following is an auto-immune disease ?
(a) Polymyositis
(b) Guill–Barre syndrome
(c) CREST syndrome
(d) All

104. Which one of the following proteins is associated with the pathogenesis of neonatel multisystem inflammatory disease?
(a) Cryptdin
(b) Cryopyrin
(c) Cathelicidin
(d) Zein

105. Mononuclear phagocytes act as a host for:
(a) *Leishmania*
(b) *Salmonella*
(c) *Toxoplasma*
(d) All

106. Which one of the following is correct about DiGeorge syndrome?
(a) Normal immunoglobulins
(b) Lack of cell-mediated immunity
(c) In some cases there is microdeletion of chromosome 22
(d) All

107. Granulysin is secreted by:
(a) Cytotoxic T cells
(b) B cells
(c) Both (a) and (b)
(d) Stromal cells

117. Which one of the following syndromes is a sex-linked recessive syndrome and is characterised by predominantly IgM deficiency as well as poor antigenic response to blood groups:
(a) Lesch–Nyhan syndrome
(b) Wiscott–Aldrich syndrome
(c) Zellweger syndrome
(d) Rett syndrome

119. The reaction of antibody with antigen causes:
(a) Agglutination and precipitation
(b) Cytotropic effects
(c) Complement-dependent interaction
(d) All

110. Which one of the following is not applicable to gamma interferon?
(a) Antiviral protein
(b) Inhibition of viral replication
(c) Heat stable and nondialysable
(d) Skin reactive factor

111. Bare lymphocyte syndrome is due to:
(a) Deletion of short arm of chromosome 9
(b) Deletion of short arm of chromosome 15
(c) Deficiencies in major histocompatibility complex
(d) Deficiency of estrogen

112. RAG1 and RAG2 are purified proteins isolated from:
(a) Nervous tissue
(b) Lymphoid tissue
(c) Stromal cells
(d) All

113. The variable region of the immunoglobulin is:
(a) Agglutination
(b) Precipitation
(c) Neutralisation of antigens
(d) All

114. The cells not affected by AIDS:
(a) B lymphocytes (b) CD_4^+ T helper cells
(c) CD_4^+ macrophages and CD_4^+ monocytes (d) CD_8^+ T cells

115. Which one of the following is an incorrect match?
(a) Adhesion molecules – CD2, CD8 (b) Co-receptor – CD19
(c) Antigen receptor – B-cell receptor chain (d) Ig binding receptors – Polymeric immunoglobin molecules

116. Regulatory T-cell activity occurs during infection of:
(a) *Leishmania* (b) *Plasmodium* (c) HIV (d) All

117. Individual immunoglobulin-producing, cells of heterozygous individuals produce immunoglobulins of a single allotype. This is known as:
(a) Genetic erosion (b) Genetic suppression
(c) Allelic exclusion (d) A type of rare mutation

118. At which end of the transcript does substitution of one exon by another take place, during expression of immunoglobulins genes:
(a) 5′ (b) 3′ (c) Both 3′ and 5′ (d) 1′

119. Which one of the following is not applicable to monoclonal antibody?
(a) High useful antibody content (b) Homogeous in composition
(c) Specificity variable from animal to animal (d) Cross reaction with other antigen is generally absent

120. Kaposi's sarcoma is associated with:
(a) Fever and shock syndrome (b) Loeffler's syndrome
(c) Goodpasture's syndrome (d) AIDS

121. Which one of the following contributes greatly to an increase in the diversity of antibody synthesis by an organism?
(a) Linkage (b) Somatic recombination
(c) Mutation (d) Mutation and somatic recombination

122. It an immunoglobulin is treated with papain, it will be unable to promote:
(a) Opsonisation (b) Agglutination (c) Precipitation (d) All

123. The immune system has the ability to recognise and respond to about:
(a) 1000 antigens (b) 10^5 antigens (c) 10^7 antigens (d) 10^{20}

124. CD3 molecule is the surface component of:
(a) β lymphocytes (b) T lymphocytes
(c) Cytotoxic T cells (d) Helper T cells

125. Light chain of immunoglobulins lack gene segment:
(a) V (variable) (b) J (Joining) (c) D (Diversity) (d) C (Constant)

126. Genes involved in autoimmune diseases:
(a) Immunoglobulins (b) T-cell receptors
(c) Major histocompatibility complexes (d) All

127. Which class of immunoglobulins is active for antibody-dependent cell-mediated cytolysis (ADCC) mechanism?
(a) IgD (b) IgE (c) IgG (d) IgM

128. Which one of the following complement components of the complement system is not synthesised mainly in the macrophages?
(a) C9 (b) C8 (c) C7 (d) C6

129. The main site of synthesis of complement component C2 of the complement system is the:
(a) Epithelial cells of small intestine (b) Reticuloendothelial cells
(c) Hepatocytes (d) Macrophages

130. Macrophage Migration Inhibition Factor (MIF) is secreted by:
(a) T cells (b) Macrophages (c) Monocytes (d) Stromal cells

131. In human beings, the genes of Major Histocompatibility Complex (MHC) are present on chromosome:
(a) 4 (b) 6 (c) 8 (d) 14

132. The tissue damage due to soluble immune complexes has been called type-III hypersensitivity reaction, and when it is localised, it is known as:
(a) Arthus pheomenon (b) Serum sickness
(c) Immunoinflammation (d) Immediate hypersensitivity reaction

133. Most of our information about immunoglobulin comes from:
(a) Human beings (b) Rabbit
(c) *Drosophila* (d) Human beings and mice

134. In the light chain of immunoglobulin, the C region is coded by:
(a) Single exon (b) Two exons (c) Three exons (d) Several exons

135. Which one of the following is not used in cell fusion?
(a) Sendai virus (b) Polyethlene glycol (c) Lysolecithin (d) SV_{40}

136. Which one of the following is applicable to both B lymphocytes and T lymphocytes?
(a) Formation of memory cells (b) Secretory product (antibodies)
(c) Short lifespan (d) Cell-mediated immunity

137. What is wrong about B lymphocytes?
(a) Origin and maturation occurs in bone marrow (b) From memory cells
(c) Long lifespan (d) Antibody mediated immunity

138. Interleukin 2(IL2) is secreted by:
(a) B lymphocytes (b) Cytotoxic T cells
(c) Suppressor T cells (d) Helper T cells

139. Interferon is secreted by:
(a) Naturally killed cells and cytotoxic cells (b) Virus-infected cells
(c) Macrophages (d) All

140. Plasma cells are derived from:
(a) B cells (b) Cytotoxic T cells (c) Helper T cells (d) All

141. T cells are specialised in recognising and destroying:
(a) Transplanted tissue cells (b) Virus-infected cells
(c) Cancer cells (d) All

142. Which of the following incorrect about interferons?
(a) Stimulate production of antibodies (b) Accelerate cell division
(c) Enhance macrophage phagocytic activity (d) Stimulate activity of special T lymphocytes

143. Which one of the following is the largest lymphoid organ?
(a) Liver (b) Spleen (c) Appendix (d) Kidney

144. Acute phase proteins are secreted by:
(a) Liver (b) Kidney (c) B lymphocytes (d) T lymphocytes

145. What is incorrect about HLA antigens?

(a) First discovered in leukocytes.
(b) They are present in all body cells.
(c) They are not present in erythrocytes.
(d) The exact patterns of HLA antigens is similar in all individuals.

146. Multiple myeloma is not characterised by:
(a) Lytic bone lesions and proteinuria
(b) Praproteianemia
(c) Malabsorption
(d) Anaemia

147. The first antihistaminic drug prepared:
(a) Isoprenaline (b) Mepyramine (c) 2, 4-dinitrophenol (d) Dolichol

148. Granulysin is secreted by:
(a) Cytotoxic T cells (b) Helper T cells (c) Suppressor T cells (d) B cells

149 HIV infects:
(a) Macrophages (b) CD_4^+ T cells (c) Dendritic cells (d) All

150. Which one of the following set of components of complement system have anaphylatoxin activity?
(a) C6, C7 and C8 (b) C8 and C9 (c) C3a and C5a (d) C3a and C3b

151. Consider the following statements:
(A) The immunoglobulin heavy and light chains are coded by three separate gene families, each on a separate chromosome
(B) Each of these gene families has one V region and one C region gene
(C) The lambda gene family of lambda light chains contains 4C region
(D) The kappa light chain gene family contains only one C region

The incorrect statements are:
(a) None (b) A, B and D (c) B (d) B and D

152. The scientist not associated with transplantation of organs:
(a) Christian Bernard (b) Edward Thomas (c) Subumu Tonegawa (d) Joseph Murray

153. What is common between the Wiskott–Aldrich syndrome and ataxia telangectasia?
(a) Both are autosomal recessive
(b) Both are autosomal dominant
(c) Both are sex-linked recessive
(d) In both lymphoreticular malignancies are common

154. Malabsorption syndrome is due to the deficiency of:
(a) IgA (b) IgD (c) IgG (d) IgM

155. The genes for lambda, kappa and heavy chains are genetically unlinked; it becomes clear by following the inheritance of____________ determinants on antibody molecules:
(a) Idiotypic (b) Allotypic (c) Isotypic (d) All

156. The different behaviour of immunoglobulin molecule is due to their:
(a) Combinatorial association
(b) Variable regions
(c) Heavy chain constant regions
(d) Folding

157. In heavy chain genes, junctional diversity:
(a) Does not occur
(b) Occurs at $V_{H\text{-}D}$ Joints
(c) Occurs at D-J_H joints
(d) Occurs both at $V_{H\text{-}D}$ joints and D-J_H joints

158. The earliest cells to show evidence of antibody gene rearrangements:
(a) Nurse cells (b) Null cells (c) Helper T cells (d) B cells

159. Cancers of B cells give rise to a large population of cells that produce:
(a) No antibodies
(b) Only single type of antibody molecule

(c) Two types of antibody molecules
(d) Several types of antibody molecules

160. *Tla* and *Qa* genes encode:
(a) Transplantation antigens
(b) Components of complement system
(c) Proteins present as cell surface antigens on lymphocytes
(d) Class II MHC proteins

161. In an immunoglobulin molecule, any type of light chain can combine with any type of heavy chain. This phenomenon is known as:
(a) Arthus phenomenon
(b) Antagonastic association
(c) Combinatorial association
(d) Junctional diversity

162. The immunoglobulin classes differ from each other in respect to kinds of:
(a) Membranes they can cross
(b) Activities they can perform after binding the antigen
(c) Cells they can enter
(d) All

163. The embryonic sequence for each immunoglobulin gene family is found in:
(a) All tissues
(b) Tissues that produce immunoglobulin proteins
(c) Tissues that do not produce immunoglobulin proteins
(d) Tissues of thymus gland

164. Which one of the following statements is correct?
(a) Immunoglobulins are absent in DiGeorge syndrome.
(b) Immunoglobulins are exception to one-gene–one-polypeptide theory.
(c) Isoagglutinins are predominantly IgG.
(d) None

165. The HLA antigens are lacking in:
(a) Erythrocytes
(b) Sperms
(c) Ovum and sperms
(d) Erythrocytes and sperms

166. The HLA antigen associated with Reiter syndrome:
(a) B8 (b) DR4 (c) DR3 (d) B27

167. Which one of the following is incorrect about monoclonal antibody?
(a) Homogenous composition
(b) Specificity highly constant
(c) Typical mixture of all class and subclass of immunoglobulins
(d) Cross reaction with other antigens generally absent

168. Which one of the following proteins is diverse?
(a) Antigens
(b) Antibodies
(c) Histocompatibility antigens
(d) All

169. Which one of the following is incorrect about innate immune system?
(a) Found in all plants
(b) Found in all animals
(c) Provide immediate defence against infection
(d) None

170. Some hyper IgM syndromes are characterised by the deficiency of:
(a) IgA (b) IgG (c) IgE (d) All

171. HLA antigens play an important role in:
(a) Infectious diseases
(b) Diabetes
(c) Rheumatoid
(d) All

172. Which one of the following is incorrect about interferons?
(a) They are natural proteins formed by the cells of the immune system.
(b) They have the ability to cross the blood-brain barrier.

(c) They have the ability to cross the placenta.
(d) They rarely cross the placenta.

173. Interferons were first described by:
(a) Isaacs and Lindemann (1957) (b) Lawrence and Landy (1969)
(c) Marks and Shay (1960) (d) Hawk and Oser (1965)

174. Which one of the following is incorrect with reference to natural killer cells?
(a) Large granular lymphocytes (b) May have T-cell lineage
(c) Have mature T-cell markers (d) Activity is increased by gamma interferon

175. Match column I with column II and select correct the answer using answer codes:

Column I	Column II
(A) Function as receptors for antigens on B cells	1. IgG and IgA
(B) Fixation to mast cells	2. IgM and IgD
(C) Opsonisation via Fc cell receptors	3. IgG
(D) Viral neutralisation	4. IgE

Answer codes:

	A	B	C	D
(a)	2	4	3	1
(b)	2	4	1	3
(c)	4	2	3	1
(d)	2	1	3	4

176. Molecules resembling immunoglobulin make their first appearance in:
(a) *Amphioxus* (b) *Petromyzon*
(c) Higher sharks and paddle fishes (d) *Seymouria*

177. Which one of the following ions is required for complement dependent interactions?
(a) Na^+ and K^+ (b) Ca^{++} (c) Ca^{++} and Mg^{++} (d) Mg^{++} and Cu^{++}

178. Which one of the following is a vasoactive amine?
(a) Bradykinin (b) Histamine (c) Serotonin (d) All

179. Activation of complement system releases vasoactive amines resulting in:
(a) Increased vascular permeability (b) Activation of kalikranin system
(c) Activation of coagulation mechanism (d) All

180. Bence Jones protein is applicable to______________ of immunoglobulin:
(a) Heavy chain (b) Light chain (c) Constant segment (d) Variable segment

181. These antibodies posses the ability to bind to cells of the same species:
(a) Xenogenic (b) Antitoxin
(c) Homocytotropic (d) Monoclonal antibodies

182. Which one of the following statements is correct?
(a) Monoclonal antibodies destroy merozoite-infected red blood cells.
(b) Monoclonal antibodies are used as enzymes and are called abzymes.
(c) The amount of antibody production can be increased by ascites fluid production method.
(d) Antigens cannot be purified using purified chromatography with monoclonal as antibodies.

183. The tissue damage due to soluble immune complexes is known as type:
(a) I hypersensitivity (b) II hypersensitivity (c) III hypersensitivity (d) IV hypersensitivity

184. Macrophage migration factor:
(a) Is a species specific factor

(b) Acts upon the macrophages derived from the same species
(c) Increases glucose metabolism accumulated macrophages
(d) All

185. S Tonegawa was awarded the 1987 Nobel Prize in Physiology and Medicine for his work on:
(a) The arrangement of DNA sequences that result in a large variety of sequences for production of antibodies
(b) Gene transfer techniques in animals
(c) Hybridoma technology
(d) Discovery of luxury genes

186. Consider the following statements:
(A) Immunoglobulins are coded by genes which undergo DNA rearrangements during cell differentiation
(B) The C and V coding sequences of immunoglobulin molecule are far apart from each other
(C) In genes for immunoglobulin molecule, the cutting and rejoining of DNA occurs only within a chromosome.
(D) The genes for immunoglobulin are located on the same chromosome.

The correct statements are:
(a) All (b) A, B and D
(c) A and C (d) B and C

187. Generation of diversity of immunoglobulin occurs by simple permutation and combination of:
(a) One V, J and C genes (b) Two V, J and C genes
(c) Many V, J and C genes (d) Many V, J, C and D genes

188. Null cells lack:
(a) Identifying surface markers only for T lymphocytes
(b) Identifying surface markers only for B lymphocytes
(c) Identifying surface markers either for T lymphocytes or B cells
(d) Adaptive traits and antibodies

189. Phytohaemagglutinin stimulates:
(a) Predominantly T lymphocytes (b) Predominantly B lymphocytes
(c) Both T and B lymphocytes (d) Predominantly formation of IgG

190. The predominant class of immunoglobulin present on B lymphocytes:
(a) IgG (b) IgE (c) IgD (d) IgM

191. Which one of the following is not applicable to Icterus Gravis Neonatorum?
(a) Manifestation of haemolytic anaemia (b) Oedema
(c) Increasing jaundice (d) Anaemia

192. The number of domains in an IgG, molecule:
(a) 2 (b) 3 (c) 6 (d) 7

193. An antigen-binding site contains_______________ domains:
(a) 2 (b) 3 (c) 5 (d) 6

194. The classical pathway of complement system can be activated by immune complexes having:
(a) Antibodies (b) IgG and IgM (c) IgA and IgD (d) IgE and IgM

195. Which one of the following is not capable of synthesising complement components?
(a) Fibroblasts (b) Granstein cells
(c) Hepatocytes (d) Epithelial cells of gastrointestinal tract

196. Which one of the following is not due to genetic deficiency of component?
    (a) Chronic vasculities (b) Anaphylactoid purpura
    (c) Ataxia telangiectasia (d) Raynaud's phenomenon
197. Deficiencies of complement components are inherited as:
    (a) Sex-linked recessive (b) Autosomal dominant
    (c) Co-dominant or sex-linked recessive (d) Autosomal recessive or co-dominant
198. The number of domains in heavy chain of immunoglobulin:
    (a) 2 (b) 3 (c) 4 (d) 5
199. The properties of antibody combining site is determined by the composition of the:
    (a) Variable region VL (b) Variable region VH
    (c) Variable region VL and VH (d) Heavy chain and light chain domains
200. On destruction, a virus from virus-infected cell enters the extracellular fluid where it is attacked by:
    (a) Activated complement components (b) Macrophages
    (c) Antibodies (d) All
201. Delayed allergy reaction is caused by:
    (a) Antibodies (b) Activated T cells (c) B cells (d) Memory cells
202. Consider the following statements with reference to T-cell receptor:
    (A) It is very similar in structure to the immunoglobulin
    (B) It consists of two chains: alpha and beta
    (C) The two chains are linked by disulphide bridge
    (D) The gene clusters for both these chains are on chromosome 14

    The correct statements are:
    (a) All (b) A, B and C (c) B and C (d) B, C and D
203. Junctional diversity results in:
    (a) Allelic exclusion (b) Null cell formation (c) Antibody diversity (d) All
204. T-cell receptors are:
    (a) Homologous to antibodies (b) Analogous to antibodies
    (c) Null cells (d) Haptens
205. Theory of clonal selection is applicable to:
    (a) T cells (b) B cells
    (c) Equally to both T and B cells (d) Partially to B cells and fully to T cells
206. The immunoglobulin having the highest percentage of carbohydrate:
    (a) IgM (b) IgG (c) IgD (d) Ig
207. Antibody molecules are classified into different classes depending on the amino acids sequence of:
    (a) Constant region of their heavy chains (b) Constant region of their light chains
    (c) Variable regions of their heavy chains (d) Junctional diversity
208. The most important HLA antigens for transplantation:
    (a) HLA-A (b) HLA-B (c) HLA-DR (d) All
209. Which one of the following is not applicable to the complement system?
    (a) Zymogens (b) Acquired immunity
    (c) Biochemical cascade (d) Mannose binding lectin pathway
210. Complement system does not involve:
    (a) Opsonisation (b) Immune clearance

(c) Lysis of bacteria and viruses (d) Allelic exclusion

211. Activation Induced Cytidine Deaminase is an important catalyst for:
(a) Gene conversion (b) Somatic conversion
(c) Class switch recombination (d) All

212. The human VH locus contains about:
(a) 100 V segments (b) 30 D segments
(c) 6 J segments (d) All

213. Which one of the following lacks the capacity to cross the placenta?
(a) IgG_1 (b) IgG_2 (c) Ig_3 (d) IgG_4

214. Patients having high frequency of antibody to milk have predominantly_______antibodies:
(a) IgM (b) IgG (c) IgE (d) IgD

215. Which one of the following is unique to antibody?
(a) Null cells (b) Allelic exclusion (c) Annealing of DNA (d) Gene amplification

216. The β-gene family resembles heavy chain genes of immunoglobulin in having:
(a) V segment (b) D segment (c) J segment (d) All

217. Two-genes–one-polypeptide model is applicable to:
(a) Epidermal proteins (b) Antibodies
(c) Australian aborigines (d) Autoimmune diseases

218. Which one of the following is a lymphokine?
(a) Macrophage migration inhibition factor (b) Interleukin
(c) Interferon (d) All

219. Which one of the following is not applicable to immunoglobulins?
(a) Combinatorial association (b) Substraction hybridisation
(c) Allelic exclusion (d) Silent mutation

220. Active immunity does not include:
(a) Humoral immunity (b) Cellular immunity
(c) Combination of both (a) and (b) (d) Normal human Ig

221. M spike in urine is related to:
(a) Bence Jones protein (b) AIDS
(c) DiGeorge's syndrome (d) Wiscott–Aldrich syndrome

222. Haemolytic disease of a newborn is due to:
(a) IgM (b) IgG (c) IgE (d) IgA

223. Which one of the following is different in classes of immunoglobulins?
(a) Light chains (b) VH regions (c) CH regions (d) None

224. The number of hypervariable regions in light and heavy chains are:
(a) 1 and 3 (b) 2 and 4 (c) 3 and 4 (d) 4 each

225. Transcription of the V region gene segments takes place when the cells undergo:
(a) V, C, D and J recombination (b) Final V and J recombination
(c) Final V, D and J recombination (d) V and J recombination

226. In which one of the following diseases, the mode of inheritance is both autosomal recessive as well as as X-linked recessive:
(a) Chediak–Higashi syndrome (b) DiGeorge syndrome
(c) Chronic granulomatous disease (d) Severe combined immunodeficiency

227. Tissue transplantation between identical twins is known as:
(a) Allograft (b) Xenograft (c) Isograft (d) Allograft

228. Which one of the following is an important level of control in the expression of immunoglobulin genes?
(a) Clonal selection (b) RNA processing
(c) Somatic mutation (d) Junctional diversity

229. In immunoglobulin, diverse amino acids codons are generated at the site of:
(a) V–JL recombination (b) V–D–JH recombination
(c) Both (a) and (b) (d) Stem and 100p structure

230. Which one of the following is the largest immunoglobulin in size?
(a) IgM (b) IgG (c) IgD (d) IgE

231. Which one of the following is responsible for immunodeficiency?
(a) Infection (b) Malnutrition (c) Gene mutation (d) All

232. Which one of the following is known as effector cell?
(a) Helper T cell (b) Killer T cell (c) Suppressor T cell (d) All

233. Who is a pioneer of immunogenetics?
(a) M R Irwin (b) Feodor Lynen (c) Robert F Furchgott (d) Ferid Murad

234. The rearrangement of kappa and lambda chains of the immunoglobulin light chain loci is similar to that of heavy chain except the light chains lack:
(a) D gene segment (b) V J gene segment
(c) C gene segment (d) D and C gene segments

235. The term 'immunogenetic' was coined by:
(a) Archilbald Vivian Hill (1895) (b) Irwin and Cole (1936)
(c) Roger Gullimin (1904) (d) Niels K Jerne (1960)

236. Alloimmunisation Gm markers is common in:
(a) Branchio-oto-renal syndrome (b) Waardenburg syndrome
(c) Wilson disease (d) Rheumatoid arthiritis

237. The diversity of T-cell receptor does not include:
(a) V J joining of gene segments (b) VDJ joining of gene segments
(c) Combinatorial association (d) Somatic mutation

238. Co-dominant expression of allotypes in the organism as a whole occurs in:
(a) Doehle body (b) Antibodies (c) Giant cells (d) Neuron cells

239. The first event to occur in developing B cell is between:
(a) One D and one J segment of the heavy chain locus
(b) One D and one J segment of the light chain locus
(c) VH and CH segment of the heavy chain
(d) One C one D and one V segments of heavy chain locus

240. Which one of the following structure is found in neutrophils in times of leukemoid reaction?
(a) Giant body (b) Adoptoti body (c) Doehle body (d) Tangible body

241. The earliest cells to show evidence of antibody DNA arrangements:
(a) Blast cells (b) Memory cells (c) Null cells (d) Suppressor T cells

242. The function of variable region of immunoglobulin:
(a) Antigen-antibody interaction (b) Precipitation and agglutination
(c) Neutralisation (d) All

243. In immunoglobulin, words light and heavy chains denote the relative:
(a) Density (b) Sedimentation coefficient
(c) Affinity to antigens (d) Size of polypeptide chains

244. Which one of the following is an incorrect match?
(a) Mitogenic factor – Blast transformation
(b) Gamma interferon – Antiviral protein
(c) Chemotoxic factor – Antigen stimulated lymphoid cells
(d) Macrophage migration inhibitory factor – Decrease the rate of proliferation

245. The two types of light chains, kappa and lambda, characteristically differ in their residues at position:
(a) 95 (b) 105 (c) 155 (d) 191

246. Consider the following statements with reference to immunoglobulins:
(A) Switching involves only the C_H gene
(B) The switching sites are called S regions
(C) Cells expressing IgM have the germ line arrangement of C_H genes
(D) The VJ joining rearrangement is capable of starting expression of the C_H gene

The incorrect statements are:
(a) A (b) B (c) C (d) D

247. In HLA class some loci are:
(a) Highly polymorphic (b) Monomorphic
(c) Oligomorphic (d) All

248. The antigen and antibody binding does not involve:
(a) Covalent bonds (b) Ionic bonds (c) Vander Waals forces (d) Hydrophobic bonds

Answers to Multiple-Choice Questions

1. (a)	2. (d)	3. (b)	4. (c)	5. (a)	6. (d)	7. (b)	8. (a)
9. (a)	10. (a)	11. (c)	12. (b)	13. (c)	14. (d)	15. (d)	16. (d)
17. (a)	18. (a)	19. (d)	20. (a)	21. (b)	22. (d)	23. (d)	24. (a)
25. (c)	26. (c)	27. (a)	28. (c)	29. (b)	30. (d)	31. (d)	32. (a)
33. (d)	34. (d)	35. (b)	36. (a)	37. (d)	38. (c)	39. (b)	40. (a)
41. (b)	42. (d)	43. (a)	44. (c)	45. (b)	46. (c)	47. (b)	48. (d)
49. (d)	50. (c)	51. (d)	52. (d)	53. (c)	54. (d)	55. (d)	56. (b)
57. (c)	58. (b)	59. (c)	60. (d)	61. (a)	62. (a)	63. (b)	64. (d)
65. (d)	66. (c)	67. (c)	68. (b)	69. (d)	70. (c)	71. (d)	72. (d)
73. (a)	74. (d)	75. (c)	76. (a)	77. (a)	78. (b)	79. (b)	80. (a)
81. (b)	82. (d)	83. (b)	84. (d)	85. (a)	86. (d)	87. (a)	88. (c)
89. (a)	90. (d)	91. (d)	92. (c)	93. (d)	94. (c)	95. (a)	96. (b)
97. (a)	98. (d)	99. (b)	100. (d)	101. (b)	102. (a)	103. (d)	104. (b)
105. (d)	106. (d)	107. (a)	108. (b)	109. (d)	110. (d)	111. (c)	112. (b)
113. (d)	114. (d)	115. (c)	116. (d)	117. (c)	118. (b)	119. (c)	120. (d)
121. (d)	122. (d)	123. (c)	124. (b)	125. (c)	126. (d)	127. (c)	128. (d)
129. (b)	130. (a)	131. (b)	132. (a)	133. (d)	134. (a)	135. (d)	136. (a)
137. (c)	138. (d)	139. (d)	140. (a)	141. (d)	142. (b)	143. (b)	144. (a)
145. (d)	146. (c)	147. (b)	148. (a)	149. (d)	150. (c)	151. (c)	152. (c)
153. (d)	154. (a)	155. (d)	156. (c)	157. (d)	158. (b)	159. (b)	160. (c)
161. (c)	162. (d)	163. (c)	164. (b)	165. (d)	166. (d)	167. (c)	168. (d)
169. (d)	170. (d)	171. (d)	172. (c)	173. (a)	174. (c)	175. (a)	176. (c)

177. (c)	178. (d)	179. (d)	180. (a)	181. (c)	182. (d)	183. (c)	184. (d)
185. (a)	186. (c)	187. (c)	188. (c)	189. (a)	190. (c)	191. (b)	192. (c)
193. (a)	194. (b)	195. (b)	196. (c)	197. (d)	198. (a)	199. (c)	200. (d)
201. (b)	202. (b)	203. (c)	204. (b)	205. (c)	206. (c)	207. (a)	208. (d)
209. (b)	210. (d)	211. (d)	212. (d)	213. (d)	214. (a)	215. (b)	216. (d)
217. (b)	218. (d)	219. (d)	220. (d)	221. (a)	222. (b)	223. (c)	224. (c)
225. (c)	226. (d)	227. (c)	228. (b)	229. (c)	230. (a)	231. (d)	232. (d)
233. (a)	234. (a)	235. (b)	236. (d)	237. (d)	238. (b)	239. (a)	240. (c)
241. (c)	242. (d)	243. (d)	244. (d)	245. (d)	246. (d)	247. (d)	248. (a)

Fill in the Blanks

1. Application of immunological methods to study inheritance of character is known as ________________.
2. Immunology is the study of ________________ and antibodies.
3. Antibodies are produced against ________________.
4. The shape of an antibody is determined by an ________________.
5. An antigen that binds to a specific antibody is called________________.antigen.
6. Immune response is of two types, viz., ________________ immunity and ________________.
7. Immunoglobulins bind to ________________ and destroy them.
8. Two principal methods used in immunogenetical study are ________________ and ________________.
9. Antibodies occur in two forms ________________ and membrane ________________.
10. The five antibody isotypes found in mammals are ________________.
11. The ________________ is identical in all antibodies of the same isotype.
12. Haemagglutination test is used to detect ________________.
13. The antigen-binding site is present in the ________________ region of the body.
14. In the serum IgA is present as a ________________ ________________monomere.
15. ________________ antigens are not of human origin.
16. All antibodies are globular proteins and are produced by ________________.
17. ________________ increase considerably in infected animals.
18. ________________ is a well-known example of drug hapten.
19. ________________ are the main structures involved in cell-mediated immunity.
20. Helper T cells activate ________________ ________________ in contact with antigen.
21. The kappa light chain gene is located on the short area of chromosome ________________.
22. The locus for the heavy chain gene is on chromosome ________________.
23. There are three immunoglobulin families called ________________, ________________and ________________.
24. Each immunoglobulin is composed of two identical ________________and two identical ________________ chains held together by ________________.

25. Each immunoglobulin chain has three regions called ________________ region, ________________ region and ________________ region.
26. In mammalian system, a group of ________________ are known to form the complement system.
27. The immunoglobulin involved in local defence is the ________________.
28. ________________ cells are important in a cell-mediated immunity.
29. ________________ is a mitogen.
30. ________________are antigenic determinants on variable regions.
31. Antigenic determinants that differentiate between different types of constant regions are called __________.
32. Antigenic determinants involved in differentiating allelic genes are called ________________.
33. Lymphokines are produced by ________________.
34. Any molecule that acts as binding enhancer during phagocytosis is known as ________________.
35. The immunoglobulin heavy and light chains are coded for by the ________________ separate gene families; each on a separate chromosome, one for heavy chain and one for each of the light chain.
36. The kappa light chain gene family contains only one ________________ region.
37. ________________ is the process, which makes microorganisms more susceptible to the action of phagocytosis.
38. Monoclonal antibodies are products of ________________ cells created by fusion of the two cells.
39. ________________ of immunoglobulin gene are required by the cells of the immune system to form antibodies.
40. In immunoglobulin, the 'J' segments are separated from the constant coding regions by an ____________.
41. The fragment antigen binding (Fab region) is a region on immunoglobulin which binds to __________.
42. Papayan cleaves the ________________ region of immunoglobulin from the fragment-binding region.
43. ________________ regions of light and heavy chains form antibody combining site.
44. The C region of heavy chain of immunoglobulin is coded by ________________ exons.
45. The IgG contains a ________________ polypeptide in the C_1 region.
46. Kininogens are synthesised by ________________.
47. Kallikrein are released by ________________.
48. Leukocyte endogenous mediator is secreted by ________________.
49. A complex antigen may have many ________________ determinant sites.
50. Certain low molecular weight organic substances which are not antigenic but become antigenic, if they become attach to proteins, they are known as ________________ proteins.
51. The exact pattern of HLA antigen varies from individual to individual except in ________________.
52. During transplantation of an organ, donar and receiver are matched for ________________ system.
53. The first antihistaminic drug, mepyramine, was prepared in 1994 by ________________.
54. ________________ is the most well-known disease that affects the immune system itself and is caused by ________________.
55. The main primary lymphoid organs of the immune system are ________________ and ______________.
56. Deficiency of major histocompatability complex causes ________________ syndrome.
57. The macromolecules which mimic the structure of an epitope are known as ________________.

58. Mutations in __________________ gene causes X-linked hyper IgM syndrome.
59. C5a is an important __________________ protein.
60. __________________ is the best marker for B cells.
61. __________________ antigens cause rapid rejection of skin and organs graft between individuals.
62. Cancers of B cells are called __________________.
63. Nucleus of an ordinary lymphocyte is fused to those of myeloma cell in order to get a __________________.
64. Within a family, on an average, one in __________________ of the siblings will be HLA identical.
65. The process by which an organism alters its surface protein is known as __________________.
66. __________________ is a family genetic disorder characterised by relatively high level of IgM.
67. Histocompatability antigens in humans are called __________________.
68. Interferon may cause increased __________________ activity in virus-infected cells.
69. The B cell are fixed and located in the __________________.
70. __________________ is the second largest class of immunoglobulin.
71. The number of different immunoglobulin is in fact limited by the number of __________________ in a given species.
72. __________________ different antibody isotypes are known in mammals.
73. Human immunoglobulin molecules are coded by __________________ gene families.
74. Major histocompatability genes specify cell proteins except for the __________________.
75. The variable region of heavy chain immunoglobulin is generated by joining the __________________ with __________________ and a __________________ segment.
76. The __________________ of the immunoglobulin molecules are responsible for the class-specific functions of the different immunoglobulin molecules.
77. In immunoglobulin molecule number, __________________ variable regions from different humans have been found to have identical amino acid sequences.
78. Recombination of __________________ and __________________ genes results in the formation of new functional immunoglobulin genes.
79. The chain in an IgG molecule is known as __________________ chain.
80. The region of antibody that combines with an antigen is called __________________.
81. HIV attacks __________________ cells.
82. Helper T cells regulate immune functions by secreting __________________.
83. Assembly of the IgG heavy chain and one of the __________________ results in the formation of IgM.
84. The L, V, J and C genes of lambda light chains are separated by __________________.
85. A large granular lymphocyte without surface marker is known as __________________.
86. The allele of the antibody chains found in the immunoglobulin of an individual is called __________________.
87. Class Switch Recombination (CSR) occurs by an intrachromosomal deletion by which light constant region gene is replaced by downstream __________________.
88. The gene segments encoding a complete antibody are widely separated in germ cells and all somatic cells except for __________________.
89. The segment CH2 of the constant region of the immunoglobulin has receptor for binding of the complement, while segment CH2 binds to __________________.

90. The T cells of a mouse contains ____________________ antigen.
91. Phytohemagglutinin (a mitogen) is extractable from ____________________.
92. ________________ is the bulk protein of complement protein.
93. Sequences of nucleotides which are internally ____________________ are called inverted repeat sequences.
94. The loop structure in immunoglobulin contains DNA sequences that are____________________.
95. In heavy chain genes of immunoglobulin, junctional diversity occurs at ____________________ joints and ____________________ joints.
96. The antibodies secreted by multiple myeloma tumour contain ____________________ type of light chain but ____________________ type of heavy chains.
97. ____________________ is responsible for two-genes–one-polypeptide model for antibodies.
98. ____________________ provides the major explanation for production of vast majority of antibodies by an organism having only a relatively small number of genes.
99. J chain is a polypeptide chain normally present in immunoglobulin, particularly in ______________ and ____________________.
100. The area of an antigen which determines the specificity of an antigen antibody is called ____________________
101. ____________________is a test for the detection of blocking antibodies.

Answers to Fill in the Blanks

1. Immunogenetics
2. Antigens
3. Antigens
4. Antigens
5. Cognate
6. Humoral, cellular immunity
7. Antigens
8. Serology, histogenetical
9. Soluble, bound
10. IgA, IgD, IgE, IgG, IgM
11. Constant region
12. Antibodies
13. Variable
14. 7S
15. Xenogenic
16. B lymphocytes
17. Gamma globulin
18. Penicillin
19. T lymphocytes
20. B cells
21. 2
22. 22
23. Kappa, lambda, rho
24. light, heavy, disulphide bonds
25. Variable, junctional , constant
26. Nine protein ($C_1 - C_9$)
27. IgA
28. T
29. Phytohemagglutinin
30. Idiotypes
31. Isotypes
32. Allotypes
33. Lymphocytes
34. Opsonin
35. Three
36. C
37. Opsonisation
38. Hybrid
39. Multiple programmed rearrangements
40. Intron
41. Antigen
42. Fragment crystallisable
43. Variable
44. Several
45. Kappa
46. Liver
47. Neutrophils
48. Macrophages
49. Antigenic determinant
50. Hapten
51. Identical twins
52. HLA
53. Daniel Bovet
54. AIDS, HIV
55. Thymus, bone marrow
56. Bare lymphocyte
57. Mimotope
58. *CD40LG*
59. Chemotoxic
60. Surface immunoglobulin
61. Transplantation.
62. Myelomas
63. Hybridoma
64. Four
65. Antigenic variation
66. Hyper IgM syndrome
67. HLA(Human Leukocyte Antigen)
68. P^{53}
69. Secondary lymphoid organs
70. IgA
71. B cells

72. Five
73. Three unlinked
74. Complement components
75. VH, D, JH
76. Constant regions
77. 78 V, C
79. γ
80. Paratope
81. T helper
82. Lymphokines
83. Light chains
84. Introns
85. Null cell
86. Allotype
87. Constant region
88. B lymphocytes
89. Macrophages
90. Theta
91. Jack bean
92. C3
93. Complementary
94. Spliced out
95. VH–D, D–JH
96. Single, Two
97. Class switching
98. Combinatorial association
99. IgM, IgG
100. Antigen determinant
101. Coomb test

True or False

1. Antibodies are produced by plasma cells.
2. The antigen-antibody reaction leads to the activation of the complement system.
3. Phagocytosis is a specific mechanism.
4. Majority of immunoglobulin have gamma mobility.
5. Overreaction of immune system is known as hypersensitivity.
6. The heavy chain of IgM globulin is longer than the size of heavy chains in other immunoglobulin.
7. Generally macrophages are mobile cells.
8. Memory cells are not formed in some diseases.
9. IgG acts as opsonin.
10. Natural killer cells nonspecifically destroy virus-infected cells and tumour.
11. Antibodies are not opsonins.
12. HLA antigens are very polymorphic.
13. During the lifespan of an individual, the complement system undergoes changes.
14. Kallikrein is released by neutrophils.
15. Antigen binding causes activated B-cell clone to stop multiplication.
16. T cells and natural killer cells secrete perforin.
17. Gamma globulins are involved in immune response.
18. Antibody is a bivalent molecule.
19. Complement factors are present in lymphocytes.
20. Most of the immunoglobulin have gamma mobility.
21. In human beings, 60 per cent light chains are of kappa variety and 40 per cent are of lambda type.
22. In cows, placental crossing of immunoglobulin does not occur.
23. Human clostrum is rich in antibodies.
24. The life of complement system at room temperature is only a few hours.
25. Ataxia telangectasis is a sex-linked recessive disease.
26. Macrophages cannot digest bacteria.

27. Bacteria may multiply inside the macrophages.
28. In multiple myeloma, there is an increase in the total immunoglobulin but biologically active immunoglobulin is suppressed.
29. In heavy chain defective diseases, heavy chains are excreted in urine.
30. IgA can bind with macrophages.
31. Avian antibodies can fix to mammalian complement systems.
32. Human IgE can fix to the skin of monkeys but the sensitivity is much lower.
33. Macrophages migration inhibitory factor increases stickiness.
34. The light and heavy chains vary with the class of immunoglobulin.
35. The crystallisable fragment (FC) of immunoglobulin has receptors for binding to all classes of immunoglobulin.
36. The allotypes are genetically inherited.
37. The constant part of the heavy chain determines the properties related to the elimination of antigen.
38. Light chains of immunoglobulin lack D segment.
39. The variable and the constant regions of light and heavy chains of immunoglobulin are the same for all types of antibody specificities.
40. All cells co-express IgM and IgD.
41. In myeloma genome, Vc segment and the C_L segment of immunoglobulins are located in the same chromosome.
42. Memory cells secrete antibodies.
43. T4 antigen is present in the helper T cells of humans.
44. Hypervariable regions are the most variable portions of the V regions.
45. A single DNA recombination is responsible for the assembling of light chain V region genes.
46. Some of the IgM responses are the cell independent.
47. The V and C genes immunoglobulin are expressed as an independent unit.
48. The structure of kappa loci of immunoglobulin is similar in humans and mice.
49. In an immunoglobulin molecule, the number of V and C genes are almost equal.
50. In an immunoglobulin molecule, in the formation of light chain gene, an additional segment called the diversity (d) segment, is present between V and J segments.
51. IgE are cytophilic antibodies.
52. The T cells secrete antibodies and are responsible for the recognition of antigen.
53. IgG are largely extravascular.
54. The presence IgG may be indicative of a recent infection.
55. The light chains are alike in all classes of immunoglobulin.
56. A hapten becomes antigen on combining with a body protein.
57. Passively transferred antibodies are generally broken down within a month.
58. The carbohydrate residues present in immunoglobulin contribute charge on the molecule.
59. Transcriptional activation of the v-gene promoter in immunoglobulin is a side effect of DNA rearrangement.
60. The arrangements of DNA in B and T cells create diversity and also affect the gene expression.

61. The type of H chain determines the class of immunoglobulin.
62. Somatic recombination leads to the formation of a large number of Ig genes.
63. Haptens have molecular weight more than 10,000.
64. Gamma globulins have positive charge.
65. The genes within each family of immunoglobulin are linked but the families are not linked.
66. Cancer is an autoimmune disease.
67. Segments of antigens that are recognised by antibodies are called epitopes.
68. Spleen is the shock organ of allergy.
69. Gamma globulins are synthesised in liver.
70. IgG function in innate immunity.
71. Complement proteins are components of innate immunity.
72. Memory cells are responsible for active immunity
73. Some antibodies function as antitoxins.
74. Spleen is a primary lymphoid organ.
75. HLA system is highly polymorphic.
76. Helper T cells recognise product of class IMHC genes.
77. Generally natural killer cells are null cells.
78. Bence Jones protein is found in blood urine.
79. IgA has three subclasses.
80. Insects lack true adaptation immunity.
81. Immunoglobulin is glycoprotein.
82. Antibodies are present in saliva.
83. Null cells are characterised by producing cytotoxic activity.
84. IgG is secreted by T cells.
85. Plasma cells have a long lifespan.
86. Humoral immunity is carried out by B cells.
87. IgD is present in milk.
88. Both kappa and lambda light chains present in a given antibody.
89. T cells lack power of synthesis of immunoglobulin.
90. Natural killer cells may have T cell lineage and have mature T cell markers.
91. The Fc part of the immunoglobulin bears receptors for the fixation of complements for all classes of immunoglobulin.
92. In macrophages mitochondria are numerous, Golgi body is well developed and lysosomes are abundant.
93. Mitosis is rare in macrophages.
94. The frequency of severe combined immunodeficiency is higher in Apache Indians.
95. Gamma globulin increases considerably in the infected animal.
96. IgD are present in human serum.
97. Gene cloning has provided the genetic bases of immunoglobulin diversity and specificity.
98. The number of four chain units in IgE is five.

99. Cyclosporin A has antilymphocyte activity.
100. Antibodies are exception to one-gene–one-polypeptide hypothesis.
101. There are sequence homologies between the variable regions of light chains and those of heavy chains of immunoglobulin.

Answers to True or False

1. True	2. True	3. False	4. True	5. True	6. True	7. True	8 True
9. True	10. True	11. False	12. True	13. False	14. True	15. False	16. True
17. True	18. True	19. False	20. True	21. True	22. True	23. False	24. True
25. False	26. True	27. False	28. True	29. True	30. False	31. False	32. True
33. True	34. False	35. False	36. True	37. True	38. True	39. False	40. False
41. True	42. False	43. True	44. True	45. True	46. True	47. False	48. True
49. False	50. False	51. True	52. False	53. True	54. True	55. True	56. True
57. True	58. True	59. True	60. True	61. True	62. True	63. False	64. False
65. True	66. False	67. True	68. True	69. True	70. False	71. True	72. True
73. True	74. False	75. True	76. False	77. True	78. True	79. False	80. False
81. True	82. False	83. True	84. False	85. False	86. True	87. False	88. False
89. True	90. False	91. False	92. True	93. True	94. True	95. True	96. False
97. True	98. False	99. True	100. True	101. True			

Give Reasons

1. Neutrophils arrive first at the inflammatory site.
   – Because they are more mobile in comparison to monocytes.
2. Humoral immune mechanisms seem to have evolved for the protection of the exposed surfaces of the body.
   – Because most cells are present generally in the exposed surfaces of the body such as skin, naso-pharyngeal mucosa, respiratory tract and gastrointestinal tract.
3. Cell-mediated immune response is also known as delayed type hypersensitivity.
   – Because cell-mediated immune reactions are delayed reactions, attaining a peak in 48 to 72 hours.
4. Each cell produces only one antibody.
   – Because in a cell there is only one rearrangement of heavy chain genes and light chain genes.
5. IgA is called secretary immunoglobulin.
   – Because it is present in all body secretions.
6. Antibodies are called gamma globulins.
   – Because they have globular structure and are capable of migration in electric field.
7. Monoclonal antibodies are identical.
   – Because they are produced by one type of immune cells, which are clones of a single parent cell.
8. X-linked hyper IgM syndrome is more susceptible to infections.
   – Because they lack a proper functioning immune system.

9. The HLA antigenic constitution of a person is entirely different from another.
   - Because the genes located on chromosomes having A, B, C, D and DR loci containing more than 150 alleles, undergo permutation and combination producing innumerable variations.
10. IgG can diffuse to interstitial fluid.
    - Because of its low molecular weight (1,160,000).
11. Being antigenically different from the mother, yet rejection of foetus does not occur.
    - Because the absence of HLA antigens from the outer layer of placental cells may be involved in this, as may the presence of foetal white cells in maternal circulation.
12. Probably cancer cells are recognised by the immune system.
    - Because they contain new and different antigens.
13. HLA antigens are similar in identical twins.
    - Because they share the same major histocompatibility encoded HLA antigens.
14. In adult men, thymus gland gradually atrophies and becomes less important.
    - Because of migration and differentiation of T cells, which occurs mostly during early development.
15. *Leishmania*, *Salmonella* , *M. tuberculosis* and *M. leprae* are protected from antibodies and nonspecific defence mechanism of the body.
    - Because of their intracellular location.
16. Antigens can be called immunogens.
    - Because they stimulate a specific immune response.
17. The first complete Ig synthesised by the B lymphocyte is an IgM.
    - Because in the pre-B-lymphocyte, a mu chain is first synthesised as constant IHGM gene is located near the V-D-J arrangement. This mu chain is associated with pseudo light chain and this combination constitutes the pre-B-receptor. So, the first complete Ig synthesised by the B lymphocyte is an IgM in which mu chain is associated with a kappa or lambda chain.
18. In the genome of B cell, there are two chromosomes (two alleles) for each Ig locus, yet it synthesises single type of heavy chain and light chain.
    - Because of allelic exclusion which is in part by the rearrangements and partly by the surface expression of the functional immunoglobulin which inhibit the rearrangements. As a result the expression of a second chain only one 14 chromosome and one 2 or 22 chromosome are, therefore, productive.
19. It is difficult to isolate a single molecular species of immunoglobulin from the blood serum of a normal person but it can be isolated from the multiple myeloma person.
    - Because the blood serum of a normal person contains a mixture of many different IgG immunoglobulin having the same basic structure. But in multiple myeloma, certain cells multiply, resulting in the excessive production of a single type of immunoglobulin which appears in the blood and can be easily isolated.
20. It is difficult to make a direct estimate of the number of Vk genes in the germ line.
    - Because of the varying degree of divergence between germ line genes.
21. Antibodies are produced against antigens.
    - Because antigens, if not destroyed, will produce diseases causing harm to the body.
22. Many cancer cells express high levels of FasL and can kill any cytotoxic cells that try to kill them.
    - Because cytotoxic T cells also express Fas, but are protected from their own FasL.

PRIONS

Multiple-Choice Questions

1. The term 'prion' was coined by:
   (a) Gross (b) Prusiner (c) Rotman (d) Zuckerkandle
2. What one of the following is correct about prions?
   (a) Devoid of nucleic acid
   (b) Transmissible particles
   (c) Composed exclusively of modified proteins
   (d) All
3. Human prion gene contains ________________ codons.
   (a) 253 (b) 353 (c) 450 (d) 575
4. The human prion gene is located on the:
   (a) Short arm of chromosome number 15
   (b) Long arm of chromosome number 15
   (c) Short arm of Chromosome number 20
   (d) Long arm of chromosome number 20
5. Prions:
   (a) Are normal proteins
   (b) Have correct primary structure
   (c) Have abnormal primary structure
   (d) All
6. Prions cause:
   (a) Zellweger syndrome
   (b) Ehlers–Danios syndrome
   (c) Alper's syndrome
   (d) Gillbert's disease
7. Which one of the following is not applicable to prions?
   (a) Infectious agent (b) Misfolded protein (c) PrP^{SC} (d) Disease not fatal
8. Prp^{C} binds has high affinity:
   (a) Copper ions (b) Zinc ions (c) Cobalt ions (d) Calcium ions
9. Prion diseases are:
   (a) Genetic (b) Infectious (c) Sporadic (d) All
10. Prion proteins in yeast were discovered by:
    (a) James Herrick (b) Reed Wickner (c) Harold E Varmus (d) Peter D Michell
11. PrP^{SC} refers to:
    (a) Misfolded form of PrP^{C}
    (b) Causes formation of amyloid plaques
    (c) Neurodegeneration
    (d) All
12. Which one of the following is incorrect about scrapie?
    (a) It is not a genetic disease.
    (b) It is not caused by a particular disease.
    (c) It is caused by an infectious agent.
    (d) Susceptible genotype is not need to be present in order to have this disease.
13. Prions are:
    (a) Bacteria (b) Fungi (c) Viruses (d) None

14. In which one of the following diseases are there two mutations in prion gene?
    (a) Kippel–Feil syndrome
    (b) Gerstmann–Straussler–Scheinker Syndrome (GSSS)
    (c) Atagille syndrome
    (d) Holt–Oram syndrome
15. Kuru was once found among the Fore tribe in:
    (a) Papua New Guinea
    (b) Kenya
    (c) Zambia
    (d) Tanzania
16. Which one of the following is incorrect about SVp^{35}?
    (a) In is a eukaryotic translation factor.
    (b) It is the yeast eukaryotic release factor.
    (c) It propagates in a prion form.
    (d) Its loss is not fatal.
17. Prions are resistant to:
    (a) Radiation and heat
    (b) Formalin
    (c) Proteases
    (d) All
18. Which one of the following is an inherited prion disease?
    (a) Fatal Familial Insomnia (FFI)
    (b) Kuru
    (c) Kearns–Sayre syndrome
    (d) Kippel–Feil syndrome
19. Prions are inactivated by:
    (a) Sterilisation
    (b) Diethyl pyrocarbonate
    (c) Ethyl ethane sulphonate
    (d) Potassium Cyanide
20. Prions accumulate in:
    (a) Adipose tissue (b) Liver tissue (c) Nervous tissue (d) Cardiac tissue
21. Consider the following statements:
    (A) It is incurable disease of nervous system
    (B) It is characterised by progressive dementia, hallucination, hyper-reflexia, gait abnormalities and tremor
    (C) There is hypertrophy of glial cells
    (D) This disease was first studied by Klatze et al. (1959)

    The name of this disease is:
    (a) Kuru
    (b) Crevtzfeldt–Jakob disease (CJD)
    (c) Ehlers–Danios syndrome
    (d) Kartagener syndrome
22. The infectious prion induces the nearby normal protein molecules to unfold to abnormal form. This is known as:
    (a) Starling's hypothesis
    (b) Two hit hypothesis
    (c) Seeding Model
    (d) Null hypothesis
23. Abnormal folding of proteins may lead to:
    (a) Prions diseases
    (b) Kartagener syndrome
    (c) Kippel–Feil syndrome
    (d) All
24. Which one of the following diseases had resulted in an economic disaster in Great Britain in 1996?
    (a) Bovine spongiform encephalopathy
    (b) Fatal familial insomnia
    (c) Potato spindle tuber disease
    (d) Anthrax
25. Which one of the following is incorrect about Prion Protein (*PrP*)?
    (a) It is a normal protein containing 253 amino acids.
    (b) It can undergo structural conformational changes.
    (c) It can be broken by lysosomal enzymes.
    (d) None

26. Unconventional virus is applicable to:
    (a) Prions (b) Transposons (c) Digoxin (d) Interleukins
27. In which one of the following disease does the cerebral cortex becomes sponge like:
    (a) Alzheimer's disease (b) Creutzfeldt–Jakob disease
    (c) Pompe's disease (d) Gillbert's disease
28. Which one of the following statements is incorrect?
    (a) The prion is a product of a human gene called *PrP* gene.
    (b) *PrP* gene contains two exons separated by a single intron.
    (c) Prion contains either DNA or RNA.
    (d) The *PrP* protein is a precursor of the prion protein.
29. Which one of the following is not a sign of diseases caused by prions?
    (a) Astrocyte gliosis (b) Amyloidosis
    (c) Inflammation or fever (d) Numerous vacuoles in the cerebellar cortex
30. What is incorrect about the infection form of the prion protein?
    (a) Infectious (b) Beta sheeted (c) Nonsoluble (d) Monomeric
31. Which one of the following diseases is related with the diagnosis of prion diseases?
    (a) West blot analysis of blood tissue (b) Capillary electrophoresis of blood
    (c) Immunofluorescence test on tonsil biopsies (d) All
32. Bovine Spongiform Encephalopathy (BSE) was first diagnosed in:
    (a) United States of America in 1980 (b) United Kingdom in 1986
    (c) France in 1998 (d) Australia in 2000
33. Alpers' syndrome is a prion disease in:
    (a) Infants (b) Young (c) Old (d) Women
34. Fatal familial insomnia is characterised by:
    (a) Cerebellar ataxia (b) Severe selective atrophy of the thalamus
    (c) Pneumonia (d) Decreased immune response
35. Deposition of amyloid occurs in:
    (a) Leprosy (b) Creutzfeldt-Jakob disease and Kuru
    (c) Down syndrome (d) All
36. Prion diseases are:
    (a) Infectious (b) Hereditary
    (c) Both infectious and hereditary (d) None
37. Which one of the following is applicable to infectious nature of prions?
    (a) Occurrence of variations due to mutation (b) Metastasis
    (c) Protein folding (d) Resistant to heat and radiation
38. Which one of the following is not an inherited prion disease?
    (a) Kuru (b) Catal familial insomnia
    (c) Creutzfeldt–Jakob disease (d) Gerstmann–Straussler–Scheinker disease
39. Prions are:
    (a) Viruses (b) Bacteria or fungi (c) Parasites (d) None
40. Prion diseases:
    (a) Destroy brain tissue
    (b) Are transmissible from host to host of a single species

(c) Sometimes they are transmissible even from one species to another
(d) All

41. Brain eater disease is applicable to:
(a) Gillbert disease (b) Andersen disease (c) Prion disease (d) Alzheimer's disease

42. Laughing death is related with:
(a) Transposons (b) Exons (c) Mutagens (d) Prions

43. Which one of the following is not applicable to prion diseases?
(a) Horizontal transmission (b) Slow infection
(c) Fever (d) Amyloidosis

44. Which one of the following statements is incorrect?
(a) Prions infect the central nervous system. (b) Each of the different prions have different incubation period.
(c) Prions attack different parts of the brain. (d) None

45. These days ozone sterilisation method is being used to deactivate:
(a) Cancer cells (b) Viruses (c) Prions (d) Transposons

46. The infection of Creutzfeldt–Jakob disease does not occur through:
(a) Growth hormone injection (b) Corneal transplant from dead persons
(c) Blood transfusions (d) Accidental surgery

47. The most common form of human disease having mean age of onset as 60–65 years:
(a) Creutzfeldt–Jakob disease (b) Kuru
(c) Fatal familial insomnia (d) Gerstamann–Straussler–Scheinker disease

48. In humans, the sporadic prion diseases are:
(a) Kuru and Fatal familial insomnia (b) Kuru and Creuztfeldt–Jackob disease
(c) Fatal familial insomnia and variant Creutzfeldt–Jackob of disease (d) Creuztfeldt–Jackob disease and fatal familial insomnia

49. Transmissible Spongiform Encephalopathies (TSEs):
(A) Are a family of rare progressive neurodegenerative disorders (B) Affect both humans and animals
(C) Have long incubation periods (D) Have no induced inflammatory response
The incorrect statements are:
(a) None (b) A, B and C (c) B and D (d) A and D

50. Cerebellar ataxia is the characteristic of:
(a) Alper syndrome (b) Gerstmann–Straussler–Scheinker syndrome
(c) Kuru (d) Fatal familial insomnia

51. Which one of the following is a correct match?
(a) Bovine spongiform encephalopathy – Mad cow disease
(b) GSS – An autosomal dominant form of ataxia
(c) FFI – An autosomal recessive sleep disorder
(d) CJD – Sporadic, familial introgenic

52. What is incorrect about prions?
(a) Infectious agent (b) Virus-like particle
(c) Lacks nucleic acid (d) Lacks replication power

53. Which one of the following is characterised by the deposition of abnormal proteins; often with large amyloid fibrils?

(a) Transmissible spongiform encephalopathies (b) Alzheimer's disease
(c) Parkinson disease (d) All

54. Prions proteins have been detected in the milk of:
(a) Humans (b) Cows (c) Sheep and goats (d) All

55. Which one of the following is not a visible result of post-mortem of prion diseases?
(a) Inflammatory lesions (b) Vacuoles
(c) Amyloid protein deposits (d) Astrogliosis

56. Gerstmann–Straussler–Scheinker syndrome differs from Creutzfeldt–Jackob disease in:
(a) It is characterised by cerebellar ataxia and concomitant motor problems
(b) Dementia is less common
(c) It occured typically in 4th to 5th decades
(d) All

57. It is an extinct disease of Pappua New Guinea, transmitted by eating the brain of dead persons who had this disease. The name of this disease is:
(a) Gerstmann–Straussler–Scheinker syndrome (b) Alper syndrome
(c) Variant Creutzfeldt–Jackob disease (d) Kuru

58. Chronic wastage disease does not occur in:
(a) Mink (b) Mule (c) Deer (d) Elk

59. A change in the folded shape of a prion protein changes its:
(a) Infectious properties (b) Its inability to jump species barrier
(c) Both (a) and (b) (d) None

Answers to Multiple-Choice Questions

1. (b)	2. (d)	3. (a)	4. (c)	5. (d)	6. (c)	7. (d)	8. (a)
9. (d)	10. (b)	11. (d)	12. (d)	13. (d)	14. (b)	15. (a)	16. (d)
17. (d)	18. (a)	19. (b)	20. (c)	21. (b)	22. (c)	23. (a)	24. (a)
25. (c)	26. (a)	27. (b)	28. (c)	29. (c)	30. (d)	31. (d)	32. (b)
33. (a)	34. (b)	35. (d)	36. (c)	37. (c)	38. (a)	39. (d)	40. (d)
41. (c)	42. (d)	43. (c)	44. (d)	45. (c)	46. (c)	47. (a)	48. (d)
49. (a)	50. (b)	51. (c)	52. (d)	53. (d)	54. (d)	55. (a)	56. (d)
57. (d)	58. (a)	59. (c)					

Fill in the Blanks

1. ___________________ are the only known diseases that can be genetic or infectious.
2. Prion protein is a normal protein found on the ___________________of cells.
3. Prusiner has named prion proteins as___________________.
4. Prion protein is easily digested by ___________________.
5. Prion protein has a secondary structure dominated by___________________.

6. In humans, prion is encoded by a gene designated as ________________ located on the chromosome____________.
7. The abnormal disease producing prion is designated as __________________.
8. Creutzfeldt–Jackob disease is inherited as an autosomal __________________.
9. The deposits of PrP^{Sc} in the brain are called __________________.
10. The discovery of prion proteins as infectious agents began in the 1980s with an outbreak of __________________ in the United Kingdom.
11. Prions are normal proteins with abnormal __________________.
12. Prion proteins have been described by __________________.
13. Mad cow disease is caused by __________________.
14. __________________ is affected in fatal familial insomnia.
15. __________________ becomes sponge like in Creutzfeld–Jackob disease.
16. Prions are produced by __________________ in the gene coding for a normal cell protein.
17. Familial form of prion diseases are caused by inherited mutations in the __________________ gene.
18. Mutations in the __________________ codon of normal prion protein have been linked to neurodegeneration.
19. The incidence of sporadic CJD is about __________________ per million per year.
20. __________________ is the name given to prion diseases in infants.
21. PrPs exist in two forms, viz., a common harmless ________________ form and a rare ________________ form that causes fatal prion diseases.
22. Humans are thought to contract __________________ diseases, most commonly by eating prion-contaminated flesh.
23. A protein called __________________ is elevated in the patients of with CJD.
24. Mutations of the *PRNP* gene and the codon 129 polymorphism can be detected by __________________ analysis of blood and tissues.
25. Abnormal prions can be detected in brain tissue extracts by ELISA and in tissue section by __________________.

Answers to Fill in the Blanks

1. Prion diseases	2. Membranes	3. PrP
4. Proteases	5. Alpha helixes	6. *PRNP*, 20
7. PrP^{Sc} (Sc for scrapie)	8. Dominant	9. Amyloid
10. Mad cow disease	11. Tertiary structure	12. Stanley Prusiner (1982)
13. Prions	14. Thalamus	15. Cerebral Cortex
16. Mutations	17. *PRNP*	18. 102nd
19. 1	20. Alper's syndrome	21. Alpha helical, beta sheet
22. Prion	23. 14-3-3	24. DNA
25. Immunohistochemistry		

True or False

1. Prions are a product of human genes which accumulate in the tissue as amyloid.
2. Prions induce no immune reactions within the humans .
3. Prion diseases have a long incubation period.
4. Prions are solely made up of glycoprotein.
5. All mammals have prion protein genes and the gene sequences are similar, but not identical in related species.
6. Some prions can be inherited.
7. Prion diseases involve modification of the prion protein.
8. Neither prions-specific nucleic acid nor like virus like particles have been detected.
9. Prion diseases are also referred to as Transmissible Spongiform Encephalopathies (TSEs).
10. The properties of prions do not change when they are passed from one species to another.
11. Prions resist inactivation by nucleases.
12. Infectious and non-infectious prions differ in their molecular structures.
13. There is no effective treatment for any of the prion diseases.
14. Prions have no ability to reproduce themselves.
15. Fungal prions are toxic to their host.
16. Prions like proteins may not always be harmful.
17. According to recent researches, small prions are much more efficiently infectious than large ones.
18. Some prions diseases can be inherited.
19. Prion diseases are usually rapidaly progressive and always fatal.
20. Variant Creutzfeldt–Jakob disease is inheritable.
21. In their normal harmless form, prions are believed to be involved in cell-to-cell communication.
22. A change in the shape transforms a harmless prion into its infectious prion form. This is due to a point mutation.
23. Baruch Blumberg showed that Kuru was transmitted by eating the brain of an infected person.
24. In Kuru disease, the cerebellum is affected.
25. In Creutzfedt–Jakob disease, aspartic acid 178 is replaced by aspargine, while valine is present at amino acid 129.
26. Prions can mutate genetically.
27. The shape change accounts for strain differences in prions.
28. Mad cow disease is also caused by genetic mutation.

Answers to Fill in the Blanks

1. True	2. True	3. True	4. True	5. True	6. True	7. True	8. True
9. True	10. False	11. True	12. True	13. True	14. True	15. True	16. True
17. True	18. True	19. True	20. False	21. True	22. True	23. True	24. True
25. False	26. False	27. True	28. True				

Give Reasons

1. Prions diseases are called Transmissible Spongiform Encephaopathies (TSFs).
   – Because in these diseases the brain becomes riddled with small holes like a sponge.
2. Prions are spread by eating contaminated meat and use of contaminated surgical instruments.
   – Because they are resistant to normal sterilisation procedures.
3. Scrapie is so named.
   – Because it is characterised by constant scratching.

TRANSPOSONS

Multiple-Choice Questions

1 The stretches of DNA, generally having repeated DNA segments at their ends and are capable of moving from one place to another, on the same or different chromosomes of a genotype, are called:
(a) Transposons (b) Mobile elements or translocatable elements
(c) Selfish genes or jumping genes (d) All

2. The term transposon was used by:
(a) Shine and Delgarno (1974) (b) Hedges and Jacob (1974)
(c) McClintock (1961) (d) Olins and Olins (1974)

3. Consider the following statements:
(a) Transposons can occupy different sites in the genome
(b) Transposons cannot be transposed between a plasmid and a bacterial cell
(c) Transposable elements have no effect on gene structure and expression
(d) Mutations caused by transposable elements are useful

The correct statements are:
(a) All (b) A, B and C (c) B, C and D (d) A

4. The enzyme responsible for transposition is the:
(a) *Translocase* (b) *Transposase* (c) *Transferase* (d) *Transketolase*

5. The elements responsible for hybrid dysgenesis in *Drosophila melanogaster* are:
(a) *Copia* like elements (b) *P* and *I* elements
(c) *FB* elements (d) *Ty* elements

6. Transposons were first discovered by McClintock (1940) in:
(a) Maize (b) *Drosophila* (c) *E. coli* (d) *Neurospora*

7. Which one of the following statements is incorrect?
(a) The insertion of transposons is random.
(b) The ends of transposons are essential for conducting the process of transposition.
(c) The deletion of ends of transposons results in transposition defect.
(d) The enzyme required for transposition is encoded by the normal transposase gene.

8. Transposons cause:
(a) Deletion (b) Insertion
(c) Complicated rearrangements (d) All

9. The first transposon to be molecularly isolated was from:
(a) Snapdragon (b) *Zea mays* (c) *Neurospora* (d) Yeast

10. About__________ per cent of total genome of maize is transposons:
(a) 10 (b) 25 (c) 50 (d) 75

11. All are examples of repeated sequence, except:
    (a) Transposons (b) Pseudogenes (c) SINEs (d) LINEs
12. Which one of the following statements is incorrect with reference to transposase?
    (a) Catalyses insertion of transposon into new sites (b) Carries out DNA cutting
    (c) Carries out DNA ligation (d) None
13. Transposons range in length from:
    (a) 1,000 to 4,000 base pairs (b) 20,000 to 25,000 base pairs
    (c) 750 to 40,000 base pairs (d) 450 to 25,000 base pairs
14. McClintock (1948) noticed all of these caused by transposons except:
    (a) Deletion (b) Insertion (c) Inversion (d) Translocation
15. An example of the most abundant type of repeated DNA sequence is:
    (a) Pseudogenes (b) Telomeres (c) Transposons (d) Centeromeres
16. Resolvase enzyme is encoded by some:
    (a) Pseudogenes (b) Transposons (c) Telomerase (d) Centromeres
17. Transposons generally cause mutations due to:
    (a) Insertion in structural region (b) Insertion in regulatory region
    (c) Addition, deletion or substitution of bases (d) Both (a) and (b)
18. Dotted (*Dt*) element was discovered by:
    (a) Marcus Rhoads (1938) (b) Hartman and Suskind (1965)
    (c) Rosenberg and Court (1979) (d) R I Sinsheimer (1959)
19. Consider the following statements:
    (A) Transposons are mobile elements and appear to play an important functional role in a cell
    (B) Transposons account for at least 30 per cent of *Drosophila* DNA
    (C) Transposons are more diverse in mammalian genome in comparison to other eukaryotes
    (D) Transposons are more numerous than satellite DNA but are shorter in length

    The correct statements are:
    (a) All (b) None (c) A, B and D (d) C and D
20. Which one of the following is not applicable to hybrid dysgenesis?
    (a) High mutation rate (b) High frequency of chromosomal aberrations
    (c) Nondisjunction (d) Fertility
21. Transposons in prokaryotes were discovered by:
    (a) McClintock (1951) (b) Hedges and Jacob (1974)
    (c) Sonneborn (1943) (d) Rubin and Spardling (1982)
22. The first transposon discovered was *Zea mays* and is named as:
    (a) Tn_{10} (b) Dissociator (c) *P* element (d) *Copia*
23. Which one of the following is not applicable to LINEs?
    (a) Lack LTRs (b) Transcribed by RNA polymerase II
    (c) Encode reverse transcriptase (d) Long terminal repeats
24. The ends of transposable elements:
    (a) Represent recognition sites (b) Define the segment of DNA undergoing transposition
    (c) Both (a) and (b) (d) None
25. Transposons move around different positions in the genome of a cell which does not cause:

(a) Increased amount of DNA in the genome (b) Decreased amount of DNA in the genome
(c) Mutations (d) Neither increased or decreased amount of DNA in the genome

26. The most abundant SINEs are the:
(a) *Alu* element (b) *P* elements (c) *Copia* elements (d) None
27. Miniature Inverted Transposable Elements (MITEs) applicable to:
(a) Class I transposons (b) Class II transposons
(c) Class III transposons (d) All
28. Reverse transcriptase is applicable to:
(a) Class I tranposons (b) Class II transposons (c) Class III transposons (d) All
29. Transposable elements may change a gene product by:
(a) Altering the base sequences in DNA (b) Altering the structure of chromosome
(c) Altering the way pre-mRNAs are spliced (d) Amplifying DNA sequences within a genome
30. Which one of the following is an incorrect match?
(a) IS_8 ⟶ RP_4 plasmid
(b) In_1 and In_2 ⟶ *Klebsiella* plasmid JR67
(c) IS_1 to IS_5 ⟶ *E. coli*
(d) *ISRI* ⟶ Geome of *Rhizobium lupini*
31. Transposon mariner was identified in 1986 in:
(a) *Drosophila melanogaster* (b) *D. malerkotliana*
(c) *D. mauritiana* (d) *D. bipectinata*
32. The best known example of replicative transposition is:
(a) *Ty* element (b) *Mu* phage (c) *P* elements (d) *Alu*
33. Which one of the following is not a transposable element?
(a) Tn_{10} (b) Tn_9 (c) Tn_5 (d) None
34. Miniature inverted repeats Transposable Elements (MITEs) have been reported in the genome of:
(a) *Xenopus* (b) Humans (c) Apples (d) All
35. Transposons are normal constituents of many:
(a) Plasmids (b) Bacterial genes (c) Bacteriophages (d) All
36. Mobile genetic elements include:
(a) Transposons (b) Plasmids and bacteriophage elements
(c) Group II introns (d) All
37. Which one of the following is a useful marker for DNA fingerprinting?
(a) LINEs (b) SINEs (c) Pseudogenes (d) Retrotransposons
38. Transposons can multiply in the genome of:
(a) Archaea (b) Eubacteria (c) Eukaryotes (d) All
39. Which one of the following is a moderately repetitive DNA?
(a) Transposons (b) Telomeres (c) Centromeres (d) All
40. Which one of the following diseases is applicable to transposons?
(a) Porphyria (b) Haemophilia A and B
(c) Duchenne muscular dystrophy (d) All
41. Retrotransposons lack:
(a) LTR (b) RT (c) Integrase (d) None

42. The first transposon to be identified was the:
    (a) Insertion sequence in *E. coli*
    (b) *P* elements from *D. melanogaster*
    (c) *Alu* in mammals
    (d) *Ty* in yeast
43. Which one of the following is not a synonym of the jumping gene?
    (a) Insertion sequence element
    (b) Transposable genetic element
    (c) Intervening sequence
    (d) Transposons
44. Which one of the following is not applicable to McClintock?
    (a) Transposons (b) Overlapping genes (c) Maize (d) Nobel Prize
45. Silenced transposons have been reported in:
    (a) *Neurospora crassa*
    (b) *Aspergillus nidulans*
    (c) Both (a) and (b)
    (d) Maize and sugarcane
46. The two transposable controlling elements discovered by McClintock are:
    (a) *P* and *I* elements
    (b) *Copia* and *FB* elements
    (c) Dissociation (*Ds*) and Activator (*A(c)*
    (d) *Ac* and *P*
47. Terminal repeats are lacking in:
    (a) *Alu* elements (b) *FB* elements (c) *Ty* elements (d) *Copia* elements
48. Which one of the following has been used to produce transgenic tobacco plants?
    (a) *dSpm* (b) *Spm* (c) *Tam* (d) *Tc1*
49. Which one of the following statements is incorrect with reference to transposons?
    (a) Changed pattern of gene expression
    (b) Cause gross chromosome rearrangements
    (c) Can change the gene sequence
    (d) Can not amplify DNA sequence
50. Noncomposite transposons lack:
    (a) Inverted repeats
    (b) Insertion sequence
    (c) Transposase gene
    (d) Antibiotic resistance gene(s)
51. Consider the following statements:
    (A) Bacteriophage mu is the longest known transposable element
    (B) Bacteriophage mu contains repeated sequences at its termini
    (C) Mu insertions are usually accompanied by deletion, duplication and inversion
    (D) Mu causes different kinds of mutation

    The correct statements are:
    (a) All (b) A, B and C (c) A, C and D (d) C and D
52. Telomerase is closely related to:
    (a) LINEs (b) SINESs (c) Junk DNA (d) Pseudo genes
53. Rare cases of haemophilia are caused by insertion of a:
    (a) LINEs-1 element from chromosome 22 into the gene for clotting factor VIII located on the X-chromosome
    (b) LINEs-1 element from chromosome 22 into the gene for clotting factor VIII located on the Y- chromosome
    (c) LINEs-1 element from chromosome 20 into the gene for clotting factor VIII located on the X-chromosome
    (d) LINEs-1 element from chromosome 20 into the gene for clotting factor VIII located on the Y-chromosome
54. The transposase promoter of *Tn10* contains DNA sequence GATC which is methylated in *E. coli* at the:
    (a) Adenine nucleotide
    (b) Guanine nucleotide

(c) Thymine nucleotide (d) Cytosine nucleotide

55. Match column I with column II and select the correct answer using answer codes:

Column I	Column II
(A) Retrons	1. SINEs, LINEs
(B) Retroposons	2. msDNA
(C) Retrotransposons	3. *Ty*, *Copia*
(D) Retrosequence	4. cDNA genes

Answer codes:

	A	B	C	D
(a)	2	4	1	3
(b)	2	1	3	4
(c)	1	2	3	4
(d)	4	3	2	1

56. Transposition can cause:
(a) Mutations (b) Change in the amount of DNA in genome
(c) Both (a) and (b) (d) None

57. Transposons cause changes in gene expression by:
(a) Shutting off genes (b) Causing insertion mutations
(c) Both (a) and (b) (d) None

58. Transposition can facilitate:
(a) Insertional activation (b) Gene amplification
(c) Gene mobilisation (d) Gene immobilisation

59. In moderately repetitive DNAs, the main constituent of the middle repetitive sequences is:
(a) Satellite DNAs (b) Centromeres (c) Telomeres (d) Transposons

60. Transposons:
(a) Are mobile genetic elements (b) Cause breakdown of the reading frame
(c) Appear to play no functional role in the cell (d) Can insert themselves at one or more sites in the genome

61. Which one of the following is incorrect with reference to resolvase?
(a) It is encoded by some transposons. (b) It catalyses the second stage of transposition.
(c) It is a repressor of its own gene but not to that of the transposase gene. (d) It represses genes by binding a site between them in transposon DNA.

62. Transposable elements can:
(a) Join unrelated segments (b) Promote rearrangements of genetic material
(c) Promotes deletion of the genetic material (d) All

63. The frequencies of many transposons-promoted genetic rearrangements range from _____ per generation:
(a) 10^5 to 10^{10} (b) 10^5 to 10^5 (c) 10^{-17} to 10^{-10} (d) 10^{-25} to 10^{-15}

64. Which one of the following is not applicable to non-autonomous elements of maize?
(a) Can be derived from autonomous element of the same family (b) Unstable
(c) Do not transpose (d) Stable

65. What is incorrect about insertion of IS-elements in bacterial operons?
(a) Inactivate gene (b) Stop transcription (c) Stop translation (d) None

66. Reverse transcription is applicable to:
(a) Retro-elements (b) Transposons (c) *IS* elements (d) Both (b) and (c)

67. Retrosequences include:
(a) cDNA or pseudogenes (b) B-DNA
(c) Z-DNA or pseudogenes (d) msDNA

68. Retroposons lack:
(a) LTR (b) Integrase (c) RT (d) None

69. The main component of moderately repetitive DNA is:
(a) Telomeres (b) Centromeres (c) Transposons (d) Retrons

70. Which one of the following is applicable to transposons?
(a) Mutagens (b) Transformation vectors
(c) Radiation (d) Both (a) and (b)

71. Cointegrate is applicable to:
(a) Crossing over (b) Gene regulations
(c) Transcription (d) Transposition process

72. Which one of the following is a correct match?
(a) McClintock – Central dogma (b) Transposons – Jumping genes
(c) Muller – Pleiotropic gene (d) Morgan – Lethal gene

Answers to Multiple-Choice Questions

1. (d)	2. (b)	3. (a)	4. (b)	6. (b)	7. (a)	8. (a)	9. (d)
10. (b)	11. (c)	12. (b)	13. (d)	14. (c)	15. (c)	16. (c)	17. (b)
18. (d)	19. (a)	20. (b)	21. (d)	22. (b)	23. (b)	24. (d)	25. (c)
26. (d)	27. (a)	28. (a)	29. (a)	30. (c)	31. (b)	32. (c)	33. (b)
34. (d)	35. (d)	36. (d)	37. (d)	38. (a)	39. (d)	40. (d)	42. (d)
43. (d)	44. (a)	45. (c)	46. (b)	47. (c)	48. (c)	49. (a)	50. (b)
51. (d)	52. (b)	53. (c)	55. (a)	56. (a)	57. (a)	58. (b)	59. (c)
60. (c)	61. (c)	62. (d)	63. (c)	64. (c)	65. (d)	66. (c)	67. (d)
69. (d)	70. (a)	71. (a)	72. (a)	73. (c)	74. (d)	75. (d)	76. (b)

Fill in the Blanks

1. The *P* element gives rise to a phenotype called ____________________.
2. Class II transposons consist only of ____________________.
3. The *P* element is a class II transposon and is found in ____________________.
4. The *P* element can be identified by its terminal ____________________ bp inverted repeats and ____________________ bp direct repeats.
5. *P* elements seem to have first appeared in ____________________ about 50 years ago.
6. ____________________ per cent of the entire human genome consists of transposons.

7. Thousands of our *Al4* elements occur in the ______________________ of structural genes.
8. Short Interspersed Elements (SINEs) are short ______________________ sequences.
9. The process of excision and reintegration of the transposable element at another site in the genome of a cell is known as ______________________.
10. In yeast, transposons are termed as ______________________ elements.
11. ______________________ elements are probably the most widespread DNA-transposon in nature.
12. In *E. coli*, there are ______________________ and ______________________ conservative methods of transposition.
13. In the ______________________ method of transposition, a new copy of the transposable element appear at a new site.
14. The eggs of *P*-strain females contain a high amount of ______________________ protein which prevents ______________________ of transpoase gene.
15. Retrotransposons move by a ______________________ mechanism.
16. ______________________ transposition is the best-known example of replicative transposition.
17. ______________________ is the total of all mobile genetic elements in a genome.
18. Class II mobile genetic elements move directly from one position to another within the genome using a ______________________.
19. Eukaryotic organisms may have developed a ______________________ mechanism as a way of reducing transposon activity.
20. Junk DNA is mostly mutated ______________________ and ______________________.
21. *Alu* sequences are interspersed in the human genome, one every ______________________ nucleotides of DNA or less.
22. The process of movement of transposons is known as ______________________.
23. ______________________ is responsible only for resistance to ampicillin.
24. In bacteria, two types of transposons are found, viz., ______________________ and ______________________.
25. The hybrid dysgenesis phenotype is effected by the transposition of ______________________ within the germ line cells of offspring of *P* strain males with M strain females.
26. Retrotransposons copy themselves to RNA and then via ______________________ back to DNA.
27. The transposons which move by cut and paste, may duplicate themselves if the transposition occurs in ______________________ phase of the cell cycle.
28. Insertion sequence contains ______________________ and ______________________.
29. Transposons form bulk of the genome which is evident through ______________________ of eukaryotic species.
30. Composition transposons contain two insertion elements and ______________________ genes.
31. Transposons that replicate via an RNA intermediate are called ______________________.
32. Resolvase catalyses recombination between the two ______________________.
33. For a particular transposon, the number of repeated nucleotides is generally ______________________ and ______________________.
34. Simple transposons of bacteria contain no genetic information except that which is necessary for ______________________.
35. Simple transposons are also called ______________________.
36. ______________________ family is the best-characterised mammalian SINE family.

37. *Alu* family descended from ______________________ gene.
38. SINEs are typically less than ______________________ bp long, while LINEs are more than 500 bp long.
39. The most abundant interspersed repeated DNA family of primates is the ______________________.
40. The fused structure called cointegrate is formed during ______________________.
41. Antibiotic resistant segments of plasmids have evolved as a collection of ______________________.
42. The mu is a ______________________ element.
43. The ______________________ can pick transposons from their hosts.
44. ______________________ of *E. coli* was the first operon where *IS* elements were detected.
45. Retrotransposons resemble transposons except that they have ______________________.
46. The first transposons were discovered in ______________________ by McClintock (1948).
47. DNA elements which can jump to a new location are called ______________________.
48. Mutations caused by *IS* elements are called ______________________ mutations.

Answers to Fill in the Blanks

1. Hybrid dysgenesis	2. DNA	3. *Drosophila*
4. 31, 8	5. *Drosophila melanogaster*	6. 42
7. Introns	8. DNA	9. Transposition
10. *Ty*	11. *Tc1*/mariner	12. *E. coli*, replicative, conservative
13. Replicative	14. Repressor, transcription	15. Copy and paste
16. Mu phage	17. Mobilome	18. Transposase
19. RNA interference	20. Viruses, transposons	21. 10,000
22. Transposition	23. *Tn3*	24. Simple, complex
25. *P* elements	26. Reverse transcriptase	27. S
28. Inverted repeats, transposase gene	29. C-values	
30. Antibiotic resistance gene(s)	31. Retrotransposons	32. Internal resolution sites (IRSs)
33. Uneven, fixed	34. Transposition	35. Insertion sequences
36. *Alu*	37. 7S-L RNA	38. 500
39. *Alu* family	40. Transposition	41. Transposons
42. Transposable	43. Baculoviruses	44. gal-operon
45. RNA origin	46. *Zea mays*	47. Transposons
48. Polar		

True or False

1. The genes that move from one location to another on a chromosome are known as transposons.
2. Transposable elements are restricted to prokaryotes.
3. The terminal inverted repeats are characteristic for each transposable element.
4. Transposons are examples of mobile genetic elements.
5. Transposons are not found in all organisms.

6. McClintock first discovered segments of DNA that can jump to new places in the genome.
7. Autonomous elements do not excise and transpose.
8. All class II (DNA) transposons move by copy and paste.
9. Tn_{10} is a composite transposon.
10. Transposons may lead to gene duplications.
11. *Alu* family is the most abundant mobile element in humans.
12. *Ty* elements of yeast transcribed first into DNA.
13. Retroelements involve reverse transcription.
14. Transposons are unable to produce new genes.
15. Simple transposition is a conservative process.
16. Tn_{10} is flanked by insertion sequences.
17. In Tn_{10} insertion sequences are asymmetrical.
18. Naturally occurring *P* elements lack coding sequence for the enzyme transposase.
19. The discovery of transposons showed that genetic information is fixed in the genome.
20. Many retrotransposons have Long Terminal Repeats (LTRs) at their ends.
21. Transposons are mutagens.
22. Integrase serves the same function as the transposases of DNA-transposons.
23. Transposons have a profound effect on embryonic development and formation of tumour in animal cells.
24. During the course of transposition, the transposons leave their original site.
25. Transposition involves the relationship between the sequences at the donor and recipient sites.
26. Mutations caused by transposable elements are deleterious.
27. Mutation in the transposon gene may stop the transposition process.
28. Transposable elements are found only in germinal tissues.
29. Sometimes, transposons cause damage to DNA.
30. Transposons form almost half of the human DNA.
31. *TC1* is the first *C. elegans* transposon to be identified.
32. *IS* elements only codes for proteins involved in transposition.
33. The coding region in an insertion sequence is generally not flanked by inverted repeats.
34. Retrotransposons are more abundant in plants.
35. Transposons are mobile segments of DNA that can not exist independent of a replicon.

Answers to True or False

1.	True	2.	False	3.	True	4.	True	5.	False	6.	True	7.	False	8.	False
9.	True	10.	True	11.	True	12	False	13	True	14.	False	15.	True	16.	True
17.	False	18.	False	19.	False	20.	True	21	True	22.	True	23.	True	24.	False
25.	False	26	True	27	True	28.	False	29.	True	30	True	31.	True	32.	True
33.	False	34.	True	35.	True										

Give Reasons

1. Transposition only occurs in germ line cells.
   – Because a splicing event needed to make transposase in mRNA does not occur in somatic cells.
2. *P* elements do little harm.
   – Because expression of their transposase gene is usually repressed.
3. Transposons are called selfish genes.
   – Because their only function seems to be making more copies of themselves.
4. Transposons are called junk.
   – Because there is no obvious benefit to their host.
5. Mutations induced by retrotransposons are relatively stable.
   – Because the sequence at the insertion site is retained as they transpose via the replication mechanism.
6. Bacterial elements may be remnants of a retrovirus.
   – Because they contain genes for reverse transcriptase.
7. Transposons are considered natural agents for genetic engineering.
   – Because they have the ability to transfer genes from one locus to another.
8. *P* element is autonomous.
   – Because it encodes a functional transposase.
9. In replicative transposition, the donar is not destroyed.
   – Because a transposon does not leave the donar DNA.
10. Transposons may be used as genetic markers.
    – Due to their ability to change the pattern of restriction fragment analysis.
11. McClintock called mobile elements as controlling elements.
    – Because of their effect on other genes.
12. Most LINEs-1 elements are not functional in humans.
    – Because of mutated-RT genes.
13. Mutator (mu) gene causes different kinds of mutation.
    – Because it can insert at multiple sites in the host bacterium.
14. McClintock was unable to map *Ac*.
    – Due to its different positions in different chromosomes, in different plants, or in different positions in the same chromosome.
15. *Ac/Ds* transposons are so named.
    – *Ac* (for activator) because it activates *Ds* and *Ds* (for dissociation) because it is associated with chromosome breaks.

APOPTOSIS

Multiple-Choice Questions

1. Which one of the following is applicable to apoptosis?
   (a) Cell suicide (b) Self-destruction (c) Genetically controlled (d) All
2. The process in which cells play an active role in their own death is known as:
   (a) Apolysis (b) Autolysis (c) Apoptosis (d) Necrosis
3. Apoptosis does not cause:
   (a) Cell shrinkage (b) Traumatic cell death
   (c) Loss of membrane symmetry (d) Nuclear fragmentation
4. Which one of the following is not a programmed cell death?
   (a) Apoptosis (b) Anoiks (c) Cornification (d) Necrosis
5. Which of the following cells are marked for apoptosis?
   (a) Cells having no function (b) Cells that develop at improper places
   (c) Cells who have completed their functions and cells treated with care inorganic chemicals (d) All
6. Granazyme B is produced by:
   (a) B cell (b) T cells (c) Lysosomes (d) Natural killer cells
7. Apoptosis does not involve:
   (a) Condensation of chromatin material (b) Shrinkage of cytoplasm
   (c) Shrinkage of cytoplasm causes damage to plasma membrane (d) The cell contents are broken down into small pieces.
8. Which one of the following is anti-apoptosis?
   (a) Ced3 (b) Ced4 (c) Ced9 (d) Hsp70
9. Consider the following statements:
   (A) Apoptosis ensures homeostasis in all tissues
   (B) Apoptosis is responsible for removal of a large number of vestigial structures
   (C) Apoptosis causes swelling and rupturing of mitochondria
   (D) In apoptosis, cell contents are released

   The correct statements are:
   (a) All (b) A, B and D (c) A and B (d) C and D
10. Apoptosis does not involve:
    (a) Cell swelling (b) Loss of membrane integrity
    (c) Inflammation (d) All
11. Genetic studies in which one of the following animals have given much information about cell death?
    (a) *C. elegans* (b) *D. melanogaster*
    (c) *Trypanosoma gambience* (d) *D. bipectinata*

12. Activity of Ced3 and Ced4 is regulated by:
    (a) Ced9 (b) Ced8 (c) Ced7 (d) Ced1
13. The caspases, cysteine proteases are homologous to *C. elegans*:
    (a) Ced9 (b) Ced7 (c) Ced5 (d) Ced3
14. The first example of an oncogene that inhibits cell death:
    (a) *BCl2* (b) *BH*1 (c) *BH*2 (d) *Apaf*-1
15. Excessive apoptosis contributes to:
    (a) Alzheimer's disease (b) Parkinson's disease
    (c) Huntington's disease (d) All
16. Ced9 has a mammalian homolog called:
    (a) DR4 (b) BC12
    (c) DR5 (d) Tumour Necrosis Factor (TNF)
17. Which one of the following is an apoptosis inducing factor?
    (a) Cytochroma b5 reductase (b) NADH dehydrogenase
    (c) NADPH oxidase (d) All
18. Principle of apoptosis was first described by:
    (a) Carl Vogt (1842) (b) Walther Flemming (1845)
    (c) Paul Ehrlich (2000) (d) Robert Horvitz (2000)
19. Tumour Necrosis Factor (TNF) is mainly activated by:
    (a) CD95 (b) Phosphatidylserine
    (c) Inflammatory response (d) Macrophages
20. Defects in an apoptosis may result in:
    (a) Autoimmune diseases (b) Cancer
    (c) Spreading of viral infections (d) All
21. In *C. elegans*, generally______________ out of every eight body cells produced is eliminated by a programmed cell death:
    (a) 1 (b) 2 (c) 4 (d) 3
22. Anoik:
    (a) Is a formation of P^{53} (b) Is the shrinkage of cell
    (c) Is a type of programmed cell death (d) Is the creation of apoptosome
23. The X-linked inhibitor of apoptosis protein is over expressed in cells of the:
    (a) NCI-H460 (b) H460 cell line (c) Cytochrome C (d) All
24. The term 'apoptosis' was introduced by:
    (a) Taylor (1978) (b) Kerr (1972) (c) Gilbert (1972) (d) Garner (1981)
25. Which one of the following is incorrect about BCl2?
    (a) The first component of the cell death mechanism to be cloned in any organism (b) Anti-apoptotic
    (c) Tumourigenic (d) Discovered by A R Currie (1972)
26. Apoptosome is applicable to:
    (a) Autolysis (b) Apolysis (c) Apoptosis (d) None
27. Apoptotic processes are involved in:
    (a) Development and differentiation (b) Regulation and development of the immune system

(c) Removal of harmful cells. (d) All

28. Excessive apoptosis may cause:
(a) AIDS (b) Ischemic diseases
(c) Neurodegenerative disorders (d) All

29. The central regulators of intrinsic apoptosis pathways are:
(a) Nuclei (b) Lysosomes (c) Mitochondria (d) Ribosomes

30. The most abundant protein on the inner membrane of mitochondria is the:
(a) Adenine nucleotide translocator (b) Voltage dependent anion channel
(c) Calmodulin (d) CoQ reductase

31. In multicellular organisms, apoptosis plays a key role in:
(a) The elimination of unwanted cells (b) Maintaining tissue integrity
(c) Tissue function (d) All

32. The intrinsic pathway of apoptosis may be triggered by:
(a) ApO2L/TRAIL (b) ApO2L/TRAIL and DNA damage
(c) DNA damage and severe cell stress (d) Cell stress and Apo2L/TRAIL

33. Which one of the following is applicable to apoptosis?
(a) Cells get injured (b) Cells get punctured
(c) Cells lyse extruding injurious compartments (d) None

34. Caspases are a hallmark of:
(a) Paraptosis (b) Apoptosis (c) Necrosis (d) Oncosis

35. Paraptosis does not involve:
(a) Swelling of cells (b) Appearance of vacuoles or large bubbles
(c) Employment of caspases (d) Appearance of liquid

36. Which one of the following may cause apoptosis?
(a) DNA damage (b) Oxidative stress
(c) Improper protein folding (d) All

37. Apoptosis inducing factor is normally present in the _________ of mitochondria.
(a) Matrix (b) Inter membrane space
(c) Elementary particles (d) Stalk of elementary particles

38. Upon release from the mitochondria, apoptosis factor migrates to:
(a) Nucleus (b) Lysosome (c) Cytoplasm (d) Outside cell

39. Apoptosis inducing factor:
(a) Binds to DNA (b) Causes fragmentation of DNA
(c) Causes cell death (d) All

40. Consider the following statements:
(A) Apoptosis inducing factor pathway of apoptosis is more common in neurons
(B) It is a caspase independent pathway
(C) BCl2 activates Bax, which creates holes in the mitochondrial membrane
(D) The wholes caused by Bax in mitochondrial membranes allow entry of cytochrome C into the cytoplasm
The correct statements are:
(a) All (b) A, B and C (c) B and C (d) C and D

41. Which one of the following is not applicable to apoptosis?
(a) An active genetically regulated process (b) Requires coordinated expression of many genes

(c) Irreversible process
(d) Inflammation

42. Cellular necrosis is:
(a) A passive process
(b) Caused by cell damage
(c) Does not require further gene activity
(d) All

43. Which one of the following is common between apoptosis and necrosis?
(a) Gene expression (b) Differentiation (c) Passive process (d) Cell death

44. Cell death is regulated by steroid hormones in:
(a) Yeast (b) *Drosophila* (c) *E. coli* (d) Humans

45. Which one of the following is incorrect about tumour necrosis factor?
(a) It is a cytokine.
(b) It is secreted by activated macrophages.
(c) It is the major extrinsic mediator of apoptosis.
(d) There are four receptors for TNF1 in human cells.

46. Apoptosis may occur:
(a) During starvation
(b) When a cell becomes infected with a virus
(c) When a cell becomes damaged beyond repair
(d) All

47. In thymocytes, apoptosis can be induced by:
(a) Glucocorticoids
(b) Irradiation
(c) Glucocorticoids and irradiation
(d) Growth hormone and estrogen

48. CD95 is applicable to:
(a) Tumour necrosis factor
(b) Lymphotoxin
(c) Fas
(d) *Apaf*-1

49. BCl2 located in the membranes of:
(a) Nuclear envelope
(b) Endoplasmic reticulum
(c) Outer membrane of mitochondria
(d) All

50. Which one of the following cells is constitutively programmed to undergo cell death by apoptosis?
(a) Lacis cells
(b) Mott cells and lacis cells
(c) Neutrophils
(d) Neutrophils and eosinophils

51. Consider the following statements:
(A) BCl2 is a human proto-oncogene
(B) BCl2 is located on chromosome 18
(C) In the cancerous B cells, the part of chromosome containing BCl2 locus has undergone reciprocal translocation with the portion of chromosome 16
(D) High levels of BCl2 protein protect the cell from death by apoptosis
The correct statements are:
(a) A, B and D (b) B, C and D (c) B and C (d) All

52. Which one of the following ageing cells kills themselves by apoptosis?
(a) Neuron cells (b) B cells (c) α cells and β cells (d) β cells

53. BCl2 protein suppresses apoptosis by
(a) Activation of caspases
(b) Preventing activation of caspases
(c) Releases of *Apaf*-1
(d) Release of tumour necrosis factor

54. Apoptosome does not include:
(a) BCl2
(b) BCl2 and *Apaf*-1
(c) Caspase 9 and cytochrome C
(d) Fas and TNF

55. The formation of synapses between neurons the brain requires removal of surplus cells by:
(a) Necrosis (b) Apoptosis (c) Apolysis (d) Autolysis

56. Apoptosis is under the influence of:
(a) Growth factors (b) Hormones (c) Cytokines (d) All

57. Which one of the following is anti-apoptotic?
(a) BCl2 (b) BClXL (c) Bax (d) MCl 1

58. High rate of apoptosis is found in:
(a) Liver cells (b) Kidney cells
(c) Male germ cells and cells of the nervous system (d) Haematopoietic cells and male germ cells

59. Very little apoptosis is shown by:
(a) Heart (b) Kidney
(c) Cells of the nervous system (d) All

60. Inhibitors of Apoptosis Proteins (IAPs) were first found in:
(a) Baculo viruses (b) Herpes viruses (c) SV40 (d) HIV

61. Apoptosis inducing factor is generally present in:
(a) Nucleus (b) Lysosomes
(c) Mitochondria (d) Endophasmic reticulum

62. Which one of the following is considered to be a hallmark of apoptosis?
(a) Breakdown of genomic DNA
(b) Breakdown of genomic DNA into multiples of approximately 180 bp
(c) Breakdown of genomic DNA into multiples of approximately 500 bp
(d) Breakdown of genomic DNA into multiples of approximately 10 bp

63. Apoptosis does not involve:
(a) Cell shrinkage (b) Pykonosis (c) Karyorrhexis (d) Karyolysis

64. Caspases have proteolytic activity and are able to cleave proteins at __________ residues:
(a) Glutamic acid (b) Aspartic acid (c) Tryptophan (d) Isoleucine

65. Which one of the following caspases is highly expressed in embryonic tissues but not in adult tissues?
(a) Caspase 14 (b) Caspase 12 (c) Caspase 10 (d) Caspase 8

66. Apoptosis inducing factor is encoded by a single gene located on the:
(a) 5th chromosome (b) 9th chromosome (c) 21st chromosome (d) X chromosome

67. The BCl2 family of proteins:
(a) Governs mitochondrial membrane permeability (b) Is pro-apoptotic
(c) Is anti-apoptotic (d) All

68. Insufficient apoptosis of auto-aggressive T cells, results in:
(a) Excess production of immunoglobulin (b) Autoimmune Lympho Proliferative Syndrome (ALPS)
(c) Alzheimer's disease (d) All

69. Excessive apoptosis does not cause:
(a) AIDS (b) Cancers
(c) Neurodegenerative diseases (d) Myocardial ischemia

70. Which one of the following is released by mitochondria during apoptosis?
(a) AIF (b) Endonuclease G (c) CAD (d) All

71. Apoptosis and necrosis can occur:
(a) Independently (b) Sequentially (c) Simultaneously (d) All

72 Generally apoptosis occurs during:
(a) Development (b) Ageing
(c) Homeostatic mechanism (d) All

73. Death ligand is:
(a) FacL (b) FADD (c) CD95 (d) *Aapf*-1

74. Granazyme B can directly activate caspases:
(a) 3 and 7 (b) 7 and 8 (c) 3 and 10 (d) All

75. Caspases:
(a) Activate DNAases (b) Inhibit DNA repair enzymes
(c) Breakdown structural protein in the nucleus (d) All

76. Nitric acid inhibits apoptosis in:
(a) Leukocytes (b) Hepatocytes and trophoblasts
(c) Endothelial cells (d) All

77. Caspases play a key role in:
(a) Necrosis (b) Inflammation (c) Apoptosis (d) All

78. Which one of the following is not involved in apoptosis?
(a) CASP-3 (b) CASP-7 (c) CASP-9 (d) CASP-14

79. The importance of caspases in apoptosis was established by:
(a) Korsmeyer (b) Andrew Wyllie (c) Robert Horvitz (d) All

80. Apoptosome activates caspase:
(a) 3 (b) 5 (c) 9 (d) 10

81. Death receptor domain is applicable to:
(a) Caspases 5 and 7 (b) Caspases 7 and 9 (c) Caspases 8 and 9 (d) Caspases 8 and 10

82. Which one of the following is involved in extrinsic death receptor pathways?
(a) Apoptosome (b) Mitochondria
(c) Both apoptosome and mitochondria (d) None

83. DNA laddering is the characteristic of:
(a) Apoptosis (b) Toxic cell death (c) Necrosis (d) All

84. DNA fragmentation caused by apoptosis can be assayed by:
(a) Flow cytometry (b) Fluorescent assays (c) Both (a) and (b) (d) None

85. Which one of the following statements is incorrect about X-linked inhibitor of apoptosis protein (XLAP)?
(a) It inhibits apoptosis.
(b) It is the most potent human inhibitor of apoptosis protein.
(c) It stops cell death induced either by viral infection or overproduction of caspases.
(d) It cannot inhibit action of caspases 3, 7 and 9.

86. Cell death is:
(a) Genetically determined (b) Carried out by physiological response
(c) Triggered by extracellular signals (d) All

87. What is correct about tumour necrosis factor?
(a) Secreted by activated macrophages (b) Pro-apoptotic
(c) Anti-apoptotic (d) Serine protease

88. Match column I with column II and select the correct answer using answer codes:

Column I	Column II
(A) Pro-apoptotic	1. Activated macrophages
(B) Anti-apoptotic	2. Bax
(C) Serine protease	3. Granazyme B
(D) Tumour necrosis factors	4. BCl2

Answer codes:

	A	B	C	D
(a)	2	4	3	1
(b)	2	1	4	3
(c)	3	2	4	1
(d)	4	3	1	2

89. During apoptosis, a cell first:
(a) Swells
(b) Shrinks
(c) Produces enzymes needed for destruction
(d) Produces blebs on its outer surface

90. Which one of the following statements is correct?
(a) All forms of programmed cell death have similar shape.
(b) All forms of programmed cell death have the same sequences of apoptosis.
(c) All forms of programmed cell death are highly regulated processes.
(d) All

91. A cell commits suicide due to:
(a) The withdrawal of signals needed for continued survival
(b) The receipt of negative signals
(c) The balance between (a) and (b)
(d) No definite reason

92. Upon recognition of their target, cytotoxic T cells:
(a) Bind with Fas
(b) Produce more FasL at their surface
(c) Produce caspase 8
(d) Bind with tumour necrosis factor

93. Allografts engineered to express FasL, have resulted in increased survival of:
(a) Kidneys
(b) Islets of Langerhans
(c) Hearts
(d) All

94. Mutation is more common in:
(a) Fas
(b) FasL
(c) Caspases
(d) TNF

95. Death Inducing Signaling Complex (DISC) contains:
(a) Caspase 8
(b) Caspase 10
(c) FADD
(d) All

96. Which one of the following statements is incorrect?
(a) T lymphocytes undergo apoptosis.
(b) The lymphocytes have receptors for self-antigens.
(c) About 90 per cent of B cell lymphocytes produced in bone marrow are eliminated by the process of apoptosis.
(d) Apoptosis does not occur in certain parasites.

97. CED-4 participate in:
(a) Cytochrome C dependent activation of caspases
(b) Cytochrome C independent activation of caspases
(c) Both (a) and (b)
(d) None

98. Glyphosate formation induces apoptosis in human:
(a) Placental cells (b) Umbilical cells
(c) Embryonic cells (d) All

99. Bystander cell damage is applicable to:
(a) Apoptosis (b) Necrosis (c) Zeiosis (d) All

100. Which one of the following is incorrect about perforin?
(a) It is a peptide. (b) It is related to complement component 9.
(c) On release, perforin is polymerises to polyperforin. (d) Polyperforin does not form trans membrane channels.

101. Kostman syndrome is caused by:
(a) Excessive necrosis (b) Excessive apoptosis of granulocytes
(c) Deletion of chromosome 15th (d) Translocation between chromosomal 14th and 19th

102. Which one of the following can induce apoptosis?
(a) Tumour necrosis factor (b) Glucocorticod
(c) Serum starvation (d) All

103. Gene directed cell death is applicable to
(a) Necrosis (b) Zeiosis (c) Apolysis (d) Apoptosis

Answers to Multiple-Choice Questions

1. (d)	2. (c)	3. (b)	4. (d)	5. (d)	6. (b)	7. (c)	8. (c)
9. (c)	10. (d)	11. (a)	12. (a)	13. (d)	14. (a)	15. (d)	16. (b)
17. (d)	18. (a)	19. (d)	20. (d)	21. (a)	22. (c)	23. (b)	24. (b)
25. (d)	26. (c)	27. (d)	28. (d)	29. (c)	30. (b)	31. (d)	32. (c)
33. (d)	34. (b)	35. (c)	36. (d)	37. (b)	38. (a)	39. (d)	40. (a)
41. (d)	42. (d)	43. (d)	44. (b)	45. (d)	46. (d)	47. (c)	48. (c)
49. (d)	50. (d)	51. (a)	52. (b)	53. (b)	54. (d)	55. (b)	56. (d)
57. (c)	58. (d)	59. (d)	60. (a)	61. (c)	62. (b)	63. (d)	64. (b)
65. (a)	66. (d)	67. (d)	68. (b)	69. (b)	70. (d)	71. (d)	72. (d)
73. (a)	74. (d)	75. (d)	76. (d)	77. (d)	78. (d)	79. (c)	80. (c)
81. (d)	82. (d)	83. (a)	84. (c)	85. (d)	86. (d)	87. (a)	88. (a)
89. (c)	90. (c)	91. (c)	92. (b)	93. (a)	94. (a)	95. (d)	96. (d)
97. (a)	98. (d)	99. (b)	100. (d)	101. (b)	102. (d)	103. (d)	

Fill in the Blanks

1. Apoptosis is an ________________ cell death.
2. Apoptosome is a key connection between mitochondria and ________________ activation.
3. BCl2 was discovered in ________________ neoplasm.
4. The duration of apoptosis is about ________________ hours but in cell culture visible morphologic changes are completed in ________________ hours.

5. ________________ cells have the ability to induce apoptosis directly.
6. ________________ proteins can promote or inhibit apoptosis.
7. The Fas receptor binds to ________________ ligand.
8. Caspase ________________, caspase ________________ and caspase ________________ function as initiator of cell death.
9. In *C.elegans,*1090 somatic cells are generated in the formation of an adult worm, of which ________________ cells undergo apoptosis.
10. ________________ uptake of apoptotic cells is the last component of apoptosis.
11. A family of proteins called ________________ are typically activated in early stages of apoptosis.
12. Death receptors are cell surfaces that transmit ________________ initiated by specific ligands.
13. Binding of TRAIL (TNF-related apoptosis inducing ligand to its receptors ________________ and ________________ triggers apoptosis in many cells.
14. Caspases belong to a group of enzymes known as ________________.
15. Release of cytochrome C from mitochondria may lead to activation of caspase ________________ and then of caspase ________________.
16. Caspases are regulated at a ________________ level.
17. For maturation of cytokines some ________________ are needed.
18. ________________ is the programmed death of an organism.
19. In Autoimmune Lymphoproliferative Syndrome (ALPS), the mutation is present in ________________.
20. During viral infection, certain cells of immune system called ________________, bind to infected cells and trigger them to undergo apoptosis.
21. Apoptosis can occur in as little as ________________ minutes.
22. A foetus eyelids form an opening by the process of ________________.
23. p^{53} is a potent inducer of ________________.
24. Apoptosome is formed by the process of ________________.
25. The formation of apoptosome is triggered by the release of ________________ from the mitochondria.
26. The mitochondrial ________________ is an early marker of the onset of apoptosis.
27. Apoptosis inducing factor is neutralised by ________________.
28. Apoptosis inducing factor is a mitochondrial intermembrane ________________.
29. Apoptosis is controlled by number of tightly regulated ________________.
30. Planned cellular death is known as ________________.
31. Granazyme B activates many caspases by cleaning ________________ residues.
32. Cell divisions and ________________ are two antagonistic processes that maintain the proper number of somatic cells.
33. Violent membrane blebbing occurring during apoptosis is called ________________.
34. Fas and tumour necrosis factor receptors are ________________ proteins.
35. Melanoma avoid apoptosis by inhibiting the expression of the genes encoding ________________.
36. The cellular mechanism that ends in cell death in a programme-controlled manner is known as ________________ .
37. The first component of cell death mechanism to be cloned in any organism was ________________.

38. CED-4 is similar to ________________.
39. In histological sections, apoptotic cells can be detected by the ________________ and labelling assay.
40. Over-expression of ________________ is associated with a specific type of lymphoma.
41. Loss of function of normal ________________ has been reported in many human cancers.
42. In humans, ________________ is involved in the activation of apoptosis.
43. The two possible ways of cell death are ________________ and ________________.
44. Kostmann's syndrome results from a deficiency in one of the three proteins called ________________.
45. A damaged cell may undergo apoptosis if it is unable to repair ________________.

Answers to Fill in the Blanks

1. Active
2. Caspase
3. B cell
4. 12 to 24, <2
5. Cytotoxic T
6. BC12
7. Fas
8. 8, 9, 10
9. 131
10. Phagocytic
11. Caspases
12. Apoptotic signals
13. DR4, DR5
14. Cysteine proteases
15. 9, 3
16. Post transcriptional
17. Caspases
18. Phenoptosis
19. Germ line
20. Cytotoxic T lymphocytes
21. 20
22. Apoptosis
23. Apoptosis
24. Apoptosis
25. Cytochromie C
26. Apoptosis-induced channel
27. Heat shock protein70
28. Flavoprotein
29. Signaling pathways
30. Apoptosis
31. Asparate
32. Apoptosis
33. Zeiosis
34. Integral membrane
35. Aapf-1
36. Apoptosis
37. BC12
38. *Aapf*-1
39. Terminal deoxyuridine nucleotide
40. BC12,
41. p^{53}
42. *Aapf*-1
43. Apoptosis, necrosis,
44. HaX1
45. Genetic errors

True or False

1. Apoptosis is a normal cellular process.
2. Apoptotic processes are observed during the regression of tumour.
3. Apoptosis is always pathological.
4. The induction of apoptosis is an active genetically regulated process.
5. Necrosis is an active process.
6. Apoptotic cell death requires gene activity.
7. Apoptosis requires a functional energy producing system.
8. Morphological alterations of the cells are earliest indications of apoptotic cell death.
9. DNA fragmentation can be used to identity apoptotic cells.
10. Cells dying by apoptosis shrink and eventually break up into small vesicles called apoptotic bodies.

11. In adult human beings, on an verage 50–70 billion cells die each day due to apoptosis.
12. Tumour necrosis factor plays a key role in auto-immune diseases.
13. Macrophages ingest apoptotic cells with release of pro-inflammatory cytokines.
14. Many growth factors and cytokines act as cellular survival factors by preventing apoptosis.
15. BCl2 is a negative regulator of apoptosis.
16. In *C. elegans*, the early steps of apoptosis are reversible.
17. High levels of BCl2 have been found in some B cell leukaemias and lymphomas.
18. During apoptosis, the cell membrane remains intact.
19. Apoptosis eliminates both excess normal cells as well as damaged cells.
20. Apoptosis plays an opposite role of mitosis.
21. Apoptosis is a disturbed process.
22. Most cells in apoptosis have its genetic material.
23. Apoptosis can be stimulated by various stimuli from outside or inside the cell.
24. Removal of apoptotic bodies from the cell does not cause an inflammatory response.
25. Necrosis causes damage to neighbouring cells.
26. In the cells, caspases are synthesised as inactive procaspases.
27. Gene regulating apoptosis is not found in all metazoan organisms.
28. Genotoxic compounds can induce apoptosis
29. Many viruses encode genes that prevent apoptotic destruction of the host cell.
30. TAS-103 induces apoptosis.
31. Apoptosis is a controlled and energy-independent process.
32. Flavoprotein is an apoptosis-inducing factor.
33. Nitric acid is involved in the regulation of apoptosis.
34. Enough apoptosis can lead to cancer.
35. During apoptosis, the cytoplasm undergoes shrinkage damaging the plasma membrane.
36. The phenomenon of apoptosis is pathological.
37. Different caspases have different substrate recognition.
38. Apoptosis involves clumping of chromatin, swelling and rupturing of mitochondria.
39. The Ced genes are responsible for all programmed cell death.
40. Apoptosis is an abnormal physiological process.
41. Apoptosis is an inherent part of cell life.
42. In humans, genetic defects in apoptosis are rare.
43. Viral infections may trigger an apoptotic response.
44. Many viruses have evolved mechanisms to repress apoptosis.
45. Natural killer cells can induce apoptosis by expression of Fas ligand on their surfaces.

Answers to True or False

1. True	2. True	3. False	4. True	5. False	6. True	7. True	8. True
9. True	10. True	11. True	12. True	13. False	14. True	15. True	16. True

17. True	18. True	19. True	20. True	21. False	22. True	23. True	24. True
25. True	26. True	27. False	28. True	29. True	30. True	31. False	32. True
33. True	34. False	35. False	36. False	37. True	38. False	39. True	40. False
41. True	42. True	43. True	44. True	45. False			

Give Reasons

1. The process of cell death is also called active cell death or programmed cell death.
   – Because it involves controlled gene expression, which is activated by a variety of external or internal stimuli or by their absence.
2. Apoptosis provides a defence mechanism against viruses.
   – Because it reduces the spread of viruses through rapid death of virus-infected cells.
3. Apoptosis is regarded as an injury-limiting mode of cell disposal.
   – Because in apoptosis, intracellular contents are not released from apoptotic cells and their fragments and the process does not result in inflammation.
4. BCl2 is so named.
   – Because the gene was discovered as a translocated locus in B cell leukaemia.
5. Apoptosis is also known as programmed cell death.
   – Because the pattern of death by suicide is an orderly event.
6. Apoptosis should be tightly regulated.
   – Because too little or too much cell death may lead to pathology.
7. Cell undergoing apoptosis shrink.
   – Due to breakdown of the protein aceous cytoskeleton by caspases.
8. Cells commit suicide.
   – Because it is essential for proper development and for the destruction of unwanted cells that threaten the integrity of the organism.
9. Any orphan cells without adequate connections to neighbouring cells are removed by the process of apoptosis.
   – Because their survival depends on the availability of neurotropic factors secreted by the target cells they innervate.
10. Caspase enzymes have been mentioned as ICE like by many authors.
    – Because of their similarity with Interleukin Converting Enzyme (ICE).

GENETIC ENGINEERING

Multiple-Choice Questions

1. Who among the following scientists is associated with discoveries in genetic engineering?
   (a) H Morgan (b) z H C Crick
   (c) J D Watson and F H C Crick (d) H G Khorana
2. The technique of artificial synthesis of a new gene and its successful transplantation in the genome of a competent organism is called:
   (a) Biotechnology (b) Genetic counseling (c) Genetic engineering (d) DNA fingerprinting
3. The first human hormone synthesised with the help of biotechnology was:
   (a) Pituitary hormone (b) Somatostatin (c) Prostaglandin (d) Interferon
4. Who amongst the following synthesised a gene for somatostatin?
   (a) Rick and Boyer (b) Ullrich et al. (c) Seeberg et al. (d) A M Chakravorty
5. Which one of the following techniques is used for the production of monoclonal antibodies?
   (a) Transformation (b) Hybridisation
   (c) Hybridoma (d) Site specific recombination
6. Super bug is the name applied to a genetically altered bacterium called:
   (a) *Clostridium pastorianum* (b) *Pseudomonas putida*
   (c) *Escherichia coli* (d) *Azotobacter agile*
7. 'Super bug' combines the genetic elements of:
   (a) 2 different bacteria (b) 4 different bacteria
   (c) 8 different bacteria (d) 12 different bacteria
8. Which type of environmental pollution is being controlled by using 'super bug'?
   (a) Radioactive pollution (b) Herbicide pollution
   (c) Pesticide pollution (d) Oil pollution
9. In genetic engineering, the most commonly used vector is a:
   (a) Plastid DNA (b) Plasmid DNA (c) Cosmid DNA (d) Chromatid DNA
10. When a foreign genetic segment is introduced on the bacterial plasmid, the plasmid DNA is most commonly referred to as:
    (a) Passenger DNA (b) Recombinant DNA
    (c) Vector DNA (d) Complementary DNA
11. Recombinant DNA is also called:
    (a) Combined DNA (b) Chimeric DNA (c) Crossover DNA (d) Criss-cross DNA
12. *Taq polymerase* enzyme is extracted from the bacterium:
    (a) *E. coli* (b) *Bacillus thurengenesis*
    (c) *Thermus aquaticus* (d) *Treponema pallidum*

13. Which one of the following *DNA polymerase* is also called thermo-stable *DNA polymerase* enzyme?
    (a) The *DNA polymerase* extracted from the bacterium *E. coli*
    (b) The *DNA polymerase* extracted from the bacterium *Bacillus thurengenesis*
    (c) The *DNA polymerase* extracted from the bacterium *Thermus aquaticus*
    (d) The *DNA polymerase* extracted from the bacterium *Treponema pallidum*
14. Which one of the following enzyme acts best at 72°C and the de-naturation temperature of 90 °C does not destroy its enzymatic activity?
    (a) *DNA polymerase -I* (b) *DNA polymerase- II*
    (c) *RNA polymerase* (d) *Taq polymerase*
15. Which one of the following polymerase enzyme is frequently used in the field of genetic engineering?
    (a) *DNA polymerase -I* (b) *DNA polymerase- II*
    (c) *RNA polymerase* (d) *Taq polymerase*
16. Which one of the following enzymes recognises short sequences of double-stranded DNA as the target for cleavage?
    (a) *DNA polymerase* (b) *RNA polymerase*
    (c) *Restriction endonuclease* (d) *Taq polymerase*
17. Some *Restriction endonuclease* enzymes also bring about modification in the DNA by methylation, these enzymes are called:
    (a) *Methylase* (b) Modification enzymes (c) Both (d) None
18. Which one of the following statements is correct about the *Taq polymerase* enzyme?
    (a) *Taq polymerase* enzyme is thermo-sensitive. (b) *Taq polymerase* enzyme is thermo-stable.
    (c) *Taq polymerase* enzyme is thermo-immune. (d) *Taq polymerase* enzyme is thermo-active.
19. In genetic engineering, which one of the following *Restriction endonuclease* enzymes are the most important and commonly used for cloning purposes?
    (a) *Type-II restriction enzyme* systems (e.g., *Eco*RI)
    (b) *Type-I restriction enzyme* systems (e.g., *Eco*K and *Eco*B)
    (c) *Type-III restriction enzyme* systems (e.g., *Eco*P1 and *Eco*I5)
    (d) All of the above
20. The DNA is a negatively charged molecule, which moves towards the anode (positive electrode). This property of DNA is utilised in:
    (a) Gel electrophoresis (b) DNA fingerprinting
    (c) Southern blotting (d) All of the above
21. The northern blotting and southern blotting techniques are used in blotting (transfer) of:
    (a) DNA molecules and RNA molecules, respectively
    (b) DNA molecules and protein molecules, respectively
    (c) RNA molecules and DNA molecules, respectively
    (d) Protein molecules and RNA molecules, respectively
22. In southern blotting, the gel is stained with the intercalcating ethidium bromide, which gives visible fluorescence on illumination of the gel with:
    (a) Infrared light (b) Ultraviolet light (c) Visible white light (d) Visible red light
23. The DNA fingerprint (VNTRs) of all individuals on the planet earth is different:
    (a) Except in monozygotic twins (b) Except in identical twins
    (c) Except in fraternal twins (d) More than one is correct
24. In recombinant DNA technology, a plasmid vector must be cleaved by:

(a) An enzyme different from what cleaves the donor gene
(b) The same enzyme that cleaves the donor gene
(c) The four different enzymes that cleave the donor gene
(d) The modified *DNA ligase*

25. The great advantage of using yeast in place of bacteria as the recipient cells for recombination of the eukaryotic DNA is that the yeast can:
(a) Excise introns from the RNA transcripts (b) Excise exons from the RNA transcripts
(c) Remove methyl groups (d) Remove formyl groups

26. When a cut caused by the *restriction endonuclease* enzyme in the DNA, produces short complementary single-stranded sticky ends in the DNA, this type is called:
(a) Sequencing cut (b) Palindromic cut (c) Replicate cut (d) Staggered cut

27. The enzyme that recognises and cleaves DNA at specific interval sites is termed as:
(a) *Endonuclease* enzyme (b) *Restriction Endonuclease* enzyme
(c) *Restriction* enzyme (d) *Restriction Exonuclease* enzyme

28. The different *Restriction Endonuclease* enzymes recognise different but specific short sequences of double-stranded DNA as the target site. The specific base sequence at the target sites has:
(a) 4 to 6 base pairs (b) 4 to 8 base pairs (c) 6 to 8 base pairs (d) 4 to 20 base pairs

29. Transfer of DNA bands from agarose gel to nitrocellulose or nylon membrane is termed as:
(a) Southern transfer (b) Northern transfer (c) Western transfer (d) Gene transfer

30. The common reporter gene used in plant expression vector is:
(a) TAC (b) GAT (c) CAT (d) TAG

31. Which one of the following processes is a key requirement in DNA fingerprinting?
(a) ELISA (b) Electron microscopy (c) Ultra centrifugation (d) Gel electrophoresis

32. The DNA element having the ability to change its position is termed as:
(a) Cistron (b) Transposon (c) transfer DNA (tDNA) (d) Intron

33. A sequential expression of a set of human genes occurs when a steroid molecule binds to:
(a) DNA sequence (b) tRNA (c) mRNA (d) rRNA

34. The enzyme called *Hind III,* isolated and purified by Smith and Wilcox from the bacterium – *Haemophilus influenzae*, belongs to which one of the following categories?
(a) *Ligases* (b) *Restriction Endonucleases*
(c) *Polymerases* (d) *Nitrogenases*

35. For the introduction of foreign DNA into higher plant cells, the two successful and best known vectors used are:
(a) Tiplasmid of *Agrobacterium tumefaciens* and Riplasmid of *A. rhizogenes*
(b) The bacterium *Agrobacterium tumefaciens* or its modified Tiplasmids and the double-stranded DNA virus – Cauliflower mosaic virus
(c) Riplasmid of *A. rhizogenes* and the double-stranded DNA virus – Cauliflower mosaic virus
(d) Riplasmid of *A. rhizogenes* and TMV

36. The use of bactriophages as cloning vectors has a limitation on the size of the foreign DNA to be cloned. This is due to:
(a) Its intracellular life cycle in the host cell (b) Lysogenic life cycle of the bacteriophage
(c) Size of the phage head (d) Size of the tail plate

38. Plasmids, phages and cosmids are used as vectors in:
(a) Gram-negative bacteria (b) Gram-positive bacteria

(c) In both gram-positive bacteria and gram-negative bacteria
(d) None

38. The small DNA segments (genomic DNA, cDNA or synthetic oligonucleotides) or RNA segments (often synthesised on DNA template) that recognise the complimentary sequences in DNA or RNA molecules and are used in identification and isolation of specific DNA sequences from an organism are called:
(a) Marker molecule (b) Selector molecule (c) Molecular template (d) Molecular probe

39. In DNA probe assay, the probes are labelled either with the radioactive isotopes (e.g., ^{32}P dCTP) or with nonradioactive signal molecules (e.g., biotin or digoxigenin). Presently the most popular method of labelling the nucleic acid is done by:
(a) ^{32}P dCTP (b) Biotin (c) Digoxigenin (d) All of the above

40. When the entire genome is cloned in the form of a library of random genomic clones without identifying them, it is called:
(a) Gunshot experiment
(b) Shotgun experiment
(c) Replication experiment
(d) Hybridisation

41. A set of DNA fragments cloned by the shotgun experiment is called:
(a) Genome (b) DNA fingerprinting (c) Genomic library (d) Gene pool

42. The mRNAs, isolated from the actively protein synthesising cells (like root and leaf cells in plants, and ovaries and reticulocytes in mammals) are used in making:
(a) Gene bank (b) Genomic library (c) DNA library (d) cDNA library

43. Which one of the following enzyme is used in the synthesis of cDNA?
(a) *Exonuclease*
(b) *Restriction Endonuclease*
(c) *Reverse Transcriptase*
(d) *Polymerase*

44. Which one of the following enzyme is used in the Polymerase Chain Reaction (PCR)?
(a) *DNA polymerase* (b) *Taq polymerase* (c) *DNA ligase* (d) *Helicasel*

45. The *Taq polymerase* enzyme acts best at a temperature of:
(a) 48°C (b) 58°C (c) 35°C (d) 72°C

46. The natural genetic engineer of plants is:
(a) *Agrobacterium tumefaciens*
(b) *Rhizobium leguminosarium*
(c) *Pseudomonas fluorescence*
(d) *Bacillus thuringiensis*

47. Modification in the Tiplasmid used in genetic engineering has been done by:
(a) Modification of introns
(b) Addition of tumour-forming genes
(c) Deletion of tumour-forming genes
(d) Modification of *endonuclease* enzyme

48. In genetic engineering, which one of the following are commonly used as vector for transfer of genes?
(a) YAC (b) BAC (c) Plasmid (d) All

49. Which one of the following enzymes inhibits the process of self-ligation in DNA fragments with sticky ends?
(a) *Endonuclease*
(b) *Restriction endonuclease*
(c) *Taq polymerase*
(d) *Alkaline phosphatase*

50. Which one of the following bacterium gene has been incorporated in the *Bt* cotton plant that gives resistance to the plant against the ball worm pest?
(a) *Bacillus thermos*
(b) *Bacillus thuringiensis*
(c) *Bacillus thrombophila*
(d) *Bacillus tumefaciens*

51. Flavr savr variety of tomato is an improved variety, which remains fresh for a long period because it has:
(a) Gene for antibiotic resistance
(b) A little amount of *polygalacturonase*

(c) A large amount of *polygalacturonase* (d) 'Cry' protein

52. *Restriction endonuclease* enzymes cuts:
    (a) Only single-stranded DNA (b) Only double-stranded DNA
    (c) Both single-stranded DNA.and double-stranded DNA (d) All single-stranded DNA/RNA.and double-stranded DNA/RNA
53. Which one of the following is a commonly used plasmid in genetic engineering?
    (a) pBR 322 (b) *Eco R I* (c) *Al Ul* (d) *Hind III*
54. DNA is generally methylated at:
    (a) A base (b) G base (c) T base (d) C base
55. Genomic DNA library is:
    (a) Packing of donor DNA in a collection of vectors (b) A collection of gene vectors
    (c) Collection of organisms for extracting DNA (d) All of the above
56. Which one of the following genetically modified organisms has been used in recent years to control fungal diseases of plants?
    (a) *Bacillus thuringiensis* (b) *Bacillus tumefaciens*
    (c) *Trichoderma* (d) *Klebsiella*
57. Polymerase chain reaction is useful in:
    (a) DNA replication (b) DNA synthesis (c) DNA amplification (d) Protein synthesis
58. Thermal cycle is used in:
    (a) Polymerase chain reaction (b) DNA replication
    (c) Organogenesis (d) Protoplast hybridisation
59. Introduction of genetically modified food is not desirable because:
    (a) It will affect the economy of the developing countries
    (b) Genetically modified foods are less tasty
    (c) Genetically modified foods are costly
    (d) There is danger of entry of toxins and virus in the food and food chain
60. Genetically engineered bacteria are being widely used in the commercial production of:
    (a) Melatonin (b) Testosterone (c) Human insulin (d) Thyroxin
61. What is incorrect about plasmids?
    (a) Plasmid DNA is small in size. (b) Plasmids are mostly circular.
    (c) They can be easily isolated. (d) They contain sex factors.
62. What is not true about *restriction endonuclease* enzyme?
    (a) Restriction enzymes are important tools for the study of DNA.
    (b) Several restriction enzymes were first isolated from bacteria.
    (c) They recognise specific sites where cuts are to be made.
    (d) They have not been found effective in eukaryotic DNA.
63. Which one of the following acts as vectors in genetic engineering?
    (a) Plasmids (b) BAC (c) YAC (d) All
64. Plasmid pBR–322 contains which one of the following gene(s)?
    (a) Ampicillin resistance (b) Tetracycline resistance
    (c) Nitrogen fixation (d) Both (a) and (b)
65. Which one of the following artificial chromosomes are efficient vectors for eukaryotic genes?
    (a) YACs (b) MACs (c) BACs (d) All

66. First protoplasmic fusion was performed by:
(a) Smith and Warner (b) Smith and Nathan (c) Harris and Watkins (d) Khorana and Holley

67. Which one of the following restriction enzyme is not able to recognise purines and pyrimidines at a certain position?
(a) *EcoR I* (b) *EcoR II* (c) *Hind II* (d) *Alu I*

68. Match column I with column II and select the correct answer using answer codes:

Column I	Column II
(A) Nitrogen fixation	1. *E. coli*
(B) Toxin production	2. *Rhizobium* species
(C) Antibiotic resistance	3. *Bacillus anthracis*
(D) Tumour formation	4. *Agrobacterium tumefaciens*

Answer codes:

	A	B	C	D
(a)	3	2	4	1
(b)	4	3	1	2
(c)	2	3	1	4
(d)	2	4	1	3

69. Which one of the following is not associated with gene manipulation?
(a) Vector DNA (b) Recombinant DNA
(c) Palindromic sequence of DNA (d) Shine–Delgarno sequence of DNA

70. The most important advancement in the area of genetic engineering was the discovery of:
(a) *Restriction endonuclease* (b) Genetic vector
(c) DNAase (d) None of these

71. Restriction enzymes are widespread in:
(a) *Drosophila* (b) Microorganisms (c) Certain plants (d) All of the above

72. Simian virus 40 (SV 40) has _________ EcoRI recognition site(s) in its single circular chromosome:
(a) Only one (b) Only two (c) Only three (d) More than three

73. The first hormone produced by the bacteria using recombinant DNA technology was:
(a) Testosteron (b) Thyroxin (c) Insulin (d) Relaxin

74. Which one of the following statements is incorrect about a good cloning vehicle?
(a) There should be unique cleavage sites for several *restriction endonucleases.*
(b) The position of these cloning sites should be such that insertion of a segment of DNA in these segments of DNA would not bring about phenotypic changes in the characteristics of plasmids.
(c) Plasmids are found in most bacteria, and are self-replicating circular, double-stranded DNA molecules.
(d) There should be marker genes in the plasmids.

75. The T_2 plasmid part when transferred into the plant cell DNA is termed as:
(a) Junk DNA (b) cDNA (c) zDNA (d) tDNA

76. T_2 plasmid is found in:
(a) *Agrobacterium tumefaciens* (b) *Rhizobium leguminosarium*
(c) *E. coli* (d) *Bacillus thuringiensis*

77. Match column I with the column II and select the correct answer using answer codes:

Column I	Column II
(a) Gene gun	1. Used as host for the recombinant DNA
(b) Tiplasmid	2. Vehicle DNA

(c) *E. coli*
(d) Bacteriophage DNA

3. Vector-less direct gene transfer
4. Crown gall tumour

Answer codes:

	A	B	C	D
(a)	3	4	1	2
(b)	1	4	3	2
(c)	2	4	1	3
(d)	4	2	1	3

78. The restriction enzymes used frequently in genetic manipulation, commonly recognise:
(a) AAAGCCCT sequence
(b) TTTTCAGG sequences
(c) Short palindromic sequences having an axis of symmetry
(d) All

79. The first experiment in genetic engineering, conducted in 1973, used plasmid vector:
(a) p15A (b) $p^{SC}101$ (c) R 6–5 (d) pMB_1

80. Which restriction enzyme does not produce a blunt end?
(a) *Alu* 1 (b) *Sma* I (c) *Pst* 1 (d) *Hae III*

81. The largest number of bacteriophage cloning vectors have been made from:
(a) SV_{40} (b) Lambda (λ) phage (c) Bacteria (d) Yeast

82. Which one of the following statement is correct?
(a) PCR works very efficiently on RNA.
(b) Northern blotting technique involves hybridisation of protein bound to filter.
(c) Southern blotting technique is used for the separation and study of RNA fragments.
(d) Restriction Fragment Length Polymorphisms (RELPs) are inherited as simple mendelian co-dominant markers.

83. The complex formed due to the binding of restriction enzyme *EcoRI* with the DNA was first revealed by:
(a) Agrose gel electrophoresis
(b) Electron microscopy
(c) NMR study
(d) X-ray analysis

84. Restriction enzymes and their corresponding *methylase* have:
(a) Common sequences but are synthesised by different genes
(b) No common sequences and are encoded by overlapping genes
(c) No common sequences and are encoded by different genes
(d) Common sequences

85. Lysogeny is the:
(a) Process of addition of phage DNA into the genetic material of the host
(b) Cleavage of DNA segment with *exonuclease*
(c) Secretion of pathogenic protein by bacteria
(d) Breakdown of the lysosome for intracellular digestion

86. Who among the following scientists is regarded as the 'Father of Genetic Engineering'?
(a) H G Khorana (b) P A Sharp (c) E B Lewis (d) Paulberg

87. Which one of the following statements is incorrect?
(a) Larger fragments of DNA move slowly through agrose gel.
(b) During movement through agrose gel, the DNA fragments are damaged.
(c) The DNA fragments are separated by agrose gel electrophoresis by their differential rates of movement.
(d) The DNA fragments produced by the restriction enzyme cuts can be separated by agrose gel electrophoresis.

88. The characteristics of the cosmids are similar to:
(a) Plasmid and yeast
(b) Yeast and lambda (λ) phage
(c) Plasmid and lambda (λ) phage
(d) All

89. A given restriction enzyme within the cell may function to degrade:
(a) DNA of the cell
(b) Protein of the cell
(c) Foreign DNA that may have entered into the cell
(d) Both (a) and (b)

90. Which one of the following DNA acts as passenger DNA, i.e., the DNA that is transferred from one organism to another by combining with vehicle DNA:
(a) Random DNA (b) SDNA (c) cDNA (d) All

91. The detailed list and sequences of restriction enzymes were provided by:
(a) Kornberg (1982)
(b) Mutz and Davis (1972)
(c) Kornberg and Nirenberg (1982)
(d) Roberts (1983)

92. The original order of the DNA fragments along with the chromosomal segment is termed as:
(a) DNA map (b) Chromosome map (c) Linkage map (d) Restriction map

93. The first restriction maps were obtained by digesting viral chromosomes with the help of the enzyme:
(a) *Hind II* (b) *Hind III* (c) *EcoRI* (d) All of the above

94. Which one of the following bases has dominancy at sites recognised by the restriction enzymes?
(a) A and G (b) A and T (c) T and C (d) G and C

95. Cleavage of DNA to create a nick by the restriction enzymes take place at a:
(a) 5 ψ phosphoryl and 3 ψ hydroxyl terminus
(b) 3 ψ phosphoryl and 5 ψ hydroxyl terminus
(c) Both (a) and (b) are correct
(d) None

96. The first transgenic animal produced was:
(a) Super pig
(b) Super mouse
(c) Super goat
(d) Large sized *Drosophila*

97. Which one of the following scientist(s) produced the first transgenic animal?
(a) Brigitte Boisslier (2002)
(b) Richard Palmiter and Ralph Brinsler (1981)
(c) Ian Wilmut (1999)
(d) Francis Collin (1990)

98. Transgenic mice are commonly used in the study of:
(a) Fertility
(b) Sterility
(c) Gene interaction
(d) Regulation of gene expression

99. Cells of a clone are identical:
(a) Morphologically (b) Physiologically (c) Genetically (d) All of the above

100. The first successful experiment of animal cloning was performed by:
(a) Bateson and Punnet (1908)
(b) Kornberg and Nirenberg (1982)
(c) Gurdon (1962)
(d) Ian Wilmut (1999)

101. Which one of the following experiments was successfully conducted by Donwolf (1996)?
(a) Microbial cloning
(b) Cloning of Asian gaur
(c) Cloning of calves
(d) Simian cloning from an eight-cell embryo

102. The first cloned mammal named 'Dolly' was born on:
(a) 5th December 1997
(b) 13th February 1997
(c) 13th July 1997
(d) 20th November1997

103. Which one of the following is the first genetically altered primate?

(a) Andi (b) Molly (c) George (d) Dolly

104. Which one of the following is statements is incorrect?
   (a) In animals, cell cloning is based on the property of Pleuriopotency.
   (b) In plants, cell cloning is based on the property of Totipotency.
   (c) All plants are Totipotent.
   (d) In animals, unfertilised eggs are Totipotent.
105. Cloned genes can be inserted only into:
   (a) Microorganisms (b) Cultured mammalian cells
   (c) The genome of a multicellular plants and animals (d) All of the above
106. Restriction maps are prepared by:
   (a) Methylation of DNA segments
   (b) Locating the position of the inversion sites in the DNA segment
   (c) Locating the position of the cleavage sites of a number of restriction enzymes in DNA segments
   (d) All of the above
107. In animal cells like mouse cells, special animal viruses were used as a cloning vehicle, in which globin gene could be integrated, one such virus is:
   (a) T 4 (b) ϕX 174 (c) SV40 (d) Sendai virus
108. Which one of the following is used as vector in gram-positive bacteria?
   (a) Cosmids only (b) Plasmids only (c) Both (d) None
109. Which one of the following is used as vector in gram-negative bacteria?
   (a) Cosmids only (b) Plasmids only (c) Phages (d) All of the above
110. Which one of the following is not used in polymerase chain reaction?
   (a) Eppendorf tube (b) Thermal cyclers
   (c) *Taq DNA polymerase* (d) Deoxygenin
111. The single-stranded free ends of DNA fragments are called:
   (a) Primer ends (b) Cut ends (c) Sticky ends (d) Recombinant ends
112. Which one of the following statements is incorrect?
   (a) Eve was the first genetically altered primate of the world.
   (b) The milk of Molly and Polly (two cloned lambs) contain blood-clotting factors.
   (c) Dolly has two genetic mothers.
   (d) Genotype can be preserved by cloning.
113. In which one of the following organisms, a clone of 4–8 identical young ones of the same sex is produced from a single zygote?
   (a) *Icthyophis* (b) Cobra (c) Krait (d) Armadillo
114. The cloning of plasmid sequences can be done up to a length of:
   (a) 40–50 Kb (b) 30–40 Kb (c) 20–25 Kb (d) 10–15 Kb
115. Match column I with the column II and select the correct answer using answer codes:

Column I	Column II
(a) pBluescript IIS	1. *Agrobacterium rhizogenes*
(b) Lambda (λ) phage	2. Expression vector
(c) Riplasmids	3. Sequences up to 22 Kb can be cloned
(d) Cosmids	4. Sequences up to 40 Kb can be cloned

Answer codes:

	A	B	C	D
(a)	3	4	2	1
(b)	2	1	3	4
(c)	2	3	1	4
(d)	2	3	4	1

116. Phagemids have combined features of:
(a) Phages with bacteria
(b) Phages with *P* elements of *Drosophila*
(c) Phages with plasmids
(d) Transposons of higher plants with bacteria

117. The first cloned member of the equine family is:
(a) Andi (b) Tetra (c) Prometea (d) Milly

118. The nucleotide sequence of first IFN-a cDNA derived from cloned IFN-a cDNA was carried out by:
(a) Edge (1981) (b) Harrison (1980) (c) Hill (1992) (d) Weiseman (1980)

119. Match column I with column II and select the correct answer using answer codes:

Column I	Column II
(A) Andi	1. First cloned member of Equine family
(B) Tetra	2. Cloned cat
(C) Prometea	3. First cloned monkey
(D) Cc	4. Cloned cat

Answer codes:

	A	B	C	D
(a)	4	1	2	3
(b)	3	4	1	2
(c)	2	3	1	4
(d)	4	2	1	3

120. Which one of the following viruses acts as a vector for introducing a foreign gene?
(a) TMV (b) Herpes virus (c) Retrovirus (d) All of the above

121. Insertion of cloned genes can be done into:
(a) Genome of an intact multicellular plant and animal
(b) Microorganisms
(c) Cultured mammalian cells
(d) All of the above

122. The cDNA are directly used in:
(a) Gene manipulation
(b) Transformation experiments
(c) Both (a) and (b)
(d) None of the above

123. If a gene has been cloned, its DNA can be used anywhere:
(a) To make a restriction map
(b) To determine whether similar DNA exists anywhere
(c) Both (a) and (b)
(d) None of the above

124. The cloning vehicle is introduced into the host by:
(a) Transformation (b) Transfection (c) Conjugation (d) All of the above

125. Transgenic pigs have been provided human genes with an aim that their organs contain:
(a) Human enzymes (b) Human antigens (c) Both (a) and (b) (d) None of the above

126. Original order of DNA fragments along with a chromosomal segment is termed as:
(a) Chromosomal map
(b) Linkage map

(c) DNA map (d) Restriction map

127. Hybridoma is an example of:
(a) Gene cloning (b) Somatic cell hybridisation
(c) Genetically modified organism (d) Cell cloning

128. The virus used in the production of human antibodies is:
(a) Epstein–Barr virus (b) Sendai virus
(c) Adeno virus (d) Simian virus 40 (SV40)

129. Which one of the following statements is incorrect?
(a) Restriction enzymes cause break in palindromic sequences.
(b) The breaks caused by restriction enzymes are usually directed opposite to one another.
(c) Restriction enzymes generate DNA fragments having complementary ends.
(d) None

130. The DNA polymerase generally used is:
(a) *DNA Pol I* from *E. coli* (b) T_4 *DNA polymerase* encoded by $T_{4\ gene}$
(c) Either (a) or (b) (d) None

131. Which one of the following is applicable to plasmids?
(a) Production of bacteriocin and toxin (b) Nitrogen fixation
(c) Antibiotic and heavy metal resistance (d) All

132. Which one of the following is incorrect with reference to the structure of Tiplasmids?
(a) Regions A – Responsible for tumour suppression
(b) Region B – Responsible for origin of replication
(c) Region C – Responsible for conjugation
(d) Region D – Responsible for virulence

133. Electroporation is only efficient on:
(a) Nuclear membrane (b) Cell membrane
(c) Cell wall (d) All

134. Polyethylene glycol causes:
(a) Changes in the pore size of the cell membrane (b) Mutation
(c) Inactivation of tumour suppressor gene (d) Activation of proto oncogenes

135. Match the given columns I, II and III and select the correct answer using answer codes:

Column I (Crops)		Column II (Trait)		Column III (Gene product)
A) Rice	p.	Male sterility	i.	Ribonuclease
B) Tomato	q.	Insect tolerance	ii.	Bt toxins
C) Cotton	r.	Virus tolerance	iii.	Coat proteins
D) Mustard	s.	Better nutritional quality	iv.	Vitamins

Answer codes:

	A	B	C	D
(a)	s, iv	r, ii	q, iii	p, i
(b)	s, iv	r, iii	q, ii	p, i
(c)	r, iv	s, iii	q, i	q, ii
(d)	p, iv	r, ii	q, I	p, iii

136. The vir region of T DNA comprises:
(a) Two operons (b) Four operons (c) Six operons (d) Eight operons

137. Molecular mechanism is done through:

(a) In Situ Hybridisation (ISH)
(b) Restriction fragment length polymorphism
(c) Restriction mapping and Random amplified polymorphic DNA
(d) All

138 Which one of the following is not a molecular probe?
(a) cDNA (b) Genomic DNA (c) RNA segments (d) None

139. Which one of the following is not applicable to polymerase chain reaction?
(a) Taq *DNA polymerase*
(b) Peoples choice reaction
(c) Eppendorf tube
(d) Shotgun experiment

140. Which one of the following is the correct sequence of a polymerase chain reaction?
(a) Primer extension→ Annealing of primers → Denaturation of DNAfragment
(b) Annealing of primers → Denaturation of DNA fragment
(c) Denaturation of DNA fragment → Primer extension → Annealing of primers
(d) Denaturation of DNA fragments → Annealing of primers → Primer extension

141. Which one of the following is an incorrect match?

	Marker/Reporter gene	Source of gene
(a)	Luciferase	*E. coli*
(b)	Bleomycin resistance	*E. coli*
(c)	β glucuronidase	*E. coli*
(d)	Neomycin phosphotransferase II	*E. coli*

142. A vector must have some qualities to be an efficient agent for:
(a) Maintenance
(b) Amplification of the passenger DNA
(c) Transfer
(d) All

143. What is incorrect about reverse transcriptase?
(a) Used in the production of Klenow fragment
(b) Used in the synthesis of DNA
(c) Used in the construction of cDNA clone bank in making of short labelled probes
(d) None

144. Match column I with column II and select the correct answer using answer codes:

Column I	Column II
(A) Southern blotting	1. Requires radioactively DNA probe
(B) Northern blotting	2. Identification of RNA band
(C) Western blotting	3. Identification of DNA band
(D) Colony hybridisation technique	4. Identification of protein band

Answer codes:

	A	B	C	D
(a)	3	2	4	1
(b)	4	1	3	2
(c)	2	4	1	3
(d)	3	4	1	2

145. Recombinant clones can be identified, selected and characterised by:
(a) Nucleic acid hybridisation
(b) Antibiotic resistance
(c) Immuno chemicals
(d) All

146. Which one of the following is applicable to hybridoma technology?
(a) Hybrid cell
(b) Monoclonal antibodies
(c) Myeloma cells
(d) All

147. Match column I with column II and select the correct answer using answer codes:

Column I	Column II
(a) Cloning	1. DNA amplification
(b) Gene splicing	2. Construction of gene bank
(c) Polymerase chain reaction	3. Use of restriction endonuclear to produce small fragments of the intergenome
(d) Shotgun technique	4. Linking of foreign DNA to the vector DNA

Answer codes:

	A	B	C	D
(a)	2	4	1	3
(b)	2	3	4	1
(c)	3	4	1	2
(d)	4	1	3	2

148. Which one of the following transposable elements can be used for cloning vector?
(a) *MU*-1 of maize
(b) *P* element of *Drosophila*
(c) *Ac-Ds* of maize
(d) All

149. Which one of the following technique is particularly useful for studying gene expression?
(a) Inverted polymerase chain reaction
(b) Asymmetrical polymerase chain reaction
(c) Reverse transcriptase polymerase chain reaction
(d) Nested polymerase chain reaction

150. Monoclonal antibodies are used:
(a) In the identification of different types of leukaemia and lymphomas
(b) To attack cancer metastases
(c) To track cancer antigen
(d) All

Answers to Multiple-Choice Questions

1 (d)	2 (c)	3 (c)	4 (a)	5 (c)	6 (b)	7. (b)	8 (d)
9 (b)	10. (c)	11. (b)	12. (c)	13. (c)	14 (d)	15. (d)	16 (c)
17. (c)	18. (b)	19 (a)	20 (d)	21 (c)	22. (b)	23. (d)	24. (b)
25. (a)	26. (d)	27 (b)	28. (b)	29. (a)	30. (c)	31. (d)	32. (b)
33. (a)	34. (b)	35. (a)	36. (c)	37. (a)	38. (d)	.39. (b)	40. (b)
41. (c)	42. (d)	43. (c)	44. (b)	45. (d)	46. (a)	47. (c)	48. (d)
49. (d)	50. (b)	51. (b)	52. (b)	53. (a)	54. (d)	55. (a)	56. (c)
57. (c)	58. (a)	59. (d)	60. (c)	61. (d)	62. (d)	63. (d)	64. (d)
65. (d)	66. (c)	67. (c)	68. (c)	69. (d)	70. (a)	71. (b)	72. (a)
73. (c)	74. (b)	75. (d)	76. (a)	77. (a)	78. (c)	79. (b)	80. (b)
81. (b)	82. (d)	83. (d)	84. (c)	85. (a)	86. (d)	87. (b)	88. (c)
89. (c)	90. (d)	91. (d)	92. (d)	93. (d)	94. (d)	95. (a)	96. (b)
97. (b)	98. (d)	99. (d)	100. (c)	101 (d)	102. (b)	103. (a)	104. (d)
105. (d)	106. (c)	107. (c)	108. (b)	109. (d)	110. (d)	111. (c)	112. (a)
113. (d)	114. (d)	115. (c)	116. (c)	117. (c)	118. (d)	119. (b)	120. (d)
121. (d)	122. (c)	123. (c)	124. (d)	125. (b)	126. (d)	127. (d)	128. (a)
129. (b)	130. (c)	131. (d)	132. (a)	133. (b)	134. (a)	135. (b)	136. (c)
137. (d)	138. (d)	139. (d)	140. (d)	141. (a)	142. (d)	143. (a)	144. (a)
145. (d)	146. (d)	147. (a)	148. (d)	149. (c)	150. (d)		

Fill in the Blanks

1. In the southern blotting technique, the DNA is visible under _________light after staining with the ethidium bromide stain.
2. Only plasmids act as a vector in cloning, in the gram ____________bacteria.
3. The extraction of genomic DNA, its breakage into fragments of reasonable size with the help of a restriction *endonuclease* enzyme and then inserted into a cloning vector for producing a population of chimeric vectors is called __________________experiment.
4. Plasmids are used as vectors in genetic engineering because of their ability to carry__________ genes.
5. Recombinant DNA or rDNA technology was discovered by _______ and __________.
6. Tumour-inducing plasmid used in producing transgenic plants is obtained from the bacterium ________________________.
7. Humulin is a ______________.
8. The toxic component of the bacterium *Bacillus thuringienesis* is ______________.
9. Transgenic pigs having genes for human antigens have opened a new dimension in medical science for __________________.
10. The *Hirudin* gene inserted in the plant *Brassica napus* is taken from the __________.
11. _________________________ is a commonly used phage mid derived from PUC-19.

Answers to Fill in the Blanks

1. Infrared	2. Positive	3. Shotgun
4. Foreign	5. Cohen, Boyer	6. *Agrobacterium tumefaciens*
7. Protein	8. 'Cry' protein	9. Organ transplantation
10. Leech	11. p Bluescript IIKs.	

True or False

1. The DNA is a negatively charged molecule, which moves towards the anode (positive electrode).
2. Majority of restriction *endonuclease* enzymes cleave only unmethylated targets but a few restriction *endonuclease* enzymes can cleave both methylated and unmethylated targets.
3. In the southern blotting technique, the DNA is visible under infrared light after staining with ethidium bromide stain.
4. Majority of restriction endonuclease enzymes used have six base pairs (bp) target sites.
5. *mi*RNA (micro RNA) are small sized RNAs which can attach to *m*RNAs and stop their complete expression.

6. Ttransposons of the higher plants (e.g., *Ds*, *Ac* or *Mu*-1 of maize) and transposons of *Drosophila* (*P* elements) are also used as vectors.
7. In case of the gram-positive bacteria, only plasmids act as the vector in cloning.
8. Antibodies are also occasionally used as molecular probes to recognise specific protein sequences.
9. DNA/RNA and antibody probe assay are faster and sensitive technique in comparison to many conventional diagnostic tests.
10. Flavr savr variety of tomato is an improved variety, which remains fresh for a long period because it has a little amount of *polygalacturonase.*
11. *Restriction* enzyme *EcoRl*, cuts the DNA at the sequence 5′ GAATTC 3′.
12. In transgenics, the expression of transgene in the target tissue is termed as 'reporter'.
13. Golden rice plant is a transgenic plant, which is rich in vitamin A content.
14. Formation of transient pores in the cell membrane to introduce gene construct is called 'electroporation'.
15. Plasmids are used as vectors in the genetic engineering because of their ability to cause infection in the host cell.

Answers to True or False

1. True	2. True	3. False	4. True	5. True	6. True	7. True	8. True
9. True	10. True	11. True	12. True	13. True	14. True	15. False	

Give Reasons

1. RNA viruses are not suitable for gene therapy.
   – Because the RNA which are not able to integrate into DNA of human cells, are degraded rapidly.
2. Plasmids are used for indirect gene transfer.
   – Because plasmids can move from one cell to another and can make several copies of themselves.
3. Restriction enzymes are important tools for the study of DNA.
   – Because:
   (a) They cut the DNA from specific sites.
   (b) They recognise specific sites where cuts have to be made.
   (c) They permit isolation of small DNA fragments.
   (d) They convert circular DNA into a linear DNA with well-defined ends.
4. Genetic engineering is termed as recombinant DNA technology.
   – Because the techniques used in genetic engineering result in a new combination of heritable materials.
5. Plasmids are used as cloning vectors.
   – Because of their increased yield potential.
6. The use of cauliflower mosaic virus (DNA) as a cloning vector is restricted.
   – Because it has a narrow host range as it infects only the members of family Crucifer.

7. EMBL3 and EMBL4 are known as replacement vectors.
   - Because in these vectors a central non-essential part of 44 Kb long phage can be replaced by a foreign DNA.
8. Restriction enzymes create double-stranded breaks.
   - Because the same sequence is found on both the strands.
9. Restriction enzymes are prevented from acting on the DNA native to their organisms.
   - Because specific *DNA methylase* add methyl groups only to the specific adenine or cytosine residues within the recognition sequences which prevent the restriction enzymes from acting on DNA native to the cell.
10. The functional forms of *methylases* are monomers.
    - Probably because their substrates are single chain segments.
11. Prokaryotes (particularly *E. coli*) are favoured materials for genetic engineering techniques.
    - Because they can be easily cultured and possess rapid proliferation ability.
12. Vectors are called replicons.
    - Because of their inherent capacity to clone independently.

GENETICS OF BACTERIOPHAGES

Multiple-Choice Questions

1. The term 'bacteriophage' was coined by:
   (a) Felix Hubert d' Herelle (b) Frederick W Towort
   (c) W Igo-Kemenes (d) D E Dykhuisen
2. Bacteriophages were discovered by:
   (a) W F Bodmer (b) K M Brinkhous (c) T Maniatis (d) Frederick W Towrt
3. The genetic material of bacteriophage is:
   (a) dsDNA (b) dsRNA (c) ssDNA or ssRNA (d) All
4. Which one of the following cycles is found in bacteriophages?
   (a) Lytic (b) Lysogenic
   (c) Both lytic and lysogenic (d) None
5. Which one of the following is a temperate phage?
   (a) Lamda (b) P1 (c) mu (d) All
6. Consider the following statements:
   (A) Bacteriophages typically carry genetic information needed for replication of their nuclei and synthesis of their protein coats
   (B) For replication of their nuclei acids and synthesis of protein coat, bacteriophages require precursor energy generation and ribosomes supplied by their host cells
   (C) The first phage of replication of the phage in its host cell is called adsorption
   (D) In propage state, all genes are repressed except one
   The correct statements are:
   (a) All (b) A, B and C (c) B and C (d) A and D
7. Which one of the following is applicable to tail fibres of bacteriophages?
   (a) Lysis of the host cell (b) Peneleration
   (c) Attachment (d) Protection
8. Which one of the following phages structurally and biochemically resemble human pathogenic enteric viruses?
   (a) PRD1 (b) φX174 (c) MS2 (d) All
9. Which one of the following is a specific receptor in bacteria to which bacteriophages attach themselves?
   (a) Proteins or flagella (b) Cipopolysaccharides
   (c) Teichoic acid (d) All
10. Sexuality in bacteria was demonstrated by:
    (a) Beadle (1958) (b) Lederberg and Tatum (1946)
    (c) Jacob and Wollman (1961) (d) Delbruck and Luria (1969)
11. Recombination in T4 has been studied by:
    (a) Hershey (b) Rotman (c) Bailey (d) Benzer

12. Bacteriophages are released:
(a) By extrusion (b) By budding (c) Via cell lysis (d) All

13. Which one of the following statements is incorrect?
(a) T4 infects *E. coli.*
(b) The DNA of T4 is one of the longest DNAs in phages.
(c) The tail of T4 is hollow.
(d) T4 is capable of undergoing a lysogenic cycle.

14. The Nobel laureate not associated with T4 bacteriophage:
(a) Roberts (b) Watson (c) Max Delbruck (d) Hershey

15. Consider the following statements:
(A) F factor is composed of circular DNA
(B) Generally the number of F factor is one per cell
(C) F factor is dependent of transmission of chromosomal genes
(D) F factor remains free in the cytoplasm in F^+ strains and Hfr strains

The correct statements are:
(a) All (b) A, B and C (c) A and B (d) B and D

16. The lambda gene repressor system does not comprise:
(a) *cl* gene (b) *cro* gene (c) *P* element (d) OR3

17. What is incorrect about T4?
(a) Terminally redundant genome
(b) Introns like eukaryote
(c) Low speed DNA copying mechanism
(d) Special DNA repair mechanism

18. Bacteriophage φX174 contains:
(a) 9 genes (b) 11 genes (c) 15 genes (d) 21 genes

19. Lambda (λ) phage was discovered by:
(a) Esther Lederberg (1950)
(b) G W Beadle (1958)
(c) E L Wollmann (1961)
(d) Kimber and Riley (1963)

20. The only protein expressed during lysogenic phage of lambda is the:
(a) Cl (b) Q (c) N (d) None

21. Which one of the following is an RNA binding protein and RNA polymerase factor?
(a) cII (clear2) (b) cIII (clear3) (c) N (a ntiterminator) (d) Int (integration)

22. Which one of the following is applicable to T7 phage?
(a) *E. coli 0157 H7*
(b) Synthetic biology
(c) Gp5 DNA polymerase
(d) All

23 Lyses by tailed phages occurred through by an enzyme called:
(a) Methylase (b) Endolysin (c) Terminase (d) Lysolase

24. A type of regulation called antitermination is unique to:
(a) Lambda phage (b) φX174 phage (c) T7 phage (d) T4 phage

25. Match column I with column II and select the correct match using answer codes:

Column I	Column II
(A) x is (eXcsion)	1. DNA binding protein and RNA polymerase cofactor
(B) c I (clear I)	2. Excisonase and int protein regulator
(C) cII (clear 2)	3. Transcription inhibitor
(D) Q	4. Transcription activator

Answer codes:

	A	B	C	D
(a)	2	3	4	1

(b)	3	1	4	2
(c)	4	3	2	1
(d)	2	3	1	3

26. Which one of the following statements is incorrect?
    (a) PBR322 is the best-known plasmid rector.
    (b) Mu phage is a large transposable element.
    (c) Bacteriophages are parasitic at the genetic level.
    (d) Bacteriophages lack the ability to mutate.
27. Which one of the following is not applicable to lambda phage?
    (a) Transposon
    (b) Lytic and lysogenic cycles
    (c) Model organism
    (d) Temperate bacteriophage
28. In lambda phage, cI dimmer bind to operator:
    (a) *OR*1 (b) *OR*2 (c) *OR*3 (d) All
29. Which one of the following is a popular vehicle for cloning of complementary DNA?
    (a) T7 phage
    (b) Some strains of *Pasteurella*
    (c) Some strains of *Shigella*
    (d) All
30. The genome of which one of the following phages exists as an autonomous self-replicating plasmid?
    (a) P1 (b) Mu (c) φX174 (d) Lambda
31. Which one of the following statements is incorrect?
    (a) The ribosomal RNA sequences of mitochondria are similar to those of sulphur bacteria.
    (b) The rRNA of chloroplast is closely related to those *cyanobacteria*.
    (c) A variety of Hfr strains exists in *E. coli.*
    (d) The bacterial genetic map is linear.
32. The scientist associated with cross feeding is:
    (a) Bernard Davis (b) Lederberg (c) Adelberg (d) Zinder
33. An episome:
    (a) Is a genetic factor in bacteria
    (b) Exists as an element in the cytoplasm
    (c) As an integrated part of chromosome
    (d) All
34. Consider the following statements:
    (A) The lambda repressor is a dimer and is also known as *cI* protein
    (B) The life cycle of lambda phages is controlled by *cI* Cro proteins
    (C) In the presence of *cI* proteins, only *cI* gene may be transcribed, while in the absence of cI proteins, the *cro* gene may be transcribed
    (D) Whenever there is predominance of *cI* proteins, the lambda remain in the lysogenic state, but on predominance of Cro proteins, it is transformed into the lytic cycle

    The correct statements are:
    (a) All (b) A and D (c) B and D (d) B and C
35. 'Holliday Junction' is a mobile junction between:
    (a) Two strands of DNA
    (b) Four strands of DNA
    (c) Eight strands of DNA
    (d) Two strands of DNA
36. cI protein is a:
    (a) Repressor
    (b) Binds to operator sequences
    (c) Prevent transcription
    (d) All
37. Which one of the following statements is correct?
    (a) A cell containing F in the first state is called F^+ cell.

(b) A cell containing F in the second state is called Hfr cell.
(c) A cell lacking F is called F^- cell.
(d) All

38. Which one of the following is incorrect about φX174?
(a) Contains very small amount of DNA (b) Contains II genes in 5386 bases
(c) 70 per cent nucleotides are coding genes (d) GC content is 44 per cent

39. Which one of the following must be timed appropriately for the functional phage to be released, after phage DNA is injected?
(a) Replication (b) Morphogenesis (c) Lysis (d) All

40. Which one of the following is not applicable to Holliday junction?
(a) Homologous recombination (b) Robin Holliday
(c) Yeast (d) Highly conserved

41. The fate of the viral genetic material is controlled by a completion between the:
(a) cro and cI proteins (b) cro and N proteins
(c) cI and Q proteins (d) N and Q proteins

42. With reference to lambda phage mutant characterisation, which one of the following is not essential?
(a) Clone of genes (b) Mapping
(c) Dominance test (d) Complementation test to determine number of genes

43. Which one of the following is not applicable to bacteriophages?
(a) Peptidoglycan (b) Endolysin (c) Prophage (d) Lysogenic cycle

44. In lambda phage, which one of the following protects cII from degradation by proteases?
(a) cI (b) cIII (c) xis (d) int

45. Plasmids are found in:
(a) Archaea (b) Bacteria (c) Eukarya (d) All

46. Which one of the following is incorrect with reference to plasmids?
(a) Contain circular DNA molecules (b) Contain genes
(c) Do not respond to cellular replication signals (d) Used in genetic engineering

47. Which one of the following is incorrect with reference to phage genetics?
(a) Rapid replication cycle (b) Mutants are available
(c) Auxotrophs are available (d) Selection strategies are possible

48. Which one of the following is an incorrect match?
(a) F-Plasmids – Production of bacterocins (b) R-Plasmids – Resistance against antibiotics
(c) Col-Plasmids–Turn a bacterium into a pathogen (d) Virulence plasmids – Digestion of toluene

49. Consider the following statements:
(A) Bacteriophages do not undergo intergenic recombination
(B) Conjugation is one means of recombination in bacteria
(C) Rec proteins are essential for bacterial recombination
(D) Transduction is virus mediated DNA transfer

The correct statements are:
(a) B, C and D (b) A, B and C (c) B and C (d) C and D

50. Generally F factors are lacking in species groups:
(a) *Salmonella* (b) *Serratia* and *Shigella* (c) *Pasteurella* (d) All

51. The first example of linkage between two genes in transformation was shown by:
(a) Hotchkiss and Marmur (1954)
(b) Taylor and Thoman (1964)
(c) Benzer (1950)
(d) Hayes (1968)

52. The lambda repressor gene system consists of:
(a) *cI* gene – *OR*3 – *OR*2 – *OR*1– *cro* gene
(b) *cI* gene – *OR*1 – *OR*2 – *OR*3 – *cro* gene
(c) *OR*1 – *OR*2 – *OR*3 – *cI* gene – *cro* gene
(d) *cI* gene – *cro* gene – *OR*1 – *OR*2 – *OR*3

53. Match column I with column II and select the correct answer using answer codes:

Column I	Column II
(A) Hershey and Chase	1. Genetic transduction
(B) Zinder and Lederberg	2. Mutations arise in the absence of selection
(C) Luria and Delbruck	3. DNA is genetic material
(D) Meselson and Weigle	4. Molecular basis of DNA recombination

Answer codes:

	A	B	C	D
(a)	4	1	2	3
(b)	3	1	2	4
(c)	4	3	1	2
(d)	2	4	1	3

54. Holin is a protein used by __________ phage to make lesions in the host's cytoplasmic membrane:
(a) T4
(b) T7
(c) Lambda
(d) φX174

55. Fertility (F) factor exists as:
(a) F+ (Plus)
(b) Hfr
(c) F prime
(d) All

56. Plasmids are found in:
(a) Bacteria
(b) Fungi
(c) Mitchondria
(d) All

57. Which one of the following is not a competent bacteria for DNA transformation?
(a) *Haemophilus influenzae*
(b) *Bacillus subtilis*
(c) *E. coli*
(d) *Neisseria gonorrhoeae*

58. Which one of the following viruses contains only four genes?
(a) MS21
(b) Qβ
(c) R17 and f2
(d) All

59. Which one of the following statements is correct?
(a) Mark Ptashne purified lambda repressor
(b) When a phage infects a cell, no repressor is present
(c) The product of *N* gene interacts with RNA polymerase
(d) cI mutation does not produce any defective repressor in the phage control system

60. Blender experiment to prove DNA as the genetic material was conducted by:
(a) Griffith
(b) Lederberg and Zinder
(c) Hershey and chase
(d) Jacob and Wollman

61. Hershey and Chase performed their experiment to prove DNA as the genetic material on:
(a) T2
(b) SV40
(c) Lambda
(d) φX174

62. Bacteria used in first transformation experiment was:
(a) *Bacillus subtilis*
(b) *E. coli*
(c) *Diplococcus pneumoniae*
(d) *Nesseria gonorrhoea*

63. Consider the following statements with reference to lambda phase:
(A) The gene *cI*, lies between PR/OR and PL/OL control regions
(B) When active *cI* inhibits expression of *cro* and *N* genes

(C) When active, *cI* enhances its expression (D) *cro* and *N* genes are delayed early genes

The incorrect statement is:

(a) D (b) A (c) B (d) C

64. Which one of the following proteins only acts as a repressor but never as activator?

(a) λ-cro (b) λ-repressor (cI) (c) CAP (d) None

65. Cro protein consists of:

(a) Two distinct domains and a connector
(b) Two distinct domains with three α-helices and a β-sheet
(c) Single domains with three α-helices and a β-sheet
(d) Three distinct domains with three α-helices and two β-sheets

66. Which one of the following is not a virulent bacteriophage?

(a) SPO1 (b) T2 (c) T4 (d) None

67. In lambda phage, each operator has ________ binding sites:

(a) One (b) Two (c) Three (d) Five

68. Which one of the following is not applicable to repressor of lambda?

(a) Dimer (b) N-terminal domain (c) C-terminal end (d) Lack of connector

69. In SPO1, the consensus sequence -35 and -10 base pair resembles those of:

(a) *E. coli* (b) *E. coli* and *B. subtilis*
(c) *E. coli* and *S. typhimurium* (d) *N. gonorrohoe* and *B. subtilis*

70. Which one of the following is incorrect about plasmids?

(a) Found in bacteria (b) Are circular DNA
(c) Possesses genes for sexuality (d) Lack genes for antibiotic resistance

71. In T4 phage, synthesis of late proteins occurs:

(a) Before replication of DNA (b) At the time of replication of DNA
(c) After replication of DNA (d) All

72. Which one of the following is incorrect with reference to T4 phage?

(a) T4 bacteriophage contains 5–35 genes
(b) 18 genes are required for the formation of the head
(c) 21 genes are required for the formation of the tail
(d) 6 genes are required for the formation of tail fibres

73. F factor:

(a) Is a cytoplasmic factor (b) Generally separated from the chromosome
(c) Multiplies independently (d) All

74. Concatemeres have been reported in:

(a) φX174 (b) T5 and T7 (c) P22 (d) All

75. Consider the following statements:

(A) Phage genome can be mapped by analysis of recombinant frequency in double infections (B) Virulent phages may become prophages
(C) Virulent phages are always lytic (D) Prophages always exist in bacterial chromosome

The incorrect statements are:

(a) None (b) B, C and D (c) B and D (d) A and C

76. Which one of the following is incorrect in the cross of Hfr(λ) × F(λ)?

(a) Conjugation proceeds normally (b) Bacterial recombinants are readily recovered
(c) Lysis takes place (d) No lysis takes place

77. The integration of phage to become a prophage ________ the genetic distance between flanking bacterial markers:
   (a) Increases
   (b) Decreases
   (c) Does not affect
   (d) May increase or decrease, depending upon the physiological conditions
78. Bacteriophages:
   (a) Are intracellular parasites
   (b) Multiply inside bacteria
   (c) Use some or all of the host biosynthetic machinery
   (d) All
79. What is incorrect about bacteriophages?
   (a) Have a very high degree of gene diversity
   (b) The genonome architectures is mosaic
   (c) The genomes contain a very high proportion of novel genetic sequences
   (d) None
80. What is incorrect about endolysin?
   (a) Antibacterial agent
   (b) Degrade peptidoglycan
   (c) dsDNA bacteriophage encoded enzyme
   (d) Have a bad effect on the normal microflora
81. Which one of the following is essential for bacterial lysis?
   (a) Endolysin
   (b) Collagenase and holing
   (c) Endolysin and holing
   (d) Holin, endolysin and endopeptidase
82. The statement: 'Bacteria do not die, they just phage away' is given by:
   (a) S Spiegelman (b) Mark Muller (c) R Rotman (d) E Zuekerkandl
83. Stability of which one of the following determines the lifestyle of a lambda phage?
   (a) cI (b) cII (c) cIII (d) None
84. Which one of the following is a horizontal gene transfer?
   (a) Transformation (b) Transduction (c) Bacterial conjugation (d) All
85. Bacterial artificial chromosome:
   (a) Is the large segment of DNA
   (b) Contains 100,000 to 200,000 bases
   (c) Cloned from another species into bacteria
   (d) All
86. Which one of the following is not applicable to plasmid?
   (a) Autonomous molecules
   (b) Alkaline lysis
   (c) Extrochromosomal
   (d) None
87. Which one of the following enables a crossover event to occur between λ and the host chromosome?
   (a) int product (b) Q product (c) cI product (d) cII product
88. In lambda, lysogeny is increased:
   (a) Starved cells
   (b) When there is high multiplicity of infection
   (c) Both (a) and (b)
   (d) When cells are nutritionally rich
89. Which one of the following is incorrect about plasmids?
   (a) Extrachromosomal genetic material
   (b) Not essential for survival growth
   (c) Has no extracellular form
   (d) Similar to genome of bacteria
90. Upon infection in the bacterial cell, the lambda DNA:
   (a) Becomes linear
   (b) Becomes single stranded
   (c) Becomes double stranded
   (d) Circularises
91. The *N* gene product is:
   (a) A protein
   (b) Enhances transcription of genes
   (c) Both (a) and (b)
   (d) A protein that represses transcription

92. The gene Q anti-terminator is responsible for:
(a) Late gene expression
(b) Enhancement of transcription of genes
(c) Activation of cro protein
(d) Early gene expression
93. Which one of the following is an RNA bacteriophage of *E.coli*?
(a) f2
(b) MS2
(c) R17
(d) All
94. Integration and replication of mu are examples of:
(a) Simple transcription
(b) Replicative transposition
(c) Both (a) and (b)
(d) Mutation
95. Nearly all bacterial genes are responsible for the synthesis of specific polypeptide chain except for those DNA segments that code for:
(a) mRNA
(b) rRNA
(c) mRNA and rRNA
(d) rRNA and tRNA
96. Cro protein turns down synthesis of:
(a) Q protein
(b) P protein
(c) O protein
(d) All
97. In lambda phage, the genes critical in determining the development of phage are:
(a) cI and cII
(b) cII and cIII
(c) N and Q
(d) cro and cII
98. RNA polymerase binds to PR when:
(a) cro is present
(b) Repressor is present
(c) Both cro and repressor are present
(d) Neither cro nor repressor is present
99. Akins to lysogeny is:
(a) Transformation in bacteria
(b) Transduction in bacteria
(c) Conjugation in bacteria
(d) None

Answers to Multiple-Choice Questions

1. (a)	2. (d)	3. (d)	4. (c)	5. (d)	6. (a)	7. (c)	8. (d)
9. (c)	10. (b)	11. (d)	12. (d)	13. (d)	14. (a)	15. (c)	16. (c)
17. (c)	18. (b)	19. (a)	20. (a)	21. (c)	22. (d)	23. (b)	24. (a)
25. (a)	26. (d)	27. (a)	28. (d)	29. (d)	30. (a)	31. (d)	32. (a)
33. (d)	34. (a)	35. (b)	36. (d)	37. (d)	38. (c)	39. (d)	40. (d)
41. (a)	42. (b)	43. (a)	44. (b)	45. (d)	46. (c)	47. (c)	48. (b)
49. (a)	50. (d)	51. (a)	52. (a)	53. (b)	54. (c)	55. (d)	56. (d)
57. (c)	58. (d)	59. (d)	60. (b)	61. (a)	62. (c)	63. (a)	64. (a)
65. (c)	66. (d)	67. (c)	68. (d)	69. (b)	70. (d)	71. (c)	72. (a)
73. (d)	74. (d)	75. (c)	76. (c)	77. (a)	78. (d)	79. (d)	80. (d)
81. (c)	82. (b)	83. (a)	84. (d)	85. (d)	86. (a)	87. (a)	88. (c)
89. (d)	90. (d)	91. (c)	92. (a)	93. (d)	94. (c)	95. (d)	96. (d)
97. (d)	98. (d)	99. (b)					

Fill in the Blanks

1. A virus that infects bacteria is called________________.
2. The bacteriophages that establish lytic and lysogenic cycles are called______________ phage.

3. The lysogeny cycle involved a protein called ________________.
4. Analysis of the genetics of phage activity is routinely done by using a ______________.
5. When the host RNA polymerase transcribes the lambda genome, two proteins are produced called ________and ______________.
6. The phage genome varies in size from__________ kilobase pairs per strand of nucleic acids.
7. Bacteriophages that can reproduce only within a host cell may have a ________ cycle or __________ cycle or ______________.
8. When the phage genome becomes a part of the bacterium chromosome, it is called ____________.
9. If the phage genome is not incorporated into the host chromosome, and remains separate it is called a ______________.
10. The first demonstration of bacterial transformation was done with ____________, ____________________.
11. Very rarely, lysogeny permits the transfer of bacterial genes between cells by ______________.
12. Lambda phase was discovered by __________________(1950).
13. Transfer of gene from one type of microorganism to another is called genetic __________.
14. Zinder discovered the first bactericphage that contained __________ as its genetic material.
15. The virion of T2 contains ____________ stranded DNA.
16. The most intensively studied bacteriophage is the ________ phage.
17. Phage gene expression during the lytic and lysogenic cycles uses the host______________.
18. Cloning of human insulin gene in bacteria was done by using a ____________ as a vector.
19. The cII protein is constantly degraded by ________ present in the cell.
20. In bacteria, the sex difference between males and females is recognised by the presence of the ______________.
21. The F factor is an ____________.
22. The Hfr chromosome is transferred to the F⁻ cell in a_______ fashion.
23. Recent researches suggested that the key components of the mitochondrial transcription and replication apparatus are derived from the ________ lineage of bacteriophage.
24. The __________ assay method is the fundamental method to count phages.
25. There are two types of transducing phages, viz., __________ and __________ transducing phages.
26. Simultaneous transduction of two or more different markers is called ________________.
27. The DNA of many phages is ____________ redundant.
28. Bacteriophage lambda is a classic example of a ____________ bacteriophage.
29. The single circular chromosome of *E. coli* contains about _________ base pairs.
30. The first type of genetic exchange between bacteria to be observed was ______________.
31. A plasmid is an extrachromosomal __________ molecule.
32. The term 'plasmid' was coined by_______________________.
33. Plasmids used in genetic engineering are called __________________ .
34. Three major types of genetic transfer found in bacteria are ______________, _________ and ______________.
35. Fertility (F) factor contains about ________________ genes.

36. ________________ is the only plasmid chimera containing inserted DNA sequences.
37. Transformation was first demonstrated by ________________.
38. Introduction of DNA into animal cells is generally termed as ________________.
39. Transduction takes place through________________ cycle or ________________ cycle.
40. Transduction was discovered by ________________ .
41. Phage ecology is the study of the interaction of bacteriophages with their________________.
42. The bacteriophages which cause lysis of bacterial cells are called __________ phages.
43. Recombination in T4 phage has been studied by ________________ .
44. ________________ exists in two forms called ______________ and rough (R).
45. The sex difference between male and female bacteria is recognised by the presence or absence of ________________ factor.
46. When lambda DNA is mixed with lambda gal DNA, the two strands undergo complementary pairing except in the ________ region.
47. The lambda phage DNA is inserted in the *E. coli* DNA at a specific site between the ______________ operon and the ____________ operon.
48. The insertion of phage DNA requires a specific enzyme called ________________.
49. The chromosome of lysogenic viruses become ________________ .
50. The Hfr strain is produced by the integration of the ______________ within the bacterial chromosome.
51. In *E. coli*, lytic infections are caused by a group of seven phages known as ________ phages while lysogenic infections are caused by the ______________ phage.
52. Concatemeres are generally due to ________________ replication.
53. The integration of lambda phage takes place at a special attachment site in the bacterial genome called ____________.
54. The *E. coli* ________ gene can inhibit or degrade cII protein.

Answers to Fill in the Blanks

1. Bacteriophage
2. Temperate
3. cI
4. Plaque
5. *Cro, N*
6. 2 to 200
7. Lytic, lysogenic, both
8. q prophage
9. Plasmid
10. Streptococcus pneumoniae
11. Transduction
12. Andre Lwoff (1950)
13. Transduction
14. RNA
15. Double
16. Lambda
17. RNA polymerase
18. Bacteriophage
19. Proteases
20. F factor
21. Episome
22. Linear
23. T-odd
24. Plaque
25. Generalised, specialised
26. Cotransduction
27. Terminally
28. Temperate
29. 4.5×10^6
30. Transformation
31. DNA
32. Lederberg (1952)
33. Vectors
34. Transformation, conjugation, transduction
35. 100
36. Hybrid plasmid
37. Federick Griffith (1928)
38. Transfection
39. Lytic, lysogenic
40. Zinder and Lederberg (1951)
41. Environment
42. Virulent
43. Benzer
44. *Diplococcus pneumoniae, Smoth (s)*
45. Fertility (F)
46. Gal
47. Gal, bio
48. Integrase

49. Prophages
50. F factor
51. T, lambda
52. Rolling circle
53. Attλ
54. *hfl*

True or False

1. Bacteriophages are bacterial viruses.
2. Bacteria do not make true zygotes .
3. *E. coli* contains enormous amount of repetitive DNA.
4. Transposons and insertion sequences are episomes.
5. Plasmids are unstable genetic elements.
6. An episome is non-essential genetic element.
7. Plasmids are circular episomes.
8. Lambda genome exists as a linear double-stranded DNA.
9. When lambda genome is injected to bacterial cytoplasm, it is converted from linear to circular.
10. *E. coli* genes are present in only one copy per genome.
11. Cro protein favours lysogenic growth.
12. The nature of lambda DNA replication changes during the growth cycle.
13. Cro protein acts in opposition to repressor.
14. In the merozygote genetics of bacteria, reciprocal recombinant is expected.
15. Lambda phage is a good example of specialised transduction.
16. The transfer of genetic material in *E. coli* is reciprocal.
17. Lambda repressor binds to co-repressor.
18. Transfection gave birth to genetic engineering.
19. F cells lack high frequency recombination (Hfr) factor.
20. Phage SPO1 infects Bacillus subtilis.
21. For the expression of early genes in SPO1, bacterial RNA polymerase is used.
22. Proteins cII and cIII are quite stable proteins.
23. Mutants in the gene cI can maintain lysogeny.
24. Gene Q induces transcription of late genes.
25. Mutants in cI are similar to the mutants cII and cIII.
26. Product of gene cI is needed to establish lysogeny.
27. Bacteria containing prophages are called lytic bacteria.
28. Prolonged incubation of F^+ strains with acridine dyes, converts it into F^- strains.
29. Plaque morphology is a phage character that can be analysed.
30. Virulant phages are not always lytic.
31. The nucleic acids of phages often contain unusual or modified bases.
32. T4 is among the largest phages.
33. Prophage lacks potential to produce phage.

34. Lysogeny is a model system for virus transformation to animal cells.
35. Lysogenic phages have been shown to carry genes that can modify the *Salmonella* O antigen.
36. Endolysin acts on cell wall.
37. Endolysin contains secretory signals.
38. Lytic infection is characterised by the retention of the phage DNA molecule in the host bacterium.
39. Phages infect only specific bacteria.
40. Stringent plasmids replicate only when the chromosome replicates.
41. Bacteria lacking plasmids, replicate in a normal manner.
42. Plasmids are merely tiny chromosomes.
43. Plasmids allow gene amplification.
44. In gram-negative bacteria, natural transformation does not occur.
45. Transduction is very common.
46. A given gene is present only once on a given plasmid.
47. A phage chromosome may participate in several pairing and crossover events.
48. T2 phage lacks tail fibres.
49. Cro stimulates gene *cII* to synthesise repressor.
50. Cro protein binds most tightly to OR3.

Answers to True or False

1. True	2. True	3. False	4. True	5. False	6. True	7. True	8. True
9. True	10. True	11. False	12. True	13. True	14. False	15. True	16. False
17. True	18. True	19. False	20. True	21. True	22. False	23. False	24. True
25. False	26. False	27. False	28. True	29. True	30. False	31. True	32. True
33. False	34. True	35. True	36. True	37. False	38. True	39. True	40. True
41. True	42. True	43. True	44. False	45. False	46. True	47. True	48. False
49. False	50. True						

Give Reasons

1. Spherical bacteriophages are used as indicators.
   – Because their assay is much easier and less expensive.
2. Bacteriophages are useful in studying how genes function.
   – Because of their small size and simple genetic organisation.
3. Study of genetics of bacteriophages is important.
   – Because it throws light on some very basic features of gene structure and organisation for the very first time.
4. Prophage is so named.
   – Because somehow it seems to be able to induce formation of little infective phage.

5. There is no way of getting all the chromosomes of two different bacteria in the same cell.
   - Because bacteria do not mate in the usual way.
6. Plasmids may be considered as part of the mobilome.
   - Because sometimes they are associated with conjugation.
7. It is advantageous to use bacteria for genetic study.
   - Because:
   (a) Haploid cells
   (b) Short generation times
   (c) Easy manipulation
8. Lysogeny is an advantage to bacterial cells.
   - Because the presence of the prophage provides bacteria to withstand infection as well as prevent vegetative growth by virus particles of the same variety.
9. Conjugation between *E. coli* F^+ cells and *Salmonella* species is more successful.
   - Because F transferred from *E. coli* to *Salmonella* cell, may mobilise the *Salmonella* chromosome for transfer.
10. Bacteriophages are ideal organisms for experimental studies of evolution.
   - Because of:
   (a) Rapid generation time
   (b) Small size
   (c) Small genomes
   (d) Availability of their molecular genetic and biochemistry details
   (e) Ease of manipulation
11. Plasmids are different from episomes.
   - Because plasmids DNA cannot link up with chromosomed DNA.
12. Bacteriophages are similar to animal and plant viruses.
   - Because like animal and plant viruses, they cannot replicate themselves or conduct metabolic processes, if they remain outside the host cells.
13. The genes cI, cII and cIII are so named.
   - Because mutants in these genes produce clear plaques.
14. A lysogenic bacterium is resistant to subsequent infection.
   - Because of the presence of an immunity conferred by the prophage.
15. Generally it is difficult to recognise lysogenic bacteria.
   - Because lysogenic and nonlysogenic cells appear identical.
16. The most common gene to be used for constructing phylogenetic relationships in prokaryotes is the 16SrRNA gene.
   - Because its sequences tend to be conserved among members with close phylogenetic distances, but variable enough that differences can be measured.
17. Plasmids are widely used in genetic modification.
   - Because they can easily pass from one cell to another.
18. Most phages kill cells.
   - Because infection destroys the vital cell membrane that maintains equilibrium between the cell and its surroundings.

Made in the USA
Monee, IL
08 November 2026

67812150R00295